INDUCED PLANT RESISTANCE TO HERBIVORY

Induced Plant Resistance to Herbivory

Edited by

Andreas Schaller
*University of Hohenheim,
Stuttgart, Germany*

 Springer

Editor

Andreas Schaller
University of Hohenheim
Stuttgart, Germany

ISBN: 978-1-4020-8181-1   e-ISBN: 978-1-4020-8182-8

Library of Congress Control Number: 2007941936

© 2008 Springer Science+Business Media B.V.
No part of this work may be reproduced, stored in a retrieval system, or transmitted
in any form or by any means, electronic, mechanical, photocopying, microfilming, recording
or otherwise, without written permission from the Publisher, with the exception
of any material supplied specifically for the purpose of being entered
and executed on a computer system, for exclusive use by the purchaser of the work.

Cover pictures showing *Pieris brassicae* caterpillars, the parasitic wasp *Cotesia glomerata*,
and a parasitized *Manduca sexta* larva were taken by Hans Smid and Tibor Bukovinszky
(http: www.bugsinthepicture.com/), and Johannes Stratmann (University of South Carolina).

Printed on acid-free paper

9 8 7 6 5 4 3 2 1

springer.com

In Memoriam Clarence A. (Bud) Ryan

Bud Ryan left us on October 7th, 2007. His sudden passing away is felt deeply by his family and friends. Bud has left us with a flourishing field of research but we must now continue along this road without him. Throughout his long career Bud gave the community many startling insights into nature. One of the first milestones in the long and unerring path to reveal the invisible secrets of the plant defense mechanism was the discovery, published in 1972, of wound-inducible proteinase inhibitors in potato. Much of Bud's career was spent finding out how these proteins functioned in defense, how they were made, and how their genes were regulated. Constantly incorporating new ideas and technologies, Bud and his collaborators brought to light the first peptide hormone in plants (systemin) and found that jasmonates regulate proteinase inhibitor gene expression. These and other achievements initiated much if not most of the ongoing work in trying to understand the wound response in plants. All those who joined him in this endeavour were touched by his vision, his modesty, his strong sense of humour, and, above all, by his friendship. I personally feel the void he has left and know that this feeling is shared by my colleagues.

Lausanne, Switzerland, October 18th, 2007   *Edward E. Farmer*

Photo: Bud Ryan (right) with his closest colleague Gregory Pearce in 1992. The image shows Bud on one of his frequent visits to the greenhouse to examine tomato plants engineered for altered defense responses.

Contents

In Memoriam **Clarence A. (Bud) Ryan** v

Introduction 1

Section I   Basic Concepts of Plant Defense Against Insect Herbivores

1 **Direct Defenses in Plants and Their Induction by Wounding and Insect Herbivores** 7
   Gregg A. Howe and Andreas Schaller

2 **Herbivore-Induced Indirect Defense: From Induction Mechanisms to Community Ecology** 31
   Maaike Bruinsma and Marcel Dicke

3 **Induced Defenses and the Cost-Benefit Paradigm** 61
   Anke Steppuhn and Ian T. Baldwin

Section II   Induced Direct Defenses

Part A   Anatomical Defenses

4 **Leaf Trichome Formation and Plant Resistance to Herbivory** 89
   Peter Dalin, Jon Ågren, Christer Björkman, Piritta Huttunen and Katri Kärkkäinen

5 **Resistance at the Plant Cuticle** 107
   Caroline Müller

6   **Wound-Periderm Formation** 131
    Idit Ginzberg

7   **Traumatic Resin Ducts and Polyphenolic Parenchyma Cells in Conifers** 147
    Paal Krokene, Nina Elisabeth Nagy and Trygve Krekling

Part B   Production of Secondary Metabolites

8   **Insect-Induced Terpenoid Defenses in Spruce** 173
    Jörg Bohlmann

9   **Phenylpropanoid Metabolism Induced by Wounding and Insect Herbivory** 189
    Mark A. Bernards and Lars Båstrup-Spohr

10 **Defense by Pyrrolizidine Alkaloids: Developed by Plants and Recruited by Insects** 213
    Thomas Hartmann and Dietrich Ober

Part C   Anti-nutritional Enzymes and Proteins

11 **Plant Protease Inhibitors: Functional Evolution for Defense** 235
    Maarten A. Jongsma and Jules Beekwilder

12 **Defensive Roles of Polyphenol Oxidase in Plants** 253
    C. Peter Constabel and Raymond Barbehenn

13 **Action of Plant Defensive Enzymes in the Insect Midgut** 271
    Hui Chen, Eliana Gonzales-Vigil and Gregg A. Howe

14 **Plant Lectins as Part of the Plant Defense System Against Insects** ... 285
    Els J.M. Van Damme

Section III   Defense Signaling

Part A   Activation of Plant Defenses

15 **Systemins and AtPeps: Defense-Related Peptide Signals** 313
    Javier Narváez-Vásquez and Martha L. Orozco-Cárdenas

16 **MAP Kinases in Plant Responses to Herbivory** 329
   Johannes Stratmann

17 **Jasmonate Biosynthesis and Signaling for Induced Plant Defense against Herbivory** 349
   Andreas Schaller and Annick Stintzi

Part B Signals Between Plants and Insects

18 **Caterpillar Secretions and Induced Plant Responses** 369
   Gary W. Felton

19 **Fatty Acid-Derived Signals that Induce or Regulate Plant Defenses Against Herbivory** 389
   James H. Tumlinson and Juergen Engelberth

20 **Aromatic Volatiles and Their Involvement in Plant Defense** 409
   Anthony V. Qualley and Natalia Dudareva

21 **Ecological Roles of Vegetative Terpene Volatiles** 433
   Jörg Degenhardt

Abbreviations 443

Subject Index 449

Taxonomic Index 457

Contributors

Jon Ågren
Department of Ecology and Evolution, Evolutionary Biology Centre, Uppsala University, SE-752 36 Uppsala, Sweden
jon.agren@ebc.uu.se

Ian T. Baldwin
Max Planck Institute for Chemical Ecology, Department of Molecular Ecology, D-07745 Jena, Germany
baldwin@ice.mpg.de

Raymond Barbehenn
Department of Molecular, Cellular, and Developmental Biology, University of Michigan, Ann Arbor, MI 48105, USA
rvb@umich.edu

Lars Båstrup-Spohr
Freshwater Biological Laboratory, University of Copenhagen, DK-3400 Hillerød, Denmark
lars@zygomorf.dk

Jules Beekwilder
Plant Research International B.V., Wageningen University and Research Center, 6700 AA Wageningen, The Netherlands
jules.beekwilder@wur.nl

Mark A. Bernards
Department of Biology, The University of Western Ontario, London, ON, Canada N6A 5B7
bernards@uwo.ca

Christer Björkman
Department of Ecology, Swedish University of Agricultural Sciences, SE-750 07 Uppsala, Sweden
Christer.Bjorkman@ekol.slu.se

Jörg Bohlmann
Michael Smith Laboratories, University of British Columbia, Vancouver, BC,
Canada V6T 1Z4
bohlmann@msl.ubc.ca

Maaike Bruinsma
Laboratory of Entomology, Wageningen University, 6709 PD Wageningen,
The Netherlands
Maaike.Bruinsma@wur.nl

Hui Chen
DOE Plant Research Laboratory, Michigan State University, East Lansing, MI
48824, USA
chenhui@wsu.edu

C. Peter Constabel
Centre for Forest Biology and Department of Biology, University of Victoria,
Victoria, BC, Canada V8W 3N5
cpc@uvic.ca

Peter Dalin
Marine Science Institute, University of California at Santa Barbara, CA
93106-6150, USA
dalin@msi.ucsb.edu

Jörg Degenhardt
Department of Biochemistry, Max Planck Institute for Chemical Ecology, D-07745
Jena, Germany
degenhardt@ice.mpg.de

Marcel Dicke
Laboratory of Entomology, Wageningen University, 6709 PD Wageningen, The
Netherlands
Marcel.Dicke@wur.nl

Natalia Dudareva
Department of Horticulture and Landscape Architecture, Purdue University, West
Lafayette, IN 47907, USA
dudareva@purdue.edu

Juergen Engelberth
Department of Biology, University of Texas at San Antonio, San Antonio, TX
78249, USA
Jurgen.Engelberth@utsa.edu

Gary W. Felton
Department of Entomology, Pennsylvania State University, University Park, PA
16802, USA
gwf10@psu.edu

Idit Ginzberg
Agricultural Research Organization, The Volcani Center, Bet Dagan 50250, Israel
iditgin@volcani.agri.gov.il

Eliana Gonzales-Vigil
DOE Plant Research Laboratory, Michigan State University, East Lansing, MI
48824, USA
gonza260@msu.edu

Thomas Hartmann
Institute of Pharmaceutical Biology, Technical University of Braunschweig,
D-38106 Braunschweig, Germany
t.hartmann@tu-bs.de

Gregg A. Howe
DOE Plant Research Laboratory, Michigan State University, East Lansing, MI
48824, USA
howeg@msu.edu

Piritta Huttunen
Department of Biological and Environmental Sciences, University of Jyväskylä,
40350 Jyväskylä, Finland
pirhutt@st.jyu.fi

Maarten A. Jongsma
Plant Research International B.V., Wageningen University and Research Center,
6700 AA Wageningen, The Netherlands
maarten.jongsma@wur.nl

Katri Kärkkäinen
The Finnish Forest Research Institute, Muhos Research Unit, 91500 Muhos,
Finland
katri.karkkainen@metla.fi

Trygve Krekling
Department of Plant and Environmental Sciences, Norwegian University of Life
Sciences, N-1432 Ås, Norway
trygve.krekling@umb.no

Paal Krokene
Norwegian Forest and Landscape Institute, N-1432 Ås, Norway
Paal.Krokene@skogoglandskap.no

Caroline Müller
Department of Chemical Ecology, University of Bielefeld, D-33615 Bielefeld,
Germany
caroline.mueller@uni-bielefeld.de

Nina Elisabeth Nagy
Norwegian Forest and Landscape Institute, N-1432 Ås, Norway
Nina.Nagy@skogoglandskap.no

Javier Narváez-Vásquez
Department of Botany and Plant Sciences, University of California Riverside,
Riverside, CA 92521, USA
jnarvaez@ucr.edu

Dietrich Ober
Institute of Botany, University of Kiel, D-24098 Kiel, Germany
dober@bot.uni-kiel.de

Martha L. Orozco-Cárdenas
Plant Transformation Research Center, University of California Riverside,
Riverside, CA 92521, USA
mlorozco@citrus.ucr.edu

Anthony V. Qualley
Department of Horticulture and Landscape Architecture, Purdue University, West
Lafayette, IN 47907, USA
aqualley@purdue.edu

Andreas Schaller
University of Hohenheim, Institute of Plant Physiology and Biotechnology,
D-70599 Stuttgart, Germany
schaller@uni-hohenheim.de

Anke Steppuhn
Max Planck Institute for Chemical Ecology, Department of Molecular Ecology,
D-07745 Jena, Germany
asteppuhn@ice.mpg.de

Annick Stintzi
Institute of Plant Physiology and Biotechnology, University of Hohenheim,
D-70599 Stuttgart, Germany
stintzi@uni-hohenheim.de

Johannes Stratmann
Department of Biological Sciences, University of South Carolina, Columbia, SC
29208, USA
johstrat@biol.sc.edu

James H. Tumlinson
Center for Chemical Ecology, Department of Entomology, Pennsylvania
State University, University Park, PA 16802, USA
jht2@psu.edu

Els J.M. Van Damme
Laboratory of Biochemistry and Glycobiology, Department of Molecular
Biotechnology, Ghent University, 9000 Gent, Belgium
ElsJM.VanDamme@UGent.be

Introduction

The class Insecta with more than one million described species is the most diverse group of animals on Earth and outnumbers all other forms of life. Almost half of all existing insect species are herbivores, i.e. they feed on living plants. The abilitity of plants to persist in such a hostile environment relies on evolved resistance systems allowing them to escape herbivores in time or in space, to confront herbivores directly by affecting host plant preference or reproductive success, or to fight herbivores indirectly by association with other species. Plant defenses against insect herbivores were long known to change in evolutionary time and may even change during the lifetime of an individual plant. Until recently however, plant defenses were generally assumed to be constitutively expressed, i.e. independent from herbivore attack. This view changed in 1972 when Green and Ryan reported that wounding by Colorado potato beetles causes the rapid accumulation of proteinase inhibitors in potato and tomato leaves. As inducible resistance factors which interfere with the digestive system of leaf-consuming insects, the proteinase inhibitors were suggested to defend the plants against herbivores.

In the 35 years following the initial discovery by Green and Ryan, many more traits have been identified that are induced by wounding or herbivory. The full breadth of induced responses became apparent with the advent of novel techniques for transcriptome analysis, revealing large-scale changes in gene expression in response to herbivory. In fact, many of this book's chapters focus on herbivore-induced responses including the results of transcriptome analyses, and how they relate to plant defense. At this point, a cautionary note is required: Inducibility of a certain gene (or any given trait) is per se not sufficient evidence for a function in resistance or plant defense. Adopting the definitions advocated by R. Karban J.H. Myers, A.A. Agrawal, and I.T. Baldwin, the term 'induced resistance' refers to induced changes in preference, performance, or reproductive success of the attacker. Therefore, to test for a role in resistance, it is necessary to compare herbivore preference and/or performance on plants that differ in the trait of interest, which can easily be done in a laboratory setting. Induced resistance is thus defined from the point of view of the herbivore, and it does not necessarily benefit the plant. For example, the investment in induced resistance may be larger than the benefit form reduced herbivore damage, or induced herbivore resistance may render the plant more susceptible to other stresses. Defense, on the other hand is defined from the plant's perspective, and it implies a benefit for the plant. Only when they result in a gain in plant fitness are induced traits considered to contribute to plant defense. Therefore, to demonstrate a role in plant defense, field studies are required that reveal differences in fitness for plants that differ in the trait of interest. Only very few herbivore-induced responses have actually been shown to result in a fitness benefit for the plant. Nevertheless, the term 'defense' is frequently used by researchers in the field and also by the authors of this book to describe induced changes in plant architecture, metabolism, or physiology that are assumed to minimize the negative impact of herbivory.

Two types of defense are commonly distinguished (Table 1). Defense is called direct if induced responses affect the interaction of the herbivore and its host plant directly. It relies on resistance factors that by themselves or in combination impact on the preference of insect herbivores during host plant selection (antixenosis), or the performance of the insect on its host plant (antibiosis). Resistance factors include morphological features of the plant which act as physical barriers to ward off invading pests, and the chemical constitution of the plant which is the primary determinant for host plant selection. As an additional component of direct defense, plants may develop tolerance, allowing them to support a herbivore population similar to that on a susceptible host but without the concomitant reduction in plant fitness. Underlying tolerance are traits related to plant morphology and to the production or allocation of resources which – similar to resistance traits – may be constitutive or inducible by herbivory (Table 1).

For indirect defense plants rely on natural enemies of their herbivores. Morphological and chemical features support the presence, abundance, and effectiveness of predators or parasites of herbivorous insects, and may thus serve as resistance factors. Examples include hollow thorns (domatia) that provide shelter, and extrafloral nectar that is produced as a food source for ants which, in a mutualistic relationship with myrmecophilic plants, protect their hosts from herbivores. Furthermore, in response to herbivory, plants may produce odors that attract predators or parasitoids of herbivores and guide them to their prey. Table 1 includes carnivorous enemies of insect herbivores as 'ecological resistance factors' that are recruited by the plant for indirect defense.

Table 1 Classification of defensive traits. Resistance factors can be categorized as physical (including morphological and structural features of the plant), biochemical (e.g., toxic or anti-nutritive metabolites and proteins), or ecological (i.e., involving other species). They differ with respect to their modes of expression (constitutive or inducible) and action (in direct or indirect defense)

Type of trait	Mode of expression	Mode of action
Physical	Constitutive	Direct defense (antixenosis, antibiosis, tolerance)
(Bio)chemical	Induced	
Ecological		Indirect defense (tritrophic interactions)

Introduction

Direct and indirect mechanisms of plant defense against herbivores, and the *raison d'être* for inducibility as opposed to constitutive expression of defense are introduced in Section I of this book. These introductory chapters provide the background for the subsequent more focussed discussion of individual resistance factors in Section II, and the signals and signaling mechanisms for the induction of direct and indirect defenses in Section III. With emphasis on plant responses that are induced by wounding or herbivory, the progress of research is summarized in a field that continues to be highly dynamic, even 35 years after the initial discovery of proteinase inhibitor accumulation as a wound-induced anti-nutritive defense by Ryan in 1972.

Section I
Basic Concepts of Plant Defense Against Insect Herbivores

Chapter 1
Direct Defenses in Plants and Their Induction by Wounding and Insect Herbivores

Gregg A. Howe and Andreas Schaller

Resistance factors for direct plant defense against herbivorous insects comprise plant traits that negatively affect insect preference (host plant selection, oviposition, feeding behavior) or performance (growth rate, development, reproductive success) resulting in increased plant fitness in a hostile environment. Such traits include morphological features for physical defense, like thorns, spines, and trichomes, epicuticular wax films and wax crystals, tissue toughness, as well as secretory structures and conduits for latices or resins. They also include compounds for chemical defense, like secondary metabolites, digestibility reducing proteins, and antinutritive enzymes. All these traits may be expressed constitutively as preformed resistance factors, or they may be inducible and deployed only after attack by insect herbivores. The induction of defensive traits is not restricted to the site of attack but extends to non-infested healthy parts of the plants. The systemic nature of plant responses to herbivore attack necessitates a long-distance signaling system capable of generating, transporting, and interpreting alarm signals produced at the plant–herbivore interface. Much of the research on the signaling events triggered by herbivory has focused on tomato and other solanaceous plants. In this model system, the peptide systemin acts at or near the wound site to amplify the production of jasmonic acid. Jasmonic acid or its metabolites serve as phloem-mobile long-distance signals, and induce the expression of defense genes in distal parts of the plant. In this chapter, we will provide an overview of physical and chemical defense traits, and review the signaling mechanisms that account for their inducible expression after insect attack.

1.1 Introduction

Plants, flowering plants in particular, exhibit a tremendous diversity in size and shape, ranging from just a few millimeters in the tiny duckweeds to almost 100

A. Schaller
University of Hohenheim, Institute of Plant Physiology and Biotechnology, D-70599 Stuttgart, Germany
e-mail: schaller@uni-hohenheim.de

meters in giant eucalyptus trees. Some may complete their life cycle in a few weeks, while others live thousands of years. The amazing diversity results from the adaptation to different, oftentimes hostile environments, as exemplified by the early evolution of land plants. The colonization of land by plants, dating back some 480 million years according to fossil records (Kenrick and Crane 1997), marks the beginning of an evolutionary success story, with flowering plants now occupying every habitat on Earth except the regions surrounding the poles, the highest mountaintops, and the deepest oceans (Soltis and Soltis 2004). The colonization of land was a major event in the history of plant life, and at the same time, paved the way for the explosive evolution of terrestrial ecosystems. Despite the vulnerability of plants as sessile organisms to adverse biotic and abiotic conditions, they actually dominate over much of the land surface. This apparent success of flowering plants relies on the evolved ability to persist in unfavorable and variable environments by virtue of effective resistance systems that are based on a combination of physical, chemical, and developmental features (Schoonhoven et al. 2005). It was recognized by Stahl in 1888 that the great diversity of mechanical and chemical 'means of protection of plants were acquired in their struggle for existence within the animal world' leading to the conclusion that 'the animal world [...] deeply influenced not only their morphology but also their chemistry' (Stahl 1888; Fraenkel 1959). Hence, not only thorns and spines as morphological resistance traits, but also the bewildering variety of plant secondary chemicals attest to the selective pressure exerted by phytophagous animals (Fraenkel 1959; Ehrlich and Raven 1964).

It was later discovered that induced expression of resistance traits increases plant fitness in environments that harbor a variety of plant parasites. The inducibility of plant resistance was first reported for fungal and bacterial pathogens in the early 1900s (Karban and Kuc 1999) and, much later, inducible defenses were shown to exist also against insect herbivores. In their seminal paper of 1972, Green and Ryan demonstrated that tomato and potato plants accumulate inhibitors of trypsin and chymotrypsin-like serine proteinases throughout their aerial tissues, as a direct consequence of insect-mediated damage or mechanical wounding (Green and Ryan 1972). Proteinase inhibitors are present constitutively in high concentrations in plant storage organs, and a possible function as protective agents against insects was discussed at that time (Lipke et al. 1954; Applebaum and Konijn 1966). Green and Ryan suggested that the expression of proteinase inhibitors may be regulated in leaves to make the plant less palatable and perhaps lethal to invading insects. The accumulation of proteinase inhibitors in aerial tissues was proposed to constitute an inducible defense system, directly affecting the performance of leaf-consuming insects by starving them of nutrients, thus resulting in enhanced plant resistance against herbivory (Green and Ryan 1972). It is now clear that the nutritional quality of the foliage is an important determinant of herbivore growth and development (Painter 1936; Berenbaum 1995; Schoonhoven et al. 2005) and anti-nutritional defense as part of the plant's arsenal for induced resistance is well accepted (Rhoades and Cates 1976; Felton 2005).

Thirty-five years of research following the initial discovery by Ryan and coworkers established plant resistance against insect herbivores as a highly dynamic process. In addition to the proteinase inhibitors, many more inducible factors have been identified which contribute to direct defense and which have the potential to enhance host plant fitness after herbivore attack. These are aspects that will be introduced in this chapter to provide the background for a more detailed discussion of the defensive role of individual proteins in the subsequent, more focused chapters of this volume. Another aspect of induced resistance that has fascinated researchers since the seminal Green-and-Ryan-paper is the systemic nature of the response: defense proteins accumulate not only at the site of wounding but also systemically in unwounded tissues of the infested plant. Obviously, a signal must be generated locally as a consequence of insect feeding which is then propagated throughout the plant, and able to induce the expression of defense proteins at distant sites (Green and Ryan 1972; Ryan and Moura 2002). Our current understanding of systemic wound signaling for direct defense will also be summarized here.

1.2 Inducible Resistance Factors for Direct Defense

Since the initial observation of proteinase inhibitor accumulation in wounded tomato and potato plants, inducibility by herbivory has been shown for a large number of other potential resistance factors (Walling 2000; Gatehouse 2002). In the light of recent studies analyzing induced responses at the level of the entire transcriptome, we now begin to appreciate the full breadth and highly dynamic nature of plant-insect interactions. Numerous studies have shown that herbivory causes large-scale changes in gene expression (Cheong et al. 2002; Delessert et al. 2004; Reymond et al. 2004; Smith et al. 2004; Voelckel and Baldwin 2004; Zhu-Salzman et al. 2004; De Vos et al. 2005; Schmidt et al. 2005; Ralph et al. 2006b; Thompson and Goggin 2006; Broekgaarden et al. 2007). In hybrid poplar, for example, it is estimated that 11% of the transcriptome is differentially regulated by insect feeding (Ralph et al. 2006a). However, inducibility of a certain gene or enzyme per se is not sufficient evidence for a function in plant defense. Whereas the potential contribution of a given trait to plant resistance can be readily tested in a laboratory setting by comparing herbivore preference and/or performance on plants that differ in the trait of interest, a role in plant defense implies that expression of the resistance trait is associated with a gain in plant fitness; such associations must ultimately be demonstrated in field experiments that simulate 'real world' conditions (Karban and Myers 1989).

A further requisite for the evolution of inducible defense systems is heritable variation in the degree of inducibility (Karban and Myers 1989; Agrawal 1999). Genetic variation has frequently been observed in natural populations, e.g., for physical (trichome density) or chemical (glucosinolate content) resistance characters in *Arabidopsis*, and both traits are associated with fitness costs (Mauricio 1998). For induced resistance traits, however, a fitness benefit has been demonstrated in only a few cases. One example is radish plants that were induced to accumulate higher levels of glucosinolates and to produce trichomes at increased density. Compared to control plants, these induced plants exhibited both increased resistance to herbivory and increased seed mass (a correlate of lifetime fitness). This experiment confirmed a role in direct defense for trichomes and glucosinolates as inducible physical and chemical resistance factors, respectively (Agrawal 1998, 1999). Likewise, in *Nicotiana attenuata*, the induced production of nicotine as a chemical resistance factor was associated with metabolic costs, but provided a fitness benefit when plants were under attack by herbivores (Baldwin 1998; see also Steppuhn and Baldwin this volume). Although these findings should not be generalized and a defensive role should not be assumed for all plant responses to wounding and herbivory, the prevalence of inducible resistance traits in present day plant-herbivore systems implies that such responses are likely the result of natural selection imposed by insect herbivores during evolution.

Any plant trait that interferes with host plant selection, oviposition, or feeding of an insect herbivore is a potential resistance factor and may further contribute to plant defense. Most prominent among these traits are morphological features and the chemical composition of the plant, both of which have long been recognized as constitutive resistance characters (Stahl 1888; Fraenkel 1959), and were also the focus of initial studies on inducible resistance to insect herbivores. In this chapter, we provide a brief overview of inducible factors that lead to enhanced resistance through direct effects on insect preference or performance. The traditional distinction between plant defense traits that are either morphological or chemical is used throughout this volume (see also Table 1 on page 2 in the Introduction). It is important to realize, however, that this classification is often arbitrary because any morphological feature is the manifestation of a genetically regulated biochemical process and, therefore, also chemical at its very basis.

1.2.1 Morphological Features for Physical Defense

Insect herbivores from all feeding guilds must make contact with the plant surface in order to establish themselves on the host plant. It is therefore not surprising that physical and chemical features of the plant surface are important determinants of resistance. Epicuticular wax films and crystals cover the cuticle of most vascular plants. In addition to their important role in desiccation tolerance, they also increase slipperiness, which impedes the ability of many non-specialized insects to populate leaf surfaces. The physical properties of the wax layer as well as its chemical composition are important factors of preformed resistance (see Müller this volume). Whereas induced changes in wax production and surface chemistry have been observed, evidence for a role of the cuticle and epicuticular waxes in induced resistance is still scant. Wax biosynthesis and composition are known to vary during plant development, and the physico-chemical properties of the cuticle respond to changes in season and temperature (Müller this volume). Considering the ingenuity of plants in dealing with their offenders, it would thus be surprising if regulated production of wax on the leaf surface were not adopted to influence the outcome of plant-insect interactions.

Other components of the plant surface that serve a role in constitutive defense include thorns and spines directed against mammalian herbivores, and hairs (trichomes) which are effective against insects (Myers and Bazely 1991; Schoonhoven et al. 2005). Non-glandular trichomes may serve as structural resistance factors preventing small insects from contacting the leaf surface or limiting their movement. Morphological and chemical resistance factors are combined in glandular trichomes. Glands produce substances which may repel insect herbivores or deter them from feeding (antixenosis), or immobilize them on the leaf surface. Quite interestingly, trichome density in some plant species increases in response to insect feeding, and therefore constitutes an inducible resistance trait. The defensive role of trichomes is discussed in more detail by Dalin et al. (this volume).

Leaf toughness is an important physical factor for plant resistance, as it affects the penetration of plant tissues by mouthparts of piercing-sucking insects, and also increases mandibular wear in biting-chewing herbivores (Schoonhoven et al. 2005). Leaf toughness is frequently correlated with insect resistance and is a good predictor of herbivory rates (Bergvinson et al. 1995; Coley and Barone 1996; Howlett et al. 2001). Although leaf toughness is typically regarded as a physical character, this trait exemplifies the general difficulty in drawing clear distinctions between physical and chemical resistance factors. Cell wall reinforcement for enhanced leaf toughness results from the deposition of 'chemicals', including macromolecules such as lignin, cellulose, suberin, and callose, small organic molecules (e.g., phenolics), and even inorganic silica particles (Schoonhoven et al. 2005). Enhanced synthesis and/or deposition of these chemicals after wounding leads to induced physical resistance (McNaughton and Tarrants 1983; Bernards this volume; Ginzberg this volume).

Another anatomical defense found in plants of diverse phyolgenetic origin is a network of canals such as lacticifers (latex-containing living cells) or resin ducts (resin-filled intercellular spaces) that store latex or resins under internal pressure. When the canal system is severed, the contents are exuded and may entrap or even poison the herbivore. Out of more than 50 plant families for which such defense systems have been described, the well-studied milkweeds (genus *Asclepias* in the family Asclepiadacea) may serve as an example. Milkweed latices coagulate upon exposure to air and immobilize small insect larvae. As an additional chemical resistance factor, the latex may contain large amounts of toxic cardenolides (Dussourd and Hoyle 2000; Agrawal 2004). Fascinatingly, many specialist herbivores that feed on milkweed or other latex-producing plants employ feeding strategies that block the flow of latex to intended feeding sites. Such feeding behavior has evolved independently in several phylogenetic lineages, and can be viewed as a counteradaptation of herbivores to circumvent latex-based plant defenses (Carroll and Hoffman 1980; Dussourd and Eisner 1987; Dussourd and Denno 1994).

A widely appreciated and well-established form of anatomical protection are the resin-based defenses in conifers (Berryman 1972). The resin, which is a mixture of monoterpenes, sesquiterpenes, and diterpene resin acids, accumulates in resin ducts and related secretory structures. Stem-boring bark beetles and other insects that breach the resin duct system are expelled ('pitched out') from the bore hole by resin flow. Upon exposure to air, the highly volatile monoterpene fraction evaporates, leaving the insects trapped in the solidifying resin acids and the wound site sealed (Phillips and Croteau 1999; Trapp and Croteau 2001). Although this complex resin-based defense system in conifers is preformed, it is further induced in response to wounding. Among the inducible components of the system are terpene biosynthesis (Bohlmann this volume) and the formation of new resin ducts (Krokene et al. this volume).

Finally, the wound healing process itself can be considered as a wound-induced anatomical trait for enhanced resistance. Efficient sealing of the wound is important to prevent water loss and opportunistic infections by bacterial and fungal pathogens at the site of tissue damage. Wound closure may involve extensive cell division and formation of wound callus (e.g., Guariguata and Gilbert 1996). In the case of plants with resin- and latex-based defenses, coagulation of the exudates may efficiently seal the wound site. More generally, a sealing cell layer is formed by infusion of antimicrobial and water-impermeable substances, including lignin and suberin (Rittinger et al. 1987). This may be followed by the induction of cell division and the formation of a periderm as a protective tissue that is impermeable to water and resistant to pathogens. Wound periderm formation and its potential contribution to plant defense are discussed in greater detail by Ginzberg (this volume).

1.2.2 Metabolites and Enzymes for Chemical Defense

Plant chemicals that play a role in direct defense impair herbivore performance by one of two general mechanisms: these chemicals may reduce the nutritional value of plant food, or they may act as feeding deterrents or toxins. There has been considerable debate as to which of these two strategies is more important for host plant selection and insect resistance. An important part of this debate concerns the extent to which variation in the levels of primary and secondary metabolites has evolved as a plant defense (Berenbaum, 1995). Plant primary metabolism, which is shared with insects and other living organisms, provides carbohydrates, amino acids, and lipids as essential nutrients for the insect. Food quality is largely determined by the availability of these nutrients, and its importance for longevity, size, fecundity, and death rates in herbivorous insects has been recognized early on by Painter (1936). In addition, more than 100,000 plant compounds (i.e., secondary metabolites) have been identified with no apparent role in primary metabolism, and many of these have been regarded as expendable metabolic waste products. While many secondary metabolites are in fact expendable for primary metabolism, it is now widely accepted that they serve important ecological functions in the interaction of plants with their biotic and abiotic environment.

According to the paradigm put forward by Fraenkel in his seminal paper in 1959, secondary metabolites in a given plant species may act both as repellents for generalist (polyphagous) insects and as attractants for specialist (monophagous) insects, and may thus be largely responsible for host range restriction (Fraenkel 1959). In addition to these allelochemical functions, secondary metabolites also act in multiple ways as toxins, feeding deterrents, as digestibility reducers or antinutritives, as precursors for physical defense, and as volatiles in indirect defense (Bennett and Wallsgrove 1994; Karban and Baldwin 1997). Despite their diversity in structure, activity, and distribution in the plant kingdom, all secondary compounds are derived from universally available intermediates of primary metabolism, including sugar phosphates (erythrose 4-phosphate), acetyl-coenzyme A, and amino acids, and are conveniently classified according to their biosynthetic pathways as phenolics, terpenoids, and alkaloids. Each of these classes of compounds and their role in induced resistance are the focus of subsequent chapters of this volume. While the importance of secondary metabolites in plant defense remained undisputed for decades following Fraenkel's landmark paper (Fraenkel 1959), the realization that some secondary metabolites (e.g., tannins and phenolics) exert anti-nutritive activity brought greater attention to the idea that food quality, nutritional value, and variation in primary metabolism may have evolved as a plant defense (Feeny 1970; Rhoades and Cates 1976; Berenbaum 1995). The relevance of nutritional quality as a resistance trait was further supported by Ryan and coworkers (Green and Ryan 1972) who showed that induced expression of serine proteinase inhibitors contributes to plant defense by interfering with the insect's digestive processes, thus limiting the availability of essential amino acids.

Following the landmark study of Green and Ryan, many workers reported that the overall chemical composition of the plant is greatly influenced by developmental and environmental parameters, including herbivory. Induced changes in plant chemistry involve the biosynthesis of a wide variety of secondary metabolites, including phenolics, terpenoids, alkaloids, cyanogenic glucosides, and glucosinolates (Karban and Baldwin 1997; Constabel 1999; Chapters 8–10, this volume). It was further shown that the induction of anti-nutritional proteins is not limited to serine proteinase inhibitors, but includes inhibitors of other classes of proteases, oxidative enzymes, amino acid-metabolizing enzymes, and lectins (Constabel 1999; Felton 2005). Such examples of protein-based defenses are further discussed in Chapters 11–14 of this volume.

1.2.3 Metabolic Reconfiguration to Shift from a Growth- to a Defense-Oriented State

The numerous anatomical and chemical changes associated with induced resistance require massive reprogramming of gene expression. For the quantitative analysis of large-scale changes in gene expression, novel techniques have been developed in recent years. Most notable among these approaches are microarray technologies for the identification of differentially expressed transcripts, and even more recently, techniques for high-throughput proteomic analysis (Kessler and Baldwin 2002; Kuhn and Schaller 2004; Giri et al. 2006; Lippert et al. 2007). With the advent of these techniques, it is now possible to obtain a relatively unbiased account of the plant's response to herbivory. Many of the genes required for the expression of known resistance traits were in fact shown to be upregulated during plant-insect interaction. Consistent with the activation of structural defenses, genes of general phenylpropanoid metabolism and monolignol biosynthesis, lignin polymerization, and cell wall fortification are induced by wounding or herbivory in hybrid poplar (Smith et al. 2004; Lawrence et al. 2006; Major and Constabel 2006; Ralph et al. 2006a), Sitka spruce (Ralph et al. 2006b), and *Arabidopsis* (Cheong et al. 2002; Delessert et al. 2004; Reymond et al. 2004). Likewise, the activation of chemical defenses is accompanied by the induction of genes involved in secondary metabolism, including phenolics, polyamine, and alkaloid biosynthesis in *N. attenuata* (Voelckel and Baldwin 2004; Schmidt et al. 2005; Giri et al. 2006), the genes for the formation of phenolics and terpenes in spruce and poplar (Ralph et al. 2006a, b), and phenolic metabolism and glucosinolate biosynthesis in *Arabidopsis* (Cheong et al. 2002; Reymond et al. 2004). These studies also confirmed the activation of genes for antidigestive and antinutritional defenses (e.g., proteinase inhibitors, oxidative enzymes, lectins) and for the signaling of the resistance response (e.g., jasmonic acid and ethylene biosynthesis, transcription factors). Notably, however, these genes represent only a fraction of the total wound-induced changes in gene activity. The insect-responsive transcriptome was estimated to comprise 10% of all transcripts, suggesting that massive reprogramming of gene expression is required to bring about a shift from growth-oriented to defense-oriented plant metabolism (Hui et al. 2003; Ralph et al. 2006a). The latter state involves the activation of genes for general stress responses (oxidative stress, dehydration stress, heat-shock proteins), protein turnover (e.g., proteases), and transport processes (e.g., aquaporins, lipid transfer proteins, ABC transporters, sugar and peptide transporters), as well as modulation of primary metabolism (carbohydrate and lipid metabolism, nitrogen assimilation), and downregulation of photosynthesis and chloroplast function. These changes in gene expression may reflect the herbivore-induced reallocation of resources from primary processes to defense (Voelckel and Baldwin 2004; Ralph et al. 2006b).

Efficient mobilization of plant resources is likely to facilitate the expression of costly resistance traits, including the accumulation of defense proteins, the synthesis of secondary metabolites, and the formation of structural defenses. On the other hand, mobilization of resources may also contribute to plant tolerance of herbivory. Unlike resistance, tolerance does not affect herbivore preference or performance, but rather allows the host plant to minimize the fitness consequences of tissue loss. Tolerance and resistance are therefore viewed as alternative and complementary strategies for plant defense against insect herbivores (Karban and Myers 1989; Mauricio 2000; Weis and Franks 2006) Whereas tolerance is still not well-understood at the molecular level, it may include the mobilization of leaf carbon and nitrogen that is threatened by herbivory, and temporary storage of these resources for later regrowth. The induction of protein turnover, lipid and carbohydrate metabolism, and transport functions observed in microarray studies (see above) may thus be equally relevant for both tolerance and induced expression of resistance traits. Temporary storage of resources occurs in organs that are less susceptible to herbivory, e.g., the root system. Indeed, a change in sink-source relations was observed in *N. attenuata* after simulated herbivore attack, resulting in increased allocation of sugars to roots and enhanced tolerance (Schwachtje et al. 2006). Likewise, the induction of vegetative storage proteins frequently observed in response to wounding (Staswick 1994; Christopher et al. 2004; Reymond et al. 2004; Major and Constabel 2006) may allow the plant to buffer mobilized resources for later use in re-growth. Remarkably, such a role as interim storage or temporary protein depot had already been suggested for proteinase inhibitor I in tomato and potato plants, the first protein shown to be systemically induced by herbivory (Ryan and Huisman 1969; Green and Ryan 1972).

1.3 Systemic Signaling for Induced Direct Defense

An important feature of many wound-induced direct defense responses is their occurrence in undamaged tissues located far from the site of wounding. Wound-inducible serine proteinase inhibitors (PIs) represent one of the best examples of a systemically induced defense response. In tomato plants, *PI* genes are expressed in distal leaves within 1–2 hrs after insect attack or mechanical wounding (Ryan 2000; Strassner et al. 2002). The rapid and systemic nature of this response is analogous to vertebrate immune responses in which endocrine signals are delivered to target tissues via the circulatory system (Bergey et al. 1996). However, because plants lack mobile defender cells, systemic signals must be transmitted long distances via mechanisms that are specific to plants (Malone 1996; León et al. 2001; Schilmiller and Howe 2005). Ryan's pioneering work on systemic wound signaling inspired generations of plant biologists to investigate the underlying mechanisms of this fascinating response.

The widespread occurrence of systemic defense responses in the plant kingdom implies the existence of common mechanisms to generate, transport, and perceive alarm signals that are generated at the site of tissue damage. Wound-inducible PIs in tomato and other solanaceous plants have been widely used as a model system in which to study the molecular mechanism of systemic wound signaling. Green and Ryan (1972) proposed that chemical signals produced at the wound site travel through the plant and activate PI expression in undamaged leaves. Identification of these signaling compounds was facilitated by a simple bioassay in which test solutions (e.g., containing an elicitor) are supplied to tomato seedlings through the cut stem, followed by measurement of PI accumulation in the leaves. Extensive use of this assay led to the discovery of several distinct classes of PI-inducing compounds, including cell-wall-derived oligogalacturonides (OGAs), systemin, jasmonic acid (JA), and hydrogen peroxide (Ryan 2000; Gatehouse 2002). Physical signals (e.g., hydraulic forces and electrical signals) generated by tissue damage have also been implicated in the systemic signaling process (Wildon et al. 1992; Malone 1996). Currently, a major challenge is to determine how these diverse signals interact with one another to promote intercellular communication across long distances.

Farmer and Ryan (1992) established the current paradigm that extracellular signals such as OGAs and systemin (so-called primary wound signals), generated in response to wounding, trigger the intracellular production of JA via the octadecanoid pathway and that JA, in turn, activates the expression of defensive genes. Wound-induced production of OGAs is catalyzed by a family of polygalacturonases (PGs) that are expressed in various plant tissues (Bergey et al. 1999). OGAs are relatively immobile in the plant vascular system and thus are thought to act as local mediators. However, because PG activity is induced systemically in response to wounding, OGAs could also amplify defense responses in undamaged leaves (Ryan 2000). OGA-mediated signal transduction may result from direct physical effects of these compounds on the plasma membrane or may involve specific receptors (Navazio et al. 2002).

Systemin was the first bioactive peptide discovered in plants (Pearce et al. 1991). This 18-amino-acid peptide is derived from proteolytic cleavage of a larger precursor protein, prosystemin. When used in the tomato seedling bioassay, systemin is >10, 000-fold more active than OGAs in inducing PI expression. Several lines of evidence indicate that systemin serves a key role in induced defense responses in tomato. For example, transgenic plants expressing an antisense *prosystemin* (*Prosys*) cDNA are deficient in wound-induced systemic expression of PIs and, as a consequence, are more susceptible to insect herbivores (McGurl et al. 1992; Orozco-Cárdenas et al. 1993). Overexpression of prosystemin from a *35S::Prosys* transgene constitutively activates PI expression in the absence of wounding, thereby conferring enhanced resistance to herbivores (McGurl et al. 1994; Li et al. 2002; Chen et al. 2005). Forward genetic analysis has shown that genes required for systemin-mediated signaling are essential for wound-induced expression of *PI* and other defense-related genes (Howe and Ryan 1999; Howe 2004). Thus, wounding and systemin activate defense genes through a common signaling pathway.

Transcriptional activation of defense genes in response to systemin requires the biosynthesis and subsequent action of JA (Farmer and Ryan 1992; Howe, 2004). The systemin signaling pathway is initiated upon binding of the peptide to a 160-kDa plasma membrane-bound receptor (SR160) that was identified as a member of the leucine-rich repeat (LRR) receptor-like kinase family of proteins (Scheer and Ryan 1999, 2002). Binding of systemin to the cell surface is associated with several rapid signaling events, including increased cytosolic Ca^{2+} levels, membrane depolarization, and activation of a MAP kinase cascade (Felix and Boller 1995; Stratmann and Ryan 1997; Moyen et al. 1998; Schaller and Oecking 1999). The precise mechanism by which systemin activates JA synthesis remains to be determined. There is evidence indicating that a systemin-regulated phospholipase A_2 activity in tomato leaves releases linolenic acid, a JA precursor, from lipids in the plasma membrane (Farmer and Ryan 1992; Narváez-Vásquez et al. 1999). Alternatively, the role of a chloroplast-localized phospholipase A_1 in JA biosynthesis (Ishiguro et al. 2001) raises the possibility that systemin perception at the plasma membrane is coupled to the activation of a similar lipase in the chloroplast. JA synthesized in response to systemin, OGAs, and wounding acts in concert with ethylene (O'Donnell et al. 1996) and hydrogen peroxide (Orozco-Cárdenas et al. 2001; Sagi et al. 2004) to positively regulate the expression of downstream target genes.

1.3.1 Genetic Analysis of the Wound Response Pathway in Tomato

Genetic analysis provides a powerful approach to identify components of the systemic wound response pathway. The robust nature of wound-induced PI expression in tomato, together with facile assays for PIs and other biochemical markers (e.g., polyphenol oxidase) of the response, has been exploited for this purpose. Forward genetic screens identified mutants that are defective in PI expression in response to mechanical wounding or treatment with methyl-JA (MeJA) (Howe 2004; Li et al. 2004). Additional screens have been conducted to identify mutations that suppress the inductive effects of the *35S::Prosys* transgene (Howe and Ryan 1999). These screens have yielded numerous mutants that are deficient in wound-induced systemic expression of defensive genes. That most of these mutants display altered resistance to arthropod herbivores and various pathogens demonstrates the importance of induced responses to plant protection (Howe 2004).

Map-based cloning and candidate gene approaches were used to identify genes defined by forward genetic analysis. The *spr2* and *acx1* mutants, which were generated by ethylmethane sulfonate mutagenesis, are defective in genes required for JA biosynthesis. *Spr2* encodes a plastidic ω-3 fatty acid desaturase that converts linoleic acid to the JA precursor linolenic acid (Li et al. 2003). *ACX1* encodes a peroxisomal acyl-CoA oxidase that catalyzes the first step in the β-oxidation stage of JA synthesis (Li et al. 2005). The *jasmonate insensitive1* (*jai1*) mutant harbors a deletion in the tomato ortholog of the *Arabidopsis Coronatine insensitive1* (*Coi1*) gene (Li et al. 2004). *Coi1* encodes an F-box protein that is essential for expression of jasmonate-responsive genes, including many wound-responsive genes involved in anti-insect defense (Xie et al. 1998). Reverse genetic strategies identified several additional wound response mutants of tomato. For example, transgenic lines specifically engineered for defects in JA biosynthesis (Stenzel et al. 2003), ethylene synthesis (O'Donnell et al. 1996), ABA signaling (Carrera and Prat 1998), and ROS production (Sagi et al. 2004) are impaired in wound-induced systemic PI expression and other defense responses.

1.3.2 Jasmonate Performs a Key Role in Systemic Wound Signaling

Despite significant progress in the identification of genes that regulate systemic defense responses, relatively little is known about the specific role of these components in the long-distance signaling pathway. In theory, genes required for the systemic response could play a role in production of the mobile signal, translocation of the signal from damaged to undamaged leaves, signal perception by target cells in distal leaves, or subsequent signaling steps leading to expression of target genes. Classical grafting techniques provide a powerful approach to determine whether a particular mutant is defective in the production of the systemic (i.e., graft-transmissible) wound signal or the recognition of that signal in responding leaves (Li et al. 2002). Reciprocal grafting experiments performed with the

JA-insensitive *jai1* mutant showed that jasmonate perception (i.e., COI1) is essential for recognition of the mobile signal in distal responding leaves (Fig. 1.1Ad). These studies also suggest that the mobile signal is produced in the absence of COI1 (Fig. 1.1Ac). Experiments conducted with JA biosynthetic mutants (e.g., *acx1*) showed that production of the graft-transmissible signal depends on JA biosynthesis in wounded tissues (Fig. 1.1Bc). The ability of JA-deficient scions to express PIs in response to a signal emanating from wild-type rootstock leaves further indicated that de novo JA synthesis is likely not necessary for recognition of the mobile signal in the responding leaves (Fig. 1.1Bd). Based on these collective studies, it was proposed that JA (or a JA derivative) is a critical component of the systemic signal (Schilmiller and Howe 2005). These findings are also consistent with DNA microarray studies showing that local and systemic tissues undergo distinct signaling events (Strassner et al. 2002).

The plant vascular system is involved in long-distance trafficking of a wide range of signaling compounds (Lucas and Lee 2004). Recent studies provide direct evidence that jasmonates are transported in the phloem (Fig. 1.2). For example, several JA biosynthetic enzymes are located in the companion cell-sieve element complex of the vascular bundle (Hause et al. 2003; Wasternack 2007). This observation is supported by the occurrence of JA in phloem bundles from *Plantago major* (Hause et al. 2003) and the preferential accumulation of jasmonates in the tomato leaf midrib (Stenzel et al. 2003). The hypothesis that the systemic signal is translocated in the phloem is further supported by the fact that wound-induced systemic responses are strongly enhanced by the strength of vascular connections between wounded and responding leaves (Davis et al. 1991; Schittko and Baldwin 2003). The rate of movement of the endogenous signal in tomato plants is estimated between 1 and 5 cm/hr (Schilmiller and Howe 2005). The ability of the phloem to transport small molecules at rates up to 40 cm/hr (Fisher 1990) could readily accommodate such a signal. Because systemic PI expression is mediated by a signal traveling within the plant rather than a signal diffusing through the atmosphere (Farmer and Ryan 1992), it is unlikely that volatile MeJA released at the wound site is a causal factor for systemic PI expression in tomato.

The idea that JA (or a JA derivative) functions as a mobile wound signal implies that JA synthesized in damaged leaves is transported to distal undamaged leaves. In tomato and other dicots, however, systemic increases in JA levels in response to mechanical damage are generally very low (i.e., <10% of that in damaged leaves) or not significant (Strassner et al. 2002). In those cases where systemic increases in JA levels have been reported, it was not determined whether accumulation of the signal results from de novo synthesis in undamaged leaves or JA transport from wounded source leaves. Grafting experiments (see above) support the latter possibility, as does the phloem mobility and systemic signaling activity of exogenous JA (Farmer and Ryan 1992; Zhang and Baldwin 1997). Low levels of wound-induced JA in systemic leaves may reflect sequestration of the signal in specific cell types of the vasculature. An alternative, though not mutually exclusive, possibility is that the phloem-mobile pool of JA is rapidly metabolized to another bioactive derivative. JA derivatives produced by methylation, glycosylation, hydroxylation, sulfonation, amino acid

A. Grafting with a jasmonate perception mutant

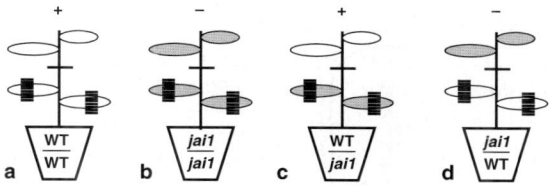

B. Grafting with a JA biosynthesis mutant

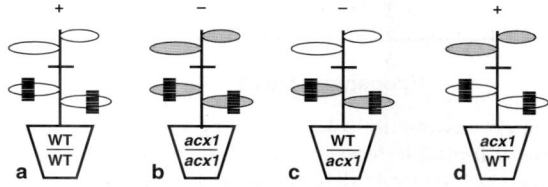

C. Grafting with a systemin-insensitive mutant

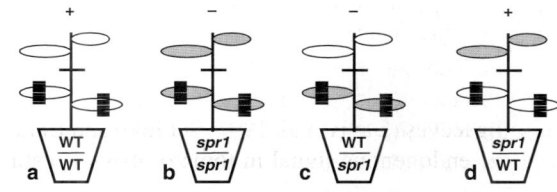

D. Grafting with a prosystemin-overexpressing mutant

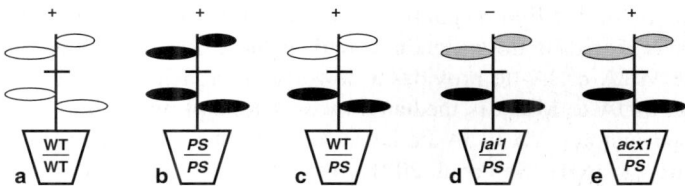

Fig. 1.1 Schematic diagram of grafting experiments used to study the role of JA in systemic wound signaling in tomato. Scions and rootstocks of the indicated genotype were joined at the graft junction (horizontal bar). For experiments shown in A, B, and C, rootstock leaves were wounded (hatched mark) and PI gene expression in the undamaged scion leaves was measured 8 hrs later. (**A**) The *jai1* mutant was used to investigate the role of jasmonate perception in systemic wound signaling. (**B**) The *acx1* mutant was used to study the role of JA synthesis in systemic wound signaling. (**C**) The *spr1* mutant was used to study the role of systemin perception in systemic signaling. For experiments depicted in panel D, no wounds were inflicted because the *35S::Prosys* (PS) transgenic line constitutively produces a systemic signal. '+' and '−' denote the expression or lack of expression, respectively, of PIs in undamaged scion leaves. Unfilled ovals correspond to wild-type (WT) leaves. Gray-shaded ovals depict leaves on mutants (*jai1*, *acx1*, or *spr1*) that are defective in systemic wound signaling. Black ovals depict leaves on the *35S::Prosys* transgenic line

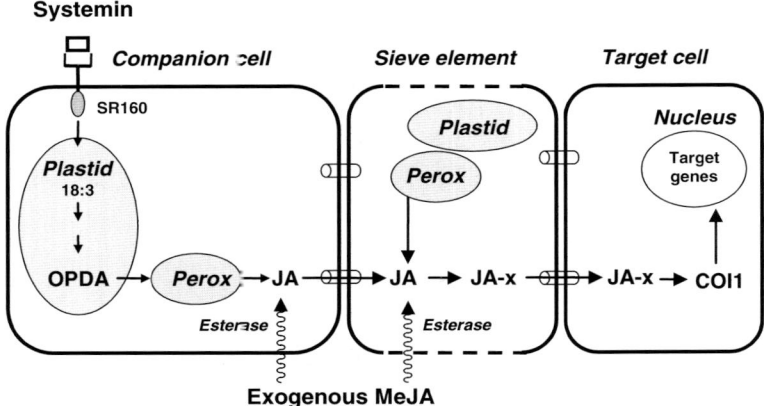

Fig. 1.2 Schematic model showing the role of JA in systemic wound signaling. Chloroplastic (Plastid) and peroxisomal (Perox) JA biosynthetic enzymes are located in vascular bundles of the leaf. Binding of systemin to its receptor (SR160) activates JA accumulation. JA synthesis in tomato leaves is also activated by systemin-independent pathways (not shown; Lee and Howe 2003). JA produced in the companion cell-sieve element complex is transported in the phloem via plasmodesmata connections between cells. JA, or a covalently modified form of JA (JA-x; such as JA-Ile), activates target gene expression in distal undamaged leaves through COI1. Esterases may convert exogenous MeJA to JA upon diffusion of MeJA across membranes

conjugation, and decarboxylation have been described (Wasternack 2007). One or more of these modifications could conceivably alter the transport, stability, or interaction of JA with target molecules (Fig. 1.2).

MeJA and certain jasmonoyl-amino acid conjugates (e.g., JA-Ile) are potent elicitors of defense gene expression (Wasternack et al. 1998). The dependence of MeJA- and JA-Ile-induced responses on COI1 indicates that both compounds are candidates for signals in the systemic wound response. Analysis of mutants that fail to produce MeJA or JA-Ile provides a powerful approach to test this hypothesis. Conversion of JA to MeJA is mediated by JA carboxyl methyltransferase (JMT), whereas conversion of JA to JA-Ile is catalyzed by the ATP-dependent adenylate-forming enzyme JAR1 (Seo et al. 2001; Staswick and Tiryaki 2004). Although the effect of loss of *JMT* function on wound-induced defense responses is not known, it is firmly established that JAR1-mediated production of JA-Ile plays a critical role in numerous jasmonate-signaled processes (Staswick and Tiryaki 2004). Moreover, recent studies have shown that JAR1 homologs in *N. attenuata* are required for wound-induced defense responses to insect attack (Kang et al. 2006). JA-Ile's key role in induced defense raises the possibility that biological responses previously attributed to JA/MeJA are in fact mediated by JA-Ile or other amino acid conjugates of JA. Consistent with this notion, physical interaction between COI1 and repressors of jasmonate-dependent gene expression, which results in proteasome-dependent degradation of the repressor proteins, was recently shown to be promoted by JA-Ile but not by JA or MeJA (Thines et al. 2007). The potency of exogenous MeJA as an elicitor of gene expression may reflect its ability to readily penetrate cellular membranes (Fig. 1.2). Once inside the cell, MeJA is likely converted to JA by specific or non-specific esterases (Stuhlfelder et al. 2004), followed by conversion to JA-Ile by JAR1 (Staswick and Tiryaki 2004). The use of *jar* mutants in grafting experiments, together with direct measurement of JA-Ile levels in phloem exudates and wounded tissues, promises to provide additional insight into the role of this bioactive conjugate in the wound signaling pathway.

1.3.3 Amplification of the Jasmonate Signal by Systemin

Activation of PI expression by systemin requires the synthesis and subsequent action of JA (Schilmiller and Howe 2005; Wasternack 2007). In the context of long-distance wound signaling, this role for systemin can be reconciled with the above-mentioned grafting studies if it is postulated that systemin activates JA synthesis at or near the site of tissue damage (Li et al. 2002; Ryan and Moura 2002). This model is consistent with grafting studies showing that a *35S::Prosys* transgenic rootstock constitutively generates a systemic signal that activates PI expression in wild-type scion leaves (Fig. 1.1Dc) (McGurl et al. 1994). Recognition of the *35S::Prosys*-derived signal in scion leaves is blocked by *jai1* but not by mutations such as *acx1* that disrupt JA biosynthesis (Fig. 1.1Dd-e) (Li et al. 2002). These findings suggest that *35S::Prosys*-expressing tissues constitutively synthesize JA, which is then mobilized to scion leaves where it initiates COI1-dependent responses in target cells. This model is consistent with the observation that *35S::Prosys* plants accumulate increased JA levels in the absence of wounding (Chen et al. 2006). Activation of PI expression in JA-deficient scions (Fig. 1.1De) indicates that the long-distance signal produced by *35S::Prosys* rootstocks is likely not systemin, but rather a signal that activates PI expression in the absence of de novo JA synthesis.

A role for systemin in localized JA production is also in agreement with results obtained from analysis of the systemin-insensitive mutant *spr1* (Howe and Ryan 1999; Lee and Howe 2003). *spr1* mutants express *PI* genes in response to elicitation by OGA and JA, but not in response to systemin and prosystemin. *Spr1* is presumably required for a signaling step that links systemin perception at the plasma membrane to activation of JA synthesis in the chloroplast. Interestingly, systemic PI expression in *spr1* plants is impaired much more than the local response (Lee and Howe 2003). This phenotype is very similar to that of *Prosys* antisense plants (Orozco-Cárdenas et al. 1993), and provides evidence that (pro)systemin functions mainly in the long-distance response. Grafting experiments provided evidence that *Spr1* function (i.e., systemin perception) is involved primarily in the generation of the systemic signal in wounded leaves and is not required for recognition of the signal in undamaged responding leaves (Fig. 1.1C). The most straightforward interpretation of these results is that (pro)systemin acts at or near the wound site to amplify JA accumulation and the strength of the systemic response.

It thus appears that numerous signals, including JA, systemin, and H_2O_2, interact through a positive feedback loop to propagate the long-distance signal via the phloem (Ryan 2000; Ryan and Moura 2002; Schilmiller and Howe 2005; Wasternack 2007). Future work is needed to understand how these signals interact with one another to promote the systemic wound response, and to determine which of these signals are functionally conserved in other plant species. The absence of *Prosys* gene homologs outside the Solanaceae suggests that systemin may have evolved in a narrow range of plants, perhaps as a mechanism to amplify systemic defense responses to insect attack (Howe 2004). The notion that systemin function is rapidly evolving is supported by recent studies indicating that a systemin homolog in *Solanum nigrum* is not involved in wound-induced direct defense responses (Schmidt and Baldwin 2006). Jasmonate-based signaling, on the other hand, appears to play a central role in regulating responses to biotic stress in all plants. Increasing evidence indicates that the role of jasmonates in promoting systemic defense may be more general than previously realized (Truman et al. 2007). These collective findings validate Ryan's original concept that chemical alarm signals produced at the plant-pest interface mediate systemic immunity to biotic stress.

1.4 Perspectives

Since the initial discovery by Ryan and coworkers of digestibility-reducing proteinase inhibitors as an inducible defense in the Solanaceae 35 years ago, inducible mechanisms for direct defense against insect herbivores have been identified throughout the plant kingdom, from unicellular green alga (Hessen and van Donk 1993; van Donk and Hessen 1993; Lampert et al. 1994) to trees (Bohlmann this volume). A plethora of inducible morphological and chemical resistance factors have been identified that reduce the availability of nutrients (e.g., incorporation of silica as structural reinforcement, antinutritive secondary metabolites and proteins), or are outright toxic to the herbivore (e.g., secondary metabolites including terpenoids, phenolics, and alkaloids). Numerous microarray studies aimed at analyzing global changes in gene expression after herbivory have confirmed the induced expression of many defensive genes. Moreover, the massive reprogramming of gene expression observed in these studies suggests that herbivory results in a shift from growth-oriented to defense-oriented plant metabolism (Hui et al. 2003; Ralph et al. 2006a, b). The number of herbivore-induced genes appears to greatly exceed the requirements for known resistance traits, suggesting that additional components of induced defense reamin to be discovered. Indeed, in addition to interfering directly with herbivore behavior or physiology, plants may use 'scorched earth' or 'escape strategies' as complementary defense measures. Valuable C and N resources are mobilized in organs threatened by herbivory, and are either used for the synthesis of resistance factors, or stored out of reach of the herbivore. Presumably, the resulting nutrient-deprived plant organs will poorly support the growth and development of attacking herbivores. Re-allocation of mobilized resources to temporary storage proteins (vegetative storage proteins and proteinase inhibitors) and/or underground storage organs (bulbs and tubers) supports later re-growth, and may allow plants to escape herbivory in time. Indeed, enhanced carbon allocation to roots in response to herbivory was recently observed in *N. attenuata*, resulting in delayed senescence and a prolonged reproductive phase. Sucrose transport to roots was found to be controlled by SNRK1, a protein kinase that was rapidly downregulated in leaves after attack by *Manduca sexta* (Schwachtje et al. 2006). Such 'civilian defenses' (Karban and Baldwin 1997) leading to enhanced tolerance of herbivory are still poorly understood at the molecular level and will be an important field for future research.

Tremendous progress has also been made with respect to the signaling events that lead to the systemic expression of defensive traits in response to herbivory. This includes the discovery of systemin as the first peptide with hormone-like activity in plants, which is now thought to act in the vicinity of the wound site to amplify the production of a long-distance signal in the vasculature. Although the systemic signal molecule remains to be identified, recent evidence suggests that JA or a JA metabolite – possibly JA-Ile – may act as a phloem-mobile signal. The perception of jasmonates and activation of defense genes in target tissues was shown to depend on COI1, which is part of an E_3 ubiquitin ligase (SCF^{COI1}) that was predicted to tag a repressor of JA signaling for degradation by the ubiquitin-proteasome pathway. Several members of the JAZ (Jasmonate ZIM domain) family of proteins have recently been identified as targets of SCF^{COI1} in tomato and *Arabidopsis*. At least two JAZ proteins are known to act as negative regulators of jasmonate-dependent transcription, and the COI1/JAZ1 complex was suggested to be the site of JA-Ile perception (Chini et al. 2007; Thines et al. 2007). Despite these exciting findings, there are still important questions to be resolved with respect to systemic wound signaling. This is particularly true for the early events in signal transduction that couple tissue damage to the activation of the octadecanoid pathway for JA production. Most notably, the events following systemin perception at the cell surface and the subsequent release of polyunsaturated fatty acids for oxylipin biosynthesis in chloroplasts remain to be elucidated. The tomato *spr1* mutant is impaired in this process and the identification of the genetic defect in *spr1* may turn out to be an important step in this direction.

Acknowledgments This work was supported in part by the National Institutes of Health Grant R01GM57795 and the US Department of Energy Grant DE-FG02-91ER20021 (G.A.H). A.S. gratefully acknowledges support from the German Research Foundation (DFG).

References

Agrawal AA (1998) Induced responses to herbivory and increased plant performance. Science 279:1201–1202

Agrawal AA (1999). Induced plant defense: evolution of induction and adaptive phenotypic plasticity. In: Agrawal AA, Tuzun S, Bent E (eds) Induced plant defenses against pathogens and herbivores: biochemistry ecology and agriculture. APS Press, Minnesota, pp 251–268

Agrawal AA (2004) Resistance and susceptibility of milkweed: competition, root herbivory, and plant genetic variation. Ecology 85:2118–2133

Applebaum SW, Konijn AM (1966) The presence of a *tribolium*-protease inhibitor in wheat. J Insect Physiol 12:665–669

Baldwin IT (1998) Jasmonate-induced responses are costly but benefit plants under attack in native populations. Proc Natl Acad Sci USA 95:8113–8118

Bennett RN, Wallsgrove RM (1994) Secondary metabolites in plant defence mechanisms. New Phytol 127:617–633

Berenbaum MR (1995) Turnabout is fair play: secondary roles for primary compounds. J Chem Ecol 21:925–940

Bergey DR, Howe GA, Ryan CA (1996) Polypeptide signaling for plant defensive genes exhibits analogies to defense signaling in animals. Proc Natl Acad Sci USA 93:12053–12058

Bergey DR, Orozco-Cardenas M, de Moura DS, Ryan CA (1999) A wound- and systemin- inducible polygalacturonase in tomato leaves. Proc Natl Acad Sci USA 96:1756–1760

Bergvinson DJ, Hamilton RI, Arnason JT (1995) Leaf profile of maize resistance factors to european corn borer, *Ostrinia nubilalis*. J Chem Ecol 21:343–354

Berryman AA (1972) Resistance of conifers to invasion by bark beetle-fungus associations. BioScience 22:598–602

Broekgaarden C, Poelman E, Steenhuis G, Voorrips R, Dicke M, Vosman B (2007) Genotypic variation in genome-wide transcription profiles induced by insect feeding: *brassica oleracea – Pieris rapae* interactions. BMC Genomics 8:239

Carrera E, Prat S (1998) Expression of the *Arabidopsis* abi1-1 mutant allele inhibits proteinase inhibitor wound-induction in tomato. Plant J 15:765–771

Carroll CR, Hoffman CA (1980) Chemical feeding deterrent mobilized in response to insect herbivory and counteradaptation by *Epilachna tredecimnotata*. Science 209:414–416

Chen H, Jones AD, Howe GA (2006) Constitutive activation of the jasmonate signaling pathway enhances the production of secondary metabolites in tomato. FEBS Lett 580:2540–2546

Chen H, Wilkerson CG, Kuchar JA, Phinney BS, Howe GA (2005) Jasmonate-inducible plant enzymes degrade essential amino acids in the herbivore midgut. Proc Natl Acad Sci USA 102:19237–19242

Cheong YH, Chang H-S, Gupta R, Wang X, Zhu T, Luan S (2002) Transcriptional profiling reveals novel interactions between wounding, pathogen, abiotic stress, and hormonal responses in *Arabidopsis*. Plant Physiol 129:661–677

Chini A, Fonseca S, Fernández G, Adie B, Chico JM, Lorenzo O, García-Casado G, López-Vidriero I, Lozano FM, Ponce MR, Micol JL, Solano R (2007) The JAZ family of repressors is the missing link in jasmonate signalling. Nature 448:666–671

Christopher ME, Miranda M, Major IT, Constabel CP (2004) Gene expression profiling of systemically wound-induced defenses in hybrid poplar. Planta 219:936–947

Coley PD, Barone JA (1996) Herbivory and plant defenses in tropical forests. Annu Rev Ecol Syst 27:305–335

Constabel CP (1999) A survey of herbivory-inducible defensive proteins and phytochemicals. In: Agrawal AA, Tuzun S, Bent E (eds) Induced plant defenses against pathogens and herbivores. Biochemistry, ecology, and agriculture. APS Press, St. Paul, Minesota, pp 137–166

Davis JM, Gordon MP, Smit BA (1991) Assimilate movement dictates remote sites of wound-induced gene expression in poplar leaves. Proc Natl Acad Sci USA 88:2393–2396

De Vos M, Van Oosten VR, Van Poecke RMP, Van Pelt JA, Pozo MJ, Mueller MJ, Buchala AJ, Metraux JP, Van Loon LC, Dicke M et al (2005) Signal signature and transcriptome changes of *Arabidopsis* during pathogen and insect attack. Mol Plant Microbe Int 18:923–937

Delessert C, Wilson IW, Van der Straeten D, Dennis ES, Dolferus R (2004) Spatial and temporal analysis of the local response to wounding in *Arabidopsis* leaves. Plant Mol Biol 55:165–181

Dussourd DE, Denno RF (1994) Host-range of generalist caterpillars – trenching permits feeding on plants with secretory canals. Ecology 75:69–78

Dussourd DE, Eisner T (1987) Vein-cutting behavior: insect counterploy to the latex defense of plants. Science 237:898–901

Dussourd DE, Hoyle AM (2000) Poisoned plusiines: toxicity of milkweed latex and cardenolides to some generalist caterpillars. Chemoecology 10:11–16

Ehrlich PR, Raven PH (1964) Butterflies and plants: a study in coevolution. Ecology 18:586–608

Farmer EE, Ryan CA (1992) Octadecanoid precursors of jasmonic acid activate the synthesis of wound-inducible proteinase inhibitors. Plant Cell 4:129–134

Feeny P (1970) Seasonal change in oak leaf tannins and nutrients as a cause of spring feeding by winter moth caterpillars. Ecology 51:565–581

Felix G, Boller T (1995) Systemin induces rapid ion fluxes and ethylene biosynthesis in *Lycopersicon peruvianum* cells. Plant J 7:381–389

Felton GW (2005) Indigestion is a plant's best defense. Proc Natl Acad Sci USA 102:18771–18772

Fisher DB (1990) Measurement of phloem transport rates by an indicator-dilution technique. Plant Physiol Biochem 94:455–462

Fraenkel GS (1959) The raison d'etre of secondary plant substances: these odd chemicals arose as a means of protecting plants from insects and now guide insects to food. Science 129:1466–1470

Gatehouse JA (2002) Plant resistance towards insect herbivores: a dynamic interaction. New Phytol 156:145–169

Giri AP, Wunsche H, Mitra S, Zavala JA, Muck A, Svatos A, Baldwin IT (2006) Molecular interactions between the specialist herbivore *Manduca sexta* (*Lepidoptera, Sphingidae*) and its natural host *Nicotiana attenuata*. VII. Changes in the plant's proteome. Plant Physiol 142:1621–1641

Green TR, Ryan CA (1972) Wound-induced proteinase inhibitor in plant leaves: a possible defense mechanism against insects. Science 175:776–777

Guariguata MR, Gilbert GS (1996) Interspecific variation in rates of trunk wound closure in a panamanian lowland forest. Biotropica 28:23–29

Hause B, Hause G, Kutter C, Miersch O, Wasternack C (2003) Enzymes of jasmonate biosynthesis occur in tomato sieve elements. Plant Cell Physiol 44:643–648

Hessen DO, van Donk E (1993) Morphological changes in *Scenedesmus* induced by substances released from *Daphnia*. Arch Hydrobiol 127:129–140

Howe GA (2004) Jasmonates as signals in the wound response. J Plant Growth Regul 23:223–237

Howe GA, Ryan CA (1999) Suppressors of systemin signaling identify genes in the tomato wound response pathway. Genetics 153:1411–1421

Howlett BG, Clarke AR, Madden JL (2001) The influence of leaf age on the oviposition preference of *Chrysophtharta bimaculata* (Olivier) and the establishment of neonates. Agric For Entomol 3:121–127

Hui D, Iqbal J, Lehmann K, Gase K, Saluz HP, Baldwin IT (2003) Molecular interactions between the specialist herbivore *Manduca sexta* (*Lepidoptera, Sphingidae*) and its natural host *Nicotiana attenuata*: V. Microarray analysis and further charaterization of large-scale changes in herbivore-induced mRNAs. Plant Physiol 131:1877–1893

Ishiguro S, Kawai-Oda A, Nishida I, Okada K (2001) The *defective in anther dehiscence1* gene encodes a novel phospholipase a1 catalyzing the initial step of jasmonic acid biosynthesis, which synchronizes pollen maturation, anther dehiscence, and flower opening in *Arabidopsis*. Plant Cell 13:2191–2209

Kang J-H, Wang L, Giri A, Baldwin IT (2006) Silencing threonine deaminase and JAR4 in N*icotiana attenuata* impairs jasmonic acid-isoleucine-mediated defenses against *Manduca sexta*. Plant Cell 18:3303–3320

Karban R, Baldwin IT (1997) Induced responses to herbivory. University of Chicago Press, Chicago

Karban R, Myers JH (1989) Induced plant responses to herbivory. Ann Rev Ecol Sys 20:331–348

Karban R, Kuc J (1999). Induced resistance against pathogens and herbivores: an overview. In: Agrawal AA, Tuzun S, Bent E (eds) Induced plant defenses against pathogens and herbivores. Biochemistry, ecology, and agriculture, APS Press, St. Paul, Minnesota, pp 1–16

Kenrick P, Crane PR (1997) The origin and early evolution of plants on land. Nature 389:33–39

Kessler A, Baldwin IT (2002) Plant responses to insect herbivory: the emerging molecular analysis. Ann Rev Plant Biol 53:299–328

Kuhn E, Schaller A (2004). DNA microarrays: methodology, data evaluation, and application in the analysis of plant defense signaling. In: Setlow JK (ed) Genetic engineering, principles and methods. Kluwer Academic Plenum, pp 49–84

Lampert W, Rothhaupt KO, von Elert E (1994) Chemical induction of colony formation in a green alga (*Scenedesmus acutus*) by grazers (*Daphnia*). Limnol Oceanogr 39:1543–1550

Lawrence SD, Dervinis C, Novak N, Davis JM (2006) Wound and insect herbivory responsive genes in poplar. Biotechnol Lett 28:1493–1501

Lee GI, Howe GA (2003) The tomato mutant *spr1* is defective in systemin perception and the production of a systemic wound signal for defense gene expression. Plant J 33:567–576

León J, Rojo E, Sánchez-Serrano JJ (2001) Wound signalling in plants. J Exp Bot 52:1–9

Li L, Li C, Lee GI, Howe GA (2002) Distinct roles for jasmonate synthesis and action in the systemic wound response of tomato. Proc Natl Acad Sci USA 99:6416–6421

Li C, Liu G, Xu C, Lee GI, Bauer P, Ling HQ, Ganal MW, Howe GA (2003) The tomato *suppressor of prosystemin-mediated responses2* gene encodes a fatty acid desaturase required for the biosynthesis of jasmonic acid and the production of a systemic wound signal for defense gene expression. Plant Cell 15:1646–1661

Li CY, Schilmiller AL, Liu GH, Lee GI, Jayanty S, Sageman C, Vrebalov J, Giovannoni JJ, Yagi K, Kobayashi Y et al (2005) Role of beta-oxidation in jasmonate biosynthesis and systemic wound signaling in tomato. Plant Cell 17:971–986

Li L, Zhao Y, McCaig BC, Wingerd BA, Wang J, Whalon ME, Pichersky E, Howe GA (2004) The tomato homolog of coronatine-insensitive1 is required for the maternal control of seed maturation, jasmonate-signaled defense responses, and glandular trichome development. Plant Cell 16:126–143

Lipke H, Fraenkel GS, Liener IE (1954) Effect of soybean inhibitors on growth of *Tribolium confusum*. J Agr Food Chem 2:410–414

Lippert D, Chowrira S, Ralph SG, Zhuang J, Aeschliman D, Ritland C, Ritland K, Bohlmann J (2007) Conifer defense against insects: proteome analysis of sitka spruce (*Picea sitchensis*) bark induced by mechanical wounding or feeding by white pine weevils (*Pissodes strobi*). Proteomics 7:248–270

Lucas WJ, Lee JY (2004) Plant cell biology – plasmodesmata as a supracellular control network in plants. Nat Rev Mol Cell Bio 5:712–726

Major IT, Constabel CP (2006) Molecular analysis of poplar defense against herbivory: comparison of wound- and insect elicitor-induced gene expression. New Phytol 172:617–635

Malone M (1996). Rapid, long-distance signal transmission in higher plants. Advances in botanical research. Academic Press, San Diego, pp 163–228

Mauricio R (1998) Costs of resistance to natural enemies in field populations of the annual plant *Arabidopsis thaliana*. Am Nat 151:20–28

Mauricio R (2000) Natural selection and the joint evolution of tolerance and resistance as plant defenses. Evol Ecol 14:491–507

McGurl B, Orozco-Cardenas M, Pearce G, Ryan CA (1994) Overexpression of the prosystemin gene in transgenic tomato plants generates a systemic signal that constitutively induces proteinase inhibitor synthesis. Proc Natl Acad Sci USA 91:9799–9802

McGurl B, Pearce G, Orozco-Cardenas M, Ryan CA (1992) Structure, expression and antisense inhibition of the systemin precursor gene. Science 255:1570–1573

McNaughton SJ, Tarrants JL (1983) Grass leaf silicification: natural selection for an inducible defense against herbivores. Proc Natl Acad Sci USA 80:790–791

Moyen C, Hammond-Kosack KE, Jones J, Knight MR, Johannes E (1998) Systemin triggers an increase of cytoplasmic calcium in tomato mesophyll cells: Ca^{2+} mobilization from intra- and extracellular compartments. Plant Cell Environ 21:1101–1111

Myers JH, Bazely D (1991). Thorns, spines, prickles, and hairs: are they stimulated by herbivory and do they deter herbivores? In: Tallamy DW Raupp MJ (eds) Phytochemical induction by herbivores. John Wiley & Sons, New York, pp 325–344

Narváez-Vásquez J, Florin-Christensen J, Ryan CA (1999) Positional specificity of a phospholipase a activity induced by wounding, systemin, and oligosaccharide elicitors in tomato leaves. Plant Cell 11:2249–2260

Navazio L, Moscatiello R, Bellincampi D, Baldan B, Meggio F, Brini M, Bowler C, Mariani P (2002) The role of calcium in oligogalacturonide-activated signalling in soybean cells. Planta 215:596–605

O'Donnell PJ, Calvert C, Atzorn R, Wasternack C, Leyser HMO, Bowles DJ (1996) Ethylene as a signal mediating the wound response of tomato plants. Science 274:1914–1917

Orozco-Cárdenas M, McGurl B, Ryan CA (1993) Expression of an antisense prosystemin gene in tomato plants reduces the resistance toward *Manduca sexta* larvae. Proc Natl Acad Sci USA 90:8273–8276

Orozco-Cárdenas ML, Narváez-Vasquez J, Ryan CA (2001) Hydrogen peroxide acts as a second messenger for the induction of defense genes in tomato plants in response to wounding, systemin, and methyl jasmonate. Plant Cell 13:179–191

Painter RH (1936) The food of insects and its relation to resistance of plants to insect attack. Am Nat 70:547–566

Pearce G, Strydom D, Johnson S, Ryan CA (1991) A polypeptide from tomato leaves induces wound-inducible proteinase inhibitor proteins. Science 253:895–898

Phillips MA, Croteau RB (1999) Resin-based defenses in conifers. Trends Plant Sci 4:184–190

Ralph S, Oddy C, Cooper D, Yueh H, Jancsik S, Kolosova N, Philippe RN, Aeschliman D, White R, Huber D et al. (2006a) Genomics of hybrid poplar (*Populus trichocarpa x deltoides*) interacting with forest tent caterpillars (*Malacosoma disstria*): normalized and full-length cDNA libraries, expressed sequence tags, and a cDNA microarray for the study of insect-induced defences in poplar. Mol Ecol 15:1275–1297

Ralph SG, Yueh H, Friedmann M, Aeschliman D, Zeznik JA, Nelson CC, Butterfield YSN, Kirkpatrick R, Liu J, Jones SJM et al (2006b) Conifer defence against insects: microarray gene expression profiling of sitka spruce (*Picea sitchensis*) induced by mechanical wounding or feeding by spruce budworms (*Choristoneura occidentalis*) or white pine weevils (*Pissodes strobi*) reveals large-scale changes of the host transcriptome. Plant Cell Environ 29:1545–1570

Reymond P, Bodenhausen N, Van Poecke RM, Krishnamurthy V, Dicke M, Farmer EE (2004) A conserved transcript pattern in response to a specialist and a generalist herbivore. Plant Cell 16:3132–3147

Rhoades DF, Cates RG (1976). Towards a general theory of plant antiherbivore chemistry. In: Wallace JW Mansell RL (eds) Biochemical interaction between plants and insects. Plenum Press, New York, pp 168–213

Rittinger PA, Biggs AR, Peirson DR (1987) Histochemistry of lignin and suberin deposition in boundary layers formed after wounding in various plant species and organs. Can J Bot 65: 1886–1892

Ryan CA (2000) The systemin signaling pathway: differential activation of plant defensive genes. Biochim Biophys Acta 1477:112–121

Ryan CA, Huisman W (1969) The regulation of synthesis and storage of chymotrypsin inhibitor I in leaves of potato and tomato plants. Plant Physiol 45:484–489

Ryan CA, Moura DS (2002) Systemic wound signaling in plants: a new perception. Proc Natl Acad Sci USA 99:6519–6520

Sagi M, Davydov O, Orazova S, Yesbergenova Z, Ophir R, Stratmann JW, Fluhr R (2004) Plant respiratory burst oxidase homologs impinge on wound responsiveness and development in *Lycopersicon esculentum*. Plant Cell 16:616–628

Schaller A, Oecking C (1999) Modulation of plasma membrane H^+-ATPase activity differentially activates wound and pathogen defense responses in tomato plants. Plant Cell 11: 263–272

Scheer JM, Ryan CA (1999) A 160 kda systemin receptor on the cell surface of *Lycopersicon peruvianum* suspension cultured cells: kinetic analyses, induction by methyl jasmonate and photoaffinity labeling. Plant Cell 11:1525–1535

Scheer JM, Ryan CA (2002) The systemin receptor SR160 from *Lycopersicon peruvianum* is a member of the LRR receptor kinase family. Proc Natl Acad Sci USA 99:9585–9590

Schilmiller AL, Howe GA (2005) Systemic signaling in the wound response. Curr Opin Plant Biol 8:369–377

Schittko U, Baldwin IT (2003) Constraints to herbivore-induced systemic responses: bidirectional signaling along orthostichies in *Nicotiana attenuata*. J Chem Ecol 29:763–770

Schmidt S, Baldwin IT (2006) Systemin in *Solanum nigrum*. The tomato-homologous polypeptide does not mediate direct defense responses. Plant Physiol Biochem 142:1751–1758

Schmidt DD, Voelckel C, Hartl M, Schmidt S, Baldwin IT (2005) Specificity in ecological interactions. Attack from the same lepidopteran herbivore results in species-specific transcriptional responses in two solanaceous host plants. Plant Physiol 138:1763–1773

Schoonhoven LM, Van Loon JJA, Dicke M (2005) Insect-plant biology, Oxford University Press, Oxford

Schwachtje J, Minchin PEH, Jahnke S, van Dongen JT, Schittko U, Baldwin IT (2006) SNF1-related kinases allow plants to tolerate herbivory by allocating carbon to roots. Proc Natl Acad Sci USA 103:12935–12940

Seo HS, Song JT, Cheong J-J, Lee J-H, Lee Y-W, Hwang I, Lee JS, Choi YD (2001) Jasmonic acid carboxyl methyltransferase: a key enzyme for jasmonate-regulated plant responses. Proc Natl Acad Sci USA 98:4788–4793

Smith CM, Rodriguez-Buey M, Karlsson J, Campbell MM (2004) The response of the poplar transcriptome to wounding and subsequent infection by a viral pathogen. New Phytologist 164:123–136

Soltis PS, Soltis DE (2004) The origin and diversification of angiosperms. Am J Bot 91:1614–1626

Stahl E (1888) Pflanzen und Schnecken: Biologische Studie über die Schutzmittel der Pflanzen gegen Schneckenfrass. Jena Z Naturwiss 22:557–684

Staswick PE (1994) Storage proteins of vegetative plant tissue. Ann Rev Plant Physiol Plant Mol Biol 45:303–322

Staswick PE, Tiryaki I (2004) The oxylipin signal jasmonic acid is activated by an enzyme that conjugates it to isoleucine in *Arabidopsis*. Plant Cell 16:2117–2127

Stenzel I, Hause B, Maucher H, Piezschke A, Miersch O, Ziegler J, Ryan CA, Wasternack C (2003) Allene oxide cyclase dependence of the wound response and vascular bundle-specific generation of jasmonates in tomato – amplification in wound signalling. Plant J 33:577–589

Strassner J, Schaller F, Frick UB, Howe GA, Weiler EW, Amrhein NA, Macheroux P, Schaller A (2002) Characterization and cdna-microarray expression analysis of 12-oxophytodienoate reductases reveals differential roles for octadecanoid biosynthesis in the local versus the systemic wound response. Plant J 32:585–601

Stratmann JW, Ryan CA (1997) Myelin basic protein kinase activity in tomato leaves is induced systemically by wounding and increases in response to systemin and oligosaccharide elicitors. Proc Natl Acad Sci USA 94:11085–11089

Stuhlfelder C, Mueller MJ, Warzecha H (2004) Cloning and expression of a tomato cDNA encoding a methyl jasmonate cleaving esterase. Eur J Biochem 271:2976–2983

Thines B, Katsir L, Melotto M, Niu Y, Mandaokar A, Liu G, Nomura K, He SY, Howe GA, Browse J (2007) JAZ repressor proteins are targets of the SCF^{COI1} complex during jasmonate signalling. Nature 448:661–665

Thompson GA, Goggin FL (2006) Transcriptomics and functional genomics of plant defence induction by phloem-feeding insects. J Exp Bot 57:755–766

Trapp S, Croteau R (2001) Defensive resin biosynthesis in conifers. Ann Rev Plant Physiol Plant Mol Biol 52:689–724

Truman W, Bennettt MH, Kubigsteltig I, Turnbull C, Grant M (2007) *Arabidopsis* systemic immunity uses conserved defense signaling pathways and is mediated by jasmonates. Proc Natl Acad Sci USA 104:1075–1080

van Donk E, Hessen DO (1993) Grazing resistance in nutrient-stressed phytoplankton. Oecologia 93:508–511
Voelckel C, Baldwin IT (2004) Herbivore-induced plant vaccination. Part II. Array-studies reveal the transience of herbivore-specific transcriptional imprints and a distinct imprint from stress combinations. Plant J 38:650–663
Walling LL (2000) The myriad plant responses to herbivores. J Plant Growth Regul 19:195–216
Wasternack C (2007) Jasmonates: an update on biosynthesis, signal transduction and action in plant stress response, growth and development. Ann Bot, 1–17
Wasternack C, Ortel B, Miersch O, Kramell R, Beale M, Greulich F, Feussner I, Hause B, Krumm T, Boland W et al (1998) Diversity in octadecanoid-induced gene expression of tomato. J Plant Physiol 152:345–352
Weis AE, Franks SJ (2006) Herbivory tolerance and coevolution: an alternative to the arms race? New Phytol 170:423–425
Wildon DC, Thain JF, Minchin PEH, Gubb IR, Reilly AJ, Skipper YD, Doherty HM, O'Donnell PJ, Bowles DJ (1992) Electrical signalling and systemic proteinase inhibitor induction in the wounded plant. Nature 360:362–65
Xie D-X, Feys BF, James S, Nieto-Rostro M, Turner JG (1998) *Coi1*: an *Arabidopsis* gene required for jasmonate-regulated defense and fertility. Science 280:1091–1094
Zhang ZP, Baldwin IT (1997) Transport of [2-C-14] jasmonic acid from leaves to roots mimics wound-induced changes in endogenous jasmonic acid pools in *Nicotiana sylvestris*. Planta 203:436–441
Zhu-Salzman K, Salzman RA, Ahn JE, Koiwa H (2004) Transcriptional regulation of sorghum defense determinants against a phloem-feeding aphid. Plant Physiol 134:420–431

Chapter 2
Herbivore-Induced Indirect Defense: From Induction Mechanisms to Community Ecology

Maaike Bruinsma and Marcel Dicke

Herbivory may induce plant defenses that promote the activity of natural enemies of the herbivores. This so-called induced *indirect* defense may involve the production of plant volatiles that attract carnivorous arthropods or extrafloral nectar that is exploited as alternative food by carnivorous arthropods. Induced indirect plant defense is mediated by different signal-transduction pathways, such as the jasmonic acid, the salicylic acid, and the ethylene pathways and may involve large-scale transcriptomic re-arrangements. Induced indirect plant defense responses result in an altered phenotype and thus can affect the interactions of the plant with various community members: attackers can be deterred, natural enemies attracted (both above- and belowground), pollinators may change flower visitation, and neighboring plants can exploit the information from the attacked plants to initiate defense responses as well. We discuss several approaches that are commonly used in molecular, chemical and ecological studies of induced indirect plant defenses and identify some remaining knowledge gaps and directions for future research. Integrating a mechanistic approach with a community ecological approach will provide important progress in understanding the selective pressures and dynamics of ecological interactions that are mediated by induced indirect plant defenses, as well as the underlying mechanisms.

2.1 Introduction

Plants face many challenges during their lifetime. Drought, flooding, high and low temperatures, and attacks by all kinds of organisms, like pathogens, nibbling insects or mammals devouring whole plants. To be able to reproduce in spite of all these difficulties plants have developed defense mechanisms to protect themselves, other than running away. The defenses of plants against herbivorous arthropods can be classified as 'direct defenses' that affect the physiology of the attacker (Howe and Schaller this volume) or 'indirect defenses' that promote the effectiveness of natural

M. Dicke
Laboratory of Entomology, Wageningen University, 6709 PD Wageningen, The Netherlands
e-mail: Marcel.Dicke@wur.nl

enemies of herbivores (Fig. 2.1). Indirect defense comprises (a) the provision of shelter, such as hollow thorns that are used by ants for nesting (b) the production of alternative food, such as extrafloral nectar that is used by carnivorous arthropods such as ants and parasitic wasps or (c) the emission of herbivore-induced plant volatiles that guide carnivorous arthropods such as predators or parasitoids to their herbivorous victims. All of these support the presence and abundance of carnivorous enemies of herbivorous arthropods and consequently the reduction of herbivore presence.

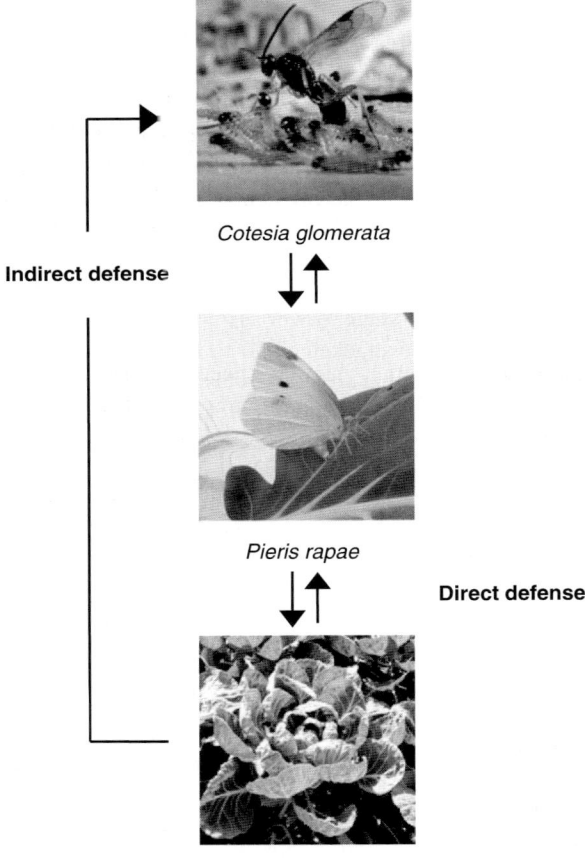

Fig. 2.1 Direct defense of a plant has a direct negative effect on the attacker of the plant. Indirect defense of a plant maintains or attracts carnivores that consume or parasitize the attacker of the plant, thereby exerting a negative effect on the attacker, which contributes to plant defense. Photographs were taken by Hans M. Smid (*C. glomerata*), Nelly Cardinel (*B. oleracea*) and Maaike Bruinsma (*P. rapae*)

Here we will address indirect defense, especially induced indirect defense. Herbivory can induce the production of extrafloral nectar (Wäckers et al. 2001) or plant volatiles (Van Poecke and Dicke 2004) and as a result the plant's phenotype changes. The extrafloral nectar can serve as food to various animals in the community and the volatiles can be used by animals to localize the plant. Throughout the plant kingdom many species have been reported to produce herbivore-induced infochemicals that attract carnivorous enemies of the herbivorous arthropods (Dicke 1999). The effect of these phenotypic changes on carnivorous enemies of herbivores has received most attention (Van Poecke and Dicke 2004; Turlings and Ton 2006). For instance, Lima bean plants that are infested with spider mites start to produce a range of volatiles, including terpenoids and methyl salicylate, several of which attract predatory mites that consume the herbivorous spider mites (Dicke et al. 1990b, 1999). Moreover, herbivory induces the production of extrafloral nectar in Lima bean plants and this results in increased numbers and duration of visits by e.g. ants and wasps (Kost and Heil 2005). The increased visitation of the plant by carnivorous arthropods results in a reduced amount of leaf damage (Heil 2004).

However, plants are at the basis of most terrestrial food webs and consequently they are members of complex communities, both below and aboveground, with a multitude of dynamic, ecological interactions (Price et al. 1980; Dicke and Vet 1999; Van Zandt and Agrawal 2004, Bezemer and van Dam 2005). Ecological interactions may involve direct interactions such as predator–prey interactions or competition among herbivores, as well as indirect interactions such as apparent competition and trait-mediated indirect effects (Holt 1977; Wootton 1994; Van Veen et al. 2006; White and Andow 2006). In many studies species interactions are considered to be fixed: all individuals within a population are considered to have the same characteristics and interact in the same way with other organisms. However, because of phenotypic plasticity, interactions within communities are context-dependent (Agrawal 2001), which implies that they are not only influenced by genotype, but also by e.g. physiological state, resource availability and interactions with community members. As a result, understanding the effects of phenotypic plasticity is important to understand community dynamics (Fig. 2.2; Agrawal 2001).

Although a phenotypic change may affect many interactions in a community, herbivore-induced indirect plant defenses have mostly been studied in a tritrophic context in simple food chains, without taking into account the effects they might have on other community members. Investigating the effects of phenotypic changes on community processes is one of the major challenges that ecologists face in the research on herbivore-induced plant defenses (Dicke and Vet 1999; Kessler and Baldwin 2001). Understanding the selection pressures and the dynamics of ecological interactions is important for understanding the ecology and evolution of communities.

To investigate the community consequences of induced indirect plant defenses, a thorough understanding of the underlying mechanisms is essential, as mechanistic knowledge allows the development of manipulative tools. Manipulative experiments are important to address the effects of induced defense on a range of individual interactions or on the total set of interactions (Kessler and Baldwin 2001; Dicke and Hilker 2003).

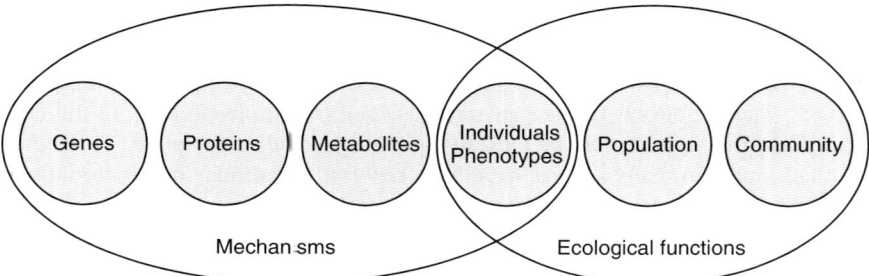

Fig. 2.2 Induced indirect defense can be investigated at different levels of biological organization: by investigating mechanisms of induction at the levels of genes up to individuals, or by investigating ecological functions at the levels of individuals up to the community. An integration of these two approaches proved to be most rewarding. Phenotypic plasticity plays a central role in understanding both, the underlying mechanisms of induced defense, and the consequences of induced defense for community dynamics

In this chapter we will address the mechanisms of induced indirect plant defenses and how information on these can be used to investigate the ecological consequences of these defenses at the level of multiple ecological interactions and the community.

2.2 Induction of Indirect Plant Defense

Herbivores can cause many types of damage to plants, according to the feeding guild to which they belong. For example, caterpillars ingest small sections of the leaves, while others feed on specific parts of the leaf material: leaf-mining insects feed on parenchymal tissue and aphids ingest phloem. Different types of damage may result in diverse defense responses in the plant. In this section we will discuss the plant responses and the various mechanisms of their induction.

2.2.1 Response of the Plant

In response to arthropod herbivory the plant may activate several major signal-transduction pathways involved in the defense response: the JA (jasmonic acid), SA (salicylic acid) and ET (ethylene) pathways (Dicke and Van Poecke 2002). These signaling pathways are differentially induced by different feeding guilds or artificial damage (Ozawa et al. 2000; Walling 2000; Dicke and Van Poecke 2002; De Vos et al. 2005; Zheng et al. 2007). The three signal-transduction pathways also interact: JA can inhibit the effect of SA, and SA can interfere with JA-mediated induction (Peña-Cortes et al. 1993; Sano and Ohashi 1995). Similarly, JA and ET synergistically affect induction of defense gene expression in tomato (O'Donnell et al. 1996), while ET inhibits the effect of JA on nicotine induction in tobacco (Kahl et al. 2000).

Table 2.1 The octadecanoid pathway and the effect of manipulation of different steps on volatile emission of Lima bean and on the attraction of natural enemies of attackers of Lima bean

Octadecanoid pathway	Elicitation ⟶ Inhibition ⊣--◆	Manipulation with	Volatile emission in response to manipulation	Carnivore attraction
Herbivory ⋮↓				
linolenic acid ⟶		Linolenic acid	DMNT, TMTT[1]	n.t.
↓ ⊣--------◆		Phenidone	No volatile production upon elicitation after pretreatment with inhibitor[1]	n.t.
13-hydroperoxylinolenic acid				
↓ ⊣--------◆		DIECA	No volatile production upon elicitation after pretreatment with inhibitor[1]	n.t.
12,13-epoxy-octadecatrienoic acid				
↓ ⊣--------◆		n-propyl gallate	No volatile production upon elicitation after pretreatment with inhibitor[1]	n.t.
12-oxo-phytodienoic acid (OPDA)				
∟⟶		OPDA	DMNT, TMTT[1,4]	Yes[2]
↓		Jasmonic acid	Similar blend as that induced by spider mite infestation[3] Hexenyl acetate, β-ocimene, linalool, DMNT, $C_{10}H_{14}$, $C_{10}H_{16}O$ indole[1]	Yes[2,3]
jasmonic acid (JA) ⟶				
↓				
methyl jasmonate (MeJA) ⟶		Methyl jasmonate	n.t.	Yes[3]

References: [1] Koch et al. 1999; [2] Dicke and Van Poecke 2002; [3] Dicke et al. 1999; [4] Boland et al. 1999; n.t. = not tested

The major signal-transduction pathway involved in plant responses to herbivorous insects is the jasmonic acid or octadecanoid pathway (Table 2.1). Octadecanoids are synthesized from the 18-carbon fatty acid linolenic acid that is released from membrane lipids in response to stimuli associated with wounding (Narváez-Vásquez et al. 1999). Through the octadecanoid pathway with 13-hydroperoxylinolenic acid, oxo-phytodienoic acid (OPDA), and other compounds as intermediates, the phytohormone jasmonic acid is produced (Liechti and Farmer 2002; Schaller and Stintzi this volume). The members of this pathway have different biological activities. JA induces a large number of genes and the emission of a volatile blend that is similar, though not identical, to the blend induced by herbivory (Dicke et al. 1999; Reymond et al. 2004; De Vos et al. 2005). JA not only affects the induction of defenses, but also developmental processes and male fertility (Creelman and Mullet 1997; Liechti and Farmer 2002). OPDA also induces gene expression and volatile emission, albeit less effectively as compared to JA (Koch et al. 1999; Stintzi et al. 2001). JA can be converted into methyl-jasmonate or cis-jasmone, both of which can induce defenses in plants (Farmer and Ryan 1990; Birkett et al. 2000).

From 13-hydroperoxylinolenic acid there is a side branch of the octadecanoid pathway that leads to the production of so-called green leaf volatiles (GLVs) such as C6-aldehydes, C6-alcohols, and their acetates (Visser and Avé 1978; Hatanaka et al. 1987). These compounds can attract both herbivorous and carnivorous arthropods on the one hand (Whitman and Eller 1990; Shiojiri et al. 2006), and prime neighboring plants for the induction of defenses on the other (Engelberth et al. 2004; Ruther and Fürstenau 2005). In addition to the octadecanoid pathway, wounding can also induce the hexadecanoid pathway, starting from 7(Z),10(Z),13(Z) hexadecatrienoic acid (Weber et al. 1997; Stintzi et al. 2001) that also leads to JA. The induction of both pathways may result in specific 'oxylipin signatures' that allow plants to fine-tune their responses to wounding or herbivory (Weber et al. 1997). The activation of the signal-transduction pathways can result in the induction of direct as well as indirect defenses. Many types of secondary metabolites involved in direct defense are produced upon herbivore damage, for example non-volatile compounds like proteinase inhibitors, glucosinolates, and alkaloids (see Howe and Schaller this volume). As indirect defense response, plants can produce volatiles such as alcohols, esters, and terpenoids, or alternative food, such as extrafloral nectar. Induced plants can produce either the same volatiles as non-induced plants but in different amounts or ratios, or they can produce new volatiles that are not emitted by intact plants (Dicke et al. 1999). However, during both types of responses plants produce the metabolites de novo; indicating an active investment in defense, rather than just a passive release of compounds (Donath and Boland 1994; Paré and Tumlinson 1997; Mercke et al. 2004).

The contribution of the JA pathway to indirect defense has been well studied in the context of tritrophic interactions. JA in itself is not attractive or deterring, but induces biosynthetic processes in the plant (e.g., the production of volatile infochemicals), that cause behavioral responses of the insects (Avdiushko et al. 1997; Thaler et al. 2002; Bruinsma et al. 2007). JA-induced defense reactions may either deter herbivores, or attract their natural enemies (Gols et al. 1999; Thaler 1999; Lou et al. 2005; Bruinsma et al. 2007). In addition to JA signaling, tritrophic interactions have been shown to also involve the SA pathway. The production of methyl salicylate (MeSA) is induced by spider mite infestation in e.g. Lima bean and tomato (Dicke et al. 1990b, 1998; Ozawa et al. 2000), and attracts the natural enemies of the spider mites (De Boer and Dicke 2004).

Another indirect defense mechanism involving JA signaling is the production of alternative food, such as extrafloral nectar (EFN) that can be used by the members of the third trophic level (Van Rijn and Tanigoshi 1999; Kost and Heil 2005; Wäckers and Van Rijn 2005). EFN is secreted from nectaries outside the flowers, which may occur on the petioles (Koptur 2005). The nectar composition can differ dramatically between floral and extrafloral nectaries of one plant (Koptur 2005). The production of EFN increases upon herbivory (Heil et al. 2001; Wäckers et al. 2001). It is highest in leaves where the herbivores are feeding, but can also be increased in systemic leaves (Wäckers et al. 2001). Wounding and JA application increase EFN production. Moreover, EFN secretion is reduced by phenidone, an inhibitor of an early step in the octadecanoid pathway (Table 2.1) supporting a role for JA in the induction of EFN secretion (Heil et al. 2001). EFN, serving as an alternative food source, may increase the duration of predator visits to plants. Predators disperse more slowly from plants with more EFN as compared to plants with little EFN (Choh et al. 2006), and plants may benefit from the presence of predators.

2.2.2 Induction by Herbivores

When herbivores are feeding on a plant, either by leaf chewing, phloem ingestion, or cell content feeding, they induce phytohormone signaling pathways and consequently elicit a plant response. The induced phytohormone signatures are attacker-specific: qualitatively, quantitatively and temporally (De Vos et al. 2005). More and more studies also show the attacker specificity at the level of global gene expression (Voelckel and Baldwin 2004; Voelckel et al. 2004; De Vos et al. 2005), although other studies recorded quite similar transcriptional responses after attack by different herbivores (Reymond et al. 2004). Herbivore-induced plant volatiles can be specific for herbivore species, and even herbivore instar, feeding on the plant (De Moraes et al. 1998; Takabayashi et al. 2006).

The main groups of plant volatiles induced by herbivory are green leaf volatiles, terpenes, and phenolics. The induction of volatile emission is mediated by the induction of the three main signal-transduction pathways – the JA, SA, and ET pathways – which may be differentially induced by insects from different feeding guilds (Walling 2000). In Lima bean for example, JA signaling is responsible for the induced production of volatiles in response to caterpillar damage, while both SA and JA mediate the response to spider mite damage (Ozawa et al. 2000). In another plant species, *Medicago truncatula*, caterpillars and spider mites induced qualitatively and quantitatively different volatile patterns. Both JA and SA accumulated in response to damage, but the accumulation differed between the attack by the chewing and by piercing–sucking insects; SA accumulation was higher in response to spider-mite as compared to caterpillar damage, and JA accumulated differently in time for the two modes of attack (Leitner et al. 2005). Likewise, induced changes in gene expression in *Arabidopsis thaliana* show differences between feeding guilds. Five attackers with different modes of attack, ranging from leaf-chewing herbivores to pathogens causing necrotic lesions, showed different degrees of relative induction of the three important signal-transduction pathways (De Vos et al. 2005). In general, it seems that the plant response to phloem-feeding herbivores is more similar to the response to pathogen attack, while leaf-chewing herbivores induce pathways also activated by wounding (Walling 2000).

2.2.3 Mechanical Wounding Versus Herbivory

When herbivores attack the plant, they inflict physical damage which in itself is sufficient to elicit a subset of the responses to herbivory. Water loss at the wound site may result in osmotic stress and therefore, there is considerable overlap between plant responses to wounding and dehydration. In addition, the plant responds to herbivore-derived compounds present in oral secretions. Physical damage and herbivore-derived elicitors are both responsible for part of the herbivore-induced response of the plant. In many plant species, however, the response to mechanical damage differs from the one elicited by herbivory (Schoonhoven et al. 2005). This may be partly due to technical difficulties in accurately mimicking herbivory. Mechanical damage differs from herbivore-inflicted damage in the amount of removed tissue, age of tissue, spatial pattern of damage, and timing (Baldwin 1990). The temporal pattern of mechanical damage was in fact shown to be an important factor influencing its effect (Mithöfer et al. 2005). Yet, when herbivore regurgitant is applied onto mechanically damaged leaves, the plant response can be similar to the response to herbivory (Turlings et al. 1990; Halitschke et al. 2001). For example, mechanical damage with subsequent regurgitant application induces herbivore-specific plant responses in *Nicotiana attenuata* (Halitschke et al. 2001). Several active compounds have been identified in the oral secretions of feeding insects. For example, the enzyme β-glucosidase and volicitin (a fatty acid–amino acid conjugate, FAC) induce volatile production when added to mechanically damaged cabbage and maize plants, respectively (Mattiacci et al. 1995; Alborn et al. 1997; Schmelz et al. 2001; Tumlinson and Engelberth this volume). FACs have been found in all lepidopteran larvae studied to date (Voelckel and Baldwin 2004). Plants can not only differentiate between mechanical wounding and herbivory, but also between herbivore species, even when they are from the same feeding guild. This has been shown for *N. attenuata*: the plant response to a specialist herbivore differed from the response to two generalist herbivores, and the different responses were correlated with the FAC composition of the regurgitant of the herbivores (Voelckel and Baldwin 2004). The role of oral secretions in induction of plant defense responses is discussed in more detail by Felton (this volume).

The difference between induction after mechanical damage and caterpillar feeding has been shown at the gene expression level in *A. thaliana*. Mechanical damage and caterpillar feeding both induce jasmonate-responsive genes and lead to accumulation of JA. However, mechanical damage induces expression of a jasmonate-responsive marker gene PDF1.2, while caterpillar feeding suppresses the induction of this gene. PDF1.2 defense gene induction was suppressed also when caterpillar regurgitant was added to mechanically damaged leaves, indicating a role for caterpillar-derived elicitors in the downregulation of plant defense responses (De Vos 2006). Three other JA-responsive genes that are induced by caterpillar feeding are not induced by mechanical wounding, demonstrating how herbivory and wounding differentially induce expression of specific genes (De Vos 2006). In *N. attenuata* plants the endogenous JA levels increase after wounding, and increase even more when oral secretion is applied to the wounds. The same pattern can be observed for ethylene emission in these plants: punctured plants treated with oral secretion temporarily emit more ethylene than do punctured plants treated with water, while ethylene emission upon herbivory increases for as long as herbivory continues (Kahl et al. 2000).

2.2.4 Priming

Apart from direct upregulation of defense signaling cascades or gene expression, the ability of the plant to rapidly activate cellular defense responses can be enhanced, a process that is called priming (Conrath et al. 2002). When exposed to a priming stimulus the plant does not respond with the immediate production of defense compounds, but rather enters a sensitized state allowing it to respond faster or stronger to subsequent challenges in the future (Turlings and Ton 2006). Priming has originally been demonstrated for plant-pathogen (Conrath et al. 2002, 2006) and plant-rhizosphere bacteria interactions (Verhagen et al. 2004), and appears to be advantageous with respect to defense-associated metabolic costs.

Priming can occur in response to different stimuli, such as plant volatiles, pathogen infestation, or herbivory. Priming of indirect defenses has been shown for induced volatile emission as well as EFN secretion. Exposure to green leaf volatiles from neighboring damaged plants results in higher endogenous JA levels and higher emission of volatiles upon herbivory or mechanical damage compared to non-exposed plants (Engelberth et al. 2004). Also exogenously applied single compounds, such as (Z)-3-hexenal, (Z)-3-hexen-1-ol, and (Z)-3-hexenyl acetate, can prime plants for responses to herbivore attack (Engelberth et al. 2004). Recently, a field study demonstrated the priming of EFN secretion. Lima bean plants were exposed to an artificial volatile blend mimicking the volatile emission from herbivore-induced Lima bean plants. Upon wounding, these plants secreted more EFN as compared to non-exposed controls indicating a priming effect of the exposure to volatiles (Heil and Kost 2006).

2.3 Responses of Community Members to Induced Indirect Plant Defense

Induced infochemicals (*sensu* Dicke and Sabelis 1988), once released from the plant, can be exploited by any of its community members, including neighboring plants, herbivores, predators, parasitoids, pollinators and other community members, both above- and belowground (Fig. 2.3). The infochemicals can function as a direct defense by repelling herbivores. For example, butterflies avoid oviposition on plants that are induced by either the presence of eggs or feeding damage (Rothschild and Schoonhoven 1977; Landolt 1993; De Moraes et al. 2001). High levels of direct defense may deter herbivores from feeding or ovipositing on a plant, and may also slow down development of larvae, rendering them more vulnerable to natural enemies (Rothschild and Schoonhoven 1977; Stout and Duffey 1996; Thaler et al. 1996). However, direct and indirect defense mechanisms can act antagonistically. Plant toxins that are ingested by the herbivore may be sequestered to affect the development of carnivores, or negatively influence carnivore fitness due to compromised host size or quality (Ode 2006). Natural populations of *Senecio jacobaea* exhibit genetic variation of pyrrolizidine alkaloid (PA) concentration. Plants infested

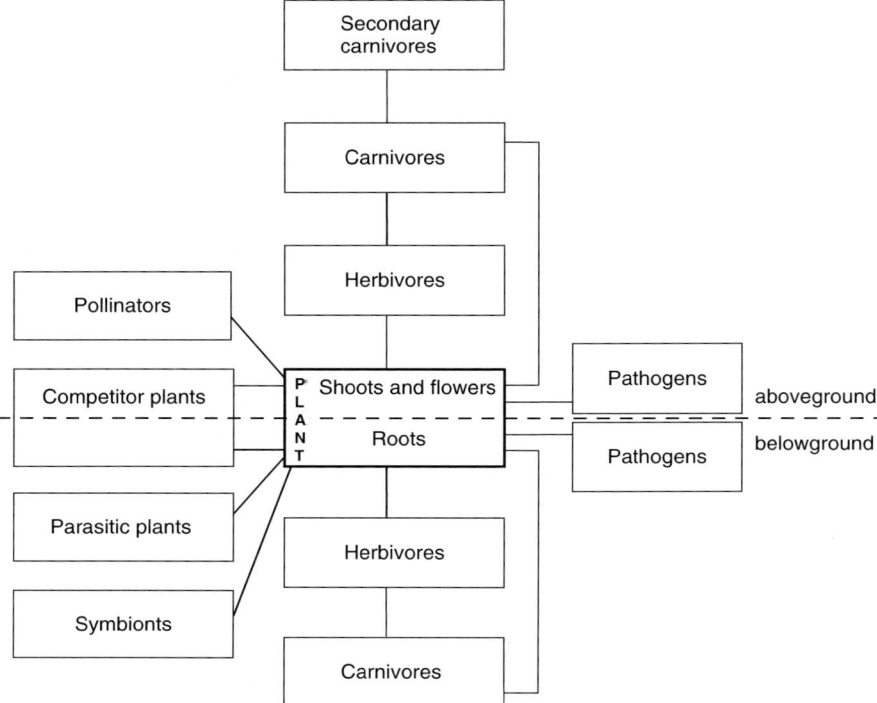

Fig. 2.3 Schematic view of the complexity of interactions above- and belowground in a multitrophic community (pathogens of herbivores, carnivores, and pollinators have not been included)

by aphids and ants that tend the aphids have lower PA-levels than plants without aphids and their ant tenders. The ants can protect the plant from complete defoliation, thus benefiting the plants in years with high pressure from the specialist herbivore *Tyria jacobaeae*, while in years with low *T. jacobaeae* abundance these plants may suffer more fitness costs from aphid herbivory than plants with higher PA-levels (Vrieling et al. 1991) The effects of induction on other community members than herbivores, and the complexity of their interactions are discussed in the following section.

2.3.1 Responses of Members of the Third Trophic Level

Plants can benefit in terms of fitness gain from carnivores that attack the herbivores (Van Loon et al. 2000; Fritzsche-Hoballah and Turlings 2001). Herbivores, however, are under selection to be inconspicuous, and are small in comparison to the plants they are feeding on. Consequently, carnivorous arthropods often depend on plant cues to locate their herbivorous victims (Turlings et al. 1990; Steinberg et al. 1992; Vet and Dicke 1992; Geervliet et al. 1994). Even though host-derived stimuli are potentially more reliable for host location, their use is often limited by low detectability, especially at longer distances (Vet and Dicke 1992). Therefore, carnivorous arthropods are usually more strongly attracted by plant-derived volatiles as compared to volatiles derived from their herbivorous victims (Turlings et al. 1990, 1991; Steinberg et al. 1993; Geervliet et al. 1994; Dicke 1999). Attraction by volatiles from host-infested plants and by EFN was shown for egg- as well as larval parasitoids and predators (e.g., Blaakmeer et al. 1994; Geervliet et al. 1997; Lou et al. 2005; Hilker and Meiners 2006; Choh et al. 2006; Mumm and Hilker 2006). Induced levels of EFN also increase the abundance of ants, wasps and flies (Kost and Heil 2005), and reduce the amount of leaf damage (Heil 2004).

The major signal-transduction pathway involved in attraction of natural enemies seems to be the JA-pathway (Dicke and Van Poecke 2002). The involvement of JA in induced attraction of members of the third trophic level has been demonstrated both by manipulation of JA on the level of the plant's phenotype and the plant's genotype. JA-treated plants attract natural enemies of herbivorous arthropods and have increased parasitism rates in field (Gols et al. 1999; Thaler 1999; Ozawa et al. 2004). Moreover, jasmonate-deficient plants are less attractive to natural enemies than control plants when attacked by herbivores (Thaler et al. 2002).

2.3.2 Responses of Members of Higher Trophic Levels

The enemies of herbivores can in turn fall victim to members from higher trophic levels, for example to hymenopterous hyperparasitoids, also called secondary parasitoids (Brodeur 2000). Induced changes in plant chemistry affect not only the development and survival of herbivores and their parasitoids, but also that of secondary parasitoids (Harvey et al. 2003). Performance of *Lysibia nana*, a hyperparasitoid with a broad host range, was shown to be negatively affected by high levels of defensive toxins in the host plant (Harvey et al. 2003; Soler et al. 2005). Secondary parasitoids are often compared to primary parasitoids, since both share common life-history strategies (Brodeur 2000). However, as compared to primary parasitoids, little is known about their host searching strategies, and whether or not they use plant volatiles to locate (parasitized) herbivores (Harvey et al. 2003; Buitenhuis et al. 2005). Plant-derived cues may be of limited value for hyperparasitoids for several reasons: firstly, although herbivore-induced plant volatiles may be reliable cues for primary parasitoids, they do not guarantee the presence of a parasitized host for the secondary parasitoid. Secondly, primary parasitoids are often more specialized than secondary parasitoids (Buitenhuis et al. 2005). Yet, a recent study demonstrated that primary parasitoids are in fact able to discriminate between herbivore-induced plant volatiles emitted from plants damaged by unparasitized or parasitized caterpillars (Fatouros et al. 2005), suggesting that reliable plant cues may be available to secondary parasitoids as well.

Several studies have been conducted on host searching behavior of secondary parasitoids with varying results. A specialized ectoparasitoid *Euneura augarus* relies on plant volatiles for long range searching (Völkl and Sullivan 2000). However,

E. augarus does not distinguish between plants with and without host mummies for long range searching. The two hyperparasitoid species *Alloxysta victrix* and *Dendrocerus carpenteri* were attracted to herbivore-induced volatiles (oat plants infested with *Sitobion avenae* aphids; Siri 1993), while in another study with another plant-host-primary parasitoid system (potato infested with *Macrosiphum euphorbiae*) they were not (Buitenhuis et al. 2005). Buitenhuis et al. (2005) did find analogies in foraging behavior of primary and secondary parasitoids in the use of contact cues while searching on a plant. Host searching behavior of hyperparasitoids may depend more on contact cues and less on plant volatiles compared to primary parasitoids (Buitenhuis et al. 2005).

2.3.3 Responses of Pollinators

Sexual reproduction of many plants species depends on pollination by honey bees, bumble bees, solitary bees, syrphid flies or moths (Klein et al. 2006). Herbivory in early stages of plant growth reduces the photosynthetic area of the plant, and may result in smaller plants and a shorter flowering period. This is possibly due to allocation of resources to defenses, rather than growth and reproduction (Poveda et al. 2003). Herbivory may affect the production of pollen and nectar, the quality of nectar, morphology of flowers and may reduce seed production (Lehtilä and Strauss 1997; Hambäck 2001; Poveda et al. 2003). Both nectar quality and quantity are parameters that determine the number and type of pollinators that are attracted to the plants (Potts et al. 2003). While extrafloral nectar is known to increase after herbivory, it remains unknown whether herbivory also affects floral nectar production (Adler et al. 2006). Just a handful of studies have addressed the effect of herbivory on floral chemistry. Leaf herbivory on tobacco plants increased alkaloid concentration in the nectar (Adler et al. 2006), while other studies have shown increases in defense compounds such as nicotine and glucosinolates in flower tissues (Euler and Baldwin 1996; Ohnmeiss and Baldwin 2000; Strauss et al. 2004; Smallegange et al. 2007). It is interesting to note that in tobacco plants nicotine levels in the corolla are lower during the scotophase, when moths are attracted for pollination (Euler and Baldwin 1996).

However, as yet there is little knowledge on the influence of herbivory on pollination. Several studies have reported an indirect effect of herbivory on pollination. Herbivory may affect flowering traits, for example reducing the number of flowers, flower size or plant height. This in turn, can affect flower visitation by pollinators (Lehtilä and Strauss 1997; Adler et al. 2001; Hambäck 2001). Lehtilä and Strauss (1997) showed that bees prefer undamaged radish plants over damaged plants, but this difference could be explained by reduced flower size and number. However, syrphid flies preferred undamaged plants over damaged ones, even when the plants were controlled for flower size and number; indicating a chemical basis for syrphid fly attraction. While in this case pollinator attraction decreased, herbivory can also enhance pollinator attraction. Root herbivory for example, can increase flower visitation, as demonstrated in mustard plants by Poveda et al. (2003, 2005). Although these examples demonstrate the influence of herbivory on pollination, the mechanism behind this phenomenon is not yet elucidated. Possibly, the effect on plant growth, number of flowers, plant volatiles or a change in nectar quantity or quality, determine the changes in pollinator attraction.

2.3.4 Responses of Neighboring Plants

Plants may gain a fitness benefit from responding to herbivore-induced volatiles from neighboring plants that are being attacked. Such a signal is indicative of the imminent danger, and the receiving plant may profit from exploiting this information by readying its defense (Dicke and Bruin 2001). Herbivore-induced plant signals can be exploited by neighboring plants of the same and of different species alike (Engelberth et al. 2004; Baldwin et al. 2006). Strong signals may immediately activate plant defenses, while lower (more common in nature) concentrations of signaling molecules can prime the plant for attack (Turlings and Ton 2006). This was nicely shown in a field study by Kessler et al. (2006) reporting that plant volatiles from damaged sagebrush plants can prime the induction of proteinase inhibitors in nearby tobacco plants. The primed tobacco plants received less damage from subsequent herbivore attack as compared to non-exposed control plants. Not only volatile emission, but also EFN secretion can be primed by exposure to volatiles from damaged plants. Plants that have been exposed to volatiles from herbivore-infested conspecific plants produce more EFN when they are attacked themselves (Heil and Kost 2006), although this effect was only observed during the early stages of herbivore attack (Choh and Takabayashi 2006).

2.3.5 Above- and Belowground Interactions

Most studies on induced defenses have focused on aboveground interactions. However, belowground interactions can have an important effect on aboveground processes. For example, root herbivory and belowground plant mutualists, such as arbuscular mycorrhizal fungi, affect pollination (Poveda et al. 2003, 2005; Wolfe et al. 2005).

Processes similar to those reported to occur aboveground do also take place belowground. Plants can release volatiles in response to root herbivory that attract natural enemies of the root herbivores. This was first reported for a coniferous host plant, *Thuja occidentalis*, that when exposed to weevil larvae, releases chemicals attractive to nematodes that can parasitize the larvae (Van Tol et al. 2001). Rasmann et al. (2005) showed the same phenomenon for maize plants, and identified a chemical attracting the entomopathogenic nematode *Heterorhabditis megidis*, a natural enemy of the attacking beetle. Plant-to-plant signaling also takes place belowground. Plants emit belowground signals that are exploited by neighboring plants for the attraction of predators (Chamberlain et al. 2001; Dicke and Dijkman 2001), or by parasitic plants to induce germination (Bouwmeester et al. 2003; Runyon et al. 2006).

In addition to belowground interactions, it is important to consider the interactions between above- and belowground communities. Cotton plants, for example, increase aboveground extrafloral nectar production upon attack by root-feeding wireworms (Wäckers and Bezemer 2003). Root herbivory, as well as arbuscular mycorrhizal fungi, can also change aboveground volatile emission, and subsequently increase the number of visits by parasitoids of aboveground herbivores (Masters et al. 2001; Neveu et al. 2002; Guerrieri et al. 2004; Soler et al. 2007). The performance of the aboveground multitrophic community associated with a plant differs between plants with and without root herbivory (Soler et al. 2005). Root herbivory changed plant quality, which in turn negatively affected the performance of an aboveground herbivore, a primary parasitoid and secondary parasitoid. Bezemer et al. (2005) found that the performance of aphids was reduced by the presence of nematodes or micro-organisms, while the primary aphid parasitoids were positively affected by these belowground community members. These studies illustrate that soil communities can influence interactions and performance of aboveground species at multiple trophic levels. To our knowledge, no such studies have been conducted to investigate possible effects of aboveground herbivory on attraction and performance of belowground carnivores.

2.4 Indirect Defense in a Complex and Variable World

Induced plant volatile blends can be very specific: parasitoids and predators can distinguish between induced blends from different plant species, herbivore species (or even herbivore instars), between feeding- and egg-induced responses, and between local and systemic damage (Table 2.2). While most studies have examined simple tritrophic systems consisting of one plant, one herbivore and one of its natural enemies, in the field an organism interacts with many more species. Host plants are frequently infested by more than one herbivore (e.g., Vos et al. 2001). For community dynamics it is important whether the organisms receiving plant-emitted signals are able to distinguish between signals indicating the presence of their host from non-host signals and other background odors. Shiojiri et al. (2002) compared the host searching behavior of two species of parasitoid wasps on host plants infested with one or two herbivores. One of the parasitoids, *Cotesia plutellae*, was more strongly attracted to plants infested solely with its host, while the other parasitoid, *Cotesia glomerata*, preferred plants with both herbivores (Shiojiri et al. 2001). Consequently, a plant infested with both herbivores provides an enemy-free space for *C. plutellae*'s host (i.e., *Plutella xylostella*), and an enemy-dense space for *C. glomerata*'s host (i.e., *Pieris rapae*). The oviposition preference of the adult herbivores corresponded to this pattern; while *P. xylostella* preferred plants with both herbivores, *P. rapae* did not show any preference (Shiojiri et al. 2002). In this

Table 2.2 Plant volatile emission can be highly specific in response to different stimuli. This table non-exhaustively illustrates the specificity of plant volatile emission and the subsequent perception and discrimination between signals by insects

Plant volatiles differ depending on:	System	Result	References
Plant species	Tobacco, cotton and maize – *Heliothis virescens* and *Helicoverpa zea* – parasitic wasp *Cardiochiles nigriceps*	All three plant species emit different volatile blends in response to their attackers.	(De Moraes et al. 1998)
Plant cultivar	Maize cultivars and wild relatives	The plant cultivars emitted qualitatively and quantitatively different volatile blends	(Gouinguene et al. 2001)
Plant with or without host	Maize *Zea mays* – beet armyworm *Spodoptera exigua* – *Cotesia marginiventris*	Undamaged and herbivore-damaged plants differ in volatile emission and *C. marginiventris* can discriminate between the odor blends.	(Turlings et al. 1990)
	Pine – aphids – *Cinara pinea* – hyperparasitoid *Euneura augarus*	Hyperparasitoid cannot discriminate at long range between plants with and without aphid mummies	(Völkl and Sullivan 2000)
Herbivore species	Tobacco, cotton and maize – *Heliothis virescens* and *Helicoverpa zea* – parasitic wasp *Cardiochiles nigriceps*	Plants produce herbivore-specific chemical signals and *C. nigriceps* females prefer host-infested over non-host-infested plants	(De Moraes et al. 1998)
	Bean plants *Vicia faba* – host: pea aphid *Acyrthosiphon pisum*, non-host black bean aphid *Aphis fabae* – parasitoid *Aphidius ervi*	Host and non-host aphid induce different volatiles, *A. ervi* can discriminate between host-infested and non-host-infested plants	(Du et al. 1996; 1998)

Table 2.2 (continued)

Plant volatiles differ depending on:	System	Result	References
Herbivore instar	Maize *Zea mays* – common armyworm *Pseudaletia separata* – *Cotesia kariyai*	*C. kariyai* discriminates between undamaged plants and plants damaged by 1st to 4th instar larvae, but not by later instar larvae. And early and late instar larvae induce qualitatively and quantitatively different volatile blends	(Takabayashi et al. 1995)
Local or systemic induction	Lima bean *Phaseolus lunatus* – spider mites *Tetranychus urticae* – predatory mites *Phytoseiulus persimilis*	Undamaged leaves of a spider mite-infested plant are induced to emit volatiles that attract the predators	(Dicke et al. 1990a)
Time since induction	Brussels sprouts *Brassica oleracea* – large cabbage white *Pieris brassicae* – *Cotesia glomerata*	Brussels sprouts plant becomes attractive to *C. glomerata* 30 minutes after caterpillar infestation, reaches a maximum after 3 hours and then remains constant for at least 14 hours	(Scascighini et al. 2005)
Time of day	Tobacco *Nicotiana tabacum* – moth *Heliothis virescens*	Difference in volatile emission during day and night, influences night-active herbivores	(De Moraes et al. 2001)
Infestation rate	Field elm *Ulmus minor* – elm leaf beetle *Xanthogaleruca luteola* –	Twigs with a low infestation rate are more attractive to beetles than uninfested or heavily infested twigs	(Meiners et al. 2005)

example, *P. xylostella* profits from a positive indirect effect (associational resistance) through the presence of *P. rapae* (White and Andow 2006). Defense responses to herbivores may also interfere with other interactions in the community, such as pollination. For tobacco plants there is a trade-off between repelling herbivores and attracting pollinators. As briefly mentioned in Section 2.3.3, these plants reduce the level of toxins and increase emission of a pollinator attractant by their flowers in the evening. This way tobacco can defend itself against herbivory during daytime and attract pollinators at night (Euler and Baldwin 1996).

While some community members are able to distinguish between different volatile blends, others cannot, or only after associative learning (Takabayashi et al. 2006). Several arthropod species, such as *Phytoseiulus persimilis*, *Cotesia marginiventris* and *C. glomerata*, can learn to associate certain odors with a reward (Geervliet et al. 1997; Turlings and Fritzsche 1999; De Boer et al. 2005). Learning to associate odors with the presence of host or prey may be a way to cope with the variation in volatile blends. Learning new associations may take some time and multiple experiences. In the meantime the host or prey species may profit from its 'invisibility' to its natural enemy while the parasitoid or predator wastes time on searching on plants without its host or prey (Shiojiri et al. 2002). The temporal pattern in learning to respond to certain blends and losing this learned response will affect the risk of herbivores to fall victim to their enemies. Temporary refuges may stabilize predator–prey or parasitoid-host systems and thus affect dynamics of arthropod communities on plants (Vos et al. 2001; Takabayashi et al. 2006). To gain more insight into these subtle interactions and ecosystem stability, field studies in which one component is changed could provide an important tool to investigate the impact on the community.

2.5 Manipulation of Indirect Plant Defense

To investigate the effect of indirect defenses and their components, a manipulative approach is most rewarding. Manipulation of defense responses can provide information on both the mechanisms of plant defense and the ecological consequences of changes in the plant's phenotype. A range of manipulative approaches can be taken that are explained below.

2.5.1 Perfuming with Individual Compounds

A simple change in a plant's phenotype can be made by adding a single compound to an undamaged plant. This method allows testing the importance of individual compounds for attraction of, for example, predators and parasitoids to the plant against a background of natural odors (Dicke et al. 2006). Even though carnivorous arthropods are usually attracted to complex odor blends, increased attraction of carnivores due to addition of individual compounds has been recorded with this method (De Boer and Dicke 2004). Also in a field study, the addition of linalool to *N. attenuata* plants resulted in both increased predation of herbivore eggs and larvae, and decreased oviposition by an herbivore (Kessler and Baldwin 2001).

2.5.2 Fractionation and Filtering

Phenotypic manipulation of the volatile blends can also be accomplished by using filters to collect compounds with specific chemical properties selectively (D'Alessandro and Turlings 2006), or by fractionation of the headspace and subsequently testing different fractions of the blend for biological activity (Turlings and Fritzsche 1999; Van den Boom 2003). As compared to the testing of single compounds, these methods have the advantage of conserving the ratios of different volatiles as they are present in natural herbivore-induced volatile blends. Filters can be used to remove one or more compounds, which are suspected to show biological activity, from the total volatile blend. It is then possible to test their importance for animal behavior in olfactometer experiments (D'Alessandro and Turlings 2006). With this method D'Alessandro and Turlings (2005) showed that the selective removal of the more polar plant volatiles using silica filters decreases the attractiveness of the volatile blend to a parasitoid wasp, *C. marginiventris*, but not to another parasitoid *Microplitis rufiventris*. This study illustrates how filtering can be used as a tool to find the ecologically relevant volatiles for different species on all trophic levels using various adsorbing materials. For fractionation, the mixture of defense compounds can be trapped onto an absorbent material and subsequently recovered by chemical or thermal desorption. Subtractive fractionation and testing of the different fractions for their biological activity provides a useful tool to identify compounds involved in indirect defense (Van den Boom 2003).

2.5.3 Genetic Modification

Another way to manipulate induced defenses of plants, is by genetic modification of signal-transduction or biosynthetic pathways. Carefully modified plants that differ in a single gene can be compared with wild-type plants to gain more insight into mechanisms of induced defense and the resulting ecological effects (for review, see Snoeren et al. 2007). Otherwise identical plants can be induced by herbivory and the differences between the wild type and mutant can be studied at the levels of gene expression, volatile production, and response of the plant-associated community. The approach is limited by the availability of modified plants. For some species like *A. thaliana* however, a wide variety of mutant and transgenic plants is available. Genetic modification is an excellent tool to study the ecological relevance of individual compounds that are difficult to obtain synthetically. For example, undamaged *Arabidopsis* plants transformed with a terpene synthase from strawberry, produce the terpenoids 4,8-dimethyl-1,3(E),7-nonatriene and (E)-nerolidol which results in the attraction of predatory mites (Kappers et al. 2005). However, *A. thaliana* has its limitations, such as its very early phenology and consequently limited interaction with potentially associated organisms, as a model for community ecology studies.

Transgenic *N. attenuata* plants that have been modified in the octadecanoid signal-transduction pathway received more herbivory and by more herbivore species than wild type control plants (Kessler et al. 2004). Similarly, the tomato mutant *def-1* is deficient in JA accumulation through a mutation early in the JA pathway. As a result, the plants produce an incomplete volatile blend in response to herbivore damage, and the natural enemies of the herbivores do not discriminate between volatile blends from induced and uninduced plants (Thaler et al. 2002; Ament et al. 2004). However, for many ecological model species mutants or genetically modified plants are not (yet) available.

2.5.4 Chemical Elicitors and Inhibitors

Signal-transduction and biosynthetic pathways can be manipulated through the application of (specific) inducers or inhibitors. By artificially inducing or inhibiting different steps of the signaling pathways it is possible to study ecological interactions at different trophic levels, or in the whole community. Plant-emitted volatile blends can be manipulated by interfering with the signal-transduction pathways, and the importance of individual steps can be investigated by specifically activating or blocking individual pathway enzymes. When these changed blends are offered to (members of) the insect community, this will provide insight into the ecological relevance of these pathways.

The most extensively tested elicitors to date are the phytohormone jasmonic acid and its methyl ester, methyl jasmonate (MeJA). These elicitors were shown to play an important role in induced defense in many plant species against a wide range of herbivores. JA induces volatile blends similar to those induced by herbivore damage (Hopke et al. 1994; Dicke et al. 1999; Ozawa et al. 2000, 2004), and while herbivores are negatively influenced (Van Dam et al. 2000; Thaler et al. 2001; Bruinsma et al. 2007), carnivorous arthropods are attracted to plants that have been induced with JA or MeJA (Thaler 1999; Van Poecke and Dicke 2002; Gols et al. 2003). For many other elicitors and phytohormones, such as methyl salicylate, linolenic acid, OPDA, β-glucosidase, cellulysin, and alamethicin, similar studies demonstrated their effects on community members (Mattiacci et al. 1995; Koch et al. 1999; Dicke and Van Poecke 2002; De Boer and Dicke 2004; Ozawa et al. 2004). One of the advantages of using elicitors is the possibility to apply a controlled dose to specific plant parts, whereas it is practically impossible to control the amount of injury inflicted by insects or other biotic agents. However, it is not easy to relate the externally applied dosage to intracellular concentrations and effects.

Specific inhibitors, although not used quite as extensively as elicitors, are suitable to demonstrate the importance of different steps in signal-transduction pathways as well. However, accumulation of pathway intermediates just before the inhibited step may cause physiological side effects. Making use of inhibitors, Koch et al. (1999)

showed the importance of early steps in the octadecanoid pathway for volatile emission of Lima bean (Table 2.1). Phenidone for example, inhibits the activity of lipoxygenase, which results in incomplete volatile blends (Piel et al. 1997; Koch et al. 1999) and reduced EFN secretion (Heil et al. 2004), indicating the importance of this early step in the octadecanoid pathway for indirect defense responses.

2.6 From Lab to the Field

To gain insight into the ecology of induced defenses, field studies incorporating the complexity of natural ecosystems are indispensable. Manipulation through one of the methods described above may facilitate research on the effects of induced indirect defense on interactions between community members and the stability of the system.

Most studies addressing induced plant defense are laboratory studies, studying simple systems of one plant, one herbivore and its natural enemies. This provides detailed insight into the effects of induced defenses on individual interactions. Through greenhouse and semi-field studies with more complex set-ups, for instance by introducing background odors under controlled conditions (Janssen 1999; Dicke et al. 2003), more insight will be gained in field situations. However, in order to use knowledge gained from these studies, the relative importance of these pieces of information should be assessed in the field. Also biological control in agricultural fields may benefit from such knowledge and understanding of multitrophic interactions in the field.

2.6.1 Natural Ecosystems

For studying the ecology of herbivore-induced plant responses, it is necessary to address natural ecosystems. For instance, in wild radish induced responses to herbivory increase plant fitness in natural environments (Agrawal 1999). Herbivory increases trichome density and subsequently reduces preference and performance of several herbivores. The effectiveness of induced defenses is clear from a field study on *N. attenuata* where herbivory was reduced by as much as 90% by the addition of defense compounds (Kessler and Baldwin 2001). In a later study, Kessler et al. (2006) showed that volatiles from damaged sagebrush can prime responses in *N. attenuata* and reduce herbivore damage to the exposed plants. However, despite the use of natural populations of *N. attenuata*, the ecological relevance of this study is questionable because the sagebrush that was used for elicitation occurs in a different successional stage than *N. attenuata*. Furthermore, the effect was only detected at a short range, when the plants grew within 15 cm from each other.

In naturally occurring milkweed populations, early-season herbivory affects subsequent herbivory throughout the season, thereby affecting community structure (Van Zandt and Agrawal 2004). In a natural community, Agrawal (2004) studied the complex interactions between milkweed and competing grass, in presence or absence of root and leaf herbivory. He concluded that the genetic differences, competition, and herbivory resulted in complex interactions that may result in diffuse co-evolution between milkweed and its herbivores. To improve our understanding of the evolution of induced defenses and resistance of herbivores more field studies are needed.

The use of elicitors in the field can provide information on the ecological relevance of pathways and certain steps therein. For example, the application of jasmonic acid to tomato plants in the field increased parasitism of herbivores and thus showed the involvement of jasmonic acid-induced changes in the attraction of carnivores (Thaler 1999). The application of methyl jasmonate to tobacco plants demonstrated the costs of jasmonate-induced responses. In environments with herbivore pressure, induced plants suffered less from herbivore attack and produced more viable seeds than uninduced plants. However, undamaged plants produced more seeds when they were not induced compared to jasmonate-induced plants (Baldwin et al. 1998).

2.6.2 Agricultural Systems

Induced plant defenses can aid pest control in agricultural systems. Attracting natural enemies of herbivores to crops can help control pests in agriculture; in the field as well as in greenhouses (Dicke et al. 1990a; Turlings and Ton 2006). Therefore, understanding the mechanisms involved in plant defenses and the consequences for the community associated with the plant can aid crop protection. Manipulation may increase the effectiveness of plant defenses, by attracting natural enemies before considerable damage is done by herbivores and by deterring oviposition by herbivores. This can be achieved by inducing the plant with phytohormones like cis-jasmone or jasmonic acid (Thaler 1999; Birkett et al. 2000; Heil 2004). Another possibility is the use of genetically modified crops that produce volatile blends that are more attractive to predators than genotypes currently used. The technology is being developed (Kappers et al. 2005; Schnee et al. 2006). However, the consequences for the community, above- as well as belowground, and the effect on interactions between different community members still need to be addressed (Groot and Dicke 2002; Kowalchuk et al. 2003).

A lot of the research on genetic modification of plant defense has been done in the model plant *A. thaliana*, for which many mutants are available (Turlings and Ton 2006). Knowledge about *A. thaliana* can be extrapolated to *Brassica* species, and therefore readily be applied in agricultural settings with crop species like *Brassica oleracea* or wild *Brassica* species (Broekgaarden et al. 2007; Zheng et al. 2007). In conclusion, the step from the laboratory into the field has not been made often yet. It will be important to make this step for different plant species to gain insight into the effects of induced indirect defense on community processes.

2.7 Perspectives

The effects of induced defense-related phenotypic changes in plants on community dynamics are difficult to predict, because many aspects are involved and the variability of plant responses is enormous. While many bi- and tritrophic interactions are well studied, plants in nature are usually under the attack of a range of organisms at the same time. How this affects plant defense has only just begun to be addressed, and first results show that the effects may be an increase as well as a decrease in defense intensity (e.g., Dicke et al. 2003; Rodriguez-Saona et al. 2005; Cardoza and Tumlinson 2005).

Although most studies thus far have focused on aboveground processes, the influence of the changes in plant phenotype is not limited to the aboveground community. Aboveground interactions can change belowground root exudates and influence the soil community, and belowground damage can influence aboveground indirect defense (Bezemer and van Dam 2005). The reverse, however, i.e. the effect of aboveground interactions on belowground indirect defense, remains as yet uninvestigated (Bezemer and van Dam 2005). Incorporating these interactions in future studies will greatly enhance our insight into the effects of induced indirect defense on the functioning of complex communities. In addition to plant-pathogen and plant-herbivore interactions, plants may also be under the attack of parasitic plants (Bouwmeester et al. 2003; Runyon et al. 2006), or interact with belowground symbiotic organisms such as mycorrhiza or symbiotic bacteria (Gange et al. 2002). Furthermore, aboveground endophytic organisms can influence the plant's defensive phenotype and consequently also the interactions with community members (Omacini et al. 2001). Incorporating these interactions in the investigations of indirect defense of plants in a community ecology approach will increase complexity, yet doing so is essential to gain a meaningful understanding of the effects of indirect plant defense on plant ecology.

Another area of research that has not received a lot of attention so far is the searching behavior of members of the higher trophic levels, such as hyperparasitoids. How they find their host and whether they use plant cues remains largely unknown (Buitenhuis et al. 2005). The same applies for pollinators. Though some effects of herbivory on pollination have been reported (Lehtilä and Strauss 1997; Poveda et al. 2003), the underlying mechanisms remain to be unraveled and which signal-transduction pathways are important in this respect, waits to be investigated.

For a complete understanding of the ecology and evolution of communities, it is necessary to include all trophic levels in field studies. Manipulative studies are likely to provide the best way forward. They can be used in the laboratory to investigate individual interactions and are a valuable tool to investigate the effects of induced defenses on the community in the field (Kessler and Baldwin 2001; Kessler et al. 2004). Using an integrated approach with molecular, chemical, and behavioral methodology will significantly advance the research in this area (Baldwin et al. 2001).

Acknowledgments We thank Joop van Loon for valuable comments on the manuscript. This work has been financially supported by a VICI grant from the Netherlands Organisation for Scientific Research, NWO (865.03.002).

References

Adler LS, Karban R, Strauss SY (2001) Direct and indirect effects of alkaloids on plant fitness via herbivory and pollination. Ecology 82:2032–2044

Adler LS, Wink M, Distl M, Lentz AJ (2006) Leaf herbivory and nutrients increase nectar alkaloids. Ecol Lett 9:960–967

Agrawal AA (1999) Induced responses to herbivory in wild radish: effects on several herbivores and plant fitness. Ecology 80:1713–1723

Agrawal AA (2001) Phenotypic plasticity in the interactions and evolution of species. Science 294:321–326

Agrawal AA (2004) Resistance and susceptibility of milkweed: competition, root herbivory, and plant genetic variation. Ecology 85:2118–2133

Alborn T, Turlings TCJ, Jones TH, Stenhagen G, Loughrin JH, Tumlinson JH (1997) An elicitor of plant volatiles from beet armyworm oral secretion. Science 276:945–949

Ament K, Kant MR, Sabelis MW, Haring MA, Schuurink RC (2004) Jasmonic acid is a key regulator of spider mite-induced volatile terpenoid and methyl salicylate emission in tomato. Plant Physiol 135:2025–2037

Avdiushko SA, Brown GC, Dahlman DL, Hildebrand DF (1997) Methyl jasmonate exposure induces insect resistance in cabbage and tobacco. Environ Entomol 26:642–654

Baldwin IT (1990) Herbivory simulations in ecological research. Trends Ecol Evol 5:91–93

Baldwin IT (1998) Jasmonate-induced responses are costly but benefit plants under attack in native populations. Proc Natl Acad Sci USA 95:8113–8118

Baldwin IT, Halitschke R, Kessler A, Schittko U (2001) Merging molecular and ecological approaches in plant–insect interactions. Curr Opin Plant Biol 4:351–358

Baldwin IT, Halitschke R, Paschold A, von Dahl CC, Preston CA (2006) Volatile signaling in plant–plant interactions: 'Talking trees' in the genomics era. Science 311:812–815

Bezemer TM, De Deyn GB, Bossinga TM, van Dam NM, Harvey JA, Van der Putten WH (2005) Soil community composition drives aboveground plant–herbivore–parasitoid interactions. Ecol Lett 8:652–661

Bezemer TM, van Dam NM (2005) Linking aboveground and belowground interactions via induced plant defenses. Trends Ecol Evol 20:617–624

Birkett MA, Campbell CAM, Chamberlain K, Guerrieri E, Hick AJ, Martin JL, Matthes M, Napier JA, Pettersson J, Pickett JA, Poppy GM, Pow EM, Pye BJ, Smart LE, Wadhams GH, Wadhams LJ, Woodcock CM (2000) New roles for cis-jasmone as an insect semiochemical and in plant defense. Proc Natl Acad Sci USA 97:9329–9334

Blaakmeer A, Geervliet JBF, Van Loon JJA, Posthumus MA, Van Beek TA, De Groot A (1994) Comparative headspace analysis of cabbage plants damaged by two species of *Pieris* caterpillars: consequences for in-flight host location by *Cotesia* parasitoids. Entomol Exp Appl 73:175–182

Boland W, Koch T, Krumm T, Piel J, Jux A (1999) Induced biosynthesis of insect semiochemicals in plants. In: Chadwick DJ, Goode J (eds) Insect–plant interactions and induced plant defence (Novartis Foundation Symposium 223). Wiley, Chichester, pp 110–126

Bouwmeester HJ, Matusova R, Sun ZK, Beale MH (2003) Secondary metabolite signaling in host–parasitic plant interactions. Curr Opin Plant Biol 6:358–364

Brodeur J (2000) Host specificity and trophic relationships of hyperparasitoids. In: Hochberg ME, Ives AR (eds) Parasitoid population biology. Princeton University Press, Princeton, pp 163–183

Broekgaarden C, Poelman EH, Steenhuis G, Voorrips RE, Dicke M, Vosman B (2007) Genotypic variation in genome-wide transcription profiles induced by insect feeding. BMC Genomics 8:239

Bruinsma M, Van Dam NM, Van Loon JJA, Dicke M (2007) Jasmonic acid-induced changes in *Brassica oleracea* affect oviposition preference of two specialist herbivores. J Chem Ecol 33:655–668

Buitenhuis R, Vet LEM, Boivin G, Brodeur J (2005) Foraging behavior at the fourth trophic level: a comparative study of host location in aphid hyperparasitoids. Entomol Exp Appl 114:107–117

Cardoza YJ, Tumlinson JH (2006) Compatible and incompatible *Xanthomonas* infections differentially affect herbivore-induced volatile emission by pepper plants. J Chem Ecol 32: 1755–1768

Chamberlain K, Guerrieri E, Pennacchio F, Pettersson J, Pickett JA, Poppy GM, Powell W, Wadhams LJ, Woodcock CM (2001) Can aphid-induced plant signals be transmitted aerially and through the rhizosphere? Biochem Syst Ecol 29:1063–1074

Choh Y, Takabayashi J (2006) Herbivore-induced extrafloral nectar production in Lima bean plants enhanced by previous exposure to volatiles from infested conspecifics. J Chem Ecol 32: 2073–2077

Choh Y, Kugimiya S, Takabayashi J (2006) Induced production of extrafloral nectar in intact lima bean plants in response to volatiles from spider mite-infested conspecific plants as a possible indirect defense against spider mites. Oecologia 147:455–460

Conrath U, Beckers GJM, Flors V, Garcia-Agustin P, Jakab G, Mauch F, Newman MA, Pieterse CMJ, Poinssot B, Pozo MJ, Pugin A, Schaffrath U, Ton J, Wendehenne D, Zimmerli L, Mauch-Mani B (2006) Priming: getting ready for battle. Mol Plant Micr Int 19:1062–1071

Conrath U, Pieterse CMJ, Mauch-Mani B (2002) Priming in plant–pathogen interactions. Trends Plant Sci 7:210–216

Creelman RA, Mullet JE (1997) Biosynthesis and action of jasmonates in plants. Annu Rev Plant Physiol Plant Mol Biol 48:355–381

D'Alessandro M, Turlings TCJ (2005) In situ modification of herbivore-induced plant odors: a novel approach to study the attractiveness of volatile organic compounds to parasitic wasps. Chem Senses 30:739–753

D'Alessandro M, Turlings TCJ (2006) Advances and challenges in the identification of volatiles that mediate interactions among plants and arthropods. Analyst 131:24–32

De Boer JG, Dicke M (2004) The role of methyl salicylate in prey searching behavior of the predatory mite *Phytoseiulus persimilis*. J Chem Ecol 30:255–271

De Boer JG, Snoeren TAL, Dicke M (2005) Predatory mites learn to discriminate between plant volatiles induced by prey and nonprey herbivores. Anim Behav 69:869–879

De Moraes CM, Lewis WJ, Paré PW, Alborn HT, Tumlinson JH (1998) Herbivore-infested plants selectively attract parasitoids. Nature 393:570–573

De Moraes CM, Mescher MC, Tumlinson JH (2001) Caterpillar-induced nocturnal plant volatiles repel conspecific females. Nature 410:577–580

De Vos M (2006) Signal signature, transcriptomics, and effectiveness of induced pathogen and insect resistance in *Arabidopsis*. Utrecht University, Utrecht

De Vos M, Van Oosten VR, Van Poecke RMP, Van Pelt JA, Pozo MJ, Mueller MJ, Buchala AJ, Metraux JP, Van Loon LC, Dicke M, Pieterse CMJ (2005) Signal signature and transcriptome changes of *Arabidopsis* during pathogen and insect attack. Mol Plant Microbe Interact 18:923–937

Dicke M (1999) Evolution of induced indirect defense of plants. In: Tollrian R, Harvell CJ (eds) The ecology and evolution of inducible defenses. Princeton University Press, Princeton, pp 62–88

Dicke M, Bruin J (2001) Chemical information transfer between plants: back to the future. Biochem Syst Ecol 29:981–994

Dicke M, Bruinsma M, Bukovinszky T, Gols R, De Jong PW, Van Loon JJA, Snoeren TAL, Zheng S-J (2006) Investigating the ecology of inducible indirect defence by manipulating plant phenotype and genotype. IOBC WPRS Bull 29:15–23

Dicke M, De Boer JG, Höfte M, Rocha-Granados MC (2003) Mixed blends of herbivore-induced plant volatiles and foraging success of carnivorous arthropods. OIKOS 101:38–48

Dicke M, Dijkman H (2001) Within-plant circulation of systemic elicitor of induced defence and release from roots of elicitor that affects neighboring plants. Biochem Syst Ecol 29:1075–1087

Dicke M, Gols R, Ludeking D, Posthumus MA (1999) Jasmonic acid and herbivory differentially induce carnivore-attracting plant volatiles in lima bean plants. J Chem Ecol 25:1907–1922

Dicke M, Hilker M (2003) Induced plant defenses: from molecular biology to evolutionary ecology. Basic Appl Ecol 4:3–14

Dicke M, Sabelis MW (1988) Infochemical terminology: based on cost benefit analysis rather than origin of compounds? Funct Ecol 2:131–139

Dicke M, Sabelis MW, Takabayashi J, Bruin J, Posthumus MA (1990a) Plant strategies of manipulating predator–prey interactions through allelochemicals: prospects for application in pest control. J Chem Ecol 16:3091–3118

Dicke M, Takabayashi J, Posthumus MA, Schütte C, Krips OE (1998) Plant–phytoseiid interactions mediated by herbivore-induced plant volatiles: variation in production of cues and in responses of predatory mites. Exp Appl Acarol 22:311–333

Dicke M, Van Beek TA, Posthumus MA, Ben Dom N, Van Bokhoven H, De Groot AE (1990b) Isolation and identification of volatile kairomone that affects acarine predator–prey interactions. Involvement of host plant in its production. J Chem Ecol 16:381–396

Dicke M, Van Poecke RMP (2002) Signaling in plant–insect interactions: signal transduction in direct and indirect plant defence. In: Scheel D, Wasternack C (eds) Plant signal transduction. Oxford University Press, Oxford, pp 289–316

Dicke M, Vet LEM (1999) Plant–carnivore interactions: evolutionary and ecological consequences for plant, herbivore and carnivore. In: Olff H, Brown VK, Drent RH (eds) Herbivores: between plants and predators. Blackwell Science, Oxford, pp 483–520

Donath J, Boland W (1994) Biosynthesis of acyclic homoterpenes in higher plants parallels steroid hormone metabolism. J Plant Physiol 143:473–478

Du YJ, Poppy GM, Powell W (1996) Relative importance of semiochemicals from first and second trophic levels in host foraging behavior of *Aphidius ervi*. J Chem Ecol 22:1591–1605

Du YJ, Poppy GM, Powell W, Pickett JA, Wadhams LJ, Woodcock CM (1998) Identification of semiochemicals released during aphid feeding that attract parasitoid *Aphidius ervi*. J Chem Ecol 24:1355–1368

Engelberth J, Alborn HT, Schmelz EA, Tumlinson JH (2004) Airborne signals prime plants against insect herbivore attack. Proc Natl Acad Sci USA 101:1781–1785

Euler M, Baldwin IT (1996) The chemistry of defense and apparency in the corollas of *Nicotiana attenuata*. Oecologia 107:102–112

Farmer EE, Ryan CA (1990) Interplant communication – airborne methyl jasmonate induces synthesis of proteinase inhibitors in plant leaves. Proc Natl Acad Sci USA 87:7713–7716

Fatouros NE, Van Loon JJA, Hordijk KA, Smid HM, Dicke M (2005) Herbivore-induced plant volatiles mediate in-flight host discrimination by parasitoids. J Chem Ecol 31:2033–2047

Fritzsche-Hoballah ME, Turlings TCJ (2001) Experimental evidence that plants under caterpillar attack may benefit from attracting parasitoids. Evol Ecol Res 3:553–565

Gange AC, Stagg PG, Ward LK (2002) Arbuscular mycorrhizal fungi affect phytophagous insect specialism. Ecol Lett 5:11–15

Geervliet JBF, Posthumus MA, Vet LEM, Dicke M (1997) Comparative analysis of headspace volatiles from different caterpillar-infested or uninfested food plants of *Pieris* species. J Chem Ecol 23:2935–2954

Geervliet JBF, Vet LEM, Dicke M (1994) Volatiles from damaged plants as major cues in long-range host-searching by the specialist parasitoid *Cotesia rubecula*. Entomol Exp Appl 73:289–297

Gols R, Posthumus MA, Dicke M (1999) Jasmonic acid induces the production of gerbera volatiles that attract the biological control agent *Phytoseiulus persimilis*. Entomol Exp Appl 93:77–86

Gols R, Roosjen M, Dijkman H, Dicke M (2003) Induction of direct and indirect plant responses by jasmonic acid, low spider mite densities or a combination of jasmonic acid treatment and spider mite infestation. J Chem Ecol 29:2651–2666

Gouinguene S, Degen T, Turlings TCJ (2001) Variability in herbivore-induced odor emissions among maize cultivars and their wild ancestors (teosinte). Chemoecology 11:9–16

Groot AT, Dicke M (2002) Insect-resistant transgenic plants in a multi-trophic context. Plant J 31:387–406

Guerrieri E, Lingua G, Digilio MC, Massa N, Berta G (2004) Do interactions between plant roots and the rhizosphere affect parasitoid behavior? Ecol Entomol 29:753–756

Halitschke R, Schittko U, Pohner G, Boland W, Baldwin IT (2001) Molecular interactions between the specialist herbivore *Manduca sexta* (Lepidoptera, Sphingidae) and its natural host *Nicotiana attenuata*. III. Fatty acid–amino acid conjugates in herbivore oral secretions are necessary and sufficient for herbivore-specific plant responses. Plant Physiol 125:711–717

Hambäck PA (2001) Direct and indirect effects of herbivory: feeding by spittlebugs affects pollinator visitation rates and seedset of *Rudbeckia hirta*. Ecoscience 8:45–50

Harvey JA, Van Dam NM, Gols R (2003) Interactions over four trophic levels: foodplant quality affects development of a hyperparasitoid as mediated through a herbivore and its primary parasitoid. J Anim Ecol 72:520–531

Hatanaka A, Kajiwara T, Sekiya J (1987) Biosynthetic pathway for C-6-aldehydes formation from linolenic acid in green leaves. Chem Phys Lipids 44:341–361

Heil M (2004) Induction of two indirect defenses benefits Lima bean (*Phaseolus lunatus*, Fabaceae) in nature. J Ecol 92:527–536

Heil M, Greiner S, Meimberg H, Krüger R, Noyer JL, Heubl G, Linsenmair KE, Boland W (2004) Evolutionary change from induced to constitutive expression of an indirect plant resistance. Nature 430:205–208

Heil M, Koch T, Hilpert A, Fiala B, Boland W, Linsenmair KE (2001) Extrafloral nectar production of the ant-associated plant, *Macaranga tanarius*, is an induced, indirect, defensive response elicited by jasmonic acid. Proc Natl Acad Sci USA 98:1083–1088

Heil M, Kost C (2006) Priming of indirect defenses. Ecol Lett 9:813–817

Hilker M, Meiners T (2006) Early herbivore alert: insect eggs induce plant defense. J Chem Ecol 32:1379–1397

Holt RD (1977) Predation, apparent competition, and structure of prey communities. Theor Popul Biol 12:197–229

Hopke J, Donath J, Blechert S, Boland W (1994) Herbivore-induced volatiles: the emission of acyclic homoterpenes from leaves of *Phaseolus lunatus* and *Zea mays* can be triggered by a beta-glucosidase and jasmonic acid. FEBS Lett 352:146–150

Janssen A (1999) Plants with spider-mite prey attract more predatory mites than clean plants under greenhouse conditions. Entomol Exp Appl 90:191–198

Kahl J, Siemens DH, Aerts RJ, Gabler R, Kuhnemann F, Preston CA, Baldwin IT (2000) Herbivore-induced ethylene suppresses a direct defense but not a putative indirect defense against an adapted herbivore. Planta 210:336–342

Kappers IF, Aharoni A, Van Herpen TWJM, Luckerhoff LLP, Dicke M, Bouwmeester HJ (2005) Genetic engineering of terpenoid metabolism attracts bodyguards to *Arabidopsis*. Science 309:2070–2072

Kessler A, Baldwin IT (2001) Defensive function of herbivore-induced plant volatile emissions in nature. Science 291:2141–2144

Kessler A, Halitschke R, Baldwin IT (2004) Silencing the jasmonate cascade: induced plant defenses and insect populations. Science 305:665–668

Kessler A, Halitschke R, Diezel C, Baldwin IT (2006) Priming of plant defense responses in nature by airborne signaling between *Artemisia tridentata* and *Nicotiana attenuata*. Oecologia 148:280–292

Klein AM, Vaissière BE, Cane JH, Steffan-Dewenter I, Cunningham SA, Kremen C, Tscharntke T (2007) Importance of pollinators in changing landscapes for world crops. Proc Roy Soc B-Biol Sci 274:303–313

Koch T, Krumm T, Jung V, Engelberth J, Boland W (1999) Differential induction of plant volatile biosynthesis in the Lima bean by early and late intermediates of the octadecanoid-signaling pathway. Plant Physiol 121:153–162

Koptur S (2005) Nectar as fuel for plant protectors. In: Wäckers FL, Van Rijn PCJ, Bruin J (eds) Plant-provided food for carnivorous insects: a protective mutualism and its applications. Cambridge University Press, Cambridge

Kost C, Heil M (2005) Increased availability of extrafloral nectar reduces herbivory in Lima bean plants (*Phaseolus lunatus*, Fabaceae). Basic Appl Ecol 6:237–248

Kowalchuk GA, Bruinsma M, Van Veen JA (2003) Assessing responses of soil microorganisms to GM plants. Trends Ecol Evol 18:403–410

Landolt PJ (1993) Effects of host plant leaf damage on cabbage looper moth attraction and oviposition. Entomol Exp Appl 67:79–85

Lehtilä K, Strauss SY (1997) Leaf damage by herbivores affects attractiveness to pollinators in wild radish, *Raphanus raphanistrum*. Oecologia 111:396–403

Leitner M, Boland W, Mithöfer A (2005) Direct and indirect defenses induced by piercing–sucking and chewing herbivores in *Medicago truncatula*. New Phytol 167:597–606

Liechti R, Farmer EE (2002) The jasmonate pathway. Science 296:1649–1650

Lou YG, Du MH, Turlings TCJ, Cheng JA, Shan WF (2005) Exogenous application of jasmonic acid induces volatile emissions in rice and enhances parasitism of *Nilaparvata lugens* eggs by the parasitoid *Anagrus nilaparvatae*. J Chem Ecol 31:1985–2002

Masters GJ, Jones TH, Rogers M (2001) Host-plant mediated effects of root herbivory on insect seed predators and their parasitoids. Oecologia 127:246–250

Mattiacci L, Dicke M, Posthumus MA (1995) Beta-glucosidase: an elicitor of herbivore-induced plant odor that attracts host-searching parasitic wasps. Proc Natl Acad Sci USA 92:2036–2040

Meiners T, Hacker NK, Anderson P, Hilker M (2005) Response of the elm leaf beetle to host plants induced by oviposition and feeding: the infestation rate matters. Entomol Exp Appl 115:171–177

Mercke P, Kappers IF, Verstappen FWA, Vorst O, Dicke M, Bouwmeester HJ (2004) Combined transcript and metabolite analysis reveals genes involved in spider mite induced volatile formation in cucumber plants. Plant Physiol 135:2012–2024

Mithöfer A, Wanner G, Boland W (2005) Effects of feeding *Spodoptera littoralis* on Lima bean leaves. II. Continuous mechanical wounding resembling insect feeding is sufficient to elicit herbivory-related volatile emission. Plant Physiol 137:1160–1168

Mumm R, Hilker M (2006) Direct and indirect chemical defence of pine against folivorous insects. Trends Plant Sci 11:351–358

Narváez-Vásquez J, Florin-Christensen J, Ryan CA (1999) Positional specificity of a phospholipase A activity induced by wounding, systemin, and oligosaccharide elicitors in tomato leaves. Plant Cell 11:2249–2260

Neveu N, Grandgirard J, Nenon JP, Cortesero AM (2002) Systemic release of herbivore-induced plant volatiles by turnips infested by concealed root-feeding larvae *Delia radicum* L. J Chem Ecol 28:1717–1732

O'Donnell PJ, Calvert C, Atzorn R, Wasternack C, Leyser HMO, Bowles DJ (1996) Ethylene as a signal mediating the wound response of tomato plants. Science 274:1914–1917

Ode PJ (2006) Plant chemistry and natural enemy fitness: effects on herbivore and natural enemy interactions. Annu Rev Entomol 51:163–185

Ohnmeiss TE, Baldwin IT (2000) Optimal defense theory predicts the ontogeny of an induced nicotine defense. Ecology 81:1765–1783

Omacini M, Chaneton EJ, Ghersa CM, Müller CB (2001) Symbiotic fungal endophytes control insect host–parasite interaction webs. Nature 409:78–81

Ozawa R, Arimura G, Takabayashi J, Shimoda T, Nishioka T (2000) Involvement of jasmonate- and salicylate-related signaling pathways for the production of specific herbivore-induced volatiles in plants. Plant Cell Physiol 41:391–398

Ozawa R, Shiojiri K, Sabelis MW, Arimura GI, Nishioka T, Takabayashi J (2004) Corn plants treated with jasmonic acid attract more specialist parasitoids, thereby increasing parasitization of the common armyworm. J Chem Ecol 30:1797–1808

Paré PW, Tumlinson JH (1997) Induced synthesis of plant volatiles. Nature 385:30–31

Peña-Cortes H, Albrecht T, Prat S, Weiler EW, Willmitzer L (1993) Aspirin prevents wound-induced gene-expression in tomato leaves by blocking jasmonic acid biosynthesis. Planta 191:123–128

Piel J, Atzorn R, Gabler R, Kuhnemann F, Boland W (1997) Cellulysin from the plant parasitic fungus *Trichoderma viride* elicits volatile biosynthesis in higher plants via the octadecanoid signaling cascade. FEBS Lett 416:143–148

Potts SG, Vulliamy B, Dafni A, Ne'eman G, Willmer P (2003) Linking bees and flowers: how do floral communities structure pollinator communities? Ecology 84:2628–2642

Poveda K, Steffan-Dewenter I, Scheu S, Tscharntke T (2003) Effects of below- and above-ground herbivores on plant growth, flower visitation and seed set. Oecologia 135:601–605

Poveda K, Steffan-Dewenter I, Scheu S, Tscharntke T (2005) Effects of decomposers and herbivores on plant performance and aboveground plant–insect interactions. OIKOS 108: 503–510

Price PW, Bouton CE, Gross P, McPheron BA, Thompson JN, Weis AE (1980) Interactions among three trophic levels: influence of plants on interactions between insect herbivores and natural enemies. Annu Rev Ecol Syst 11:41–65

Rasmann S, Köllner TG, Degenhardt J, Hiltpold I, Toepfer S, Kuhlmann U, Gershenzon J, Turlings TCJ (2005) Recruitment of entomopathogenic nematodes by insect-damaged maize roots. Nature 434:732–737

Reymond P, Bodenhausen N, Van Poecke RMP, Krishnamurthy V, Dicke M, Farmer EE (2004) A conserved transcript pattern in response to a specialist and a generalist herbivore. Plant Cell 16:3132–3147

Rodriguez-Saona C, Chalmers JA, Raj S, Thaler JS (2005) Induced plant responses to multiple damagers: differential effects on an herbivore and its parasitoid. Oecologia 143:566–577

Rothschild M, Schoonhoven LM (1977) Assessment of egg load by *Pieris brassicae* (Lepidoptera: Pieridae). Nature 266:352–355

Runyon JB, Mescher MC, De Moraes CM (2006) Volatile chemical cues guide host location and host selection by parasitic plants. Science 313:1964–1967

Ruther J, Fürstenau B (2005) Emission of herbivore-induced volatiles in absence of a herbivore – response of *Zea mays* to green leaf volatiles and terpenoids. Z Naturforsch [C] 60:743–756

Sano H, Ohashi Y (1995) Involvement of small GTP-binding proteins in defense signal-transduction pathways of higher plants. Proc Natl Acad Sci USA 92:4138–4144

Scascighini N, Mattiacci L, D'Alessandro M, Hern A, Rott AS, Dorn S (2005) New insights in analysing parasitoid attracting synomones: early volatile emission and use of stir bar sorptive extraction. Chemoecology 15:97–104

Schmelz EA, Alborn HT, Tumlinson JH (2001) The influence of intact-plant and excised-leaf bioassay designs on volicitin- and jasmonic acid-induced sesquiterpene volatile release in *Zea mays*. Planta 214:171–179

Schnee C, Köllner TG, Held M, Turlings TCJ, Gershenzon J, Degenhardt J (2006) The products of a single maize sesquiterpene synthase form a volatile defense signal that attracts natural enemies of maize herbivores. Proc Natl Acad Sci USA 103:1129–1134

Schoonhoven LM, Van Loon JJA, Dicke M (2005) Insect–plant biology, 2nd edn. Oxford University Press, Oxford

Shiojiri K, Kishimoto K, Ozawa R, Kugimiya S, Urashimo S, Arimura G, Horiuchi J, Nishioka T, Matsui K, Takabayashi J (2006) Changing green leaf volatile biosynthesis in plants: an approach for improving plant resistance against both herbivores and pathogens. Proc Natl Acad Sci USA 103:16672–16676

Shiojiri K, Takabayashi J, Yano S, Takafuji A (2001) Infochemically mediated tritrophic interaction webs on cabbage plants. Popul Ecol 43:23–29

Shiojiri K, Takabayashi J, Yano S, Takafuji A (2002) Oviposition preferences of herbivores are affected by tritrophic interaction webs. Ecol Lett 5:186–192

Siri N (1993) Analysis of host finding behavior of two aphid hyperparasitoids (Hymenoptera: Alloxystidae, Megaspilidae). Christian-Albrechts University, Kiel

Smallegange RC, Van Loon JJA, Blatt SE, Harvey JA, Agerbirk N, Dicke M (2007) Flower vs. leaf feeding by *Pieris brassicae*: glucosinolate-rich flower tissues are preferred and sustain higher growth rate. J Chem Ecol 33:1831–1844

Snoeren TAL, De Jong PW, Dicke M (2007) Ecogenomic approach to the role of herbivore-induced plant volatiles in community ecology. J Ecol 95:17–26

Soler R, Bezemer TM, Van der Putten WH, Vet LEM, Harvey JA (2005) Root herbivore effects on above-ground herbivore, parasitoid and hyperparasitoid performance via changes in plant quality. J Anim Ecol 74:1121–1130

Soler R, Harvey JA, Kamp AFD, Vet LEM, Van der Putten WH, Van Dam NM, Stuefer JF, Gols R, Hordijk CA, Bezemer TM (2007) Root herbivores influence the behavior of an aboveground parasitoid through changes in plant-volatile signals. OIKOS 116:367–376

Steinberg S, Dicke M, Vet LEM (1993) Relative importance of infochemicals from 1st and 2nd trophic level in long-range host location by the larval parasitoid *Cotesia glomerata*. J Chem Ecol 19:47–59

Steinberg S, Dicke M, Vet LEM, Wanningen R (1992) Response of the braconid parasitoid *Cotesia* (= *Apanteles*) *glomerata* to volatile infochemicals – effects of bioassay set-up, parasitoid age and experience and barometric flux. Entomol Exp Appl 63:163–175

Stintzi A, Weber H, Reymond P, Browse J, Farmer EE (2001) Plant defense in the absence of jasmonic acid: the role of cyclopentenones. Proc Natl Acad Sci USA 98:12837–12842

Stout MJ, Duffey SS (1996) Characterization of induced resistance in tomato plants. Entomol Exp Appl 79:273–283

Strauss SY, Irwin RE, Lambrix VM (2004) Optimal defence theory and flower petal colour predict variation in the secondary chemistry of wild radish. J Ecol 92:132–141

Takabayashi J, Sabelis MW, Janssen A, Shiojiri K, van Wijk M (2006) Can plants betray the presence of multiple herbivore species to predators and parasitoids? The role of learning in phytochemical information networks. Ecol Res 21:3–8

Takabayashi J, Takahashi S, Dicke M, Posthumus MA (1995) Developmental stage of herbivore *Pseudaletia separata* affects production of herbivore-induced synomone by corn plants. J Chem Ecol 21:273–287

Thaler JS (1999) Jasmonate-inducible plant defenses cause increased parasitism of herbivores. Nature 399:686–688

Thaler JS, Farag MA, Paré PW, Dicke M (2002) Jasmonate-deficient plants have reduced direct and indirect defenses against herbivores. Ecol Lett 5:764–774

Thaler JS, Stout MJ, Karban R, Duffey SS (1996) Exogenous jasmonates simulate insect wounding in tomato plants (*Lycopersicon esculentum*) in the laboratory and field. J Chem Ecol 22:1767–1781

Thaler JS, Stout MJ, Karban R, Duffey SS (2001) Jasmonate-mediated induced plant resistance affects a community of herbivores. Ecol Entomol 26:312–324

Turlings TCJ, Fritzsche ME (1999) Attraction of parasitic wasps by caterpillar-damaged plants. In: Chadwick DJ, Goode JA (eds) Insect–plant interactions and induced plant defence. John Wiley & Sons, Chichester, pp 21–38

Turlings TCJ, Ton J (2006) Exploiting scents of distress: the prospect of manipulating herbivore-induced plant odors to enhance the control of agricultural pests. Curr Opin Plant Biol 9:421–427

Turlings TCJ, Tumlinson JH, Heath RR, Proveaux AT, Doolittle RE (1991) Isolation and identification of allelochemicals that attract the larval parasitoid, *Cotesia marginiventris* (Cresson), to the microhabitat of one of its hosts. J Chem Ecol 17:2235–2251

Turlings TCJ, Tumlinson JH, Lewis WJ (1990) Exploitation of herbivore-induced plant odors by host-seeking parasitic wasps. Science 250:1251–1253

Van Dam NM, Hadwich K, Baldwin IT (2000) Induced responses in *Nicotiana attenuata* affect behavior and growth of the specialist herbivore *Manduca sexta*. Oecologia 122:371–379

Van den Boom CEM (2003) Plant defence in a tritrophic context. Chemical and behavioral analyses of the interactions between spider mites, predatory mites and various plant species. Wageningen University, Wageningen

Van Loon JJA, De Boer JG, Dicke M (2000) Parasitoid-plant mutualism: parasitoid attack of herbivore increases plant reproduction. Entomol Exp Appl 97:219–227

Van Poecke RMP, Dicke M (2002) Induced parasitoid attraction by *Arabidopsis thaliana*: involvement of the octadecanoid and the salicylic acid pathway. J Exp Bot 53:1793–1799

Van Poecke RMP, Dicke M (2004) Indirect defence of plants against herbivores: using *Arabidopsis thaliana* as a model plant. Plant Biol 6:387–401

Van Rijn PCJ, Tanigoshi LK (1999) The contribution of extrafloral nectar to survival and reproduction of the predatory mite *Iphiseius degenerans* on *Ricinus communis*. Exp Appl Acarol 23:281–296

Van Tol RWHM, Van der Sommen ATC, Boff MIC, Van Bezooijen J, Sabelis MW, Smits PH (2001) Plants protect their roots by alerting the enemies of grubs. Ecol Lett 4:292–294

Van Veen FJF, Morris RJ, Godfray HCJ (2006) Apparent competition, quantitative food webs, and the structure of phytophagous insect communities. Annu Rev Entomol 51:187–208

Van Zandt PA, Agrawal AA (2004) Community-wide impacts of herbivore-induced plant responses in milkweed (*Asclepias syriaca*). Ecology 85:2616–2629

Verhagen BWM, Glazebrook J, Zhu T, Chang HS, Van Loon LC, Pieterse CMJ (2004) The transcriptome of rhizobacteria-induced systemic resistance in *Arabidopsis*. Mol Plant Micr Int 17:895–908

Vet LEM, Dicke M (1992) Ecology of infochemical use by natural enemies in a tritrophic context. Annu Rev Entomol 37:141–172

Visser JH, Avé DA (1978) General green leaf volatiles in the olfactory orientation of the Colorado beetle, *Leptinotarsa decemlineata*. Entomol Exp Appl 24:538–549

Voelckel C, Baldwin IT (2004) Generalist and specialist lepidopteran larvae elicit different transcriptional responses in *Nicotiana attenuata*, which correlate with larval FAC profiles. Ecol Lett 7:770–775

Voelckel C, Weisser WW, Baldwin IT (2004) An analysis of plant–aphid interactions by different microarray hybridization strategies. Mol Ecol 13:3187–3195

Völkl W, Sullivan DJ (2000) Foraging behavior, host plant and host location in the aphid hyperparasitoid *Euneura augarus*. Entomol Exp Appl 97:47–56

Vos M, Berrocal SM, Karamaouna F, Hemerik L, Vet LEM (2001) Plant-mediated indirect effects and the persistence of parasitoid-herbivore communities. Ecol Lett 4:38–45

Vrieling K, Smit W, Van der Meijden E (1991) Tritrophic interactions between aphids (*Aphis jacobaeae* Schrank), ant species, *Tyria jacobaeae* L., and *Senecio jacobaea* L. lead to maintenance of genetic variation in pyrrolizidine alkaloid concentration. Oecologia 86:177–182

Wäckers FL, Bezemer TM (2003) Root herbivory induces an above-ground indirect defence. Ecol Lett 6:9–12

Wäckers FL, Van Rijn PCJ (2005) Food for protection: an introduction. In: Wäckers FL, PCJ Van Rijn, Bruin J (eds) Plant-provided food for carnivorous insects: a protective mutualism and its applications. Cambridge University Press, Cambridge

Wäckers FL, Zuber D, Wunderlin R, Keller F (2001) The effect of herbivory on temporal and spatial dynamics of foliar nectar production in cotton and castor. Ann Bot 87:365–370

Walling LL (2000) The myriad plant responses to herbivores. J Plant Growth Regul 19:195–216

Weber H, Vick BA, Farmer EE (1997) Dinor-oxo-phytodienoic acid: a new hexadecanoid signal in the jasmonate family. Proc Natl Acad Sci USA 94:10473–10478

White JA, Andow DA (2006) Habitat modification contributes to associational resistance between herbivores. Oecologia 148:482–490

Whitman DW, Eller FJ (1990) Parasitic wasps orient to green leaf volatiles. Chemoecology 1:69–76

Wolfe BE, Husband BC, Klironomos JN (2005) Effects of a belowground mutualism on an aboveground mutualism. Ecol Lett 8:218–223

Wootton JT (1994) The nature and consequences of indirect effects in ecological communities. Annu Rev Ecol Syst 25:443–466

Zheng S-J, Van Dijk J, Bruinsma M, Dicke M (2007) Sensitivity and speed of induced defense of cabbage (*Brassica oleracea* L.): dynamics of BoLOX expression patterns during insect and pathogen attack. Mol Plant Micr Int 20:1332–1345

Chapter 3
Induced Defenses and the Cost-Benefit Paradigm

Anke Steppuhn and Ian T. Baldwin

Defense costs are thought to be the *raison d'etre* for inducibility, by which costs are only incurred when a defense is needed. Costs can arise when resources are allocated to defenses and consequently not available for growth and reproduction, from the havoc that defenses might wreck with primary metabolism or the plant's ability to respond to other stresses, and from the damage caused by herbivores whose ability to resist plant defenses have been honed by constitutive defense expression. Though the cost-benefit paradigm is widely accepted, empirical evidence is rather slim. Elicitation studies, which elucidate fitness consequences of activating defense responses with elicitors, presented first conclusive evidence, but in these studies disentangling the costs of co-regulated responses is not possible. Quantitative genetics correlate changes in resistance and fitness among different genotypes, but defense-related traits can be genetically linked to other traits unrelated to defense. Mutant and transformant studies provide the strongest evidence for cost-benefit tradeoffs, because confounding effects of genetic linkage are minimized in isogenic lines. In addition to genetic linkage, defenses are linked to various aspects of primary and secondary metabolism and defense functions can be altered by other metabolites. The diverse linkages among responses to herbivory together with the complexity of internal and external factors that influence plant fitness suggest that cost-benefit functions are best examined as shifting in a multi-dimensional space bracketed by environmental conditions. Methodological advances allow the connections of the metabolic grid that shape phenotypes to be visualized and for the falsification of hypotheses about tradeoffs.

3.1 Introduction

Plants employ chemical defenses to protect themselves from attack by a variety of herbivores and pathogens. In many plants chemical defenses are deployed

---

I.T. Baldwin
Max Planck Institute for Chemical Ecology, Department of Molecular Ecology, D-07745 Jena, Germany
e-mail: baldwin@ice.mpg.de

inducibly, that is, their production and accumulation increase dramatically after attack (Ryan 1983; Karban and Baldwin 1997). Different elicitation systems allow plants to elicit defenses that are most efficient against particular attackers (e.g., pathogens or herbivores of different feeding guilds; Kessler and Baldwin 2004; Voelckel and Baldwin 2004a, b). But, due to the time lag between the first attack and the activation of the defense, plants remain unprotected for hours or even days until the defense is established. So why do plants put up with this risk? The selective advantage of induced over constitutively expressed defense traits, which lack the detrimental delay, is commonly explained by the high cost of defenses. Consequently, the fitness benefit of forgoing these costs when resistance is not needed may outweigh the disadvantages of the delay.

Defenses can have physiological, ecological, and evolutionary costs. Physiological costs may involve allocation and autotoxicity. Allocation costs occur when fitness-limiting resources are tied up in defenses, which may not be readily recycled, and hence are unavailable for processes relevant to fitness such as growth or reproduction. Autotoxicity costs may occur because certain secondary metabolites can be toxic to the plant itself. Their constitutive expression may therefore impose a significant metabolic burden. Metabolites that are universally toxic can be very effective defenses, but they also represent a significant liability for the producer. Plants solve this 'toxic waste dump' problem associated with chemical defense deployment by elaborating defenses that target tissues that do not occur in plants, such as neurotoxic alkaloids Ecological costs result from the negative fitness effects of defenses on the complex interactions that plants have with their environment. These costs include the liability that results from producing toxins that are sequestered by adapted herbivores, protecting them from their own enemies. But the ecological costs can also be seen as opportunity costs: activating a particular defense response may compromise the activation of other defenses vis-à-vis other attackers or tolerance responses. Inducibility may minimize evolutionary costs, because constitutively deployed defenses may provide a stronger selection pressure for herbivores to evolve counter resistance. Given that the inducible defenses of plants greatly increase the complexity of the chemical environments that a plant's natural enemies are exposed to, the minimization of these evolutionary costs may be important.

Putative allocation costs play a central role in theories that attempt to explain the patterns of defense metabolite distribution within and among plants, but the assumption that defenses incur significant allocation costs has been difficult to test empirically. The optimal defense theory (McKey 1974; Zangerl and Bazzaz 1992) argues that the distribution of defense metabolites within the plant reflects an optimization of the fitness benefits over the costs of defense deployment: tissues with high fitness value to the plant and a high probability of attack receive preferential allocation. For example, the toxic alkaloid nicotine that is produced in the roots of *Nicotiana attenuata* is concentrated in young leaves, stems, and reproductive parts, whereas roots and old leaves have low levels (Ohnmeiss and Baldwin 2000). Similarly, the distribution of induced defenses within the plant kingdom is commonly explained by the cost-benefit tradeoffs involved in defense traits. The apparence theory (Feeny 1976) predicts that slow-growing plants with long-lived tissues that are likely to be attacked by herbivores invest more in constitutive defenses compared to fast-growing species with short-lived tissues that are more likely to escape herbivory; such species optimize their fitness by expressing induced defenses, as the costs of defense are expected to be particularly onerous for fast-growing species. Consistent with this expectation, most species for which induced resistance has been reported are fast growers (Karban and Baldwin 1997). Other theories, such as the carbon/nutrient (C/N) balance hypothesis (Bryant et al. 1983) and the substrate/enzyme theory (Bryant et al. 1983), and the growth/differentiation theory (Herms and Mattson 1992), attempt to explain the inducibility of secondary metabolites in terms of nutrient (or substrate) availability either within the plant, throughout its ontogeny, or in its local environment. These 'supply-side hypotheses' (Lerdau et al. 1994) postulate that the increases in secondary metabolites that commonly occur after herbivory do not result from a signal-mediated regulation of secondary metabolism, but rather as an indirect consequence of changes in overall metabolic balance. Although these theories are of little value as mechanistic models, their underlying concepts might help to explain the distribution of plant defenses with regard to their allocation costs. For example, carbon-intensive defenses indeed tend to increase when growth is constrained due to nutrient limitations which are thought to make carbon more available than nutrients; however, such changes may also result from adaptive responses by the plant to conditions which hinder the replacement of damaged tissues.

The hallmark of ecological costs is their environmental contingency, that is, they can only be seen in particular environments. Ecological costs may result from tradeoffs between resistance and susceptibility to different natural enemies and mutualists; between resistance and tolerance to herbivory; or between resistance to herbivores and the resulting effects on the plant's competitiveness. The latter is exemplified by the fact that large allocation costs can often only be observed when plants are grown with conspecific competitors, not without (e.g., in *N. attenuata*; Baldwin and Preston 1999). Tradeoffs between resistance and susceptibility to different natural enemies may be particularly important for some indirect defenses, such as the release of volatile 'alarm' signals to attract the third trophic level, which in *N. attenuata* has been shown to be an exceptionally effective defense that can decrease herbivore abundance by up to 90% (Kessler and Baldwin 2001). Although such small investments in volatile organic compounds (VOCs) are likely to exact only minor allocation costs (Dicke and Sabelis 1992), a release of volatiles may increase a plant's apparency if herbivores co-opt the VOCs as signals when foraging for hosts (Takabayashi and Dicke 1996; Halitschke et al. 2007). Additionally, the simultaneous elicitation of direct and indirect defenses may incur ecological costs: for example, a toxic direct defense may poison organisms at the third trophic level which are attracted by the release of volatiles. Parasitoids have been found to be attracted to herbivore-induced VOCs, but as these predators spend their entire larval life bathed in the hemolymph of their hosts, they are exposed to all the plant-produced toxins ingested by the herbivore. For example, the tobacco hornworm *Manduca sexta* feeds on species of *Nicotiana* known to induce both VOCs and high levels of nicotine. *M. sexta* retains quantities of the toxin in its hemolymph that are sufficient to cause substantial mortality to its parasitoid wasp, *Cotesia congregata* (Barbosa et al. 199_). Although the herbivore-induced release of volatiles has been demonstrated to attract parasitoids in many laboratory settings, their real-world significance as a plant defense has only rarely been elucidated (but see Kessler and Baldwin 2001; Halitschke et al. 2007). The potential for these ecological costs to influence the net fitness benefit of an induced response is great, underscoring the importance of testing the cost-benefit paradigm under environmentally realistic conditions.

In the past two decades, many studies have tested the fundamental assumption of the cost-benefit paradigm: that resistance traits are intrinsically costly. Various approaches have been used with differing degrees of success to control the genetic, physiological, and ecological complexity that may confound the relationship between defense expression and fitness.

3.2 The Evidence for Defense Costs

3.2.1 Types of Study

The cost-benefit paradigm, the foundation for most theories about the evolution of plant defenses, has been difficult to test, first, because of the complexity of internal and external processes that influence how defense traits may compromise male and/or female reproductive function. Second, the costs of a defense trait must be separated from those of genetically linked traits. The most conclusive evidence that plant defenses exact fitness costs originates from three types of studies: (1) elicitation studies, which measure the fitness consequences of activating defenses by applying signal molecules known to elicit defenses; (2) quantitative genetic studies, which correlate fitness with genetic variation in levels of defense metabolites within plant populations; and (3) mutant and transformant studies, which evaluate the fitness consequences for plants that have been altered (overexpression or loss of function) in the expression of specific genes.

3.2.2 Elicitation Studies

Most costs studies have elicited secondary metabolites production by wounding or herbivore attack or applied signaling molecules that mediate induced responses, and compared the fitness or performance of these elicited plants with that of unelicited control plants. Herbivore feeding, though biologically realistic, is difficult to standardize and the costs of induced responses are hard to separate from the fitness costs resulting from tissue loss. Mechanical wounding, which induces resistance in most plants, enables damage to be quantitatively, qualitatively, and temporally controlled. Puncture wounds, in particular, allows leaf area loss to be kept to a minimum, while still effectively eliciting defense responses. The first study which was able to detect the costs of induced defenses used different wounding regimes to elicit plant defense responses to different degrees, but all resulted in the same loss of tissue (Baldwin et al. 1990). However, the responses induced by mechanical wounding are not always the same as those induced by herbivore feeding (Halitschke et al. 2001, 2003; but see Mithöfer et al. 2005).

Recently, our understanding of the different signaling molecules mediating induced plant responses has increased, enabling the costs of induced defense to be uncoupled from the costs of tissue loss. Since the jasmonic acid (JA) signaling cascade is known to mediate many herbivore-induced responses (Creelman and Mullet 1997; Wasternack et al. 1998; Halitschke and Baldwin 2004; Schaller and Stintzi this volume), numerous studies of the costs of induced responses have applied JA or its methyl ester (MeJA; van Dam and Baldwin 1998; Agrawal et al. 1999; Cipollini and Sipe 2001; Redman et al. 2001) to elicit responses, and this approach has enabled the cost-benefit paradigm to be tested in nature. The lifetime seed production of MeJA-treated *N. attenuata* plants grown in natural populations was lower than that of untreated plants if herbivores were absent; however, when plants were exposed to moderate or high levels of herbivory, fitness benefits exceeded the costs of JA elicitation (Baldwin 1998). In wild radish, Agrawal et al. (1999) detected fitness costs of JA induction not in terms of lower number or mass of seeds, but rather the time to first flowering and amount of pollen produced. However, JA applications do not faithfully mimic the responses to herbivore attack, as this phytohormone functions substantially downstream of the herbivore-specific elicitors which activate JA signaling in *N. attenuata* (Wu et al. 2007) and it is known to interact with other herbivore-elicited signaling molecules in complex ways (Thomma et al. 1998; Mur et al. 2006). Moreover, exogenous treatments are not likely to realistically imitate the frequently highly tissue-specific and transient changes in endogenous pools of signal molecules that are elicited by herbivore attack.

The identification of herbivore-specific elicitors, such as the fatty acid–amino acid conjugates (FACs) in the oral secretions of lepidopteran larvae (Alborn et al. 1997; Halitschke et al. 2001), has made it possible to elicit those plant responses that are most similar to the responses elicited by actual herbivore attack (Halitschke et al. 2003). But elicitation studies are usually unable to demonstrate the fitness costs of any particular biochemically characterized induced defense because many of the responses include physiological and morphological changes, which although unrelated to resistance, nevertheless affect fitness parameters. For example, as well as mediating resistance to herbivores, JA is involved in numerous processes such as fruit ripening, pollen fertility, root growth, tendril coiling, and resisting pathogens (Creelman and Mullet 1997). Treatment with MeJA inhibits the gene transcription of proteins essential for growth, for example, RubisCO and chlorophyll a/b-binding proteins (Reinbothe et al. 1994; Halitschke et al. 2001); and as a consequence photosynthesis is reduced (Metodiev et al. 1996). This suppression can decrease fitness when plants are grown under resource-limited conditions. The down-regulation of growth and the degradation of photosynthetic proteins may be necessary to free up the resources required for the de novo production of resistance traits but may otherwise not be related to resistance traits. To date we know very little about the traits that are necessary and sufficient for establishing resistance. These pleiotropic effects may cause the fitness costs of resistance to be overestimated in elicitation studies or they may accurately represent the true fitness costs of defense elicitation by including the metabolic readjustments required for defense activation.

3.2.3 The Quantitative Genetics Approach

Elicitation studies make use of the phenotypic plasticity of inducible plants to compare the fitness of plants with and without activated defenses; another approach uses the genetic variation in resistance. Early attempts to estimate the genetic tradeoff involved in allocating resources to defense compared the fitness parameters of offspring from crosses of species, cultivars, or populations that differ in the degree of their defense investment (Krischik and Denno 1983). Because genetic differences within populations encompass many traits in addition to those that control resistance, sib analysis and artificial selection experiments have been used to examine negative correlations between the genetic variation in resistance and in fitness. In a pioneering study using half-sib families of *Pastinaca sativa*, Berenbaum et al. (1986) provided strong evidence for such a genetic tradeoff, showing that the genetically controlled ability to produce an anti-herbivore defense is negatively associated with fitness parameters. The production of furanocoumarins, which explains 75% of the variation in resistance to a specialist herbivore, correlated negatively with seed production in the absence of herbivores. Others studies have also found negative correlations between inherited defense traits and fitness parameters; for example, higher glucosinolate production in *Brassica rapa* is associated with lower female fitness (Mitchell-Olds et al. 1996; Stowe 1998) and fewer pollinator visits (Strauss et al. 1999), but many studies have also failed to detect any significant correlations between the level of resistance and reproductive success (for example, Simms and Rausher 1987, 1989; Agren and Schemske 1993; Vrieling et al. 1996; Juenger and Bergelson 2000; Agrawal et al. 2002). A meta-analysis conducted in 1996, examining conditions under which costs were detected, revealed that only about one-third of the studies found significant costs and that greater control over the genetic background increased the probability that costs of resistance traits would be detected (Bergelson and Purrington 1996). That the fraction of studies that were able to detect costs subsequently more than doubled can be attributed to increased experimental control in both elicitation studies and genetic correlations (Heil and Baldwin 2002; Strauss et al. 2002; Cipollini et al. 2003). However, without control over the genetic background, the presence or absence of fitness costs cannot be ascribed to the expression of resistance (Bergelson et al. 1996).

3.2.4 Mutant and Transformant Studies

Advances in molecular techniques during the past decade have enabled specific defense genes to be altered and therefore obviated the problems of confounding factors that vary with defense expression. The use of mutants and transformed plants that constitutively express resistance genes, or that are hindered in the expression of such genes, allows the fitness consequences to be examined against a common genetic background. Initial support for the hypothesis that the constitutive expression of resistance lowers fitness came from the phenotypes of mutants that constitutively express genes of the signaling cascades, which mediate pathogen resistance; these mutants typically have stunted growth and development (references in Heil and Baldwin 2002). Subsequent studies actually measured the fitness consequences of such mutations. Constitutive activation of systemic acquired resistance (SAR) by the mutations *cpr1*, *cpr5*, and *cpr6* decrease seed production in *Arabidopsis thaliana* (Heidel et al. 2004). An elegant study that rigorously controlled for potential differences in genetic background unambiguously revealed the fitness costs of a particular *R* gene conferring resistance to the pathogen *Pseudomonas syringae* (Tian et al. 2003). Sense-expression of this *R* gene, *rpm1*, which encodes a peripheral plasma membrane protein that functions as a receptor for the pathogen elicitors, decreased reproductive output in *A. thaliana* by 9%. However, the specific responses elicited by this pathogen recognition system which are responsible for the decrease in reproductive output are unknown.

Only a few studies have investigated the fitness consequences of mutations or transformations in the signal transduction of herbivore defense. In *Solanum tuberosum* antisense-silencing a biosynthetic gene of the jasmonate signaling cascade (*lox*) increased tuber yield (Royo et al. 1999). Surprisingly, a study that used the *jar1* mutant in *A. thaliana*, which is deficient in JA signaling and in the expression of protease inhibitors (PIs), found greater reductions in seed production after elicitation with JA in the mutant line than in the wild type (WT) lines (Cipollini 2002). However, transgenic *A. thaliana* plants that overexpress carboxyl methyltransferase (*jmt*), and thereby increase MeJA production, have decreased seed production (Cipollini 2007).

As in elicitation studies, the interpretation of signaling mutants or plants transformed to constitutively express signaling molecules is confounded by the various roles the signaling molecules play, especially with regard to controlling growth and photosynthesis. Therefore, the fitness costs of specific defenses are best determined in isogenic lines that differ in a biosynthetic gene of a given defense trait. Furthermore, if plants are transformed to over-express a gene, the chemical and also the ecological (if the gene is heterologously expressed) contexts in which the trait evolved are altered. Expression of a defense trait in a particular tissue or cell compartment may be essential for its defensive function as well as for minimizing of its fitness costs. Hence ectopic over-expression of a defense in all cells may provide spurious results. Primary metabolites and the chemical milieu are known to affect the function of secondary metabolites (Broadway and Duffey 1988; Govenor et al. 1997; Green et al. 2001). Additionally, secondary metabolites can affect each other (Felton et al. 1989; Steppuhn and Baldwin 2007) and in response to herbivore attack, most plants increase their concentrations of a diverse bouquet of secondary metabolites. The potentially confounding effects of this chemical diversity can be minimized by silencing endogenous genes that only influence the production of a particular defense metabolite. This approach allows the ecological function as well as costs of a particular trait to be studied in its natural context, as has been shown for the case of inducibly produced PIs in *N. attenuata* (Zavala et al. 2004a, b).

To be able to attribute altered fitness or resistance to the targeted gene, the genetic consequences of the transformation or mutation procedure need to be characterized. Transgenic lines should harbor a single insertion of the transgene and be homozygous; the success of the silencing or sense-expression should be verified by measuring transcript levels of the targeted gene; and the observed phenotypes should not be due to either the effect of interrupting the genetic code at the site of insertion or the vector used for transformation. The likelihood that an insertion site is the basis of an observed fitness effect can be excluded by backcrossing (for example, Tian et al. 2003, who used site-specific recombination to excise the *rpm1* transgene), or by using multiple lines of independent transformation events (Zavala et al. 2004b). Similarly, it should be verified that the phenotype of a certain mutant is not caused by other mutations that occurred during its creation. To control for the transformation vector and the procedure, plants transformed with empty vectors should be included in the analysis.

3.2.5 Protease Inhibitors – A Case Study

A variety of plants produce JA-elicited PIs as a direct defense. PIs are antidigestive proteins that can decrease the performance of many herbivores by suppressing gut protease activity (Birk 2003). To determine whether *N. attenuata*'s production of trypsin PIs incurs a fitness cost, Zavala et al. (2004b) transformed two natural genotypes: in one, endogenous PI production was silenced by introducing a fragment of the *pi* gene in antisense orientation; in the other, a genotype that had a nonsense mutation in the *pi* gene, PIs were constitutively expressed. Plants deficient in PI activity grew faster and taller, flowered earlier, and produced more seed capsules (25%–53%) than did the isogenic PI-producing genotypes against which they were competing (Fig. 3.1). Results were similar regardless of whether PI activity was suppressed in two independent lines or restored. The difference in seed capsule production between two competing neighbors correlated with the difference in PI activity. Clearly, PI production exacts a large fitness cost when plants are grown under herbivore-free competitive conditions, which is consistent with the hypothesis that inducibility evolved as a cost-saving mechanism.

A subsequent study with the same isogenic lines, some of which were unattacked and on others of which *M. sexta* larvae were allowed to feed freely, demonstrated that the PI-mediated decreases in *M. sexta* performance translate into a fitness benefit for the plant that outweighs the costs of PI production under greenhouse

Fig. 3.1 Competitive growth reveals the costs of protease inhibitor (PI) production in *N. attenuata*. When isogenic pairs of plants that differ only in their expression of PIs are grown in the same pots, PI-producing plants are out-competed by PI-free plants and produce significantly fewer seeds. (**A**) Natural genotypes harboring a mutation resulting in a premature stop codon in the PI gene; (**B**) isogenic genotypes transformed to silence PI production by RNAi. (Photographs: (**A**) G. Glawe, (**B**) H. Wünsche)

conditions (Zavala and Baldwin 2004). Again, PI production of unattacked plants was negatively correlated with seed capsule production and caterpillar attack not only reduced seed capsule production but also reversed this pattern of seed capsule production among genotypes; high-PI-producing genotypes had higher fitness than those which produced few or no PIs. The use of isogenic plants that are altered in the expression of a particular defense genes enable the cost-benefit-paradigm to be examined under various natural conditions: field studies with transgenic *N. attenuata* plants silenced for genes involved in the biosynthesis of the alkaloid nicotine or in the JA cascade that elicits it revealed their function as anti-herbivore defenses in nature (Kessler et al. 2004; Steppuhn et al. 2004).

3.3 The Impact of Linkage

3.3.1 Types of Linkage

The expression of defense traits can be linked to other traits by different mechanisms. Genes mediating resistance are linked to other genes located on the same chromosome. Even though this linkage is disrupted by crossing over during sexual reproduction, genes that are close to each other may not segregate independently; this dependency can be quantified as linkage disequilibrium. In *A. thaliana* polymorphisms are typically independent if separated by more than fifty kilo bases, whereas below this distance linkage disequilibrium is significant (Plagnol et al. 2005). However these are only rough estimates, as a gene's location in particular chromosomal regions can dramatically influence its recombination rate. The advances in minimizing the genetic linkage to a manipulated gene have decreased many of the potential confounding effects that have plagued tests of the cost-benefit paradigm, but none of the previously described approaches excludes another confounding effect, which is the linkage of defense traits to the plant's metabolism.

Eliminating one component of the metabolic machinery is bound to cause changes in metabolite flow, which can in turn regulate metabolism through complicated feedback and feed-forward controls. The extent to which metabolism is altered depends on how the trait is regulated and how deeply it is involved in the metabolic machinery. The large extent to which the various signaling pathways that regulate resistance genes also profoundly influence metabolism is clear (Halitschke and Baldwin 2003; Reymond et al. 2004), but changes at the end branches of metabolic pathways can also have consequences due to the accumulating precursors or side products. Hence a comparison of two plants that differ genetically only in the expression of a single resistance gene may include the consequences of significantly altered metabolisms. And these alterations may be the cause or consequence of defense costs.

In addition to genomic and metabolic linkage, functional linkage also influences the costs and benefits of defense traits. How a defense trait functions can depend on the presence or absence of other metabolites. The ecological costs of a broadly toxic direct defense that interferes with indirect defenses when tolerant herbivores co-opt the plant's resistance trait for their own defense is one example already mentioned. Moreover, the functions of different defense traits might be directly linked: synergistic interactions, for example, increase the resistance above the sum of the resistance provided by each defense alone (Berenbaum and Neal 1985; Nelson and Kursar 1999). By increasing the benefits more than the costs of defenses, functional linkage may be a very efficient cost-saving mechanism.

3.3.2 Metabolic Linkage

The difficulties of disentangling the fitness costs of allocating resources to a specific defense trait from the fitness consequences of shifts in allocation resulting from other processes activated during elicitation have already been mentioned. Furthermore, it is not clear whether in isogenic genotypes, disabled in the expression of a specific trait, the resources that are not used for defenses are redirected to be available for growth and reproduction. The resource allocation shifts required for a defense may occur far up-stream of the portion of biosynthetic pathway committed to the actual metabolite production, especially if a defense is produced by a long, multi-step pathway. A physiologically well-studied example of such a defense is nicotine accumulation in leaves of attacked *Nicotiana* species. This accumulation involves a large spatial separation between the site of synthesis and the site of herbivore attack and a long-distance signaling system is required to bridge the gap.

The induced increases of nicotine in the shoots of *Nicotiana sylvestris* and *N. attenuata* are mediated by JA-signaling and result from increased de novo nicotine biosynthesis, which takes place in the roots (Baldwin et al. 1994, 1998). In both species, the elicitation of leaves with MeJA increases the root transcripts of putrescine N-methyl transferase (PMT), which is the enzyme that regulates nicotine biosynthesis (Voelckel et al. 2001; Winz and Baldwin 2001). The signal that moves from the site of attack to the roots to induce nicotine biosynthesis is transported via the phloem (Baldwin 1989), whereas the nicotine is transported into the shoot via the xylem (Baldwin 1991). Pulse-chase experiments with ^{15}N revealed that nitrogen reduction and assimilation are unaffected after induction but the proportion of reduced nitrogen that is allocated to nicotine doubles (Baldwin et al. 1993, 1994). This suggests large allocation costs of nicotine production may be due to a shift in nitrogen allocation, especially because nicotine is not sufficiently catabolized to recycle its nitrogen for use in reproductive processes (Baldwin et al. 1994, 1998). Eliciting *N. attenuata* plants with MeJA applied to the roots in order to specifically stimulate nicotine synthesis reduced seed production by 43%–71% in glasshouse experiments and by 17%–26% in field plantations and native populations of plants that had not been attacked by herbivores (Baldwin 1998; Baldwin et al. 1998). MeJA elicitation of roots will elicit processes in addition to those required for nicotine biosynthesis. To circumvent these confounding secondary effects of MeJA elicitation, the *pmt* gene was silenced by RNAi, and isogenic lines of *N. attenuata* were generated which accumulated less than 10% of the nicotine of WT plants. However, these lines accumulated levels of an alkaloid not detected in WT plants, anatabine, that were about a quarter of the nicotine levels in WT plants (Chintapakorn and Hamill 2003; Steppuhn et al. 2004). Silencing PMT disrupts nicotine biosynthesis at the formation of the five-membered pyrrolidine ring. Pulse-chase experiments with deuterated nicotinic acid, the precursor of the six-membered pyridine ring of nicotine, revealed that the excess of nicotinic acid dimerizes to form anatabine. Though nicotine biosynthesis is specifically addressed by silencing the regulatory enzyme, the elicited increase in nitrogen-containing metabolites in the nicotine biosynthetic pathway is not completely interrupted, as evidenced by the accumulation of a novel side product. No evidence was found that the precursor of the five-membered ring (putrescine) accumulates. The derivatives of putrescine that are known to accumulate after MeJA elicitation (such as caffeoylputrescine) showed no changes as a consequence of PMT silencing. However, a quarter of the nitrogen normally allocated to nicotine is still allocated to an alkaloid (anatabine) even when PMT is silenced. This example underscores the importance of a careful characterization of the metabolism of lines disabled in defense production and more generally of the importance of metabolic linkage in studying allocation costs.

Metabolic linkage may also occur if biosynthetic enzymes for secondary metabolite production have an additional catalytic function in primary metabolism or if the products themselves regulate other processes. Though the production of PIs in *N. attenuata* functions defensively when PIs are ingested by herbivores, they may also regulate proteases in the plant (Laing and McManus 2002). Whether the fitness costs of PI production result from resource-based allocation costs or from alternative physiological functions, such as the regulation of an endogenous protease, remains to be determined (Zavala et al. 2004b). To determine whether the fitness costs of a defense are due to a resource allocation shift from growth and reproduction to secondary metabolism, the flow of fitness-limiting resources into these traits warrants following. This is possible in pulse-chase experiments with isotopically labeled fitness-limiting resources. Metabolic linkage via PIs' secondary function in *N. attenuata* may explain why the lack of PIs in the natural genotype with the nonsense mutation is coupled with a lack of the predator-attracting release of cis-α-bergamotene (Glawe et al. 2003); alternatively these traits may also be functionally linked.

3.3.3 Functional Linkage

Because plants produce a cocktail of chemically and functionally diverse defenses, it has long been debated whether these mixtures of secondary metabolites have adaptive value. The selective advantage of producing combinations of different allelochemicals may be a broader range of enemies against which defense is provided, an increased toxicity of mixtures, or the delayed evolution of herbivores resistant to plant defenses. Yet, the defensive functions of allelochemicals are usually examined separately, regardless of evidence for synergistic interactions among secondary metabolites (Berenbaum and Neal 1985; Dyer et al. 2003). Functional linkage may allow plants to shift the fitness consequences of defense metabolites towards the benefits within the cost-benefit tradeoff.

The existence of a functional linkage between *N. attenuata*'s PI production and VOC release – these interestingly share the same elicitor and are therefore coordinately expressed (Halitschke et al. 2001) – can be assumed, if their defensive functions depend on each other's presence. Because PI production decreases herbivore growth rates by inhibiting insects' digestive processes (Birk 2003), the fitness benefits of PI expression may result from extending the period during which larvae can be successfully attacked by natural enemies. The predators in *N. attenuata*'s native habitat are small predatory bugs that prey only on early instars of large lepidopteran herbivores such as *Manduca* species (Fig. 3.2). Hence, the defensive benefit of PI expression should be considerably enhanced if the plants can also attract a herbivore's natural enemies (Zavala et al. 2004b). However, under glasshouse conditions, PI production increases plant fitness when plants are attacked by *M. sexta* even in the absence of predators (Zavala and Baldwin 2004), but the fitness benefits of PIs have yet to be determined in natural environments with the native herbivore community.

Herbivore attack elicits a JA-mediated and coordinated increase in both leaf PI and nicotine levels. When both of these defenses are individually silenced in *N. attenuata*, each can be shown to function effectively as defense against one of this species's major generalist herbivores, *Spodoptera exigua* (Steppuhn and Baldwin 2007). However, silencing PIs and nicotine simultaneously demonstrates that the defensive function of PIs is strongly dependent on the presence of nicotine: *S. exigua* larvae performed better and consumed more on PI-producing but nicotine-deficient plants than they did on plants silenced in both defenses. The reason for the synergism lies in the compensatory responses of the larvae fed PI-containing diets. Nicotine prevented *S. exigua* from overcompensating for PIs' antinutritive

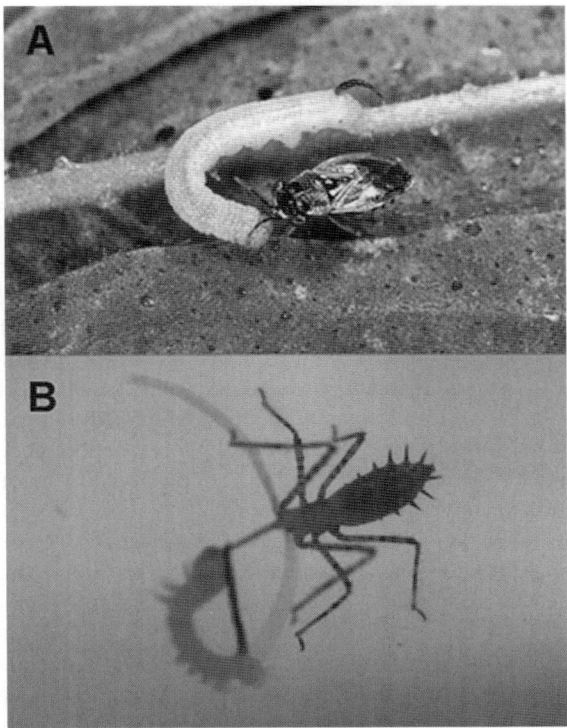

Fig. 3.2 Native predators preying on first instar larvae of *M. sexta*, which increase their body mass 10,000 fold from eclosion through 5 instars to pupation. *Geocoris pallens* (**A**) is the dominant predator in *N. attenuata*'s native habitat throughout the growing season and Reduviidae nymphs (**B**) are abundant predators during flowering and seed set; by the time *M. sexta* larvae reach the third instar, both predators are substantially smaller than their prey. Induced direct defenses that delay growth and development of the herbivores likely increase the probability of predation by prolonging the period during which larvae can be successfully attacked by these small predators. (Photographs: (**A**) A. Kessler, (**B**) A. Steppuhn)

activity by increasing consumption. Therefore, deploying both defenses provides a defensive synergism. Synergistic interactions of defenses are rarely studied but might well be the rule rather than the exception (Berenbaum 1985; Dyer et al. 2003).

3.3.4 Dynamic View of Costs and Benefits of Induced Responses

In summary, though many theories about the ecology and evolution of plant defenses are based on tradeoffs between costs and benefits, the evidence for costs is scarce. This is due to the difficulties of separating the fitness costs of a specific defense trait from genetically, metabolically, and functionally linked traits. Methodological advances allow the genetic background and genetic linkage to be controlled for, but to elucidate the full spectrum of metabolic and ecological roles of induced defenses, detailed analyses of the transcriptional and metabolic consequences of their expression are required. Elicitation with herbivore-specific elicitors is known to result in large changes of the transcriptome (Halitschke et al. 2003) and metabolomes of plants (Giri et al. 2006). The massive change in primary metabolism (for example, the down-regulation of photosynthesis) that commonly accompanies defense elicitation may be required to re-allocate resources to fuel secondary metabolism. Alternatively, this metabolic shift may itself function as a defense. The herbivore-induced response in *N. attenuata* includes a re-configuration of source–sink relationships in the plant so that recently fixed assimilates are transported to roots rather than young leaves. This shift is mediated by the rapid down-regulation of a subunit (GAL83) of a SNF1-related kinase (SnRK1) in the attacked leaves (Schwachtje et al. 2006). This carbon diversion response leads to increased root reserves, which delays senescence and prolongs flowering and seed production. The SnRK1-mediated bunkering of assimilates to the roots allows plants to better tolerate herbivory, but may also function as a 'passive' defense by decreasing the nutritional value of the remaining above-ground parts to above-ground herbivores.

The simplistic view of induced defenses – that an initially constitutively expressed defense evolves to become inducible so as to minimize the defense costs when the defenses are not needed – is difficult to harmonize with the complex and flexible nature of plant metabolism. The diverse linkages among various responses to herbivory suggests that a more dynamic view of the costs and benefits of induced defense would be more appropriate. The fitness of a plant depends on internal and external factors (Fig. 3.3A). The latter are the environmental conditions including resource, light, and water availability as well as biotic and abiotic stresses. The internal factors are first the physiological capacities such as resource uptake rates, photosynthetic rates, and metabolic efficiency, and second, the ability of metabolism to maintain these capacities under the environmental stresses which the plant is likely to experience. The resistance mechanisms of a plant are the means by which plant metabolism is adjusted to maximize reproductive output in the face of the diverse suite of stresses that plants face. The costs of the resistance mechanisms result from the tradeoffs between maximizing physiological capacity and (I) increasing the stress level and (II) expanding the range of conditions under which this capacity is optimized, and (III) from the tradeoff between the advantage of phenotypic plasticity (i.e., inducible resistance), and their drawbacks, such as the delays inherent in inducible activation (Fig. 3.3B).

If a genotype produces more of a defense against herbivores, it may be that this production results in decreased fitness under conditions at which a less defended genotype has its fitness maximum. However, small environmental changes might reverse this fitness outcome. The conditions under which the better defended genotype realizes its maximum fitness are likely altered due to metabolic and functional linkages. By deploying defenses inducibly, plants may be able to increase their fitness opportunities with respect to many stresses, as the tradeoffs with other stress resistances occur only when required. The increasing evidence of crosstalk between the signaling cascades of various induced responses suggests that plants

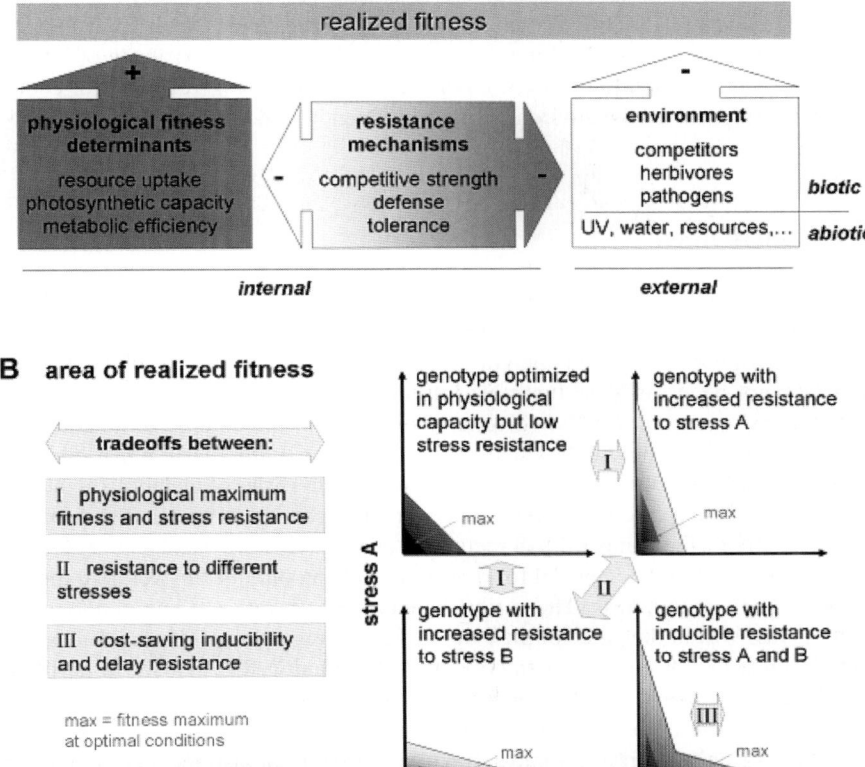

Fig. 3.3 (**A**) Conceptual framework for how internal and external factors determine the fitness a plant can achieve. Environmental conditions vary with respect to many biotic and abiotic stresses which influence a plant's realized fitness. Resistance mechanisms activated by these stresses adjusts plant metabolism so as to optimize physiological capacities in the diversity of environments in which a plant lives. (**B**) The resistance mechanisms increase the area of realized fitness when plants grow in environments with particular stresses and alter the conditions under which maximum fitness can be achieved. Increasing resistance to a certain stress results in different tradeoffs with respect to physiological capacities or resistance to other stresses and these costs can be minimized when resistance is deployed inducibly, which in turn, produces a cost resulting from the delay in the activation of resistance traits

are able to 'fine-tune' their metabolism to maximize fitness under various stress combinations. One example of such crosstalk is the concentration–dependent interaction between the pathogen-related salicylate (SA) and the herbivore-related JA signaling in *N. tabacum* (Mur et al. 2006). JA and SA signaling synergize each other at low concentration but at high concentrations they antagonize each other's biosynthesis and the subsequent signaling cascades. Thus, the induced reconfiguration of metabolism to increase herbivore resistance changes the plant's physiological capacities and its ability to maintain these capacities under other environmental stresses. This view of induced defenses does not presume the existence of intrinsically costly defenses, which can be calculated in a currency of resources or energy, but rather emphasizes the environmental contingency of defense costs.

3.4 Conclusions

The wide-spread acceptance of the assumption that plant defenses are intrinsically costly is more likely a reflection of our shared experience (or beliefs) that what benefits us is expensive, than the weight of the experimental evidence. Photosynthetic rate may not necessarily be the most important determinant of plant growth, and therefore the quantity of carbon diverted from growth into defense production may not be the best currency with which to measure costs, particularly when resources other than carbon are more limiting for growth. The divisions of metabolism into primary or secondary, into herbivore defense or pathogen resistance, and so forth, are useful in parsing the complexity of metabolic changes in more manageable parts, but such parsing may not be helpful for questions that can only be answered in the currency of Darwinian fitness. The various metabolic pathways are linked in complex ways so that – as in the game of pick-up-sticks – changing a specific trait invariably has ramifications throughout metabolism. Perhaps induced responses to herbivory do not automatically incur fitness cost but reduce the ability of the plant to optimize its fitness as the environment changes. Induced responses can be seen as responses that alter not only the physiological maximum fitness but also the size, shape, and position of the plant's opportunity to realize fitness in the multi-dimensional space of environmental conditions that each plant must successfully occupy during its life. This view of costs and benefits is depicted in Fig. 3.4 showing two genotypes, one of which evolved a greater defense against herbivores. This genotype (II) is still able to reproduce under high levels of herbivory; however, these benefits are constrained by a lower resistance to a pathogen, drought, and a competitor. The latter are those conditions which benefits the other genotype (I) but at a cost of reduced herbivore resistance. Yet, under many conditions both genotypes can achieve comparable fitness, underscoring the main point: defense costs are not absolute but highly environmentally contingent.

If the production of plant defenses is not intrinsically costly, cost minimization may not be the raison d'être of induced defenses. A model which does not assume that defense incurs costs is the moving-target model stating that inducibility evolved to present a changing metabolic phenotype to the plant's herbivores (Karban et al. 1997). Because nutritional variability decreases herbivore performance, inducibility may itself represent a defense strategy. Another hypothesis that does not rely on costs supposes that the adaptive value of inducibility arises from increasing the level of defense to highly toxic levels at a time when the herbivore is at its most

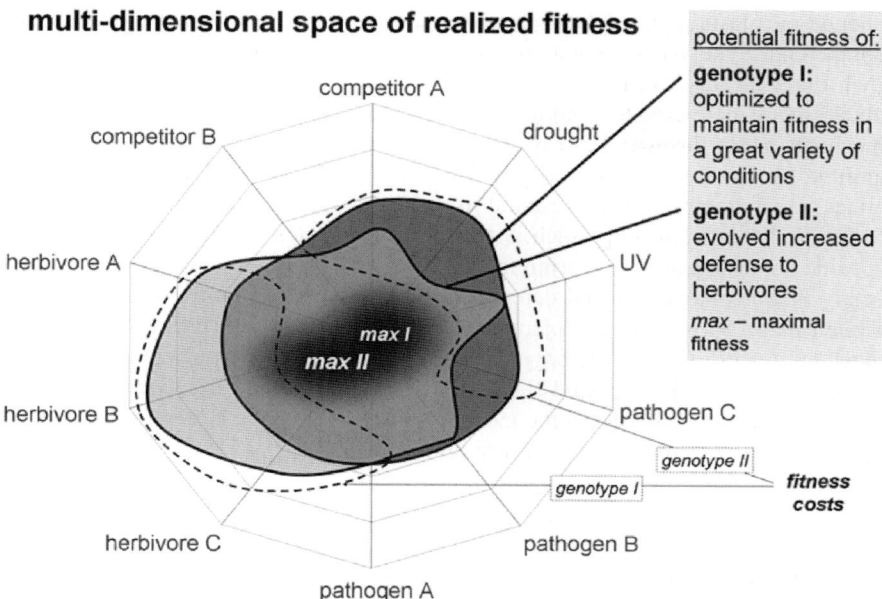

Fig. 3.4 Model of costs and benefits of induced defense in the multi-dimensional space of environmental conditions. Within this space, each plant genotype has a limited opportunity to realize a certain fitness level. The realized fitness of a genotype with increased resistance to one or more environmental stresses can be restricted under a variety of conditions. Fitness of genotype II (*light gray*) with increased anti-herbivores defense is lower than that of the less-defended genotype I (*dark gray*) in environments with high levels of competition, drought, or attack by pathogen C (*right dotted line*). However, in environments with high levels of herbivore attack (*left dotted line*), the realized fitness of genotype II is greater than that of genotype I. In intermediate stress environments (in the center of the overlapping areas), the two genotypes have similar realized fitness levels

voracious stage, thereby stimulating the herbivore to move to an adjacent competitor (van Dam et al. 2001). All these hypotheses, including the cost-benefit paradigm, assume that inducibility is a derived trait and constitutive expression of defense the pleisiomorphic condition. However, defenses may have originally all been inducibly expressed, and evolved constitutive expression under particular conditions (Karban and Baldwin 1997). If costs are not the driving force behind the evolution of induced plant defenses, they can also evolve under conditions of predictable and high herbivory, even if permanently expressed. This would be consistent with the observations that induced defense are common, despite the fact that herbivores are everywhere and their attack on plants, relentless.

On the other hand inducibility per se may be a costly trait, which is difficult to examine because such costs need to be separated from the costs of defense production and other induced traits (Cipollini et al. 2003). The costs of maintaining the required signaling systems, which are likely co-opted from developmental processes, are generally assumed to be small. Some studies suggest that genetically inherited phenotypic plasticity for defense expression is costly (such as Agrawal et al. 2002), but the underlying mechanism remains unclear. The obvious drawback of inducibility is the delayed activation of defenses, and plants may optimize the benefits of inducibility by increasing their ability to predict attack from natural enemies. Potential mechanisms for this may be 'eavesdropping' on the neighboring plants for signals emitted by attacked plants, (such as MeJA or VOC), or a 'memory' of previous attacks. For example, plants respond to pathogen attack with increased resistance to future pathogen attack, which is attributed to a faster and stronger response of specific defense genes. This priming uses specific signaling components and is less costly than activated induced defense, although providing similar resistance (van Hulten et al. 2006). Similarly, plants that experienced herbivore attack can be vaccinated by attack from a less damaging herbivore against attack from a more damaging herbivore species (Kessler and Baldwin 2004; Voelckel and Baldwin 2004b). We still have much to learn about the mechanisms by which plants rapidly adjust their phenotypes to the environments in which they live.

3.5 Perspectives

The empirical evidence for costs of induced defense remains underwhelming, and more comprehensive research is required to test the different concepts of costs and benefits of induced defense. Our understanding of the eliciting compounds, the signaling cascades, and the metabolic pathways involved has increased dramatically, and advances in molecular methods enable individual traits to be isolated and their consequences for an organism's Darwinian fitness to be determined. The simple model of intrinsically costly defenses does not jibe with the complexity with which plants adjust their metabolism in response to herbivore attack as well as to other environmental stresses. Hence, simply seeking evidence for negative correlations between defense level and realized fitness is not sufficient to adequately test the existence of defense costs. The analysis needs to proceed at many levels, examining the underlying metabolic mechanisms and their consequences for whole-organism performance in the range of environments that a plant likely experiences in nature. Advanced molecular and analytical tools in the areas of transcriptomics and metabolomics provide insights into the metabolic adjustments; together with manipulative approaches such as gene silencing, it is possible to examine the connections in the metabolic grid that are responsible for the phenotype. In addition, the methodological advances will allow for refinements of hypotheses about tradeoffs occurring at a whole-organismic level. The transgenic approach of altering specific defense trait, allows their function as well as the tradeoffs among traits to be determined under different natural conditions. Quantitative genetic studies elucidate the mechanisms underlying evolution of induced responses (Mitchell-Olds and Schmitt 2006). Extending the cost-benefit paradigm from a purely resource-based model to a dynamic view of tradeoffs among different metabolic and functional traits may improve the falsifiability of the hypotheses and broaden the experimental approaches to test them. The field of research that Clarence Ryan started more than three decades ago, when he described increases in PIs in potato leaves when they were attacked by Colorado potato beetles (Green and Ryan 1972), has blossomed into a field which is shaping our understanding of how metabolism is regulated and continues to hold the promise of developing crop plants that are as sophisticated as their wild ancestors in dodging environmental stresses.

Acknowledgments We want to thank Emily Wheeler and Jens Schwachtje for comments on earlier versions and editorial assistance, Jorge Zavala and Ashok Giri for helpful discussions, and Grit Glawe and Hendrik Wünsche for the photographs in Fig. 3.1, André Kessler for the photograph in Fig. 3.2A, and the Max Planck Society for funding.

References

Agrawal AA, Conner JK, Johnson MTJ, Wallsgrove R (2002) Ecological genetics of an induced plant defense against herbivores: additive genetic variance and costs of phenotypic plasticity. Evolution 56:2206–2213

Agrawal AA, Strauss SY, Stout MJ (1999) Costs of induced responses and tolerance to herbivory in male and female fitness components of wild radish. Evolution 53:1093–1104

Agren J, Schemske DW (1993) The cost of defense against herbivores: an experimental study of trichome production in *Brassica rapa*. Am Nat 141:338–350

Alborn HT, Turlings TCJ, Jones TH, Stenhagen G, Loughrin JH, Tumlinson JH (1997) An elicitor of plant volatiles from beet armyworm oral secretion. Science 276:945–949

Baldwin IT (1989) Mechanism of damage-induced alkaloid production in wild tobacco. J Chem Ecol 15:1661–1680

Baldwin IT (1991) Damage-induced alkaloids in wild tobacco. In: Raupp MJ, Tallamy DW (eds) Phytochemical induction by herbivores. John Wiley and Sons, New York, pp 47–49

Baldwin IT (1998) Jasmonate-induced responses are costly but benefit plants under attack in native populations. Proc Natl Acad Sci USA 95:8113–8118

Baldwin IT, Gorham D, Schmelz EA, Lewandowski CA, Lynds GY (1998) Allocation of nitrogen to an inducible defense and seed production in *Nicotiana attenuata*. Oecologia 115:541–552

Baldwin IT, Karb MJ, Ohnmeiss TE (1994) Allocation of N-15 from nitrate to nicotine – production and turnover of a damage-induced mobile defense. Ecology 75:1703–1713

Baldwin IT, Oesch RC, Merhige PM, Hayes K (1993) Damage-induced root nitrogen metabolism in *Nicotiana sylvestris* – testing C/N predictions for alkaloid production. J Chem Ecol 19:3029–3043

Baldwin IT, Preston CA (1999) The eco-physiological complexity of plant responses to insect herbivores. Planta 208:137–145

Baldwin IT, Sims CL, Kean SE (1990) The reproductive consequences associated with inducible alkaloidal responses in wild tobacco. Ecology 71:252–262

Barbosa P, Gross P, Kemper J (1991) Influence of plant allelochemicals on the tobacco hornworm and its parasitoid *Cotesia congregata*. Ecology 72:1567–1575

Berenbaum M, Neal JJ (1985) Synergism between myristicin and xanthotoxin, a naturally co-occurring plant toxicant. J Chem Ecol 11:1349–1358

Berenbaum MR (1985) Brementown revisited: interactions among allelochemicals in plants. Rec Adv Phytochem 19:139–169

Berenbaum MR, Zangerl AR, Nitao JK (1986) Constraints on chemical coevolution: wild parsnips and the parsnip webworm. Evolution 40:1215–1228

Bergelson J, Purrington CB (1996) Surveying patterns in the cost of resistance in plants. Am Nat 148:536–558

Bergelson J, Purrington CB, Palm CJ, LopezGutierrez JC (1996) Costs of resistance: a test using transgenic *Arabidopsis thaliana*. Proc R Soc Lond B Biol Sci 263:1659–1663

Birk Y (ed) (2003) Plant protease inhibitors; significance in nutrition, plant protection, cancer prevention and genetic engineering. Springer, Berlin, p 170

Broadway RM, Duffey SS (1988) The effect of plant protein quality on insect digestive physiology and the toxicity of plant proteinase inhibitors. J Insect Physiol 34:1111–1117

Bryant JP, Chapin FS, Klein DR (1983) Carbon nutrient balance of boreal plants in relation to vertebrate herbivory. Oikos 40:357–368

Chintapakorn Y, Hamill JD (2003) Antisense-mediated down-regulation of putrescine N-methyltransferase activity in transgenic *Nicotiana tabacum* L. can lead to elevated levels of anatabine at the expense of nicotine. Plant Mol Biol 53:87–105

Cipollini DF (2002) Does competition magnify the fitness costs of induced responses in *Arabidopsis thaliana*? A manipulative approach. Oecologia 131:514–520

Cipollini D (2007) Consequences of the overproduction of methyl jasmonate on seed production, tolerance to defoliation and competitive effect and response of *Arabidopsis thaliana*. New Phytol 173:146–153

Cipollini D, Purrington CB, Bergelson J (2003) Costs of induced responses in plants. Basic Appl Ecol 4:79–89

Cipollini DF, Sipe ML (2001) Jasmonic acid treatment and mammalian herbivory differentially affect chemical defenses and growth of wild mustard (*Brassica kaber*). Chemoecology 11:137–143

Creelman RA, Mullet JE (1997) Biosynthesis and action of jasmonates in plants. Annu Rev Plant Physiol Plant Mol Biol 48:355–381

Dicke M, Sabelis MW (1992) Cost and benefits of chemical information conveyance: proximate and ultimate factors. In: Roitberg BD, Isman MB (eds) Insect chemical ecology. An evolutionary approach. Chapman and Hall, New York, pp 122–155

Dyer LA, Dodson CD, Stireman JO, Tobler MA, Smilanich AM, Fincher RM, Letourneau DK (2003) Synergistic effects of three *Piper* amides on generalist and specialist herbivores. J Chem Ecol 29:2499–2514

Feeny P (1976) Plant apparency and chemical defense. Recent Adv Phytochem 10:1–40

Felton GW, Broadway RM, Duffey SS (1989) Inactivation of protease inhibitor activity by plant-derived quinones: complications for host-plant resistance against noctuid herbivores. J Insect Physiol 35:981–990

Giri AP, Wunsche H, Mitra S, Zavala JA, Muck A, Svatos A, Baldwin IT (2006) Molecular interactions between the specialist herbivore *Manduca sexta* (Lepidoptera, Sphingidae) and its natural host *Nicotiana attenuata*. VII. Changes in the plant's proteome. Plant Physiol 142:1621–1641

Glawe GA, Zavala JA, Kessler A, Van Dam NM, Baldwin IT (2003) Ecological costs and benefits correlated with trypsin protease inhibitor production in *Nicotiana attenuata*. Ecology 84:79–90

Govenor HL, Schultz JC, Appel HM (1997) Impact of dietary allelochemicals on gypsy moth (*Lymantria dispar*) caterpillars: importance of midgut alkalinity. J Insect Physiol 43:1169–1175

Green TR, Ryan CA (1972) Wound-induced proteinase inhibitors in plant leaves: a possible defense mechanism against insects. Science 175:776–777

Green ES, Zangerl AR, Berenbaum MR (2001) Effects of phytic acid and xanthotoxin on growth and detoxification in caterpillars. J Chem Ecol 27:1763–1773

Halitschke R, Baldwin IT (2003) Antisense LOX expression increases herbivore performance by decreasing defense responses and inhibiting growth-related transcriptional reorganization in *Nicotiana attenuata*. Plant J 36:794–807

Halitschke R, Baldwin IT (2004) Jasmonates and related compounds in plant–insect interactions. J Plant Growth Regul 23:238–245

Halitschke R, Gase K, Hui D, Schmidt DD, Baldwin IT (2003) Molecular interactions between the specialist herbivore *Manduca sexta* (Lepidoptera, Sphingidae) and its natural host *Nicotiana attenuata*. VI. Microarray analysis reveals that most herbivore-specific transcriptional changes are mediated by fatty acid–amino acid conjugates. Plant Physiol 131:1894–1902

Halitschke R, Schittko U, Pohnert G, Boland W, Baldwin IT (2001) Molecular interactions between the specialist herbivore *Manduca sexta* (Lepidoptera, Sphingidae) and its natural host *Nicotiana attenuata*. III. Fatty acid–amino acid conjugates in herbivore oral secretions are necessary and sufficient for herbivore- specific plant responses. Plant Physiol 125:711–717

Halitschke R, Stenberg JA, Kessler D, Kessler A, Baldwin IT (2007) Shared signals – 'alarm calls' from plants increase apparency to herbivores and their enemies in nature. Ecol Lett doi: 10.1111/j.1461-0248.2007.01123.x

Heidel AJ, Clarke JD, Antonovics J, Dong XN (2004) Fitness costs of mutations affecting the systemic acquired resistance pathway in *Arabidopsis thaliana*. Genetics 168:2197–2206

Heil M, Baldwin IT (2002) Fitness costs of induced resistance: emerging experimental support for a slippery concept. Trends Plant Sci 7:61–67

Herms DA, Mattson WJ (1992) The dilemma of plants: to grow or defend. Q Rev Bio 67:283–335

Juenger T, Bergelson J (2000) The evolution of compensation to herbivory in scarlet gilia, *Ipomopsis aggregata*: herbivore-imposed natural selection and the quantitative genetics of tolerance. Evolution 54:764–777

Karban R, Agrawal AA, Mangel M (1997) The benefits of induced defenses against herbivores. Ecology 78:1351–1355

Karban R, Baldwin IT (1997) Induced responses to herbivory. University of Chicago Press, Chicago, IL

Kessler A, Baldwin IT (2001) Defensive function of herbivore-induced plant volatile emissions in nature. Science 291:2141–2144

Kessler A, Baldwin IT (2004) Herbivore-induced plant vaccination. Part I. The orchestration of plant defenses in nature and their fitness consequences in the wild tobacco *Nicotiana attenuata*. Plant J 38:639–649

Kessler A, Halitschke R, Baldwin IT (2004) Silencing the jasmonate cascade: induced plant defenses and insect populations. Science 305:665–668

Krischik VA, Denno RF (1983) Individual, population, and geographic patterns in plant defense. In: Denno RF, McClure MS (eds) Variable plants and herbivores in natural and managed systems. Academic Press, New York, pp 463–512

Laing W, McManus MT (2002) Proteinase inhibitors. In: McManus MT, Laing WA, Allan AC (eds) In protein–protein interactions in plant biology. CRC, Boca Raton, FL, pp 77–119

Lerdau M, Litvak M, Monson R (1994) Plant chemical defense: monoterpenes and the growth-differentiation balance hypothesis. Trends Ecol Evol 9:58–61

McKey D (1974) Adaptive patterns in alkaloid physiology. Am Nat 108:305–320

Metodiev MV, Tsonev TD, Popova LP (1996) Effect of jasmonic acid on the stomatal and non-stomatal limitation of leaf photosynthesis in barley leaves. J Plant Growth Regul 15:75–80

Mitchell-Olds T, Schmitt J (2006) Genetic mechanisms and evolutionary significance of natural variation in *Arabidopsis*. Nature 441:947–952

Mitchell-Olds T, Siemens D, Pedersen D (1996) Physiology and costs of resistance to herbivory and disease in *Brassica*. Entomol Exp Appl 80:231–237

Mithöfer A, Wanner G, Boland W (2005) Effects of feeding *Spodoptera littoralis* on lima bean leaves. II. Continuous mechanical wounding resembling insect feeding is sufficient to elicit herbivory-related volatile emission. Plant Physiol 137:1160–1168

Mur LAJ, Kenton P, Atzorn R, Miersch O, Wasternack C (2006) The outcomes of concentration-specific interactions between salicylate and jasmonate signaling include synergy, antagonism, and oxidative stress leading to cell death. Plant Physiol 140:249–262

Nelson AC, Kursar TA (1999) Interactions among plant defense compounds: a method for analysis. Chemoecology 9:81–92

Ohnmeiss TE, Baldwin IT (2000) Optimal defense theory predicts the ontogeny of an induced nicotine defense. Ecology 81:1765–1783

Plagnol V, Padhukasahasram B, Wall JD, Marjoram P, Nordborg M (2005) Relative influences of crossing-over and gene conversion on the pattern of linkage disequilibrium in *Arabidopsis thaliana*. Genetics 172:2441–2448

Redman AM, Cipollini DF, Schultz JC (2001) Fitness costs of jasmonic acid-induced defense in tomato, *Lycopersicon esculentum*. Oecologia 126:380–385

Reinbothe S, Mollenhauer B, Reinbothe C (1994) JIPs and RIPs: the regulation of plant gene expression by jasmonates in response to environmental cues and pathogens. Plant Cell 6: 1197–1209

Reymond P, Bodenhausen N, Van Poecke RMP, Krishnamurthy V, Dicke M, Farmer EE (2004) A conserved transcript pattern in response to a specialist and a generalist herbivore. Plant Cell 16:3132–3147

Royo J, León J, Vancanneyt G, Albar JP, Rosahl S, Ortego F, Castañera P, Sánchez-Serrano JJ (1999) Antisense-mediated depletion of a potato lipoxygenase reduces wound induction of proteinase inhibitors and increases weight gain of insect pests. Proc Natl Acad Sci USA 96: 1146–1151

Ryan CA (1983) Insect-induced chemical signals regulating natural plant protection responses. In: Denno RF, McClure MS (eds) Variable plants and herbivores in natural managed systems, Academic Press, New York, pp 43–60

Schwachtje J, Minchin PEH, Jahnke S, van Dongen JT, Schittko U, Baldwin IT (2006) SNF1-related kinases allow plants to tolerate herbivory by allocating carbon to roots. Proc Natl Acad Sci USA 103:12935–12940

Simms EL, Rausher MD (1987) Costs and benefits of plant resistance to herbivory. Am Nat 130:570–581

Simms EL, Rausher MD (1989) The evolution of resistance to herbivory in *Ipomoea purpurea*. II. Natural selection by insects and costs of resistance. Evolution 43:573–585

Steppuhn A, Baldwin IT (2007) Resistance management in a native plant: nicotine prevents herbivores from compensating for plant protease inhibitors. Ecol Lett 10:499–511

Steppuhn A, Gase K, Krock B, Halitschke R, Baldwin IT (2004) Nicotine's defensive function in nature. PLoS Biol 2:1074–1080

Stowe KA (1998) Experimental evolution of resistance in *Brassica rapa*: correlated response of tolerance in lines selected for glucosinolate content. Evolution 52:703–712

Strauss SY, Rudgers JA, Lau JA, Irwin RE (2002) Direct and ecological costs of resistance to herbivory. Trends Ecol Evol 17:278–285

Strauss SY, Siemens DH, Decher MB, Mitchell-Olds T (1999) Ecological costs of plant resistance to herbivores in the currency of pollination. Evolution 53:1105–1113

Takabayashi J, Dicke M (1996) Plant-carnivore mutualism through herbivore-induced carnivore attractants. Trends Plant Sci 1:109–113

Thomma BPHJ, Eggermont K, Penninckx IAMA, Mauch-Mani B, Vogelsang R, Cammue BPA, Broekaert WF (1998) Separate jasmonate-dependent and salicylate-dependent defense response pathways in *Arabidopsis* are essential for resistance to distinct microbial pathogenes. Proc Natl Acad Sci USA 95:15107–15111

Tian D, Traw MB, Chen JQ, Kreitman M, Bergelson J (2003) Fitness costs of *R*-gene-mediated resistance in *Arabidopsis thaliana*. Nature 423:74–77

van Dam NM, Baldwin IT (1998) Costs of jasmonate-induced responses in plants competing for limiting resources. Ecol Lett 1:30–33

van Dam NM, Hermenau U, Baldwin IT (2001) Instar-specific sensitivity of specialist *Manduca sexta* larvae to induced defences in their host plant *Nicotiana attenuata*. Ecol Entomol 26: 578–586

van Hulten M, Pelser M, van Loon LC, Pieterse CMJ, Ton J (2006) Costs and benefits of priming for defense in *Arabidopsis*. Proc Natl Acad Sci USA 103:5602–5607

Voelckel C, Baldwin IT (2004a) Generalist and specialist lepidopteran larvae elicit different transcriptional responses in *Nicotiana attenuata*, which correlate with larval FAC profiles. Ecol Lett 7:770–775

Voelckel C, Baldwin IT (2004b) Herbivore-induced plant vaccination. Part II. Array-studies reveal the transience of herbivore-specific transcriptional imprints and a distinct imprint from stress combinations. Plant J 38:650–663

Voelckel C, Krügel T. Gase K, Heidrich N, van Dam NM, Winz R, Baldwin IT (2001) Anti-sense expression of putrescine N-methyltransferase confirms defensive role of nicotine in *Nicotiana sylvestris* against *Manduca sexta*. Chemoecology 11:121–126

Vrieling K, deJong TJ, Klinkhamer PGL, van der Meijden E, van der VeenvanWijk CAM (1996) Testing trade-offs among growth, regrowth and anti-herbivore defences in *Senecio jacobaea*. Entomol Exp Appl 80:189–192

Wasternack C, Ortel B, Miersch O, Kramell R, Beale M, Greulich F, Feussner I, Hause B, Krumm T, Boland W, Parthier B (1998) Diversity in octadecanoid-induced gene expression of tomato. J Plant Physiol 152:345–352

Winz RA, Baldwin IT (2001) Molecular interactions between the specialist herbivore *Manduca sexta* (Lepidoptera, Sphingidae) and its natural host *Nicotiana attenuata*. IV. Insect-induced ethylene reduces jasmonate-induced nicotine accumulation by regulating putrescine N-methyltransferase transcripts. Plant Physiol 125:2189–2202

Wu J, Hettenhausen C, Meldau S, Baldwin IT (2007) Herbivory rapidly activates MAPK signaling in attacked and unattacked leaf regions but not between leaves of *Nicotiana attenuata*. Plant Cell 19:1096–1122

Zangerl AR, Bazzaz FA (1992) Theory and pattern in plant defense allocation. In: Fritz RS, Simms EL (eds) Plant resistance to herbivores and pathogens. Ecology, evolution, and genetics. University of Chicago Press, Chicago, IL, pp 363–391

Zavala JA, Baldwin IT (2004) Fitness benefits of trypsin proteinase inhibitor expression in *Nicotiana attenuata* are greater than their costs when plants are attacked. BMC Ecol 4:11

Zavala JA, Patankar AG, Baldwin IT (2004a) Manipulation of endogenous trypsin proteinase inhibitor production in *Nicotiana attenuata* demonstrates their function as antiherbivore defenses. Plant Physiol 134:1181–1190

Zavala JA, Patankar AG, Gase K, Baldwin IT (2004b) Constitutive and inducible trypsin proteinase inhibitor production incurs large fitness costs in *Nicotiana attenuata*. Proc Natl Acad Sci USA 101:1607–1612

Zavala JA, Patankar AG, Gase K, Baldwin IT (2004b) Constitutive and inducible trypsin proteinase inhibitor production incurs large fitness costs in *Nicotiana attenuata*. Proc Natl Acad Sci USA 101:1607–1612

Section II
Induced Direct Defenses

Part A
Anatomical Defenses

Chapter 4
Leaf Trichome Formation and Plant Resistance to Herbivory

Peter Dalin, Jon Ågren, Christer Björkman, Piritta Huttunen and Katri Kärkkäinen

Leaf trichomes contribute to plant resistance against herbivory. In several plant species, the trichome density of new leaves increases after herbivore damage. Here we review the genetic basis of trichome production and the functional and adaptive significance of constitutive and induced trichome formation. We focus on leaf trichomes and their production in response to damage caused by herbivores. The genetic basis of trichome production has been explored in detail in the model species *Arabidopsis thaliana*. Recent comparative work indicates that the regulatory networks governing trichome development vary and that trichome production has evolved repeatedly among angiosperms. Induced trichome production has been related to increased levels of jasmonic acid in *Arabidopsis*, indicating a common link to other changes in resistance characteristics. Damage from insect herbivores is oftentimes negatively related to trichome production, and enhanced trichome production may thus be advantageous as it increases resistance against herbivores. There are yet few studies exploring the costs and benefits of induced trichome production in terms of plant fitness. Trichome density affects interactions with insect herbivores, but may also affect the abundance and effectiveness of predators and parasitoids feeding on herbivores, and the tolerance to abiotic stress. This suggests that an improved understanding of the functional and adaptive significance of induced trichome production requires field studies that consider the effects of trichome density on antagonistic interactions, tritrophic interactions, and plant fitness under contrasting abiotic conditions.

4.1 Introduction

Trichomes are hair-like appendages that develop from cells of the aerial epidermis and are produced by most plant species (Werker 2000). Leaf trichomes can serve several functions including protection against damage from herbivores (Levin 1973).

P. Dalin
Marine Science Institute, University of California, Santa Barbara, CA 93106-6150, USA
e-mail: dalin@msi.ucsb.edu

While most plants produce trichomes constitutively, some species respond to damage by increasing trichome density in new leaves. The purpose of this paper is to review processes affecting trichome formation, and the importance of trichomes in plant resistance. We mainly focus on leaf trichomes and their production in response to damage caused by herbivorous insects. We begin by briefly reviewing current understanding of the genetic basis of trichome formation. Based on literature data, we explore the magnitude of damage-induced increases in trichome production and the hormonal basis of damage-induced trichome production. We then discuss the effects of induced trichome production on interactions with herbivorous insects, their natural enemies and on plant fitness. Finally, we identify problems in need of further research for a better understanding of the functional and adaptive significance of induced trichome production in plants.

The morphology and density of leaf trichomes vary considerably among plant species, and may also vary among populations and within individual plants. The structure of trichomes can range from unicellular to multi-cellular, and the trichomes can be straight, spiral, hooked, branched, or un-branched (Southwood 1986; Werker 2000). Some trichomes have glands that release secondary metabolites (e.g., terpenes and alkaloids) which can be poisonous, repellent, or trap insects and other organisms. These trichomes are commonly referred to as glandular trichomes (Duffey 1986). In some species, individual plants produce both glandular and non-glandular leaf trichomes (e.g., Hare and Elle 2002; Rautio et al. 2002).

Trichome production is an important component of resistance against herbivorous insects (Levin 1973; Southwood 1986; Ågren and Schemske 1994; Karban and Baldwin 1997; Traw and Dawson 2002a). Damage from many insect herbivores is negatively related to trichome density (Ågren and Schemske 1993; Mauricio 1998; Valverde et al. 2001; Hare and Elle 2002; Handley et al. 2005). The glabrous (non-hairy) morph of the perennial herb *Arabidopsis lyrata* is more damaged by insect herbivores than the trichome-producing morph (Løe et al. 2007), and experimental removal of leaf trichomes resulted in increased feeding and growth of herbivorous insects in bioassay studies of several species (Rowell-Rahier and Pasteels 1982; Baur et al. 1991; Fordyce and Agrawal 2001). Moreover, several plant species respond to damage caused by herbivores by producing new leaves with an increased density and/or number of trichomes (Table 4.1), and insects feeding on the induced plants often consume less foliage and grow less well compared with insects on non-induced plants (Agrawal 1999; 2000; Dalin and Björkman 2003). These studies indicate that the production of leaf trichomes contributes to the protection against herbivorous insects.

Leaf trichomes may also increase resistance to abiotic stress. They may increase tolerance to drought by reducing absorbance of solar radiation and increasing the leaf surface boundary layer (Ehrlinger 1984; Choinski and Wise 1999; Benz and Martin 2006), and by facilitating condensation of air moisture onto the plant surface (Jeffree 1986). Trichomes may further protect living cells from damage caused by solar UV-radiation (Skaltsa et al. 1994), and low temperatures (Agrawal et al. 2004). Trichomes may thus have multiple functions and trichome density may evolve in response to variation in several environmental factors. In general, plants with high

4 Leaf Trichome Formation and Plant Resistance to Herbivory

Table 4.1 Induced increases in trichome density recorded after artificial wounding and damage by insect herbivores

Plant Species (life history)[1]	Defoliation treatment			Tissue damaged	Increase in trichome density		Reference
	Damage	Environment	Amount		Magnitude	Timing	
Arabidopsis thaliana (a)	Artificial	Lab	Pinching	Leaves	25%–117%	weeks	Traw and Bergelson (2003)
Alnus incana (p)	Coleoptera	Field	30%–100%	Leaves	500%	2%–4 weeks	Baur et al. (1991)
Brassica nigra (a)	Lepidoptera	Lab	25%	Leaves	76%	weeks	Traw and Dawson (2002a)
Brassica nigra (a)	Lepidoptera	Lab	25%	Leaves	113%	weeks	Traw and Dawson (2002a)
Brassica nigra (a)	Coleoptera	Lab	17%	Leaves	no induction	weeks	Traw and Dawson (2002a)
Brassica nigra (a)	Lepidoptera	Lab	25%	Leaves	43%	4%–5 weeks	Traw and Dawson (2002b)
Brassica nigra (a)	Coleoptera	Lab	25%	Leaves	no induction	4%–5 weeks	Traw and Dawson (2002b)
Brassica nigra (a)	Lepidoptera	Lab	22%	Leaves	38%	2%–3 weeks	Traw 2002
Betula pubescens (p)	Artificial	Field	50%	Leaves	300%	2%–7 weeks (1998)	Rautio et al. (2002)
Betula pubescens (p)	Artificial	Field	100%	Leaves	1000%	2%–7 weeks (1998)	Rautio et al. (2002)
Betula pubescens (p)	Artificial	Field	50%	Leaves	no induction	2%–7 weeks (1999)	Rautio et al. (2002)
Betula pubescens (p)	Artificial	Field	100%	Leaves	no induction	2%–7 weeks (1999)	Rautio et al. (2002)
Betula pubescens (p)	Artificial	Field	≤ 25%	Leaves	10%–20%	next year	Valkama et al. (2005)
Betula pubescens (p)	Artificial	Field	10 cm	Shoots	no induction	next year	Valkama et al. (2005)

Table 4.1 (continued)

	Defoliation treatment				Increase in trichome density		
	Damage	Environment	Amount	Tissue damaged	Magnitude	Timing	Reference
Cnidoscolus texanus (p)	Livestock	Field	no info.	Shoots	38%	no info.	Pollard (1986)
Lepidium virginicum (a)	Lepidoptera	Lab	20%–25%	Leaves	60%	2 weeks	Agrawal (2000)
Raphanus raphanistrum (a)	Lepidoptera	Lab	50%	Leaves	25%–100%	3 weeks	Agrawal (1999)
Salix borealis (p)	Coleoptera	Field	100%	Leaves	51%–300%	next year	Zvereva et al. (1998)
Salix cinerea (p)	Coleoptera	Lab	5%	Leaves	72%	3 weeks	Dalin and Björkman (2003)
Salix cinerea (p)	Artificial	Lab	5%	Leaves	no induction	3 weeks	Dalin and Björkman (2003)
Salix cinerea (p)	Coleoptera	Lab	10%	Leaves	93%	3 weeks	Björkman and Ahrne (2005)
Salix cinerea (p)	Coleoptera	Lab	3%	Leaves	83%–186%	3 weeks	Björkman et al. (2008)
Salix viminalis (p)	Coleoptera	Lab	5%–25%	Leaves	13%	3 weeks	Dalin et al. (2004)
Urtica dioica (p)	Artificial	Field	>50%	Shoots	140%–790%	10 weeks	Pullin and Gilbert (1989)
Urtica dioica (p)	Artificial	Lab	>50%	Shoots	23%–106%	10 weeks	Pullin and Gilbert 1989
Urtica dioica dioica (p)	Artificial	Lab	50%	Leaves	26%	5–7 weeks	Mutikainen and Walls (1995)
Urtica dioica dioica (p)	Artificial	Lab	No info.	Apical buds	no induction	5–7 weeks	Mutikainen and Walls (1995)
Urtica dioica sondenii (p)	Artificial	Lab	50%	Leaves	no induction	5–7 weeks	Mutikainen and Walls (1995)
Urtica dioica sondenii (p)	Artificial	Lab	No info.	Apical buds	50%	5–7 weeks	Mutikainen and Walls (1995)

[1]The life history of the different plant species is indicated as (p) for perennial, or (a) for annual

density of leaf trichomes can be expected in environments that are dry or cold, where UV-radiation is intense, and in areas where the risk of being damaged by herbivorous insects is high (Ehrlinger 1984; Løe et al. 2007).

4.2 Leaf Trichome Formation

4.2.1 Molecular Basis of Leaf Trichome Formation

Trichomes are initiated in the epidermis of developing leaves: some epidermal cells will develop into trichome cells whereas surrounding cells will develop into regular epidermal cells. Present understanding of trichome formation is based on studies of a few model species. In particular, trichome formation in *A. thaliana* has been used as a model system to study various developmental and cellular mechanisms in plants. In *Arabidopsis*, trichomes are non-glandular, large single cells, surrounded by accessory epidermal cells on rosette leaves, stems, cauline leaves, and sepals. With mutagenesis screens, dozens of genes involved in trichome initiation, spacing, and shape have been identified in *Arabidopsis* (Marks 1997; Hülskamp and Schnittger 1998). Recent studies of *A. thaliana* and its close relatives have shown that variation in some of the same genes can explain differences in trichome formation in natural populations (Hauser et al. 2001; Kivimäki et al. 2007).

Studies of the genetic basis of trichome initiation have identified genes that (a) control the entry into the trichome pathway and (b) control the spacing of initiation events. Loss-of-function mutations in *A. thaliana* have suggested that the genes *GLABROUS1* (*GL1*) and *TRANSPARENT TESTA GLABRA1* (*TTG1*) are important for both the initiation and spacing of leaf trichomes. *GL1* encodes a myb-transcription factor that is expressed diffusively in developing leaves (Oppenheimer et al. 1991). *TTG1* is expressed in most major plant organs, codes for a WD-40 repeat-containing protein, and is needed for trichome cell initiation in leaves, non-hair cell type (atrichoblast) formation in roots, anthocyanin biosynthesis in leaves, and formation of mucilage layer in germinating seeds (Koornneef 1981; Walker et al. 1999). For normal trichome initiation, the genes *GLABROUS3* (*GL3*) and *ENHANCER OF GLABRA3* (*EGL3*), which code for helix-loop-helix (bHLH) proteins, are also needed (Payne et al. 2000; Zhang et al. 2003). *GL1* and *TTG1* have the capacity to limit trichome initiation together with other genes, e.g., *CAPRICE* (*CPC*) and *TRIPTYCHON* (*TRY*) that are known to promote root hair cell fate (Szymanski et al. 2000; Schiefelbein 2003). The gene *GLABROUS2* (*GL2*) quantitatively regulates the frequency of trichome initiation and is involved in determining trichome spacing and root hair development (Ohashi and Oka 2002), but is also involved in the control of seed oil accumulation (Shen et al. 2006). Thus, many genes involved in trichome initiation and spacing also regulate other important phenotypic traits, which may constrain the evolution of genes and traits.

Although current understanding of the genetic basis of trichome initiation and spacing depends to a large degree on studies of mutants in *Arabidopsis*, some information is also available for other plant species. Trichome initiation in cotton seems to be regulated by a similar multimeric complex between MYB, bHLH and WD repeat proteins as in *Arabidopsis* (Wang et al. 2004). Sequence comparisons and functional analyses suggest that genes underlying trichome formation in *Arabidopsis* (and perhaps cotton) may have evolved from duplication and neo-functionalization of genes involved in anthocyanin and flavonoid production, whereas the genes involved in the formation of analogous multicellular trichomes in *Nicotiana* and *Antirrhinum* have a different origin (Serna and Martin 2006). Thus, trichome production appears to have evolved several times suggesting convergent adaptive evolution.

4.2.2 Genetic Basis of Natural Variation in Trichome Production

Both trichome production as such (trichome production vs. glabrousness) and the type of trichomes produced (glandular vs. non-glandular) may vary within species. Many plant species are polymorphic for trichome production with trichome-producing and glabrous morphs. In several species, the frequency of the trichome-producing morph varies geographically, among and within habitats (Björkman and Anderson 1990; Westerberg and Saura 1992; Barnes and Han 1993; St Hilaire and Graves 1999; Arroyo-García et al. 2001; Morrison 2002; Kärkkäinen et al. 2004; Løe et al. 2007). Genetic analyses have shown that the inheritance of the glabrous morph is simple in many species (one gene with a recessive allele for glabrousness; e.g., Westerbergh and Saura 1992; Sharma and Waines 1994; Kärkkäinen and Ågren 2002). In *A. lyrata*, association analysis of phenotypic variation and genotypic variation in candidate genes revealed that the polymorphism in trichome production was caused by variation in *GLABROUS1* (Kivimäki et al. 2007), and sequence analysis suggested that several independent mutations in the same gene cause glabrousness in different natural populations of *A. lyrata* and *A. thaliana* (Kivimäki et al. 2007; Hauser et al. 2001). Polymorphism in trichome morphology (glandular vs. non-glandular trichomes) can also be inherited in a simple Mendelian fashion, as documented for *Datura wrightii* (van Dam et al. 1999).

The number of trichomes produced and trichome density vary genetically within several species. Quantitative genetic studies have indicated ample genetic variation for trichome number and trichome density (Ågren and Schemske 1994; Mauricio and Rausher 1997; Roy et al. 1999; Clauss et al. 2006). QTL-studies have been conducted to reveal the genetic basis of among-population differences in trichome formation. In *A. thaliana*, a QTL-analysis of a cross between two accessions identified a major QTL for trichome number (*REDUCED TRICHOME NUMBER, RTN*) that explained 70% of the variation in number of trichomes produced (Larkin et al. 1996). More recent studies of crosses among other accessions of *A. thaliana* detected additional QTLs (altogether nine QTLs in four different crosses, most of which were shared among populations; Symonds et al. 2005). Preliminary results from QTL-crosses and within-population studies in *A. lyrata* suggest that a limited number of loci contribute to variation in the number of trichomes produced, and that some genes segregate in several populations also in this species (Kärkkäinen et al. unpublished data).

4.3 Induced Trichome Production

4.3.1 Damage-Induced Trichome Production

Both artificial wounding and damage by herbivores can induce an increase in trichome density, and induction of trichome production has been observed in both annual and perennial plants (Table 4.1). The magnitude of the reported increase in trichome density is typically between 25% and 100%, but in some cases as large as 500%–1000% (Table 4.1). The responses often involve changes in trichome density expressed within days or weeks (Baur et al. 1991; Agrawal 1999, 2000; Rautio et al. 2002; Dalin and Björkman 2003). In some woody perennials, the response is delayed and not observed until the year following initial attack (Zvereva et al. 1998; Valkama et al. 2005). Damage can also induce a change in the relative proportions of glandular and non-glandular trichomes (Rautio et al. 2002).

Abiotic stress, such as drought and UV-radiation, may influence trichome formation (Nagata et al. 1999; Höglund and Larsson 2005), and abiotic conditions may modify damage-induced responses in trichome density. For example, Björkman et al. (2008) showed that the increase in trichome production induced by leaf beetle herbivory in the willow *Salix cinerea* is stronger in the shade than under direct sunlight, suggesting that the plants invested more resources into trichome defense when growing in the shade. The results from Table 4.1 also show that damage-induced increases in trichome density tend to be stronger under field conditions than in the lab. All five studies, in which plants increased the density of leaf trichomes by more than 200% following damage, were conducted in the field. However, empirical data are still few, and additional studies of how abiotic conditions affect damage-induced responses in trichome formation are clearly needed.

Damage has not been observed to induce trichome production in all studies (Table 4.1). Only about half of the 15 independent studies listed in Table 4.1, in which the experimental treatment consisted of artificial wounding (e.g., clipping of leaves) reported an increased trichome density. Also, the effect of damage on trichome production may vary depending on the identity of the herbivore; not all herbivorous insects induce a change in trichome formation (Traw and Dawson 2002a, b). Artificial wounding differs from the damage caused by the feeding of insects in several ways, which may reduce the chances of detecting induced responses (Hartley and Lawton 1987; Cipollini et al. 2003). Moreover, the feeding patterns of insects vary, especially among feeding guilds (e.g., chewers, miners and borers), which may affect how plants respond to damage. Further studies are therefore needed to examine the extent to which responses vary with damage type, and whether plants may benefit from responding differently to different kinds of damage.

4.3.2 Hormonal Regulation of Induced Trichome Production

Recent work on *Arabidopsis* indicates that several regulatory networks may influence damage-induced increases in trichome production. Jasmonic acid regulates the systemic expression of chemical defenses (Karban and Baldwin 1997; Schaller and Stintzi this volume), and herbivore damage as well as artificial wounding cause rapid increases in jasmonic acid (Bostock 1999; Reymond et al. 2000). In *A. thaliana*, artificial damage, but also application of jasmonic acid and application of giberellin increases trichome production in new leaves (Traw and Bergelson 2003). Application of salicylic acid, on the other hand, reduces trichome production and inhibits the response to jasmonic acid, which is consistent with negative cross-talk between the jasmonate- and salicylate-dependent pathways in *Arabidopsis* (Traw and Bergelson 2003). Concentrations of salicylic acid typically increase in response to infection by biotrophic pathogens (Gaffney et al. 1993; Ryals et al. 1994), but may also increase in response to damage caused by some herbivores (Stotz et al. 2002; van Poecke and Dicke 2002). This suggests that induction of increased trichome production will be affected by interactions with both herbivores and pathogens, and should vary depending on the identity of the herbivore.

4.4 Leaf Trichomes and Plant Resistance

4.4.1 Effects on Herbivore Behavior and Performance

Trichomes influence insect oviposition and/or feeding in a wide range of insects and other herbivores (Levin 1973). Non-glandular trichomes mainly function as a structural defense against small herbivores. They interfere with the movement of insects and other small arthropods over the plant surface and make it more difficult for insects to access the leaf epidermis underneath for feeding (Southwood 1986). Trichomes are often composed of cellulose and other substances that constitute low nutritional value for the insects (Levin 1973). Insects that need to feed through the trichomes before accessing the leaf epidermis might therefore gain less weight and, ultimately, show increased mortality. Glandular trichomes can be viewed as a combination of a structural and chemical defense. This is because the glandular trichomes release secondary metabolites that can be poisonous or repellent to herbivorous organisms. Stinging trichomes, such as those produced by stinging nettles (*Urtica dioica*), may deter even large herbivores (Pollard and Briggs 1984).

The presence and density of leaf trichomes can influence both host-plant selection behavior and performance (i.e., growth, survival, and fecundity) of herbivorous insects, but can be considered a relatively soft 'weapon' in plant defense compared with many other traits that are lethal to insects. It has been argued that weak defense traits might influence a wide range of herbivorous insects because they are less likely to result in counter-adaptations in insects (Feeny 1976). For example, there are a number of studies suggesting that leaf trichomes reduce the feeding of both generalist and specialist insects (Traw and Dawson 2002a; Agrawal 1999, 2004). Agrawal (1999) showed that two lepidopteran species, one specialist and one generalist, were both negatively influenced by previous damage on wild radish plants, which was correlated with an increased density and number of trichomes on the leaves. Agrawal (2004) also showed that several insect feeding guilds, both leaf chewers and leaf miners, were negatively affected by an increased density of leaf trichomes on milkweed (*Asclepias syriaca*). Thus, non-glandular trichomes might influence the feeding by a wide range of insects and therefore reduce total damage caused by herbivores.

Insects may evolve physiological or behavioral traits that allow them to cope with structural plant defenses. For example, several mirid bugs (Heteroptera: Miridae) have special structures on their legs, which facilitate movement across trichome-covered plant surfaces (Southwood 1986; van Dam and Hare 1998). The leaf beetle *Phratora vulgatissima* has been observed to remove non-glandular trichomes when feeding on leaves of the willow *Salix viminalis* (Dalin et al. 2004). This behavior was not observed when the same beetle species was feeding on *S. cinerea*, which produces shorter and broader trichomes that appeared more difficult for the larvae to handle (Dalin and Björkman 2003). The evolution of mouthparts strong enough to handle structural plant traits is thus one possible mechanism for insects to circumvent structural plant defenses such as leaf trichomes (Levin 1973; Raupp 1985).

4.4.2 Effects on Tritrophic Interactions

Leaf trichomes do not only affect herbivores, but also their natural enemies. This may indirectly affect the intensity of damage caused by herbivores. In theory, the effect on the abundance and effectiveness of natural enemies may be neutral (no effect), negative or positive (Fig. 4.1). If we assume that herbivores are affected negatively by leaf trichomes (Fig 4.1; top graph), a neutral effect on the third trophic level will result in natural enemies having an additive effect on plant damage caused by herbivores and on plant fitness. A negative effect on the third trophic level will act antagonistically. Depending on the relative strength of the negative effects of leaf trichomes on herbivores and natural enemies the correlation between trichome density and damage may vary from negative to positive. Plant damage will be (i) negatively correlated with trichome density if the effect on natural enemies is weaker than that on herbivores (dotted line) (ii) uncorrelated with trichome density if the effect on the natural enemies and herbivores balance (solid line), or (iii) positively correlated with trichome density if the effect on natural enemies is stronger than that on the herbivores (dashed line). A positive effect of leaf trichomes on the third trophic level should result in a synergistic positive effect on plant fitness. From a plant protection perspective, this last scenario would be most desirable, but there seem to be few such examples.

In most empirical studies, the abundance and effectiveness of natural enemies were found to be negatively correlated with the density of plant trichomes (Kauffman and Kennedy 1989; Farrar and Kennedy 1991; Nihoul 1993; Gange 1995; Heinz and Zalom 1996; Romeis et al. 1998, 1999; Krips et al. 1999; Rosenheim et al. 1999; Lovinger et al. 2000; Fordyce and Agrawal 2001; Stavrinides and Skirvin 2003; Mulatu et al. 2006; Olson and Andow 2006). The effectiveness of both predators

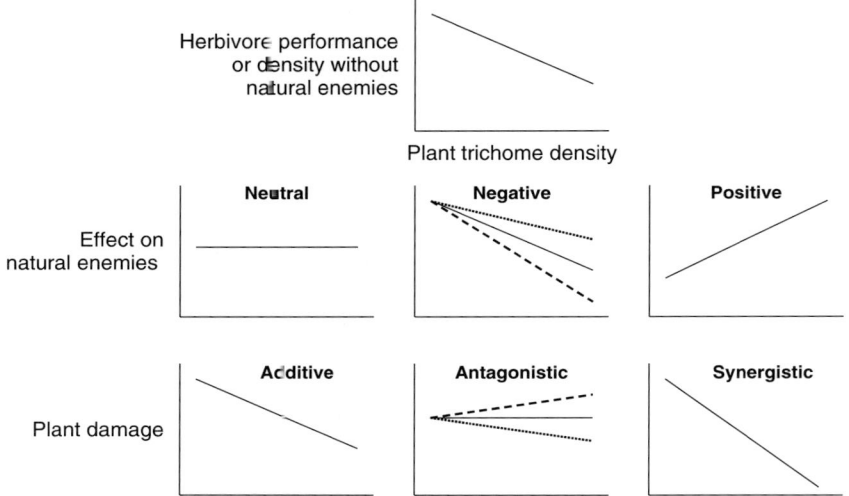

Fig. 4.1 Simplified graphical presentation of how the effect of leaf trichome density on the abundance or effectiveness of predators and parasitoids of herbivores may influence the relationship between trichome density and plant damage caused by herbivores (bottom row of graphs). The natural enemies may not be affected, or may be negatively or positively affected by plant trichome density (middle row of graphs). A basic assumption is that plant trichome density negatively affects herbivore performance or density (top graph), i.e., trichomes have the potential to function as a plant resistance trait (cf. Hare 1992)

(in 9 out of 13 reviewed studies) and parasitoids (in 12 out of 14 reviewed studies) is affected negatively because movement is inhibited and searching time is prolonged. Both non-glandular and glandular plant trichomes can have these effects. Glandular trichomes, in addition, affect natural enemies negatively by (a) the release of repellent or toxic compounds and (b) being sticky and hindering movements. However, the effect of the sticky substances produced by glandular trichomes may sometimes be of little importance in the field as the adhesive effect of the trichome exudates can be negatively affected by dust, wind, and rain (Obrycki and Tauber 1984; Romeis et al. 1999).

In a few cases, the abundance or effectiveness of natural enemies were found to be positively correlated with the density of plant trichomes. For example, Styrsky et al. (2006) recently showed that fire ant predation of herbivores on soybean was higher on isolines of soybean with high trichome density. The suggested mechanism was a functional response by the ants to a higher abundance of caterpillars on pubescent plants. Apple trees producing high density of trichomes had a higher abundance of predatory mites than apple trees with a low density of trichomes, and it has been suggested that this was because pubescent plants capture more pollen and fungal spores that serve as alternative food for the predators (Roda et al. 2003). Lucas and Brodeur (1999) showed that within potato plants, an aphidophagous midge was more abundant on leaves with a high trichome density. The authors suggested that both a decreased risk of intraguild predation by e.g., coccinelids and a higher density of food (i.e., aphids) on leaves with a high trichome density may have contributed to this pattern. Finally, leaf trichome exudates may serve as extrafloral nectar for a scelonid egg parasitoid of squash bugs (Olson and Nechols 1995). Adult females of the parasitoid lived longer when they had access to squash leaves than when only provided with water, and this increase in longevity was enough to allow maximum realized fecundity.

In even fewer cases, the abundance or effectiveness of parasitoids and predators were found not to be correlated with leaf trichome density (e.g., Sutterlin and van Lenteren 1997; Björkman and Ahrné 2005; see also reviews by Kennedy 2003; Simmons and Gurr 2005), possibly because of a reluctance to report 'negative' data.

The density of prey or hosts can affect the impact of leaf trichomes on predators and parasitoids, but relationships appear system-specific. In some studies, an effect of trichomes on natural enemies was detectable only at high prey/host densities (e.g., Sutterlin and van Lenteren 2000) while in other studies such an effect was observed only at low densities (Krips et al. 1999).

4.4.3 Adaptive Significance of Induced Trichome Production in Plants

The adaptive significance of damage-induced increases in trichome production should depend on the benefits and costs associated with this change in phenotype. Plants induced to increase leaf trichome density often receive less damage by insects than non-induced plants (Baur et al. 1991; Agrawal 1999; Dalin and Björkman 2003). This suggests that plants should benefit from induced trichome production as long as the fitness costs associated with increased trichome production do not exceed the benefits of reduced damage. Agrawal (1999) showed that induced wild radish plants (*Raphanus raphanistrum* and *Raphanus sativus*) received less damage by a variety of insects, and produced more seeds and fruits than non-induced plants. Wild radish responded to herbivory by producing new leaves with increased trichome density, but also with a systemic increase in the concentrations of secondary metabolites (Agrawal 1999). Thus, although this study clearly shows that the plants benefited from the induced responses, additional studies are needed to distinguish the effects of increased trichome production on fitness from those of other induced differences in plant quality in this and other species.

Apart from reducing overall damage, induced trichome production may divert damage away from the most valuable parts of the plant and cause insects to feed on other, less valuable leaves, or other non-induced plant individuals nearby. Because plants cannot change the density of trichomes on already existing leaves, an induced increase in trichome production can only be expressed in leaves developing during or subsequent to attack (Traw and Bergelson 2003). Induced increases in trichome production can however be relatively rapid and expressed in new leaves within days or weeks after the initial attack (Baur et al. 1991; Agrawal 1999; Dalin and Björkman 2003). A similarly rapid reduction in trichome production in new leaves when damage ceases indicates that trichomes are costly to produce (Björkman et al. 2008). Additional evidence for costs of trichome production is provided by examples of trade-offs between trichome density and other putative defense traits, such as thorns (Björkman and Anderson 1990). A damage-induced increase in trichome production can lead to variation in trichome density within plants with older, basal leaves having lower trichome density than younger, induced apical leaves. Larvae of the leaf beetle *P. vulgatissima* showed a more dispersed feeding behavior among induced willow leaves with high trichome density at the top of shoots, indicating that the insects were searching for more optimal food within the branches (Dalin and Björkman 2003). Such damage-induced within-branch variation in trichome density may reflect developmental constraints, but may also be advantageous since damage to old basal leaves is likely to be less detrimental to plant fitness than damage to the young apical leaves (cf. Feeny 1976).

4.5 Perspectives

Several plant species respond to herbivory by producing leaves with increased density and number of trichomes. The apparent benefit of this response is to reduce further damage by making it more difficult for insects to feed on the leaves. However, the magnitude of induced trichome production may vary with habitat conditions, and increased trichome density may affect predators and parasitoids negatively, which could reduce the effect on leaf damage. Moreover, trichome production may also affect tolerance to drought and other forms of abiotic stress. Taken together, this suggests that an analysis of the adaptive significance of induced trichome production requires a multidimensional view of the realized fitness (see also chapter by Steppuhn and Baldwin this volume). The effects of induced trichome production on interactions with herbivores should not be examined in isolation, but has to be considered in the context of effects on tritrophic interactions, and the extent to which fitness costs and benefits vary with abiotic conditions in natural populations. More information is also needed on the genetic and hormonal basis of induced trichome production. Such knowledge could be important for the development of sustainable pest control methods, using induced plant responses as a tool to prevent insect pest damage in agricultural systems.

References

Agrawal AA (1999) Induced responses to herbivory in wild radish: effects on several herbivores and plant fitness. Ecology 80:1713–1723

Agrawal AA (2000) Specificity of induced resistance in wild radish: causes and consequences for two specialist and two generalist caterpillars. Oikos 89:493–500

Agrawal AA (2004) Resistance and susceptibility of milkweed: competition, root herbivory, and plant genetic variation. Ecology 85:2118–2133

Agrawal AA, Conner JK, Stinchcombe JR (2004) Evolution of plant resistance and tolerance to frost damage. Ecol Lett 7:1199–1208

Ågren J, Schemske DW (1993) The cost of defense against herbivores: an experimental study of trichome production in *Brassica rapa*. Am Nat 141:338–350

Ågren J, Schemske DW (1994) Evolution of trichome number in a naturalized population of *Brassica rapa*. Am Nat 143:1–13

Arroyo-García R, Martínez-Zapater JM, Fernández Prieto JA, Álvarez-Arbesú R (2001) AFLP evaluation of genetic similarity among laurel populations (*Laurus* L.). Euphytica 122: 155–164

Barnes BV, Han FQ (1993) Phenotypic variation of Chinese aspens and their relationships to similar taxa in Europe and North America. Can J Botany 71:799–815

Baur R, Binder S, Benz G (1991) Non glandular leaf trichomes as short-term inducible defence of the grey alder, *Alnus incana* (L.), against the chrysomelid beetle, *Agelastica alni* L. Oecologia 87:219–226

Benz BW, Martin CE (2006) Foliar trichomes, boundary layers, and gas exchange in the species of epiphytic Tillandsia (Bromeliaceae). J Plant Physiol 163:648–656

Björkman C, Ahrne K (2005) Influence of leaf trichome density on the efficiency of two polyphagous insect predators. Entomol Exp Appl 115:179–186

Björkman C, Anderson DB (1990) Trade-off among antiherbivore defences in a South American blackberry (*Rubus bogotensis*). Oecologia 85:247–249

Bostock RM (1999) Signal conflicts and synergies in induced resistance to multiple attackers. Physiol Mol Plant Pathol 55:99–109

Choinski JS, Wise RR (1999) Leaf growth and development in relation to gas exchange in *Quercus marilandica* Muenchh. J Plant Physiol 154:302–309

Cipollini D, Purrington CB, Bergelson J (2003) Costs of induced responses in plants. Basic Appl Ecol 4:79–85

Clauss MJ, Dietel S, Schubert G, Michell-Olds T (2006) Glucosinolate and trichome defenses in a natural *Arabidopsis lyrata* population. J Chem Ecol 32:2351–2373

Dalin P, Björkman C (2003) Adult beetle grazing induces willow trichome defense against subsequent larval feeding. Oecologia 134:112–118

Dalin P, Björkman C, Eklund K (2004) Leaf beetle grazing does not induce willow trichome defence in the coppicing willow *Salix viminalis*. Agric For Entomol 6:105–109

Duffey SS (1986) Plant glandular trichomes: their partial role in defence against insects. In: Juniper B, Southwood SR (eds) Insects and the plant surface, Arnold, London, pp 151–172

Ehrlinger J (1984) Ecology and physiology of leaf pubescence in North American desert plants. In: Rodriguez E, Healey PL, Mehta I (eds) Biology and chemistry of plant trichomes. Plenum Press, New York, pp 113–132

Farrar RR, Kennedy G (1991) Inhibition of *Telenomus sphingis* an egg parasitoid of *Manduca* spp by trichome 2 tridecanone-based host plant resistance in tomato. Entomol Exp Appl 60:157–166

Feeny PP (1976) Plant apparency and chemical defense. In: Wallace JM, Mansell RL (eds) Biochemical interaction between plants and insects. Plenum Press, New York, pp 1–40

Fordyce JA, Agrawal AA (2001) The role of plant trichomes and caterpillar group size on growth and defence of the pipevine swallowtail *Battus philenor*. J Anim Ecol 70:997–1005

Gaffney T, Frierich L, Vernooij B, Negrotto D, Nye G, Uknes S, Ward E, Kessman H, Ryals J (1993) Requirement of salicylic acid for the induction of systemic acquired resistance. Science 261:754–756

Gange AC (1995) Aphid performance in an alder (*Alnus*) hybrid zone. Ecology 76:2074–2083

Handley R, Ekbom B, Ågren J (2005) Variation in trichome density and resistance against a specialist insect herbivore in natural populations of *Arabidopsis thaliana*. Ecol Entomol 30:284–292

Hare JD (1992) Effects of plant variation on herbivore-natural enemy interactions. In: Fritz RS, Simms EL (eds) Plant resistance to herbivores and pathogens: ecology, evolution and genetics. University of Chicago Press, Chicago, pp 278–298

Hare JD, Elle E (2002) Variable impact of diverse insect herbivores on dimorphic *Datura wrightii*. Ecology 83:2711–2720

Hartley SE, Lawton JH (1987) Effects of different types of damage on the chemistry of birch foliage. Oecologia 74:432–437
Hauser M-T, Harr B, Schlötterer C (2001) Trichome distribution in *Arabidopsis thaliana* and its close relative *Arabidopsis lyrata*: molecular analysis of the candidate gene GLABROUS1. Mol Biol Evol 18:1754–1763
Heinz KM, Zalom FG (1996) Performance of the predator *Delphastus pusillus* on Bemisia resistant and susceptible tomato lines. Entomol Exp Appl 81:345–352
Höglund S, Larsson S (2005) Abiotic induction of susceptibility in insect-resistant willow. Entomol Exp Appl 115:89–96
Hülskamp M, Schnittger A (1998) Spatial regulation of trichome formation *Arabidopsis thaliana*. Semin Cell Dev Biol 9:213–220
Jeffree CE (1986) The cuticle, epicuticular waxes and trichomes of plants, with reference to their structure, function and evolution. In: Juniper B, Southwood SR (eds) Insects and the plant surface. Arnold, London, pp 23–64
Karban R, Baldwin IT (1997) Induced responses to herbivory. The University of Chicago Press, Chicago, p 319
Kärkkäinen K, Ågren J (2002) Genetic basis of trichome production in *Arabidopsis lyrata*. Hereditas 136:219–226
Kärkkäinen K, Løe G, Ågren J (2004) Population structure in *Arabidopsis lyrata*: evidence for divergent selection on trichome production. Evolution 58:2831–2836
Kauffman WC, Kennedy GG (1989) Relationship between trichome density in tomato and parasitism of *Heliothis* spp. (Lepidoptera, Noctuidae) eggs by *Trichogramma* spp (Hymenoptera, Trichogrammatidae). Environ Entomol 18:698–704
Kennedy GG (2003) Tomato, pests, parasitoids, and predators: tritrophic interactions involving the genus *Lycopersicon*. Ann Rev Entomol 48:51–72
Kivimäki M, Kärkkäinen K, Gaudeul M, Løe G, Ågren J (2007) Gene, phenotype and function: GLABROUS1 and resistance to herbivory in natural populations of *Arabidopsis lyrata*. Mol Ecol 16:453–462
Koornneef M (1981) The complex syndrome of *ttg* mutants. Arabid Inf Serv 18:45–51
Krips OE, Kleijn PW, Willems PEL, Gols GJZ, Dicke M (1999) Leaf hairs influence searching efficiency and predation rate of the predatory mite *Phytoseiulus persimilis* (Acari: Phytoseiidae). Exp Appl Acarol 23:119–131
Larkin JC, Young N, Prigge M, Marks MD (1996) The control of trichome spacing and number in *Arabidopsis*. Development 122:997–1005
Levin DA (1973) The role of trichomes in plant defence. Q Rev Biol 48:3–15
Løe G, Toräng P, Gaudeul M, Ågren J (2007) Trichome production and spatiotemporal variation in herbivory in the perennial herb *Arabidopsis lyrata*. Oikos 116:134–142
Lovinger A, Liewehr D, Lamp WO (2000) Glandular trichomes on alfalfa impede searching behavior of the potato leafhopper parasitoid. Biol Control 18:187–192
Lucas E, Brodeur J (1999) Oviposition site selection by the predatory midge *Aphidoletes aphidimyza* (Diptera: Cecidomyiidae). Environ Entomol 28:622–627
Marks MD (1997) Molecular genetic analysis of trichome development in *Arabidopsis*. Annu Rev Plant Physiol Plant Mol Biol 48:137–163
Mauricio R (1998) Costs of resistance to natural enemies in field populations of the annual plant *Arabidopsis thaliana*. Am Nat 151:20–28
Mauricio R, Rausher MD (1997) Experimental manipulations of putative selective agents provides evidence for the role of natural enemies in the evolution of plant defense. Evolution 51:1435–1444
Morrison LW (2002) The geographic distribution of pubescence in the sea daisy, *Borrichia aborescens*, on Bahamian Islands. Glob Ecol Biogeogr 11:247–252
Mulatu B, Applebaum SW, Coll M (2006) Effect of tomato leaf traits on the potato tuber moth and its predominant larval parasitoid: a mechanism for enemy-free space. Biol Control 37:231–236

Mutikainen P, Walls M (1995) Growth, reproduction and defence in nettles: responses to herbivory modified by competition and fertilization. Oecologia 104:487–495

Nagata T, Todoriki S, Hayashi T, Shibata Y, Mori M, Kanegae H, Kikuchi S (1999) Gamma-radiation induces leaf trichome formation in *Arabidopsis*. Plant Physiol 120:113–119

Nihoul P (1993) Do light-intensity, temperature and photoperiod affect the entrapment of mites on glandular hairs of cultivated tomatoes. Exp Appl Acarol 17:709–718

Obrycki JJ, Tauber MJ (1984) Natural enemy activity on glandular pubescent potato plants in the greenhouse: an unreliable predictor of effects in the field. Environ Entomol 13:679–683

Ohashi Y, Oka A (2002) Entopically additive expression of GLABRA2 alters the frequency and spacing of trichome initiation. Plant J 29:359–369

Olson DM, Andow DA (2006) Walking patterns of *Trichogramma nubilale* Ertle & Davis (Hymenoptera; Trichogrammatidae) on various surfaces. Biol Control 39:329–335

Olson DL, Nechols JR (1995) Effects of squash leaf trichome exudates and honey on adult feeding, survival, and fecundity of the squash bug (Heteroptera, Coreidae) egg parasitoid *Gryon pennsylvanicum* (Hymenoptera, Scelionidae). Environ Entomol 24:454–458

Oppenheimer DG, Herman PL, Sivakumaran S, Eschm J, Marks MD (1991) A myb gene required for leaf trichome differentiation in *Arabidopsis* is expressed in stipules. Cell 67:483–493

Payne CT, Zhang F, Lloyd AM (2000) GL3 encodes a bHLH protein that regulates trichome development in *Arabidopsis* through interaction with GL1 and TTG1. Genetics 156:1349–1362

Pollard AJ (1986) Variation in *Cnidoscolus texanus* in relation to herbivory. Oecologia 70:411–413

Pollard AJ, Briggs D (1984) Genecological studies of *Urtica dioica* L. III. Stinging hairs and plant-herbivore interactions. New Phytol 97:507–522

Pullin AS, Gilbert JE (1989) The stinging nettle, *Urtica dioica*, increases trichome density after herbivore and mechanical damage. Oikos 54:275–280

Raupp MJ (1985) Effects of leaf toughness on mandibular wear of the leaf beetle, *Plagiodera versicolora*. Ecol Entomol 10:73–79

Rautio P, Markkola A, Martel J, Tuomi J, Härmä E, Kuikka K, Siitonen A, Riesco IL, Roitto M (2002) Developmental plasticity in birch leaves: defoliation causes shift from glandular to nonglandular trichomes. Oikos 98:437–446

Reymond P, Weber H, Damond M, Farmer EE (2000) Differential gene expression in response to mechanical wounding and insect feeding in *Arabidopsis*. Plant Cell 12:707–719

Roda A, Nyrop J, English-Loeb G (2003) Pubescence mediates the abundance of non-prey food and the density of the predatory mite *Typhlodromus pyri*. Exp Appl Acarol 29:193–211

Romeis J, Shanower TG, Zebitz CPW (1998) Physical and chemical plant characters inhibiting the searching behaviour of *Trichogramma chilonis*. Entomol Exp Appl 87:275–284

Romeis J, Shanower TG, Zebitz CPW (1999) *Trichogramma* egg parasitism of *Helicoverpa armigera* on pigonpea and sorghum in southern India. Entomol Exp Appl 90:69–81

Rosenheim JA, Limburg DD, Colfer RG (1999) Impact of generalist predators on a biological control agent, *Chrysoperla carnea*: direct observations. Ecol Appl 9:409–417

Rowell-Rahier M, Pasteels JM (1982) The significance of salicin for a Salix-feeder, *Phratora (Phyllodecta) vitellinae*. In: Visser JH, Mink AK (eds) Proceedings from the 5th international symposium on insect-plant relationships. Pudoc, Wageningen, pp 73–79

Roy BA, Stanton ML, Eppley SM (1999) Effects of environmental stress on leaf hair density and consequences for selection. J Evol Biol 12:1089–1103

Ryals J, Uknes S, Ward E (1994) Systemic acquired resistance. Plant Physiol 104:1109–1112

Schiefelbein J (2003) Cell-fate specification in the epidermis: a common patterning mechanism in the root and shoot. Curr Opin Plant Biol 6:74–78

Serna L, Martin C (2006) Trichomes: different regulatory networks lead to convergent structures. Trends Plant Sci 11:1360–1385

Sharma HC, Waines JG (1994) Inheritance of leaf pubescence in diploid wheat. J Hered 85:286–288

Shen B, Sinkevicius KW, Selinger DA, Tarczynski MC (2006) The homeobox gene GLABRA2 affects seed oil content in *Arabidopsis*. Plant Mol Biol 60:377–387

Simmons AT, Gurr GM (2005) Trichomes of *Lycopersicon* species and their hybrids: effects on pests and natural enemies. Agric For Entomol 48:51–72

Skaltsa H, Verykokidou E, Harvala C, Krabourniotis G, Manetas Y (1994) UV-protective potential and flavonoid content of leaf hairs of *Quercus ilex*. Phytochemistry 37:987–990

Southwood SR (1986) Plant surfaces and insects – an overview. In: Juniper B, Southwood SR (eds) Insects and the plant surface. Arnold, London, pp 1–22

Stavrinides MC, Skirvin DJ (2003) The effect of chrysanthemum leaf trichome density and prey spatial distribution on predation of *Tetranychus urticae* (Acari: Tetranychidae) by *Phytoseiulus persimilis* (Acari: Phytoseiidae). Bull Entomol Res 93:343–350

St Hilaire R, Graves WR (1999) Foliar traits of sugar and black maples near 43 degrees N latitude in the eastern and central United States. J Am Soc Hort Sci 124:605–611

Stotz HU, Koch T, Biedermann A, Weniger K, Boland W, Mitchell-Olds T (2002) Evidence for regulation of resistance in *Arabidopsis* to Egyptian cotton worm by salicylic and jasmonic acid signaling pathways. Planta 214:648–652

Styrsky JD, Kaplan I, Eubanks MD (2006) Plant trichomes indirectly enhance tritrophic interactions involving a generalist predator, the red imported fire ant. Biol Control 36:375–384

Sutterlin S, van Lenteren JC (1997) Influence of hairiness of *Gerbera jamesonii* leave on the searching efficiency of the parasitoid *Encarsia formosa*. Biol Control 9:157–165

Sutterlin S, van Lenteren JC (2000) Pre- and post-landing response of the parasitoid *Encarsia formosa* to whitefly hosts on *Gerbera jamesonii*. Entomol Exp Appl 96:299–307

Symonds VV, Godoy AV, Alconada T, Botto JF, Juenger TE, Casal JJ, Lloyd AM (2005) Mapping quantitative trait loci in multiple populations of *Arabidopsis thaliana* identifies natural allelic variation for trichome density. Genetics 169:1649–1658

Szymanski DB, Lloyd AM, Marks MD (2000) Progress in the molecular genetic analysis of trichome initiation and morphogenesis in *Arabidopsis*. Trends Plant Sci 5:214–219

Traw BM (2002) Is induction response negatively correlated with constitutive resistance in black mustard? Evolution 56:2116–2205

Traw BM, Bergelson J (2003) Interactive effects of jasmonic acid, salicylic acid, and gibberellin on induction of trichomes in *Arabidopsis*. Plant Physiol 133:1367–1375

Traw BM, Dawson TE (2002a) Reduced performance of two specialist herbivores (Lepidoptera: Pieridae, Coleoptera: Chrysomelidae) on new leaves of damaged black mustard plants. Environ Entomol 31:714–722

Traw BM, Dawson TE (2002b) Differential induction of trichomes by three herbivores of black mustard. Oecologia 131:526–532

Valkama E, Koricheva J, Ossipov V, Ossipova S, Haukioja E, Pihlaja K (2005) Delayed induced responses of birch glandular trichomes and leaf surface lipophilic compounds to mechanical defoliation and simulated winter browsing. Oecologia 146:385–393

Valverde PL. Fornoni J, Nunez-Farfan J (2001) Defensive role of leaf trichomes in resistance to herbivorous insects in *Datura stramonium*. J Evol Biol 14:424–432

Walker AR, Davidson PA, Bolognesi-Winfield AJ, James CM, Srinivasan N, Blundell TL, Esch JJ, Marks MD, Gray JC (1999) The TRANSPARENT TEST GLABRA1 locus, which regulates trichome differentiation and anthocyanin biosynthesis in *Arabidopsis*, encodes a WD40 repeat protein. Plant Cell 11:1337–1349

van Dam NM, Hare JD (1998) Differences in distribution and performance of two sap-sucking herbivores on glandular and non-glandular *Datura wrightii*. Ecol Entomol 23:22–32

van Dam NM, Hare JD, Elle E (1999) Inheritance and distribution of trichome phenotypes in *Datura wrightii*. J Hered 90:220–227

van Poecke RMP, Dicke M (2002) Induced parasitoid attraction by *Arabidopsis thaliana*: involvement of the octadecanoid and the salicylic acid pathway. J Exp Biol 53:1793–1799

Wang S, Wang JW, Yu N, Li CH, Luo B, Gou JY, Wang LJ, Chen XY (2004) Control of plant trichome development by a cotton fiber MYB gene. Plant Cell 16:2323–2334

Werker E (2000) Trichome diversity and development. Adv Bot Res 31:1–35

Westerberg A, Saura A (1992) The effect of serpentine and the population structure of *Silene dioica*. Evolution 46:1537–1548

Zhang F, Gonzales A, Zhao MZ, Payne CT, Lloyd A (2003) A network of redundant bHLH proteins functions in all TTG1-dependent pathways of *Arabidopsis*. Development 130:4859–4869

Zvereva EL, Kozlov MV, Niemelä P (1998) Effects of leaf pubescence in *Salix borealis* on host-plant choice and feeding behaviour of the leaf beetle, *Melasoma lapponica*. Entomol Exp Appl 89:297–303

Chapter 5
Resistance at the Plant Cuticle

Caroline Müller

Resistance against herbivores could potentially start at the plant cuticle, the first contact zone between an approaching organism and the plant. Due to its physico-mechanical and chemical properties the cuticle comprises several important physiological and ecological functions. Chemical compounds of or on the cuticle can directly act deterrent or even toxic to herbivores. The physico-mechanical cuticle characteristics influence locomotion and attachment of herbivorous insects but also of antagonists, thereby affecting direct and indirect constitutive defense properties. Little is known on induced resistance with regard to the plant cuticle. Induced direct defense could act through a production of waxes or biosynthesis of secondary metabolites that are deposited at the plant cuticle after herbivore attack or which are induced by compounds of oviduct secretion or footsteps. Furthermore, the specific cuticle properties influence the sorption of compounds originating from herbivorous species that could directly inhibit conspecifics, or affect the searching behavior of antagonists. An indirect induced defense can also be mediated by similarities between cuticular compound profiles between plants and herbivores that influence the palatability by predators. Finally, optical properties can change due to herbivory and affect the host finding behavior of predators or parasitoids. The deposition and behavior of agrochemicals used to increase crop resistance is highly determined by various cuticle characteristics. Different aspects of the role of the plant cuticle in resistance are outlined in this chapter, as well as potential future studies in this field.

5.1 Introduction

The plant cuticle secludes all above-ground plant tissue from the environment and forms an important protection barrier against water loss by transpiration but also against biotic harms from outside. For any approaching organism this outer plant

---

C. Müller
Department of Chemical Ecology, University of Bielefeld, D-33615 Bielefeld, Germany
e-mail: caroline.mueller@uni-bielefeld.de

surface thus builds the first contact zone. Before herbivorous insects can damage the plant tissue, the cuticle can highly efficiently inhibit attachment, locomotion, feeding, or oviposition. Thereby it can be generally prevented that (not-adapted) insects will come into contact with diverse plant primary and secondary metabolites, and that damage could activate enzyme activities, thereby evoking induced defense responses of the plant.

The plant cuticle is well characterized by its physico-chemical and morphological features. It forms a continuous translucent film of polymeric lipids and soluble waxes covering all aerial primary tissues of higher plants which is only interrupted by stomata. The cuticle consists of three main parts that are deposited at the outermost extracellular matrix of the epidermis: the cuticle layer forms the basis that embeds the cuticle proper which is covered by epicuticular waxes. The biopolyester cutin is composed of ω-hydroxy and hydroxyepoxy fatty acids and offers a densely networked structural support. The cuticular waxes form a complex mixture of very long-chain fatty acids and their derivatives, and additionally often various cyclic compounds such as pentacyclic triterpenoids (Riederer and Markstädter 1996; Müller and Riederer 2005). The specific chemical composition leads to a specific ultrastructure of the cuticle, which can vary highly between species. Barthlott et al. (1993) classified a total of 23 wax types, which differ in thickness and in presence and shape of local wax projections, among other features.

The physico-chemical properties of the cuticle influence features of the plant surface such as color, texture and presence of phytochemicals (volatile or non-volatile; Müller and Riederer 2005) that are important determinants for host plant acceptance by herbivorous insects (Powell et al. 1999). Generally, the cuticle will form a constitutive resistance barrier against herbivores that lack adaptations to a specific host plant surface. However, its fundamental role in plant-insect interactions is underrepresented in the scientific literature. Many more studies focused on the role of secondary plant compounds for insect acceptance or rejection (Schoonhoven et al. 2006) that are, however, often not present on the outer surface but only in inner plant tissue (Müller and Riederer 2005). Studies on resistance effects of the plant cuticle were mainly driven by investigations of crop pests (e.g., Eigenbrode and Shelton 1990; Stoner 1990; Cervantes et al. 2002). Comparisons between acceptance of plant genotypes or varieties with different levels of wax covers helped to elucidate the role of constitutive resistance parameters of plant individuals (Yang et al. 1991; Eigenbrode and Pillai 1998; Cervantes et al. 2002). Furthermore, indirect constitutive defense can be implemented at the plant cuticle, as the cuticle does not only determine features for herbivorous insects but also for predators or parasitoids, specifically by influencing their locomotion abilities (Eigenbrode 2004). The knowledge of the role of the plant cuticle in induced resistance on the other hand is extremely poor, and offers a comprehensive interesting field for future studies.

This chapter will start with a description of the physico-chemical nature of the plant cuticle which is essential to explain features that determine the role of the cuticle in interactions with herbivores. In the following, constitutive resistance characteristics of the plant against herbivores will be highlighted which can be divided in direct resistance and indirect, the latter acting by influencing the efficiency of predators or parasitoids of the herbivores. The underlying mechanisms can be based on physical and/or chemical features of the cuticle. Few examples will be given where the plant cuticle has been found to be involved in induced resistance, either direct or indirect. As agrochemicals need to intrude the plant tissue through the cuticle and thereby might change the plant surface considerably, the impact of crop protection treatments on plant-insect interactions will be shortly discussed. The chapter will end with an outlook, suggesting some future studies that might be relevant to gain a deeper understanding on the role of the plant cuticle as a plant resistance factor.

5.2 Characteristics of the Cuticle

5.2.1 Physico-Chemical Features

The plant cuticle forms a hydrophobic outer barrier of nearly all above-ground parts of terrestrial plants. Its thickness can vary from 0.01 to 200 μm (Nawrath 2006). The cuticle is a bilayered membrane. At first, the cuticle proper consisting of soluble and polymeric lipids is synthesized at the outside of the primary cell wall. During organ development, a zone of the primary cell wall and later of the secondary cell wall becomes impregnated with cutin, forming the cuticle layer beneath the cuticle proper which can vary in thickness (Jeffree 1996, 2006; Bargel et al. 2006). The cutin matrix (intracuticular) is composed predominantly of C16 ω-hydroxy and unsaturated C18 hydroxyepoxy fatty acid monomers (Holloway 1982) and contains to a lesser extent also glycerol and phenolic compounds (Nawrath 2006). Furthermore, various cell wall polysaccharides are present in the cuticle layer (Holloway 1982). Cuticular waxes are typically comprised of a complex mixture of aliphatic long-chain hydrocarbons together with cyclic compounds. The major classes of cuticular waxes are n-alkanes (C_{17}–C_{35}), smaller portions of iso- and anteiso homologues, primary alcohols (C_{22}–C_{32}), fatty acids (C_{12}–C_{32}), aldehydes (C_{22}–C_{36}), secondary alcohols (C_{21}–C_{33}) β-diketones (C_{27}–C_{35}), and n-alkyl esters (C_{36}–C_{72}) (Walton 1990; Jeffree 2006). Additionally, various triterpenoids can be found, most of them being pentacyclic triterpenoids (Walton 1990; Jetter et al. 2006). For a thorough review of the composition of plant cuticular waxes see Jetter et al. (2006). The intracuticular waxes of the cuticle proper mainly form well-packed subdomains, while the epicuticular waxes are deposited as film or form crystals (Müller and Riederer 2005). Intra- and epicuticular waxes can be sampled separately (Ensikat et al. 2000) and have been shown to have distinct qualitative and quantitative compositions of wax compounds (Jetter et al. 2000, 2006). After removal or abrasion of waxes, a constant regeneration usually follows (Neinhuis et al. 2001). While wax films appear to be ubiquitous, thick wax layers are rare. The crystals can vary from one to tens of micrometers in size and their shape is determined by the qualitative and quantitative composition of wax compounds (Bargel et al. 2006; Jeffree 2006). Among the crystalloids, platelets, tubules, stomatal wax chimneys, rodlets, needles, ribbons and others can be found (Jeffree 1986; Barthlott et al. 1998), which determine the microrelief of the plant surface.

The microrelief influences several features of the plant surface as e.g., the wettability. While smooth surfaces, lacking wax crystals, are more or less wettable, rough surfaces offer permanent water repellency (Neinhuis and Barthlott 1998). The roughness can be caused by different microstructures such as wax crystals and cuticular folds, but also by trichomes (Neinhuis and Barthlott 1997). A rough surface helps to keep the cuticle dry and clean, as mechanical adhesion of small particles is impaired and water droplets running off the surface take along particles (Barthlott and Neinhus 1997). In the plant-insect literature, the different microreliefs are usually simplified and only the terms 'glossy' and 'waxy' are used. Glossy plants have usually a reduced complexity of the microstructure and reduced amounts of epicuticular lipids in comparison to normal wax phenotypes (Eigenbrode and Espelie 1995). Generally, glossy plants allow an easier attachment for insects than normal or extremely waxy surfaces (Eigenbrode and Pillai 1998; Gorb and Gorb 2002; Eigenbrode 2004) but these features can also differ depending on the specific plant-insect system.

Furthermore, the microrelief of the plant cuticle strongly influences its optical properties (Pfündel et al. 2006). On the one hand radiation is absorbed at the plant surface, but on the other hand photons are also redirected at the transition of phases with different refractive indices causing specular reflectance (Vogelmann 1993; Barnes and Cardoso-Vilhera 1996). Electronic absorption of radiation depends on the presence of linear or cyclic compounds including several double bonds (Cockell and Knowland 1999) such as phenolics, hydroxycinnamic acids, colorless flavonoids or anthocyanins (Krauss et al. 1997; Olsson et al. 1998; Harborne and Williams 2000; Markstädter et al. 2001). Their primary role is the protection against damage caused by ultraviolet radiation, but their qualitative and quantitative occurrence in the cuticle also determines its optical behavior. The specular reflectance of the radiation depends on the angle of incident radiation and on the viewing angle of the observer, e.g., an approaching insect (Woolley 1971). Upper surfaces of glabrous leaves exhibit reflectance values in the UV-A spectral range (315–400nm) between four and 10% (Caldwell 1968; Bullas-Appleton et al. 2004), while glaucous leaf surfaces with a blue–green appearance reflect radiation up to 25% (Pfündel et al. 2006). In several succulent plants with powdery epicuticular waxes a reflectance of UV up to 80%, and of visible radiation (400–700nm) up to 70% can be reached on the upper leaf surfaces (Mulroy 1979). Furthermore, trichomes can influence the reflectance of radiation to highly variable degrees (Ehleringer and Björkman 1978; Robberecht et al. 1980; Baldiri et al. 1997). Combination of absorption and specific wavelength reflectance properties can finally influence the colored appearance of the cuticle which can be used as cue for host plant finding of herbivorous insects (Prokopy and Owens 1983; Prokopy et al. 1983; Bullas-Appleton et al. 2004).

5.2.2 Further Chemical Characteristics

Next to the typical wax compounds or in combination with these, other plant metabolites whose presence might be important in plant-insect interactions were identified in the cuticle over the last decades. Wax fatty acids can esterify with phenylpropanoids (cinnamyl alcohol), diterpenes (phytol) (Griffiths et al. 1999), benzyl and phenylethyl alcohol (Gülz and Marner 1986) or alkyl benzoates (Gülz et al. 1987). Very long-chain phenylpropyl and phenylbutyl esters have been found in needle cuticles of *Taxus baccata* L. (Taxaceae) (Jetter et al. 2002). Tocopherols have been reported in the waxes of a number of species (Griffiths et al. 2000; Gülz et al. 1992).

Soluble carbohydrates and amino acids can leak to the leaf surface (Fiala et al. 1990; Soldaat et al. 1996; Derridj et al. 1996a; Leveau 2004) and offer information about the quality of the plant as a food source available to insects (Derridj et al. 1996b). However, the leaking of carbohydrates onto the surface is size-limited, since large carbohydrates will not be able to pass the cuticle (Popp et al. 2005). Compounds located in trichomes can be exuded or released onto the plant surface when the cuticular sac of the trichomes matures. Diterpenoids, sesquiterpenes and flavonoids have been reported to be deposited on the leaf surface via this way (Severson et al. 1984; Talley et al. 2002; Roda et al. 2003; Bargel et al. 2004). But flavonoids and other phenolic compounds are also found on the surface of diverse higher plant families independent of trichomes (Wollenweber et al. 2004, 2005a, b). In ferns, flavonoids can form thick farinose layers on the fronds (Wollenweber 1989; Roitman et al. 1993).

Various secondary metabolites, which can be important infochemicals in plant-insect interactions, were described to be present in leaf surface extracts, such as pyrrolicidine alkaloids in Asteraceae (Vrieling and Derridj 2003), glucosinolates in Brassicaceae (Städler 1992; Hopkins et al. 1997; Griffiths et al. 2001; van Loon et al. 1992), terpenoids (Hubbell et al. 1984) but also phytosterols (Holloway 1971; Hubbell et al. 1984; Avato et al. 1990; Shepherd et al. 1999a). However, these findings are discussed controversially, specifically when the metabolites are rather non volatile, polar and/or charged (Müller and Riederer 2005). Surface extracts prepared with organic solvents, used in several of the mentioned studies for isolation of 'cuticle compounds', always hold the risk that compounds are eluted from underlying plant tissue and do not originate from the cuticle (Riederer and Markstädter 1996; Jetter et al. 2000, 2006). Soldaat et al. (1996) did not find pyrrolicidine alkaloids in surfaces of *Senecio* (Asteraceae) and the only alkaloids that have been confirmed as cuticle constituents with more sophisticated methods are these of fruits of *Papaver* (Papaveraceae) (Jetter and Riederer 1996). Also, the often reported presence of glucosinolates on the plant cuticle is not certain (Reifenrath et al. 2005). Keeping plants in darkness before solvent extraction lowers the risk of dissolving compounds from the inner leaf through stomata (Reifenrath et al. 2005). Preferably, extractions with organic solvents should be avoided altogether and replaced by different adhesion methods using frozen water or glycerol (Ensikat et al. 2000; Riedel et al. 2003), cellulose acetate (Müller and Hilker 2001), or gum arabic (Jetter et al. 2000; Riedel et al. 2003; Reifenrath et al. 2005), to mechanically detach the epicuticular waxes from the plant surface for chemical analyses of compounds present therein (Jetter et al. 2006; Müller 2006).

Lipophilic uncharged organic chemicals, specifically volatiles, can easily accumulate in the cuticle and be sorbed onto plant surfaces whereby the cuticle can function as fixative (Müller and Riederer 2005). Volatile compounds could reach the surface via transport from the inner leaf through the cuticle (depending on the molecule's characteristics) or by sorption from the environment. These volatiles can function as infochemicals for insects (Lambdon et al. 1998).

5.2.3 Variation of the Plant Cuticle

The plant cuticle may vary in microstructure, thickness and chemical composition within an individual due to different endogenous and exogenous factors. Adaxial and abaxial surfaces can differ dramatically in chemical composition and ultrastructure (Netting and von Wettstein 1973; Holloway et al. 1977; Baum et al. 1980) which in turn influences feeding and oviposition by herbivores (Müller 1999; Müller and Hilker 2001). Age, developmental stage and organ are important endogenous determinants of leaf and fruit cuticle characteristics (Freeman et al. 1979a, b; Atkin and Hamilton 1982; Edwards 1982; Riederer and Schönherr 1988; Bianchi et al. 1989; Neinhuis and Barthlott 1998; Jetter and Schäffer 2001; Ficke et al. 2004). During senescence, reflectance and scattering patterns change as well, influencing the optical properties of the leaf surface (Major et al. 1993), and thereby offering visual information about the plant quality. Climate conditions in a certain habitat such as season (Gülz and Boor 1992; Neinhuis and Barthlott 1998; Jenks et al. 2002), or temperature (Gülz and Boor 1992; Riederer and Schreiber 2001) are important exogenous factors affecting cuticle characteristics, as well as erosion by wind and rain (Baker 1974; Baker and Hunt 1986; Rogge et al. 1993), or certain gases (Percy et al. 2002). Endogenous and exogenous factors might interact, as for example the presence of essential oils together with high temperature may lead to melting of wax platelets (Gülz and Boor 1992). Changes of the plant cuticle can be also caused by visiting insects (see Section 5.4) or by pathogens.

5.3 Constitutive Resistance

5.3.1 Physico-Mechanical Direct Defense

Due to the specific morphology of the surface epicuticular waxes, the plant cuticle forms a constitutive resistance barrier against many visiting insects. The protection of flowers from undesired visitors has already early been described by Kerner (1879), and is realized by 'antirobbing' devices that cover the pedicles. Leaves of *Ilex aquifolium* (Aquifoliaceae) are so smooth that many insects cannot grasp them (Jolivet 1988) Attachment and locomotion of insects' eggs and tarsae, respectively, can be strongly inhibited depending on the amount and structure of epicuticular waxes (Peter and Shanower 2001). The movement and feeding of neonate larvae in particular is affected by the specific constitution of the cuticle

(Zalucki et al. 2002). The less complex epicuticular microstructure of glossy surfaces can lead to ten times lower survival of *Plutella xylostella* L. larvae (Lepidoptera: Plutellidae) compared to survival on waxy genotypes (Eigenbrode and Shelton 1990). Initiation of feeding and walking speed of these larvae are reduced on glossy as compared to waxy genotypes. *Sorghum* (Poaceae) genotypes with a smooth amorphous wax layer and sparse wax crystals are more resistant to the sorghum shoot fly *Atherigona soccata* (Rondani) (Diptera: Muscidae) than other genotypes (Nwanze et al. 1992). Thus, glossy surfaces can provide an important resistance factor against herbivores (Stoner 1990). Other resistance mechanisms might involve pleiotropic effects of the glossy trait on further plant characteristics, such as e.g., water stress (Cole and Riggal 1992; Rutledge and Eigenbrode 2003). Overall, agricultural crop species with glossy surfaces and reduced wax blooms are more often resistant to insect pests within the reported occurrences than species or varieties with waxy surfaces, whereby the resistance was mainly investigated against neonate caterpillars or smaller insects such as aphids or thrips (Eigenbrode 2004). The increased ability of these insects to walk on waxy surfaces might have evolved as part of the host plant specialization process.

On the other hand, several insects have trouble adhering properly to waxy surfaces, as e.g., flea beetles (Coleoptera: Chrysomelidae) on waxy cabbage genotypes (Bodnaryk 1992). This results in reduced feeding on such plants. Waxblooms also prevent attachment of the mustard leaf beetle, *Phaedon cochleariae* (F.) (Coleoptera: Chrysomelidae), whose traction is much better on glossy leaf surfaces (Stork 1980). Leaves can change their microstructure when maturing (Edwards 1982), turning from a more diverse array of epicuticular waxes to a glossy type, as e.g., in various eucalypt species. The young eucalypt leaves are more resistant to attack by the tortoise beetle *Paropsis charybdis* Stål (Coleoptera: Chrysomelidae) (Edwards 1982) and to oviposition and settling by the psyllid *Ctenarytaina spatulata* Taylor (Homoptera: Psyllidae; Brennan et al. 2001), whereas they are more susceptible to the autumn gum moth, *Mnesampela privata* (Guenée) (Lepidoptera: Geometridae) (Steinbauer et al. 2004). Normal or waxy surfaces of leaves and fruits can receive more (Steinbauer et al. 2004) or fewer eggs than glossy phenotypes (Neuenschwander et al. 1985; Uematsu and Sakanoshita 1989; Blua et al. 1995), the latter leading to an increased pest resistance. Different visual appearances of glossy and waxy phenotypes may contribute to acceptance by insects that use leaf color to distinguish between host plants (Prokopy et al. 1983; see Section 5.2.1).

Mechanical removal of plant epicuticular waxes often led to a higher susceptibility to herbivores, shown e.g., for several psyllids (Brennan and Weinbaum 2001), chrysomelid beetles (Edwards 1982; Reifenrath et al. 2005; Müller 2006), and moths (Hagley et al. 1980; Uematsu and Sakanoshita 1989). Cellulose acetate-stripped oat leaves were penetrated earlier by aphids than intact oat leaves (Powell et al. 1999). In contrast, monophagous herbivores might need intact epicuticular layers for host detection (Müller and Hilker 2001). The abundance of normal and reduced wax blooms can finally have effects on arthropod diversity (Chang et al. 2004b).

5.3.2 Chemical Direct Defense

The chemical composition of the cuticular lipids can cause resistance against several insect herbivores (Eigenbrode and Espelie 1995) through perception of stimuli by highly adapted sensory structures (Städler 1984, 1986) followed by avoidance behavior. The antixenotic activity of hexane and chloroform surface extracts of some grasses is higher than that of others against the Hessian fly (Diptera: Cecidomyiidae; Foster and Harris 1992; Cervantes et al. 2002). Applying epicuticular waxes of resistant on susceptible *Rhododendron* genotypes (Ericaceae) caused a reduction of feeding and oviposition of *Stephanitis pyrioides* Scott (Heteroptera: Tingidae; Chappell and Robacker 2006). Meridic diet containing cuticular lipid extracts from foliage of four maize genotypes (Poaceae) led to reduced growth of *Spodoptera frugiperda* (J. E. Smith) (Lepidoptera: Noctuidae; Yang et al. 1991). Cultivars of *Medicago sativa* that contain 50% more wax esters are more resistant to the aphid *Therioaphis maculata* (Buckton) (Homoptera: Aphididae; Bergman et al 1991). Deterrent effects of surfaces waxes of *Sorghum* seedlings against *Locusta migratoria* L. (Orthoptera: Acrididae) are attributable to *p*-hydroxybenzaldehyde, *n*-alkane, and ester fractions (Woodhead 1983). Addition of *n*-alkane-1-ols, specifically C_{24} and C_{25} alcohols, to cabbage wax components deposited on a glass slide reduced the time biting, and increased the time walking of neonate *P. xylostella* (Eigenbrode and Pillai 1998). The 3-O-fatty acid ester derivatives (C_{12}–C_{18}) of two pentacyclic triterpenic acids, ursolic acid, and oleanolic acid, together with the parent acids, were found to exhibit potent antifeedant activities against the agricultural pest *Spodoptera litura* F. (Mallavadhani et al. 2003). The triterpenols α- and β-amyrin inhibit feeding of *L. migratoria* and reduce acceptance behavior of *P. xylostella* (Eigenbrode and Espelie 1995). Cultivars of *Rubus idaeus* L. (Rosaceae) with higher levels of cycloartenyl esters, α-amyryl esters and other compounds present in the cuticular waxes are more resistant to the aphid *Amphorophora idaei* Börner (Homoptera: Aphididae; Shepherd et al. 1999b). Sesquiterpenes located in trichomes of *Lycopersicon hirsutum* f. *hirsutum* (Solanaceae) are toxic to neonate larvae of *Leptinotarsa decemlineata* Say (Coleoptera: Chrysomelidae; Carter et al. 1989). Adults of the Colorado potato beetle abandon certain wild *Solanum* species by falling to the ground after a few minutes of contact with the plant surface. While this behavior is induced by trichomes, non-volatile fractions of leaf surface extracts were identified as phagodeterrents against the beetle (Yencho et al. 1994; Pelletier and Dutheil 2006). However, these extracts might contain also other compounds than cuticular waxes (see also Section 5.2.2).

Toxic or deterrent secondary compounds can be concentrated to higher levels in glossy as compared to waxy surfaces and thus increase plant resistance (Cole and Riggal 1992). On the other hand, epicuticular waxes might also cover secondary compounds that act as feeding or oviposition stimulants and thereby increase plant resistance features (Reifenrath et al. 2005).

5.3.3 Constitutive Indirect Defense

An indirect constitutive defense of the plant against herbivorous insects can be realized by impacting the locomotion and attachment abilities of predators or parasitoids on the cuticle. Differences of leaf micro morphology and chemistry lead to different predation efficiencies of carnivores. The efficiency of predators increased several-fold on glossy genotypes compared to waxy (Eigenbrode and Espelie 1995; Eigenbrode et al. 1999). Specifically, coccinnelid larvae and beetles often fall off plants, or forage aphids less efficiently on plants with a pronounced epicuticular wax bloom (Grevstad and Klepetka 1992; Shah 1982; Rutledge et al. 2003). Thus, glossy phenotypes improve effectiveness of some predators leading indirectly to a higher resistance against herbivores. Selection and breeding of such phenotypes has been used especially for management of cabbage pests (Eigenbrode 2004).

Similar effects could be observed with parasitoids. Whereas *Diaeratiella rapae* (M'Intosh) wasps (Hymenoptera: Braconidae) foraged more slowly, groomed with a higher frequency, and fell off from leaves of cauliflower varieties with heavier wax blooms more frequently, their foraging efficacy on cabbage aphids improved highly when epicuticular waxes were removed (Gentry and Barbosa 2006). Wasps of *Aphidius ervi* Haliday (Hymenoptera: Aphidiidae) spent more time actively foraging on pea plants with reduced wax loads, which may contribute to a higher parasitism rate of the pea aphid *Acyrthosiphon pisum* Harris (Homoptera: Aphidae) observed on those varieties (Chang et al. 2004a). However, predators and parasitoids do not seem to prefer plants with reduced wax blooms over waxy leaves, they just show higher success rates of predation and parasitation on these plants due to a much better attachment on glossy leaves (Eigenbrode and Kabalo 1999; Eigenbrode et al. 1999; Eigenbrode 2004).

In contrast, several predator species can actually forage more efficiently on waxy surfaces (Eigenbrode 2004). A few syrphid species were shown to oviposit preferentially on plants with heavy epicuticular wax blooms, thereby increasing the plant resistance against herbivores (Rutledge et al. 2003). In some predator species, different developmental stages have contrasting abilities to cope with the surface structure of the plant. Parasitoids were also shown to differ in their ability to walk and forage on glossy versus waxy plant cuticles (Eigenbrode 2004). Further structures on the surface, such as trichomes, pronounced leaf veins, or smaller diameter stems or tendrils might help carnivorous insects to find a better grip to the plant (Kareiva and Sahakian 1990; Federle et al. 1997; Olson and Andow 2006).

Finally, in ant-plant mutualistic relationships the physico-chemical features of epicuticular waxes play an important role for attachment abilities of specialized and non-specialized ant species. A stable association with an efficient ant species can increase the resistance of the plant against herbivores which are foraged by the ants. Different *Macaranga* plant-ant associations, and the involvement of the particular surface characteristics of the cuticle in these interactions have been studied in detail by Federle and coworkers (Federle et al. 1997, 2000; Markstädter et al. 2000; Federle and Bruening 2005).

Whereas constitutive direct defense has been shown to be influenced by physico-mechanical characteristics of the cuticle (Section 5.3.1) and by the presence of certain chemical compounds on the surface (Section 5.3.2), indirect constitutive defense mechanisms are mainly driven by physico-mechanical properties.

5.4 Induced Resistance

5.4.1 Induced Direct Defense

Plants can respond with a number of induced defense responses to herbivore feeding. In many plant species, changes in primary composition, as e.g., the C/N ratio (Ohgushi 2005), as well as changes of secondary metabolite concentrations in response to herbivory have been found that led to increased plant performance (e.g., Bennett and Wallsgrove 1994; Agrawal 1998; van Dam et al. 2001). Also, proteinase inhibitors can be induced by damage (Bergey et al. 1996; Koiwa et al. 1997; Ryan 2000; Tamayo et al. 2000), lowering the digestibility of the plant material for herbivores and thereby increasing plant resistance. The herbivores themselves show different levels of adaptation to overcome these plant defense responses, which can also be divided into constitutive and induced (Gatehouse 2002). While considerable knowledge has accumulated on mechanisms that lead to an induction of defense traits in the plant tissue after mechanical or herbivore damage (Kessler and Baldwin 2002; Arimura et al. 2005), relatively little is known on inducible effects on the plant cuticle.

Direct effects on wax production were observed by Bystrom et al. (1968) after aphid punctures in *Beta vulgaris* L. (Chenopodiaceae). Infestation of leaves of *Sorghum halepense* (L.) (Poaceae) by *Sipha flava* (Forbes) (Homoptera: Aphididae) led to a reduction of crystalline epicuticular waxes compared to non-infested ones. However, infested leaves became even more susceptible to the aphid than intact plants (Gonzales et al. 2002).

At the leaf surface, the density of trichomes can increase in response to herbivory (Agrawal et al. 2002; Traw and Dawson 2002; Valkama et al. 2005; Dalin et al. this volume). The induced trichomes can exude various metabolites such as e.g., flavonols (Roda et al. 2003), or other lipophilic compounds (Valkama et al. 2005) that change the surface chemistry of the plants. These responses can appear quite delayed in long-living species such as trees: the density of glandular trichomes and concentrations of several lipophilic compounds on the leaf surface increased in silver birch (*Betula pendula* Roth) defoliated during the previous summer. Effects were also visible in systemic branches and affected different herbivore guilds to various degrees (Valkama et al. 2005).

Treatment of celery leaves (*Apium graveolens*, cv. *secalinum*, Apiaceae) with jasmonic acid or amino-acid conjugates of jasmonic acid stimulates the biosynthesis of a number of furanocoumarins which are at least in part deposited on the leaf surface (Stanjek et al. 1997). Whereas these act as oviposition stimulants for some specialists, they likely also have deterrent effects on generalist herbivores.

Wounding by insects and the detection of insect elicitors originating from oral secretions by plants can lead to the release of systemin and other polypeptides that systemically activate plenty of defensive genes and signal transduction pathways, including the jasmonic acid signaling pathway (Bergey et al. 1996; Ryan 2000; Kessler and Baldwin 2002; Arimura et al. 2005). But even without any damage of the plant tissue, plants can build up resistance by responding to the mere touch of a crawling insect (Bown et al. 2002). Footsteps of crawling non-wounding insect larvae stimulated the synthesis of 4-aminobutyrate and superoxide within short time which might mediate rapid local resistance mechanisms (Bown et al. 2002). The superoxide production, leading to local cell death, might also play an important function in protection against infection by opportunistic pathogens at the site of the larval footprint damage (Hall et al. 2004). The response of the plant to insect crawling could be mimicked by suction applied with a micropipette (Bown et al. 2002), suggesting a purely physico-mechanical effect of larval pads on the plant surface.

However, attachment pads on the tarsae of insects produce also fluid secretions that are needed for wet adhesion to the plant surface (Eisner and Aneshansley 2000; Gorb 1998; Federle et al. 2002; Vötsch et al. 2002; Federle and Bruening 2005). While the chemical composition of these fluids has only been studied in a small number of species (Attygalle et al. 2000; Betz 2003; Vötsch et al. 2002), they likely etch cuticular waxes. Etched wax crystals have been found that form an amorphous mass on adhesive pads influencing locomotion of insects (Gaume et al. 2004; Federle and Bruening 2005). These fluid secretions and their impact on epicuticular waxes could potentially be important stimuli for the plant to induce resistance responses.

Another elicitor for early defense responses of plants may be present in oviduct secretions which are emitted during oviposition (Hilker and Meiners 2006). When eggs are only attached to the cuticle and leaves are not visibly damaged by oviposition it is unclear how potential elicitors could enter the plant tissue. Müller and Rosenberger (2006) showed that oviposition secretion can be sorbed in the plant cuticle in such species (i.e., in *Lilioceris lilii* L., Coleoptera: Chrysomelidae) and could even penetrate through stomata (in *L. merdigera* (L.)). This way signals may reach inner plant tissues allowing the plant to respond to insect attack with the induction of direct or indirect defense responses.

Finally, sorption of compounds released from individual herbivores on the cuticle might affect conspecifics. The iridoid (epi)-chrysomelidial, which is the main component of the larval secretion of the leaf beetle *P. cochleariae*, was detectable in surface extracts of leaves on which larvae had fed (Rostás and Hilker 2002). The secretion has a deterrent effect on conspecific adults. Oviposition deterring pheromones have been reported in a number of species of different orders (Anderson 2002). Female tephritid flies mark the fruit with such pheromones after oviposition. As the fruits can only support a limited number of larvae, other females will avoid marked fruits as potential hosts for their offspring. The persistence of the pheromone varies from a few hours up to several days and depends on the nature of the pheromone and the physico-chemical properties of the cuticular surface (Müller and Riederer 2005).

5.4.2 Induced Indirect Defense

Cuticular lipids can hold important functions in multitrophic interactions, as their major constituents can be chemically very similar between cuticles of plants, herbivores, and their predators. The cuticular lipid components of larvae of *Manduca sexta* L. (Lepidoptera: Sphingidae) were shown to vary slightly depending on the surface composition of their specific host plants. This diet-related variation influences palatability by predatory ants (Espelie and Bernays 1989) and thus indirectly plant resistance features. The migratory grasshopper *Melanopus sanguinipes* (F.) (Orthoptera: Acrididae) incorporates n-alkanes of the plant cuticle in its integument (Blomquist and Jackson 1973) which is exploited for host recognition by some predators and parasitoids (Espelie et al. 1991). However, such kind of mimicry can also lead to perfect protection: larvae of *Mechanitis polymnia* (L.) (Lepidoptera: Nymphalidae) are defended against the predatory ant *Camponotus crassus* Mayr (Hymenoptera: Formicidae) by a chemical similarity between their cuticular lipids and those of the solanaceous host plant (Henrique et al. 2005). On other plant species, the larvae do not receive this chemical camouflage and experience higher predation rates. Plant cuticular hydrocarbons of fruits and seed coats of apples were detected in larvae and adults of codling moths, *Cydia pomonella* L. (Lepidoptera: Torticidae), and were furthermore identical between adult moths and their parasitoid *Ascogaster quadridentata* Wesmael (Hymenoptera: Braconidae). Even in the hyperparasitoid *Perilampus fulvicornis* Ashmead (Hymenoptera: Perilampidae) one cuticular lipid, pentacosene, was identical with its host (Espelie and Brown 1990). These relationships may have important implications for the biology of the members of the different trophic levels, for host acceptance and suitability. Indirectly, they again might affect plant resistance features.

Feeding of a leaf mining herbivore was shown to lead to increased amounts of the triterpene squalene on the leaf surface of apples, which mediated host location of a parasitoid (Dutton et al. 2002). Induced wax production after aphid punctures that has been observed in *B. vulgaris* (Bystrom et al. 1968; see also Section 5.4.1) may also be used for host location by parasitoids.

The volatility of compounds which are induced by feeding (Dicke et al. 1998; Kessler and Baldwin 2001; Mattiacci et al. 2001; Röse and Tumlinson 2004; Arimura et al. 2005) or oviposition (Meiners and Hilker 1997; Hilker et al. 2002; Colazza et al. 2004; Fatouros et al. 2005), and which are attractive to members of the third trophic level is certainly influenced by epicuticular lipids at the plant surface which thereby influence indirect plant resistance traits. The sorption of volatile organic compounds in the waxes (Welke et al. 1998) and the transport pathways through the cuticle (Riederer and Friedmann 2006) are highly determined by the lipophilicity of the compounds, by the saturation vapor pressure in the atmosphere above the cuticle (Riederer et al. 1995, 2002), and by the specific physicochemical characteristics of the plant cuticle (Müller and Riederer 2005). For studies on attraction of parasitoids by herbivore-induced volatiles it would be important to understand the kinetics of transport, and to know the concentrations of the key compounds, at the plant surface and in a distance.

Herbivore infestation can also influence the optical appearance of the plant surface. Predators with good color perception will react to slight changes and predate more efficiently. For example, willow warblers were shown to respond to changes in UV spectral maxima which resulted from induction of flavonoids in the surface due to sawfly damage (Mäntylä et al. 2004). Endophytic feeding can lead to desiccation and discoloration, influencing chromatic and achromatic cues, mainly by changes of contrasts, for parasitoids searching for pupal hosts (Fischer et al. 2004). Also, polarized reflectance, determined by specular reflectance of the epicuticular waxes and surface particle scattering (Grant et al. 1993), may be changed by mining insects and influence host finding by carnivores.

5.5 Crop Protection

The nature of the plant cuticle highly influences the deposition and subsequent behavior of pesticides, growth regulators, foliar nutrients, or other chemicals used for plant cultivation. For systemic crop protection, many agrochemicals are sprayed on the leaves where they first have to penetrate the plant cuticle whose waxy surface film and embedded cuticular waxes act as an important barrier. Organic non-electrolytes physically diffuse within and across the cuticle with varying mobility, depending on plant species and ontogenetic state, the solute size and shape, the temperature and the presence of accelerators (Buchholz 2006; Riederer and Friedmann 2006). Accelerators enhance the fluidity of cuticular waxes by acting as plasticizers (Schreiber 2006), thereby facilitating the subsequent penetration of lipophilic non-electrolytes (Riederer and Friedmann 2006). The plethora of compounds used as agrochemicals and as transport-enhancing chemicals via the cuticle has been reviewed from the plant perspective (Riederer and Schönherr 1990; Baur 1998; Zabkiewicz 2000). The specific nature of the cuticle also determines the persistence of the various agrochemicals on the plant surface (Riccio et al. 2006).

However, little is known how changes of the morphological and physico-chemical characteristics of the cuticle after treatment with insecticides or herbicides (Taylor et al. 1981; Whitehouse et al. 1982; Gouret et al. 1993) act on herbivorous insects. Given that the cuticle plays an important role for host recognition, host acceptance, and also attachment abilities of insect legs and eggs, a tremendous effect on plant-insect interactions can be expected by changing plant surface features through application of agrochemicals. Herbicide treatment of *Brassica* spp. with *S*-ethyl dipropylthiocarbamate resulted in a reduction of epicuticular waxes (Eigenbrode and Shelton 1992; Justus et al. 2000). Treated plants turned more acceptable for oviposition by adult *P. xylostella* (Justus et al. 2000). The survival of neonate larvae was significantly reduced on these plants, however (Eigenbrode and Shelton 1992). Insecticide treatment not only kills the herbivores but may also negatively affect beneficial insects such as predators or parasitoids. Deltamethrin, used as insecticide, was shown to rapidly diffuse in oilseed rape cuticles. The parasitoid *Diaeretiella rapae* (M'Intosh) (Hymenoptera: Braconidae) was only little affected by residues of this pyrethroid, and thus could limit populations of the aphid *Myzus persicae* (Sulzer) (Homoptera: Aphididae) efficiently (Desneux et al. 2005).

To summarize, a plant cuticle that offers resistance to a plant should comprise different properties. For efficient direct defense, it should have a surface microstructure on which herbivores are unable to walk, and eggs cannot attach; feeding or oviposition stimulants, and specifically nutrients, should be covered by cuticular waxes and not be present on the outer surface. For indirect defense, the microstructure should enable carnivores to hunt effectively; the sorption for attractants or stimulants for predators or parasitoids should be optimal; the visual appearance of the plant should betray the presence of feeding herbivores to the enemies. However, as the different species of herbivores and carnivores have various adaptations to specific cuticle characteristics, this optimal cuticle will probably never be feasible.

5.6 Perspectives

In almost all plant-insect relationships, the cuticle as the first contact zone can certainly be expected to play a significant role for host acceptance or rejection by herbivores. However, for only few genera of plants have the plant-insect interactions that are mediated at the cuticular surface been studied well (e.g., *Pisum, Brassica, Macaranga, Nepenthes, Eucalyptus*). In these investigations, either glossy or waxy genotypes were compared, or manipulative experiments were carried out to assess the role of specific plant cuticle structures and chemistry. The different methods for investigating the role of the plant cuticle for insects have recently been reviewed (Müller 2006).

Aspects of plant resistance mechanisms mediated by the plant cuticle are specifically important for plant breeding programs (Pelletier and Dutheil 2006). Also, for genetically modifying plants, cuticle characteristics and their barrier function need to be carefully evaluated (Kerstiens et al. 2006). However, this subject is quite complex, as the epicuticular waxes can have positive or negative effects on herbivorous and predatory insect species, or could decrease susceptibility to one pest but simultaneously increase susceptibility to another pest species.

Many questions are still open, specifically with regard to induced plant responses at the plant surface: What happens morphologically at the plant cuticle after pure contact with insect legs, mouthparts, antennae or the ovipositor? What changes chemically? Does signaling occur at/through the cuticle? Do induced changes of the cuticle finally affect herbivorous insects, either the individual itself, conspecifics, or individuals of another species? It is surprising that there is so little knowledge on the very first contact zone of the plant for an approaching insect – a goldmine for future studies!

References

Agrawal AA (1998) Induced responses to herbivory and increased plant performance. Science 279:1201–1202

Agrawal AA, Conner JK, Johnson MT, Wallsgrove R (2002) Ecological genetics of induced plant defense against herbivores: additive genetic variation and costs of phenotypic plasticity. Evolution 56:2206–2213

Anderson P (2002) Oviposition pheromones in herbivorous and carnivorous insects. In: Hilker M, Meiners T (eds) Chemoecology of insect eggs and egg deposition. Blackwell, Berlin pp 235–263

Arimura G, Kost C, Boland W (2005) Herbivore-induced, indirect plant defences. Biochim Biophys Acta-Molec Cell Biol Lipids 1734:91–111

Atkin DSJ, Hamilton RJ (1982) The changes with age in the epicuticular wax of *Sorghum bicolor*. J Nat Prod (Lloydia) 45:697–703

Attygalle AB, Aneshansley DJ, Meinwald J, Eisner T (2000) Defense by foot adhesion in a chrysomelid beetle (*Hemisphaerota cyanea*): characterization of the adhesive oil. Zool-Anal Complex Sys 103:1–6

Avato P, Bianchi G, Pogna N (1990) Chemosystematics of surface lipids from maize and some related species. Phytochemistry 29:1571–1576

Baker EA (1974) The influence of environment on leaf wax development in *Brassica oleraceae* var. *gemmifera*. New Phytol 73:955–966

Baker EA, Hunt GM (1986) Erosion of waxes from leaf surfaces by simulated rain. New Phytol 102:161–173

Baldini E, Facini O, Nerozzi F, Rossi F, Rotondi A (1997) Leaf characteristics and optical properties of different woody species. Trees-Struct Funct 12:73–81

Bargel H, Barthlott W, Koch K, Schreiber L, Neinhuis C (2004) Plant cuticles: multifunctional interfaces between plant and environment. In: Hemsley AR, Poole I (eds) The evolution of plant physiology. Elsevier Academic Press, London, pp 171–194

Bargel H, Koch K, Cerman Z, Neinhuis C (2006) Structure-function relationships of the plant cuticle and cuticular waxes – a smart material? Funct Plant Biol 33:893–910

Barnes JD, Cardoso-Vilhena J (1996) Interactions between electromagnetic radiation and the plant cuticle. In: Kerstiens G (ed) Plant cuticles: an integrated functional approach. BIOS Scientific, Oxford, pp 157–174

Barthlott W, Neinhuis C (1997) Purity of the sacred lotus, or escape from contamination in biological surfaces. Planta 202:1–8

Barthlott W, Neinhuis C, Cutler D, Ditsch F, Meusel I, Theisen I, Wilhelmi H (1998) Classification and terminology of plant epicuticular waxes. Bot J Linn Soc 126:237–260

Baum BR, Tulloch AP, Bailey LG (1980) A survey of epicuticular waxes among genera of Triticeae. 1. Ultrastructure of glumes and some leaves as observed with the scanning electron-microscope. Can J Bot 58:2467–2480

Baur P (1998) Mechanistic aspects of foliar penetration of agrochemicals and the effect of adjuvants. Rec Res Devel Agric Food Chem 2:809–837

Bennett RN, Wallsgrove RM (1994) Secondary metabolites in plant defense-mechanisms. New Phytol 127:617–633

Bergey DR, Hoi GA, Ryan CA (1996) Polypeptide signaling for plant defensive genes exhibits analogies to defense signaling in animals. Proc Natl Acad Sci USA 93:12053–12058

Bergman DK, Dillwith JW, Zarrabi AA, Caddel JL, Berberet RC (1991) Epicuticular lipids of alfalfa relative to its susceptibility to spotted alfalfa aphids (Homoptera: Aphididae). Environ Entomol 20:781–785

Betz O (2003) Structure of the tarsi in some *Stenus* species (Coleoptera, Staphylinidae): external morphology, ultrastructure, and tarsal secretion. J Morphol 255:24–43

Bianchi G, Avato P, Scarpa O, Murelli C, Audisio G, Rossini A (1989) Composition and structure of maize epicuticular wax esters. Phytochemistry 28:165–171

Blomquist GJ, Jackson LL (1973) Incorporation of labelled dietary n-alkanes into cuticular lipids of the grasshopper *Melanopus sanguinipes*. J Insect Physiol 19:1639–1647

Blua MJ, Yoshida HA, Toscano NC (1995) Oviposition preference of 2 *Bemisia* Species (Homoptera, Aleyrodidae). Environ Entomol 24:88–93

Bodnaryk RP (1992) Leaf epicuticular wax, an antixenotic factor in Brassicaceae that affects the rate and pattern of feeding of flea beetles, *Phyllotreta cruciferae* (Goeze). Can J Plant Sci 72:1295–1303

Bown AW, Hall DE, MacGregor KB (2002) Insect footsteps on leaves stimulate the accumulation of 4-aminobutyrate and can be visualized through increased chlorophyll fluorescence and superoxide production. Plant Physiol 129:1430–1434

Brennan EB, Weinbaum SA (2001) Stylet penetration and survival of three psyllid species on adult leaves and 'waxy' and 'de-waxed' juvenile leaves of *Eucalyptus globulus*. Entomol Exp Appl 100:355–363

Brennan EB, Weinbaum SA, Rosenheim JA, Karban R (2001) Heteroblasty in *Eucalyptus globulus* (Myricales: Myricaceae) affects ovipositonal and settling preferences of *Ctenarytaina eucalypti* and *C. spatulata* (Homoptera: Psyllidae). Environ Entomol 30:1144–1149

Buchholz A (2006) Characterization of the diffusion of non-electrolytes across plant cuticles: properties of the lipophilic pathway. J Exp Bo 57:2501–2513

Bullas-Appleton ES, Otis G, Gillard C, Schaafsma AW (2004) Potato leafhopper (Homoptera: Cicadellidae) varietal preferences in edible beans in relation to visual and olfactory cues. Environ Entomol 33:1381–1338

Bystrom BG, Glater RB, Scott FM, Bowler FSC (1968) Leaf surface of *Beta vulgaris*-electron microscope study. Bot Gaz 129:133–138

Caldwell MM (1968) Solar ultraviolet radiation as an ecological factor for alpine plants. Ecol Monogr 38:243–268

Carter CD, Gianfagna TJ, Sacalis JN (1989) Sesquiterpenes in glandular trichomes of a wild tomato species and toxicity to the Colorado potato beetle. J Agric Food Chem 37:1425–1428

Cervantes DE, Eigenbrode SD, Ding HJ, Bosque-Perez NA (2002) Oviposition responses by Hessian fly, *Mayetiola destructor*, to wheats varying in surfaces waxes. J Chem Ecol 28:193–210

Chang GC, Neufeld J, Durr D, Duetting PS, Eigenbrode SD (2004a) Waxy bloom in peas influences the performance and behavior of *Aphidius ervi*, a parasitoid of the pea aphid. Entomol Exp Appl 110:257–265

Chang GC, Rutledge CE, Biggam RC, Eigenbrode SD (2004b) Arthropod diversity in peas with normal or reduced waxy bloom. J Insect Sci 4. Art. No. 18

Chappell M, Robacker C (2006) Leaf wax extracts of four deciduous azalea genotypes affect azalea lace bug (*Stephanitis pyrioides* Scott) survival rates and behavior. J Am Soc Hort Sci 131:225–230

Cockell CS, Knowland J (1999) Ultraviolet radiation screening compounds. Biol Rev 74:311–345

Colazza S, Fucarino A, Peri E, Salerno G, Conti E, Bin F (2004) Insect oviposition induces volatile emission in herbaceous plants that attracts egg parasitoids. J Exp Biol 207:47–53

Cole RA, Riggal W (1992) Pleiotropic effects of genes in glossy *Brassica oleracea* resistant to *Brevicoryne brassicae*. In: Menken SBJ, Visser JH, Harrewijn P (eds) Proceedings of the 8th international symposium on insect-plant relationships. Kluwer Academic, Dordrecht, pp 313–315

Derridj S, Boutin JP, Fiala V, Soldaat LL (1996a) Primary metabolites composition of the leek leaf surface: comparative study: impact on the host-plant selection by an ovipositing insect. Acta Bot Gall 143:125–130

Derridj S, Wu BR, Stammitti L, Garrec JP, Derrien A (1996b) Chemicals on the leaf surface, information about the plant available to insects. Entomol Exp Appl 80:197–201

Desneux N, Fauvergue X, Dechaume-Moncharmont FX, Kerhoas L, Ballanger Y, Kaiser L (2005) *Diaeretiella rapae* limits *Myzus persicae* populations after applications of deltamethrin in oilseed rape. J Econ Entomol 98:9–17

Dicke M, Takabayashi J, Posthumus MA, Schütte C, Krips OE (1998) Plant-phytoseiid interactions mediated by herbivore-induced plant volatiles: variation in production of cues and in responses of predatory mites. Rev Exp Appl Aca 22:311–333

Dutton A, Mattiacci L, Amadò R, Dorn S (2002) A novel function of the triterpene squalene in a tritrophic system. J Chem Ecol 28:103–116

Edwards PB (1982) Do waxes of juvenile *Eucalyptus* leaves provide protection from grazing insects? Aust J Ecol 7:347–352

Ehleringer JR, Björkman O (1978) Pubescence and leaf spectral characteristics in a desert shrub, *Encelia farinosa*. Oecologia 36:151–162

Eigenbrode SD (2004) The effects of plant epicuticular waxy blooms on attachment and effectiveness of predatory insects. Arthr Struct Devel 33:91–102

Eigenbrode SD, Espelie KE (1995) Effects of plant epicuticular lipids on insect herbivores. Annu Rev Entomol 40:171–194

Eigenbrode SD, Kabalo NN (1999) Effects of *Brassica oleracea* waxblooms on predation and attachment by *Hippodamia convergens*. Entomol Exp Appl 92:125–130

Eigenbrode SD, Kabalo NN, Stoner KA (1999) Predation, behavior, and attachment by *Chrysoperla plorabunda* larvae on *Brassica oleracea* with different surface waxblooms. Entomol Exp Appl 90:225–235

Eigenbrode SD, Pillai SK (1998) Neonate *Plutella xylostella* responses to surface wax components of a resitant cabbage (*Brassica oleracea*). J Chem Ecol 24:1611–1627

Eigenbrode SD, Shelton AM (1990) Behavior of neonate diamondback moth larvae (Lepidoptera, Plutellidae) on glossy-leafed resistant *Brassica oleracea* L. Environ Entomol 19:1566–1571

Eigenbrode SD, Shelton AM (1992) Survival and behavior of *Plutella xylostella* larvae on cabbages with leaf waxes altered by treatment with S-ethyl dipropylthiocarbamate. Entomol Exp Appl 62:139–145

Eisner T, Aneshansley DJ (2000) Defense by foot adhesion in a beetle (*Hemisphaerota cyanea*). Proc Natl Acad Sci USA 97:6568–6573

Ensikat HJ, Neinhuis C, Barthlott W (2000) Direct access to plant epicuticular wax crystals by a new mechanical isolation method. Int J Plant Sci 161:143–148

Espelie KE, Bernays EA (1989) Diet-related differences in the cuticular lipids of *Manduca sexta* larvae. J Chem Ecol 15:2003–2017

Espelie KE, Bernays EA, Brown JJ (1991) Plant and insect cuticular lipids serve as behavioural cues for insects. Arch Insect Biochem Physiol 17:389–399

Espelie KE, Brown JJ (1990) Cuticular hydrocarbons of species which interact on four trophic levels: apple, *Malus pumila* Mill.; codling moth, *Cydia pomonella* L.; a hymenopteran parasitoid, *Ascogaster quadridentata* Wesmael; and a hyperparasite *Perilampus fulvicornis* Ashmead. Comp Biochem Physiol 95B:131–136

Fatouros NE, Bukovinszkine'Kiss G, Kalkers LA, Gamborena RS, Dicke M, Hilker M (2005) Oviposition-induced plant cues: do they arrest *Trichogramma* wasps during host location? Entomol Exp Appl 115:207–215

Federle W, Bruening T (2005) Ecology and biomechanics of slippery wax barriers and wax running in *Macaranga* ant mutualisms. In: Harrel A, Speck T, Rowe N (eds) Ecology and biomechanics: a mechanical approach to the ecology of animals and plants. CRC Press, Boca Raton, pp 163–185

Federle W, Maschwitz U, Fiala B, Riederer M, Hölldobler B (1997) Slippery ant-plants and skilful climbers: selection and protection of specific ant partners by epicuticular wax blooms in *Macaranga* (Euphorbiaceae). Oecologia 112:217–224

Federle W, Riehle M, Curtis ASG, Full RJ (2002) An integrative study of insect adhesion: mechanics and wet adhesion of pretarsal pads in ants. Integ and Comp Biol 42:1100–1106

Federle W, Rohrseitz K, Hölldobler B (2000) Attachment forces of ants measured with a centrifuge: better 'wax-runners' have a poorer attachment to a smooth surface. J Exp Biol 203:505–512

Fiala V, Glad C, Martin M, Jolivet E, Derridj S (1990) Occurrence of soluble carbohydrates on the phylloplane of maize (*Zea mays* L.): variations in relation to leaf heterogeneity and position on the plant. New Phytol 115:609–615

Ficke A, Gadoury DM, Godfrey D, Dry IB (2004) Host barriers and responses to *Uncinula necator* in developing grape berries. Phytopathology 94:438–445

Fischer S, Samietz J, Wäckers FL, Dorn S (2004) Perception of chromatic cues during host location by the pupal parasitoid *Pimpla turionellae* (L.) (Hymenoptera: Ichneumonidae). Environ Entomol 33:81–87

Foster SP, Harris MO (1992) Foliar chemicals of wheat and related grasses influencing oviposition by Hessian fly, *Mayetiola destructor* (Say) (Diptera: Cecidomyiidae). J Chem Ecol 18:1965–1980

Freeman B, Albrigo LG, Biggs RH (1979a) Cuticular waxes of developing leaves and fruit of Blueberry, *Vaccinium ashei* Reade Cv *bluegem*. J Am Soc Hort Sci 104:398–403

Freeman B, Albrigo LG, Biggs RH (1979b) Ultrastructure and chemistry of cuticular waxes of developing *Citrus* leaves and fruits. J Am Soc Hort Sci 104:801–808

Gatehouse JA (2002) Plant resistance towards insect herbivores: a dynamic interaction. New Phytol 156:145–169

Gaume L, Perret P, Gorb E, Gorb S, Labat JJ, Rowe N (2004) How do plant waxes cause flies to slide? Experimental tests of wax-based trapping mechanisms in three pitfall carnivorous plants. Arthr Struct Devel 33:103–111

Gentry GL, Barbosa P (2006) Effects of leaf epicuticular wax on the movement, foraging behavior, and attack efficacy of *Diaeretiella rapae*. Entomol Exp Appl 121:115–122

Gonzales WL, Ramirez CC, Olea N, Niemeyer HM (2002) Host plant changes produced by the aphid *Sipha flava*: consequences for aphid feeding behaviour and growth. Entomol Exp Appl 103:107–113

Gorb SN (1998) The design of the fly adhesive pad: distal tenent setae are adapted to the delivery of an adhesive secretion. Proc Roy Soc Lond Ser B-Biol Sci 265:747–752

Gorb EV, Gorb SN (2002) Attachment ability of the beetle *Chrysolina fastuosa* on various plant surfaces. Entomol Exp Appl 105:13–28

Gouret E, Rohr R, Chamel A (1993) Ultrastructure and chemical composition of some isolated plant cuticles in relation to their permeability to the herbicide, diuron. New Phytol 124:423–431

Grant L, Daughtry CST, Vanderbilt VC (1993) Polarized and specular reflectance variation with leaf surface-features. Physiol Plant 88:1–9

Grevstad FS, Klepetka BW (1992) The influence of plant architecture on the foraging efficiencies of a suite of ladybird beetles feeding on aphids. Oecologia 92:399–404

Griffiths DW, Deighton N, Birch ANE, Patrian B, Baur R, Städler E (2001) Identification of glucosinolates on the leaf surface of plants from Cruciferae and other closely related species. Phytochemistry 57:693–700

Griffiths DW, Robertson GW, Shepherd T, Birch AN, Gordon SC, Woodford JAT (2000) A comparison of the composition of epicuticular wax from red raspberry (*Rubus idaeus* L.) and hawthorn (*Crataegus monogyna* Jacq.) flowers. Phytochemistry 55:111–116

Griffiths DW, Robertson GW, Shepherd T, Ramsay G (1999) Epicuticular waxes and volatiles from faba bean (*Vicia faba*) flowers. Phytochemistry 52:607–612

Gülz PG, Boor G (1992) Seasonal variations in epicuticular wax ultrastructures of *Quercus robur* leaves. Z Naturforsch C B osci 47:807–814

Gülz PG, Marner FJ (1986) Esters of benzyl alcolhol and 2-phenyl-ethanol-1 in epicuticular waxes from *Jojoba* leaves. Z Naturforsch C Biosci 41:673–676

Gülz PG, Müller E, Schmitz K, Güth S (1992) Chemical composition and surface structures of epicuticular leaf waxes of *Ginkgo biloba*, *Magnolia grandiflora* and *Liriodendron tulipifera*. Z Naturforsch C Biosci 47:516–526

Gülz PG, Scora RW, Müller E, Marner FJ (1987) Epicuticular leaf waxes of *Citrus halimii* stone. J Agric Food Chem 35:716–720

Hagley EAC, Bronskill JF, Ford EJ (1980) Effect of the physical nature of leaf and fruit surfaces on oviposition by the codling moth, *Cydia pomonella* (Lepidoptera, Tortricidae). Can Entomol 112:503–510

Hall DE, MacGregor KB, Nijsse J, Bown AW (2004) Footsteps from insect larvae damage leaf surfaces and initiate rapid responses. Eur J Plant Pathol 110:441–447

Harborne JB, Williams CA (2000) Advances in flavonoid research since 1992. Phytochemistry 55:481–504

Henrique A, Portugal A, Trigo JR (2005) Similarity of cuticular lipids between a caterpillar and its host plant: a way to make prey undetectable for predatory ants? J Chem Ecol 31:2551–2561

Hilker M, Kobs C, Varama M, Schrank K (2002) Insect egg deposition induces *Pinus sylvestris* to attract egg parasitoids. J Exp Biol 205:455–461

Hilker M, Meiners T (2006) Eggs of herbivorous insects inducing early alert in plants. J Chem Ecol 32:1379–1397
Holloway PJ (1971) The chemical and physical characteristic of leaf surfaces. In: Preece TF, Dickinson CH (eds) Ecology of leaf surface microorganisms. Academic Press, London, pp 39–53
Holloway PJ (1982) The chemical constitution of plant cutins. In: Cutler DF, Alvin KL, Price CE (eds) The plant cuticle. Academic Press, London, pp 45–85
Holloway PJ, Hunt GM, Baker EA, Macey MJK (1977) Chemical composition and ultrastructure of epicuticular wax in four mutants of *Pisum sativum* (L). Chem Phys Lipids 20:141–155
Hopkins RJ, Birch ANE, Griffiths DW, Baur R, Städler E, McKinlay RG (1997) Leaf surface compounds and oviposition preference of turnip root fly *Delia floralis*: the role of glucosinolate and nonglucosinolate compounds. J Chem Ecol 23:629–643
Hubbell SP, Howard JJ, Wiemer D. (1984) Chemical leaf repellency to an Attine ant – seasonal distribution among potential host plant-species. Ecology 65:1067–1076
Jeffree CE (1986) The cuticle, epicuticular waxes and trichomes of plants, with reference to their structure, functions and evolution. In: Juniper BE, Southwood TRE (eds) Insects and the plant surface. Edward Arnold, London, pp 23–64
Jeffree CE (1996) Structure and ontogeny of plant cuticles. In: Kerstiens G (ed) Plant cuticles: an integrated functional approach. BIOS Scientific Publishers, Oxford, pp 33–82
Jeffree CE (2006) The fine structure of the plant cuticle. In: Riederer M, Müller C (eds) Biology of the plant cuticle. Blackwell Publishing, Oxford, pp 11–125
Jenks MA, Gaston CH, Goodwin MS, Keith JA, Teusink RS, Wood KV (2002) Seasonal variation in cuticular waxes on *Hosta* genotypes differing in leaf surface glaucousness. Hort Sci 37:673–677
Jetter R, Klinger A, Schäffer S (2002) Very long-chain phenylpropyl and phenylbutyl esters from *Taxus baccata* needle cuticular waxes. Phytochemistry 61:579–587
Jetter R, Kunst L, Samuels L (2006) Composition of plant cuticular waxes. In: Riederer M, Müller C (eds) Biology of the plant cuticle. Blackwell, London, pp 145–181
Jetter R, Riederer M (1996) Cuticular waxes from the leaves and fruit capsules of eight Papaveraceae species. Can J Bot 74:419–430
Jetter R, Schäffer S (2001) Chemical composition of the *Prunus laurocerasus* leaf surface. Dynamic changes of the epicuticular wax film during leaf development. Plant Physiol 126:1725–1737
Jetter R, Schäffer S, Riederer M (2000) Leaf cuticular waxes are arranged in chemically and mechanically distinct layers: evidence from *Prunus laurocerasus* L. Plant Cell Environ 23:619–628
Jolivet P (1988) Interrelationship between insects and plants. CRC Press, Boca Raton
Justus KA, Dosdall LM, Mitchell BK (2000) Oviposition by *Plutella xylostella* (Lepidoptera: Plutellidae) and effects of phylloplane waxiness. J Econ Entomol 93:1152–1159
Kareiva P, Sahakian R (1990) Tritrophic effects of a simple architectural mutation in *Pea* plants. Nature 345:433–434
Kerner A (ed) (1879) Schutzmittel der Blüthen gegen unberufene Gäste. Verlag der Wagner'schen Universitäts-Buchhandlung, Innsbruck
Kerstiens G, Schreiber L, Lendzian KJ (2006) Quantification of cuticular permeability in genetically modified plants. J Exp Bot 57:2547–2552
Kessler A, Baldwin IT (2001) Defensive function of herbivore-induced plant volatile emissions in nature. Science 291:2141–2144
Kessler A, Baldwin IT (2002) Plant responses to insect herbivory: the emerging molecular analysis. Annu Rev Plant Biol 53:299–328
Koiwa H, Bressan RA, Hasegawa PM (1997) Regulation of protease inhibitors and plant defense. Trends Plant Sci 2:379–384
Krauss P, Markstädter C, Riederer M (1997) Attenuation of UV radiation by plant cuticles from woody species. Plant Cell Environ 20:1079–1085

Lambdon PW, Hassall M, Mithen R (1998) Feeding preferences of woodpigeons and flea-beetles for oilseed rape and turnip rape. Ann Appl Biol 133:313–328

Leveau JHJ (2004) Leaf surface sugars. In: Decker M (ed) Encyclopedia of plant and crop science. Dekker M, New York, pp 642–645

Major DJ, McGinn SM, Gillespie TJ, Baret F (1993) A technique for determination of single leaf reflectance and transmittance in field studies. Remote Sens Environ 43:209–215

Mallavadhani UV, Mahapatra A, Raja SS, Manjula C (2003) Antifeedant activity of some pentacyclic triterpene acids and their fatty acid ester analogues. J Agric Food Chem 51:1952–1955

Mäntylä E, Klemola T, Haukioja E (2004) Attraction of willow warblers to sawfly-damaged mountain birches: novel function of inducible plant defences? Ecol Lett 7:915–918

Markstädter C, Federle W, Jetter R, Riederer M, Hölldobler B (2000) Chemical composition of the slippery epicuticular wax blooms on *Macaranga* (Euphorbiaceae) ant-plants. Chemoecology 10:33–40

Markstädter C, Queck I, Baumeister J, Riederer M, Schreiber U, Bilger W (2001) Epidermal transmittance of leaves of *Vicia faba* for UV radiation as determined by two different methods. Photosynth Res 67:17–25

Mattiacci L, Rocca BA, Scascighini N, D'Alessandro M, Hern A, Dorn S (2001) Systemically induced plant volatiles emitted at the time of 'danger'. J Chem Ecol 27:2233–2252

Meiners T, Hilker M (1997) Host location in *Oomyzus gallerucae* (Hymenoptera: Eulophidae), an egg parasitoid of the elm leaf beetle *Xanthogaleruca luteola* (Coleoptera: Chrysomelidae). Oecologia 112:87–93

Müller C (1999) Chemische Ökologie des Phytophagenkomplexes an *Tanacetum vulgare* L. (Asteraceae). Logos, Berlin

Müller C (2006) Plant-insect interactions on cuticular surfaces. In: Riederer M, Müller C (eds) Biology of the plant cuticle. Blackwell, Oxford, pp 398–422

Müller C, Hilker M (2001) Host finding and oviposition behavior in a chrysomelid specialist – the importance of host plant surface waxes. J Chem Ecol 27:985–994

Müller C, Riederer M (2005) Review: plant surface properties in chemical ecology. J Chem Ecol 31:2621–2651

Müller C, Rosenberger C (2006) Different oviposition behaviour in Chrysomelid beetles: characterisation of the interface between oviposition secretion and the plant surface. Arthr Struct Devel 35:197–205

Mulroy TW (1979) Spectral properties of heavily glaucous and non-glaucous leaves of a succulent rosette-plant. Oecologia 38:349–357

Nawrath C (2006) Unravelling the complex network of cuticular structure and function. Curr Opin Plant Biol 9:281–287

Neinhuis C, Barthlott W (1997) Characterization and distribution of water-repellent, self-cleaning plant surfaces. Ann Bot 79:667–677

Neinhuis C, Barthlott W (1998) Seasonal changes of leaf surface contamination in beech, oak, and ginkgo in relation to leaf micromorphology and wettability. New Phytol 138:91–98

Neinhuis C, Koch K, Barthlott W (2001) Movement and regeneration of epicuticular waxes through plant cuticles. Planta 213:427–434

Netting AG, von Wettstein P (1973) Physicochemical basis of leaf wettability in wheat. Planta 114:289–309

Neuenschwander P, Michelakis S, Holloway P, Berchtold W (1985) Factors affecting the susceptibility of fruits of different olive varieties to attack by *Dacus oleae* (Gmel.) (Dipt., Tephritidae). Z Angew Entomol 100:174–188

Nwanze KF, Pring RJ, Sree PS, Butler DR, Reddy YVA, Soman P (1992) Resistance in sorghum to the shoot fly, *Atherigona soccata*: epicuticular wax and wetness of the central whorl leaf of young seedlings. Ann Appl Biol 120:373–382

Ohgushi T (2005) Indirect interaction webs: herbivore-induced effects through trait change in plants. Annu Rev Ecol Syst 36:81–105

Olson DM, Andow DA (2006) Walking pattern of *Trichogramma nubilale* Ertle & Davis (Hymenoptera; Trichogrammatidae) on various surfaces. Biol Control 39:329–335

Olsson LC, Veit M, Weissenbock G, Bornman JF (1998) Differential flavonoid response to enhanced UV-B radiation in *Brassica napus*. Phytochemistry 49:1021–1028

Pelletier Y, Dutheil J (2006) Behavioural responses of the colorado potato beetle to trichomes and leaf surface chemicals of *Solanum tarijense*. Entomol Exp Appl 120:125–130

Percy KE, Awmack CS, Lindroth RL, Kubiske ME, Kopper BJ, Isebrands JG, Pregitzer KS, Hendrey GR, Dickson RE, Zak DR, Oksanen E, Sober J, Harrington R, Karnosky DF (2002) Altered performance of forest pests under atmospheres enriched by CO_2 and O_3. Nature 420:403–407

Peter AJ, Shanower TG (2001) Role of plant surface in resistance to insect herbivores. In: Ananthakrishnan TN (ed) Insects and plant defence dynamics. Science Publishers, Enfield, pp 107–132

Pfündel EE, Agati G, Cerovic ZG (2006) Optical properties of plant surfaces. In: Riederer M, Müller C (eds) Biology of the plant cuticle. Blackwell, London, pp 216–249

Popp C, Burghardt M, Friedmann A, Riederer M (2005) Characterization of hydrophilic and lipophilic pathways of *Hedera helix* L. cuticular membranes: permeation of water and uncharged organic compounds. J Exp Bot 56:2797–2806

Powell G, Maniar SP, Picket JA, Hardie J (1999) Aphid responses to non-host epicuticular lipids. Entomol Exp Appl 91:115–123

Prokopy RJ, Collier RH, Finch S (1983) Leaf color used by cabbage root flies to distinguish among host plants. Science 221:190–192

Prokopy RJ, Owens ED (1983) Visual detection of plants by herbivorous insects. Ann Rev Entomol 28:337–364

Reifenrath K, Riederer M, Müller C (2005) Leaf surface wax layers of Brassicaceae lack feeding stimulants for *Phaedon cochleariae*. Entomol Exp Appl 115:41–50

Riccio R, Trevisan M, Capri E (2006) Effect of surface waxes on the persistence of chlorpyrifos-methyl in apples, strawberries and grapefruits. Food Addit Contam 23:683–692

Riedel M, Eichner A, Jetter R (2003) Slippery surfaces of carnivorous plants: composition of epicuticular wax crystals in *Nepenthes alata* Blanco pitchers. Planta 218:87–97

Riederer M, Burghardt M, Mayer S, Obermeier H, Schönherr J (1995) Sorption of monodisperese alcohol ethoxylates and their effects on the mobility of 2,4-D in isolated plant cuticles. J Agric Food Chem 43:1067–1075

Riederer M, Daiss A, Gilbert N, Kohle H (2002) Semi-volatile organic compounds at the leaf/atmosphere interface: numerical simulation of dispersal and foliar uptake. J Exp Bot 53:1815–1823

Riederer M, Friedmann A (2006) Transport of lipophilic non-electrolytes across the cuticle. In: Riederer M, Müller C (eds) Biology of the plant cuticle. Blackwell, Oxford, pp 250–279

Riederer M, Markstädter C (1996) Cuticular waxes: a critical assessment of current knowledge. In: Kerstiens G (ed) Plant cuticles – an integrated functional approach. BIOS Scientific, Oxford, pp 189–200

Riederer M, Schönherr J (1988) Development of plant cuticles: fine structure and cutin composition of *Clivia miniata* Reg. leaves. Planta 174:127–138

Riederer M, Schönherr J (1990) Effects of surfactants on water permeability of isolated cuticles and on the composition of their cuticular waxes. Pest Sci 29:85–94

Riederer M, Schreiber L (2001) Protecting against water loss: analysis of the barrier properties of plant cuticles. J Exp Bot 52:2023–2032

Robberecht R, Caldwell MM, Billings WD (1980) Leaf ultraviolet optical-properties along a latitudinal gradient in the arctic-alpine life zone. Ecology 61:612–619

Roda AL, Oldham NJ, Svatoš A, Baldwin IT (2003) Allometric analysis of the induced flavonols on the leaf surface of wild tobacco (*Nicotiana attenuata*). Phytochemistry 62:527–536

Rogge WF, Hildemann LM, Mazurek MA, Cass GR, Simoneit BRT (1993) Sources of fine organic aerosol. 4. Particulate abrasion products from leaf surfaces of urban plants. Environ Sci Technol 27:2700–2711

Roitman JN, Wong RY, Wollenweber E (1993) Methylene bisflavonoids from frond exudate of *Pentagramma triangularis* ssp. *triangularis*. Phytochemistry 34:297–301

Röse USR, Tumlinson JH (2004) Volatiles released from cotton plants in response to *Helicoverpa zea* feeding damage on cotton flower buds. Planta 218:824–832

Rostás M, Hilker M (2002) Feeding damage by larvae of the mustard leaf beetle deters conspecific females from oviposition and feeding. Entomol Exp Appl 103:267–277

Rutledge CE, Eigenbrode SD (2003) Epicuticular wax on pea plants decreases instantaneous search rate of *Hippodamia convergens* larvae and reduces attachment to leaf surfaces. Can Entomol 135:93–101

Rutledge CE, Robinson AP, Eigenbrode SD (2003) Effects of a simple plant mutation on the arthropod community and the impacts of predators on a principle insect herbivore. Oecologia 135:39–50

Ryan CA (2000) The systemin signaling pathway: differential activation of plant defensive genes. Biochim Biophys Acta-Prot Struct Molec Enzymol 1477:112–121

Schoonhoven LM, van Loon JJA, Dicke M (2006) Insect-plant biology. Oxford University Press, Oxford, UK

Schreiber L (2006) Review of sorption and diffusion of lipophilic molecules in cuticular waxes and the effects of accelerators on solute mobilities. J Exp Bot 57:2515–2523

Severson RF, Rrendale RF, Chortyk OT, Johnson AW, Jackson DM, Gwynn GR, Chaplin JF, Stephenson MG (1984) Quantitation of the major cuticular components from green leaf of different tobacco types. J Agric Food Chem 32:566–570

Shah MA (1982) The influence of plant-surfaces on the searching behavior of coccinellid larvae. Entomol Exp Appl 31:377–380

Shepherd T, Robertson GW, Griffiths DW, Birch ANE (1999a) Epicuticular wax composition in relation to aphid infestation and resistance in red raspberry (*Rubus idaeus* L.). Phytochemistry 52:1239–1254

Shepherd T, Robertson GW, Griffiths DW, Birch ANE (1999b) Epicuticular wax ester and triacylglycerol composition in relation to aphid infestation and resistance in red raspberry (*Rubus idaeus* L.). Phytochemistry 52:1255–1267

Soldaat LL, Boutin JP, Derridj S (1996) Species-specific composition of free amino acids on the leaf surface of four *Senecio* species. J Chem Ecol 22:1–12

Städler E (1984) Contact chemoreception. In: Bell WJ, Cardé RT (eds) Chemical ecology of insects. Chapman & Hall, London, pp 3–35

Städler E (1986) Oviposition and feeding stimuli in leaf surface waxes-an overview. In: Juniper BE, Southwood TRE (eds) Insects and the plant surface. Edward Arnold, London, pp 1–22

Städler E (1992) Behavioral responses of insects to plant secondary compounds. In: Rosenthal GA, Berenbaum MR (eds) Herbivores: their interactions with secondary plant metabolites. Academic Press, San Diego, pp 45–88

Stanjek V, Herhaus C, Ritgen U, Boland W, Städler E (1997) Changes in the leaf surface chemistry of *Apium graveolens* (Apiaceae) stimulated by jasmonic acid and perceived by a specialist insect. Helv Chim Acta 80:1408–1420

Steinbauer MJ, Schiestl FP, Davies NW (2004) Monoterpenes and epicuticular waxes help female autumn gum moth differentiate between waxy and glossy *Eucalyptus* and leaves of different age. J Chem Ecol 30:1117–1142

Stoner KA (1990) Glossy leaf wax and plant resistance to insects in *Brassica oleracea* under natural infestation. Environ Entomol 19:730–739

Stork NE (1980) Role of waxblooms in preventing attachment to brassicas by the mustard beetle, *Phaedon cochleariae*. Entomol Exp Appl 28:100–107

Talley SM, Coley PD, Kursar TA (2002) Antifungal leaf-surface metabolites correlate with fungal abundance in sagebrush populations. J Chem Ecol 28:2141–2168

Tamayo MC, Rufat M, Bravo JM, San Segundo B (2000) Accumulation of a maize proteinase inhibitor in response to wounding and insect feeding, and characterization of its activity toward digestive proteinases of *Spodoptera littoralis* larvae. Planta 211:62–71

Taylor FE, Davies LG, Cobb AH (1981) An analysis of the epicuticular wax of *Chenopodium album* leaves in relation to environmental change, leaf wettability and the penetration of the herbicide bentazone. Ann Appl Biol 98:471–478

Traw MB, Dawson TE (2002) Differential induction of trichomes by three herbivores of black mustard. Oecologia 131:526–532

Uematsu H, Sakanoshita A (1989) Possible role of cabbage leaf wax bloom in suppressing Diamondback moth *Plutella xylostella* (Lepidoptera: Yponomeutidae) oviposition. Appl Entomol Zool 24:253–257

Valkama E, Koricheva J, Ossipov V, Ossipova S, Haukioja E, Pihlaja K (2005) Delayed induced responses of birch glandular trichomes and leaf surface lipophilic compounds to mechanical defoliation and simulated winter browsing. Oecologia 146:385–393

van Dam NM, Hadwich K, Baldwin IT (2001) Induced responses in *Nicotiana attenuata* affect behavior and growth of the specialist herbivore *Manduca sexta*. Oecologia 122:371–379

van Loon JJA, Blaakmeer A, Griepink FC, van Beek TA, Schoonhoven LM (1992) Leaf surface compound from *Brassica oleracea* (Cruciferae) induces oviposition by *Pieris brassicae* (Lepidoptera, Pieridae). Chemoecology 3:39–44

Vogelmann TC (1993) Plant-tissue optics. Annu Rev Plant Physiol Plant Molec Biol 44:231–251

Vötsch W, Nicholson G, Müller R, Stierhof YD, Gorb S, Schwarz U (2002) Chemical composition of the attachment pad secretion of the locust *Locusta migratoria*. Insect Biochem Molec Biol 32:1605–1613

Vrieling K, Derridj S (2003) Pyrrolizidine alkaloids in and on the leaf surface of *Senecio jacobaea* L. Phytochemistry 64:1223–1228

Walton TJ (1990) Waxes, cutin and suberin. In: Harwood JL, Boyer J (eds) Lipids, membranes and aspects of photobiology. Academic Press, London, pp 105–158

Welke B, Ettlinger K, Riederer M (1998) Sorption of volatile organic chemicals in plant surfaces. Environ Sci Technol 32:1099–1104

Whitehouse P, Holloway PJ, Caseley JC (1982) The epicuticular wax of wild oats in relation to foliar entry of the herbicides diclofop-methyl and difenzoquat. In: Cutler DF, Alvin KL, Price CE (eds) The plant cuticle. Academic Press, London, pp 315–330

Wollenweber E (1989) Exudate flavonoids in flowering plants and ferns. Naturwissenschaften 76:458–463

Wollenweber E, Christ M, Dunstan RH, Roitman JN, Stevens JF (2005a) Exudate flavonoids in some Gnaphalieae and Inuleae (Asteraceae). Z Naturforsch C Biosci 60:671–678

Wollenweber E, Dorr M, Bohm BA, Roitman JN (2004) Exudate flavonoids of eight species of *Ceanothus* (Rhamnaceae). Z Naturforsch C Biosci 59:459–462

Wollenweber E, Dorsam M, Dorr M, Roitman JN, Valant-Vetschera KM (2005b) Chemodiversity of surface flavonoids in Solanaceae. Z Naturforsch C Biosci 60:661–670

Woodhead S (1983) Surface chemistry of *Sorghum bicolor* and its importance in feeding by *Locusta migratoria*. Physiol Entomol 8:345–352

Woolley JT (1971) Reflectance and transmittance of light by leaves. Plant Physiol 47:656–662

Yang G, Isenhour DJ, Espelie KE (1991) Activity of maize leaf cuticular lipids in resistance to leaf-feeding by the fall armyworm. Florida Entomol 74:229–236

Yencho GC, Renwick JAA, Steffens JC, Tingey WM (1994) Leaf surface extracts of *Solanum berthaultii* Hawkes deter Colorado beetle feeding. J Chem Ecol 20:991–1007

Zabkiewicz JA (2000) Adjuvants and herbicidal efficacy – present status and future prospects. Weed Res 40:139–149

Zalucki MP, Clarke AR, Malcolm SB (2002) Ecology and behavior of first instar larval Lepidoptera. Annu Rev Entomol 47:361–393

Chapter 6
Wound-Periderm Formation

Idit Ginzberg

Herbivores, and particularly chewing insects, cause substantial damage to the plant. In addition to lost tissue, there are great concerns of pathogen invasion and water loss at the site of the attack. One of the plant's defense strategies is the formation of wound periderm at the boundaries of the invaded or damaged region to isolate it from non-wounded healthy tissue. The development of wound periderm following insect feeding has never been specifically examined; although studies of herbivory and wound signaling have indicated extensive overlap in the respective sets of induced genes.

The periderm protective characteristics are mainly due to the suberized walls of its outer cell layers. Suberin is composed of aromatic and aliphatic polyester domains, and associated waxy material, providing biochemical and structural barriers against pathogen infection, and contributes to water-proofing of the periderm. Most of the current knowledge on wound periderm derives from healing processes of mechanically wounded potato tubers. The review summaries these studies, in light of plant response to herbivory.

6.1 Introduction

Aerial plant organs are generally protected by the cuticle and waxes that separate the outer environment and the interior tissues, and help maintain internal homeostasis. Upon injury of plant surfaces, the inner tissues are exposed to the surroundings, and may become desiccated and/or infected by pathogens, unless impermeability is rapidly reestablished. Plants do not heal wounds in the sense that damaged tissue is completely restored; instead, wound healing involves minimizing pathogen invasion and fluid loss.

Although wound responses may differ among plant species and organs, they follow a similar pattern. The first layer of intact cells adjacent to the wound is modified by the accumulation of various antimicrobial and water-impermeable

---

I. Ginzberg
Agricultural Research Organization, The Volcani Center, Bet Dagan 50250, Israel
e-mail: iditgin@volcani.agri.gov.il

substances, including lignin and suberin (Rittinger et al. 1987), to form a sealing layer. This may be followed by the development of periderm tissue under the sealing layer (Lulai and Freeman 2001; Rittinger et al. 1987; Thomson et al. 1995).

Periderm is a protective tissue of secondary origin that replaces the epidermis when the latter is damaged. The protective characteristics of the periderm are mainly due to the deposition of suberin macromolecules on the walls of its outer cell layers. Periderm formation is a common phenomenon in stems and roots of dicotyledons and gymnosperms which increase in thickness by secondary growth, as well as in lenticels and following wounding. Actually, periderm formation and suberization processes are considered one of the generalized responses to wounding, irrespective of the nature of the natural protective polymer of the plant tissue (Bloch 1941; Dean and Kolattukudy 1976; Hawkins and Boudet 1996; Kolattukudy 2001; Rittinger et al. 1987).

Almost all herbivores, and particularly chewing insects, cause substantial injury at the site of attack. In addition to the immediate and local damage to the plant due to lost tissue, there is the great concern of water loss from the perimeter of the injury (Aldea et al. 2005). One of the plant's defense strategies is the formation of wound periderm at the boundaries of the invaded or damaged region to isolate it from non-wounded healthy tissue (Franceschi et al. 2005). The wound periderm may prevent successive pest invasions (Ichihara et al. 2000) and wall off pathogen infection. However, the development of wound periderm following insect feeding has never been specifically studied. Studies of herbivory and wound signaling have indicated extensive overlap in the respective sets of induced genes (Major and Constabel 2006; Ralph et al. 2006; Reymond et al. 2000), and insect feeding has been simulated using mechanically wounded plants, but such experiments are typically too short for periderm to form.

Most of our current knowledge on wounding periderm and suberization is derived from wound-induced healing processes in potato tuber. Wound-periderm formation is induced by removing the tuber skin or excising discs of tuber flesh using a cork borer, and allowing the exposed tissue to heal. This procedure enables the isolation of significant amounts of suberized periderm which are then available for anatomical and chemical characterization, as well as for direct measurements of water permeability. In addition to the technical advantages involved in using the potato-tuber-disc system, there is a resemblance between the wound periderm of potato and that of other plants. Tissue development and anatomy are highly similar, and chemical components of suberin from other species are also found in the suberized periderm of potato (Gandini et al. 2006; Schreiber et al. 2005). Collectively, these characteristics of induced wound periderm in potato tubers make them a good model to study suberized wound-periderm tissue in plants.

6.2 Potato Periderm: Origin and Anatomy

Periderm tissue replaces the epidermis early in potato-tuber development (Reeve et al. 1969). This type of periderm, designated natural or native periderm, is similar to wound periderm in terms of tissue origin, structure and morphology (Sabba and

Lulai 2002). Both periderm types consist of similar chemical components, although they differ in their relative quantities (Lapierre et al. 1996; Negrel et al. 1996).

The native periderm is initiated by divisions of both the epidermal and subepidermal cells, first at the stem end and later in the bud-end region of the tuber (Artschwager 1924; Reeve et al. 1969). New periderm is continuously added by cell divisions during tuber maturation. The capacity to produce new periderm upon wounding is retained by all parenchyma tissues in the tuber, but it is most pronounced close to the original, native periderm (Artschwager 1927).

Both native and wound periderms consist of three cell types: the external suberized phellem cells derive from a single-cell meristematic layer, the phellogen (cork cambium), localized underneath them; inward cell divisions of the phellogen give rise to the parenchyma-like phelloderm (Fig. 6.1). Following suberization of their walls, the phellem cells die, thus creating the outer protective layer. The term periderm and those referring to its components are derived from Greek words, among which *phellem* means cork, *gen*, to produce, *derma*, skin, and *peri*, about (Esau 1965). In non-wounded tubers, the suberized phellem constitutes the potato skin (Lulai and Freeman 2001).

In both wounded and native periderm, the phellem cells are identified by their rectangular shape and arrangement in columnar rows and by the fact that their walls autofluoresce under UV illumination (Fig. 6.1). This columnar pattern is a result of their origin from periclinal divisions in the phellogen cell layer (Artschwager 1924); the autofluorescence is due to the presence of aromatic suberin polymers in their cell walls (Bernards and Lewis 1998). In immature periderm (both native and wounding), the actively dividing phellogen is labile and prone to fracture, allowing separation of the suberized phellem from the underlying phelloderm and parenchyma cells (Lulai and Freeman 2001; Sabba and Lulai 2002). When the phellogen's meristematic activity ceases, the cell walls thicken and become more resistant to fracture. Unesterified pectins may be involved in cell–cell adhesion via calcium-pectate formation (Kobayashi et al. 1999; Parker et al. 2001), and they have been detected in maturing native periderm (Lulai and Freeman 2001; Sabba and Lulai 2002, 2005). However, maturation of wound periderm is not associated with an increase in unesterified pectin in the phellogen walls (Sabba and Lulai 2002, 2004). Nevertheless, to gain maximal protection in the wounded area, a mature wound periderm should develop in which phellogen activity ceases and the suberized phellem layers adhere strongly to the underlying phelloderm.

6.3 Chemical Composition of the Periderm

6.3.1 Suberin

6.3.1.1 Occurrence

The periderm owes its protective characteristics to the suberized walls of the phellem cells. Identifying suberized walls in a variety of plant organs has led to

A Native periderm

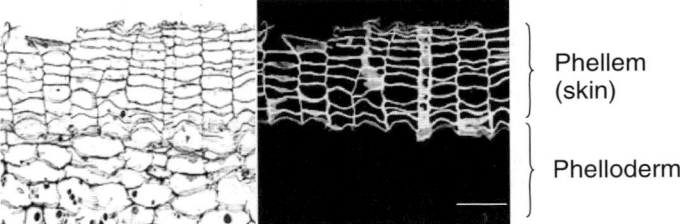

B Wound periderm

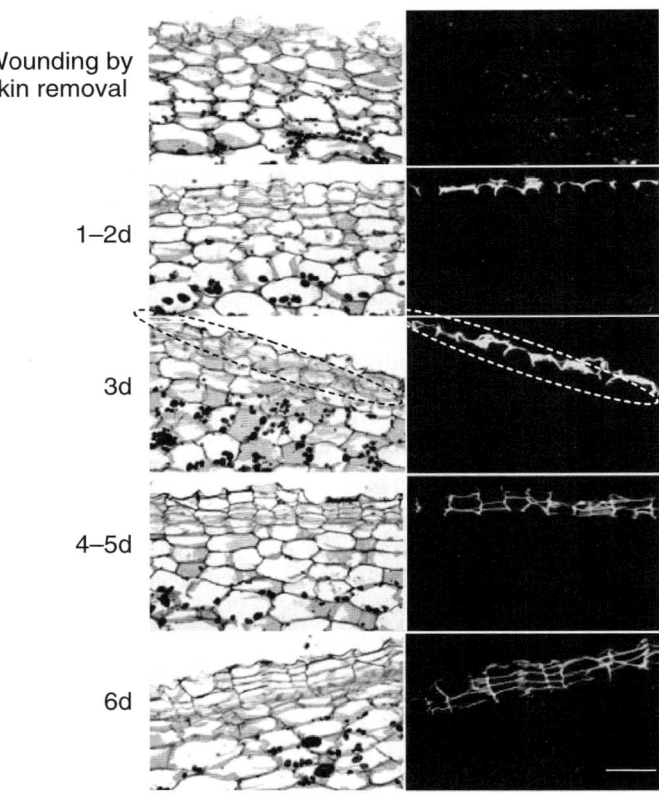

Fig. 6.1 Potato native and wound periderm. Cross sections of potato (*Solanum tuberosum* L. cv. Desiree) tuber surface were stained with Safranin O/Fast Green and viewed with light (left panel) and UV (right panel) microscopes to examine tissue morphology and autofluorescence of suberized cells, respectively. (**A**) Native periderm of mature tuber with suberized phellem cells (the 'skin') and the parenchyma-like phelloderm, (**B**) Development of wound periderm following removal of the skin. Wounding induced suberization of the exposed tuber cells (1–3d). On the third day a meristematic phellogen that appears as multinuclear region developed below these cells (circled). Outward divisions of the phellogen produced the characteristic suberized phellem layers (4–6d). The dark grains are amyloplasts. Bar represents 100 μm

the conclusion that suberization of cell walls is a universal mechanism that plants use whenever there is a need for the cell wall to become an effective diffusion barrier (Dean and Kolattukudy 1976; Kolattukudy 1984, 2001). Examples include epidermis, hypodermis and endodermis in roots (Schreiber et al. 1999; Zeier et al. 1999; Zimmermann et al. 2000) and in other underground plant organs, such as tubers and stolons (Espelie et al. 1980), as well as suberization of chalazal cells (Cochrane et al. 2000), the bundle sheath (Evert et al. 1996), and the base of secretory glands (Bakker and Baas 1993), where compartmentalization of nutrients or toxic materials is required. One specific example is the suberized walls of phenol-storing cells which are part of the plant's constitutive defense mechanism against grazing animals and feeding insects. The toxic compounds are kept compartmentalized in a reduced, poised state within the cell vacuoles, while at the same time providing the means for their rapid oxidation, should decompartmentalization occur, e.g. via wounding (Beckman 2000). These phenolic compounds can be further used as building blocks for the suberin macromolecules as one means of sealing off the immediate site of cellular penetration. If the stress persists, these same processes promote the prolonged build-up of indole acetic acid (IAA) and ethylene, resulting in a cascade of events that produce a peridermal defense at depth, several centimeters below the point of immediate danger (Beckman 2000).

In addition to its developmentally regulated biosynthesis, suberin synthesis is induced and its polymers are deposited in the walls of cells adjacent to wound sites during wound healing. Most importantly, suberization develops not only on tubers and other organs that are normally protected by suberin but also on fruits and leaves, which are normally protected by cutin (Bakker 1988; Dean and Kolattukudy 1976; Keren-Keiserman et al. 2004; Simons and Chu 1978).

6.3.1.2 Cellular Localization

Suberin is composed of aromatic and aliphatic polyester domains, as well as associated waxy material, and is localized between the primary wall and the plasmalemma (Kolattukudy 1984). The model for suberin macromolecules suggests that the aromatic domains are attached to the cell-wall carbohydrates (Yan and Stark 2000), and the aliphatic domains are attached covalently to the phenolic domain via ester bonds (Bernards and Lewis 1992; Kolattukudy 1980, 1984). Transmission electron microscopy of suberized walls shows a lamellar structure in which the light segments probably constitute the aliphatic domain (Soliday et al. 1979). The suberin in potato wound periderm is deposited from the oldest, outer cell layer inward. Deposition occurs in a segmented fashion within each cell layer, first on the outer tangential cell walls, then on the associated radial walls and finally, on the inner tangential walls (Lulai and Corsini 1998).

Suberin structure and chemical composition, as well as the enzymes involved in its biosynthesis, have been reviewed in detail (Bernards 2002; Bernards and Lewis 1998; Gandini et al. 2006; Graça and Santos 2007; Kolattukudy 2001). In the present review, this information is given in brief, to demonstrate the complexity of the suberization process and the difficulties in deciphering its structure and composition, which to date have only been partially characterized.

6.3.1.3 Purification and Detection

As suberin is deposited within the cell wall and attaches to it, it is impossible to isolate pure suberin from the cell wall. In the case of healing potato discs, the periderm tissue is peeled away from the tuber flesh and the adhering carbohydrates are removed by treatment with a mixture of hydrolases, such as cellulase and pectinase (Kolattukudy and Agrawal 1974; Vogt et al. 1983), leaving behind an insoluble material which is enriched in suberized walls. Further hydrolysis of noncellulosic polysaccharides is possible using trifluoroacetic acid (Arrieta-Baez and Stark 2006). The soluble waxes associated with the polymer are then extracted with organic solvents and analyzed by gas chromatography/mass spectrometry (GC/MS). Aliphatic components are released from the polymer by depolymerization with alkaline hydrolysis, transesterification with methanol in the presence of BF_3 or $NaOCH_3$, or exhaustive hydrogenolysis with $LiAlH_4$ in tetrahydrofuran (Kolattukudy 1984). The ether and C–C bonds in the aromatic polymer are oxidatively cleaved with alkaline nitrobenzene, CuO or thioglycolate treatment (Kolattukudy 1984). The derivatized suberin monomers are identified by GC/MS, high-pressure liquid chromatography (HPLC) or nuclear magnetic resonance (NMR; Kolattukudy 1984; Yan and Stark 2000).

6.3.1.4 Chemical Composition and Biosynthesis

The aromatic domain of potato-periderm suberin is composed of hydroxycinnamates and their amide derivatives as well as monolignols (Bernards and Lewis 1998; Yan and Stark 2000). The major components of the aliphatic domain are α, ω-diacids and ω-hydroxyacids, with minor amounts of alkan-1-ols and alkanoic acids (Graça and Pereira 2000). Glycerol has been suggested to be involved in the cross-linking between the corresponding polyesters, allowing the formation of a three dimensional network (Graça and Pereira 2000), and recently it was shown that glycerol-3-phosphate acetyltransferase5 gene play a critical role in suberized cell wall biogenesis (Beisson et al. 2007). In addition, the potato periderm contains up to 20% extractable lipids (waxes). Besides linear long-chain aliphatic wax compounds, alkyl ferulates have been detected as significant constituents; in wound periderm they amount to more than 60% of the total extracts (Schreiber et al. 2005).

It is worth noting that suberin aromatic and aliphatic domains differ from lignin and cutin, respectively. The major distinction between cutin and suberin is that the former contains polar in-chain hydroxylated and/or epoxy acids as major components (Matzke and Riederer 1991). As for lignin, it is comprised of polymerized hydroxycinnamyl alcohols, i.e. monolignols (Bernards and Razem 2001; Bernards et al. 2000).

The phenylpropanoid pathway is induced in both aromatic suberin and lignin biosynthesis. The activities of phenylalanine ammonia-lyase (PAL), 4-coumaryl-CoA ligase (4-CL) and cinnamyl alcohol dehydrogenase (CAD) increase in both suberization and lignification, while cinnamoyl-CoA oxidoreductase (CCR) activity is induced only upon initiation of lignification (Bernards et al. 2000). The lack of CCR induction in suberizing potato periderm indicates channeling of hydroxycinnamoyl-CoA derivatives away from monolignol formation and toward that of suberin derivatives (Bernards et al. 2000). Hydroxylation and methylation of hydroxycinnamic acids occur at the hydroxycinnamoyl-CoA derivatives level, and putatively suberin-specific O-methyltransferase has been cloned from maize (Held et al. 1993). Channeling of the phenylpropanoid phenolics into suberin has been suggested to involve the activity of tyramine hydroxycinnamoyl transferase (THT; Negrel et al. 1993, 1996), and its cDNA has been cloned from potato (Schmidt et al. 1999). It has further been suggested that the phenolic domain is polymerized via peroxidase activity (Cottle and Kolattukudy 1982) and requires H_2O_2 (Razem and Bernards 2003), and a suberization-associated anionic peroxidase has been cloned (Roberts et al. 1988; Roberts and Kolattukudy 1989). Cationic isoforms may also be involved, although these have not yet been cloned (Quiroga et al. 2000).

Reactions which are specific to aliphatic suberin include fatty-acid elongation (C-18 and C-24 to C-32) and oxidation. Ketoacyl-CoA synthase, catalyzing the first step in the elongation of 18:0 via malonyl-CoA, has been cloned from suberizing primary roots of maize (Schreiber et al. 2000). The activities of two NADP-dependent dehydrogenases that catalyze the oxidation of ω-hydroxyacids and ω-oxoacid to their corresponding dicarboxylic acids are considered to be key steps in aliphatic suberin biosynthesis (Agrawal and Kolattukudy 1977); however, their corresponding cDNAs have not yet been cloned. The incorporation of ferulates into suberin polymers has been suggested to be through conjugation to the hydroxyfatty acids by the hydroxyfatty acid feruloyl-CoA transferase (HHT; Lofty et al. 1994).

Extensive chemical analysis of suberin has enabled the identification of its monomeric and oligomeric fragments and the construction of a conceptual model of its structure in the cell wall. Nevertheless, only sketchy details, described herein, are known regarding the biosynthesis of suberin-specific components, the polymerization reactions, the bonding between the aromatic and aliphatic domains and the anchoring of the macromolecules to carbohydrates in the primary cell wall. This and the regulation of suberin induction and biosynthesis await further clarification.

6.4 Protective Characteristics of Wound Periderm

Following suberization of their cell walls, the phellem cells die, creating a protective layer that covers the previously exposed tissue at the wounding site. The corpses of the cells are filled with air and therefore provide thermal insulation, the suberized walls prevent invasion by microorganisms (mechanically and chemically), and wax deposits that are embedded within the suberin material prevent desiccation of the previously exposed plant tissue (Kolattukudy 1977). Mechanisms by which suberin may impart disease resistance have been suggested by Kolattukudy (1984): it may act as a barrier to the diffusion of pathogen enzymes or toxins into living tissues, as a structural barrier to pathogen ingression, or as a biochemical barrier to microbes due to the high proportion of phenolic materials incorporated into the suberin polymer.

In potato tubers, synthesis and integration into the cell walls of hydroxycinnamic acid amides of tyramine and octopamine constitute a very early response to wounding (Negrel et al. 1993). These amides exhibit antifungal activity and are postulated to contribute to the formation of a primary phenolic barrier against pathogen invasion, and to be the building blocks, with other phenolics, of the suberin polyaromatic domain which then serves as a permanent barrier (Bernards et al. 1995), replacing the cuticle. The suberin aromatic domain further strengthens the cell walls against fungal enzymatic hydrolysis (Negrel et al. 1995), and the thickness of the suberized cell layers is highly correlated with resistance to pathogens (Morris et al. 1989). These postulated roles of the suberin phenolic domain were supported by experiments with inhibitors of suberin phenolic synthesis which prevented the development of pathogen resistance (Bostock and Stermer 1989; Hammerschmidt 1984). Lulai and Corsini (1998) further suggested that the suberin phenolic domain provides resistance to bacterial (*Erwinia carotovora* subsp. *carotovora*) but not fungal (*Fusarium sambucinum*) infection, whereas deposition of the suberin aliphatic domain is responsible for the latter.

In potato-tuber wound-healing tissue, the deposition of suberin aliphatic components has also been correlated with the development of resistance at the tissue's surface to diffusion of water vapor (Kolattukudy and Dean 1974). However the waxes that associated with the suberin polymer were suggested to play the crucial role in water-proofing of the periderm. Specific inhibition of wax synthesis inhibited the development of resistance to water vapor diffusion at the periderm surface, indicating that the waxes associated with the suberin polymer, rather than the polymer itself, constitute the major diffusion barrier formed during wound healing (Soliday et al. 1979). This role of suberin-associated waxes is analogous to the contribution of waxy material to the aerial cuticle: in cuticles, water permeability is completely determined by the waxes (Schönherr 1976). Extraction of the wax from the periderm results in an increase in water permeability (Espelie et al. 1980; Schreiber et al. 2005; Vogt et al. 1983).

Interestingly, water permeability of the wound periderm is, on average, 100 times higher than that of natural periderm (Schreiber et al. 2005; Vogt et al. 1983). Schreiber et al. (2005) correlated the formation and amount of the aliphatic domain in potato-tuber periderm with its efficiency as a barrier to water transport. The differences in suberin chemical composition and amounts between native and wound periderm were minor compared to the large differences in water permeability, suggesting that the molecular arrangement and precise localization of the deposition of suberin and its associated waxes within the cell wall contribute to its efficiency as a barrier to water transport.

6.5 Erecting the Periderm Barrier

6.5.1 Time Course of Periderm Development

Following wounding, the erection of a protective barrier before pathogen invasion and establishment is fundamental to host resistance. Hence, the rate of periderm development and suberization of its outer phellem cells are important determinants in plant recovery after wounding.

Cytological events that may lead to phellogen initials and periderm development have been suggested to follow a transient increase in IAA and lipid hydroxyperoxide, both reaching maximal levels 20–30 min post-wounding of potato tubers (Fabbri et al. 2000). This is followed by mitotic activity of cells competent to produce periderm, starting from 120 min post-wounding (Fabbri et al. 2000). Cell divisions are accompanied by a biphasic increase in protein synthesis, 2 h and then between 12 and 24 h post-wounding (Morelli et al. 1994), and with an increase in polymerized actin and microfilament bundles in cells at the wounding site, 12 h post-wounding (Morelli et al. 1998).

The depositions of suberin aliphatic and aromatic compounds in the developing periderm have been found to be separate processes (Lulai and Morgan 1992). Suberin polyphenolics accumulate first during wound-induced suberization, while the suberin aliphatics accumulate last and complete the suberin barrier (Lulai and Corsini 1998; Lulai and Morgan 1992).

PAL activity, which is a necessary prerequisite for the biosynthesis of suberin aromatic monomers, was found to reach a transient peak 12 h after potato-tuber wounding; however, it was not restricted to the suberizing cell layers, and was also detected in deeper layers (Borchert 1978). Suberin deposition initiated 24 h after slicing the potato tuber (Thomas 1982). When radioactive cinnamic acid was fed to wound-healing potato-tuber discs, maximal uptake was measured during the first two days post-wounding and declined thereafter (Cottle and Kolattukudy 1982b). Consequently, the highest deposition of the phenolic suberin components onto the cell wall was measured between three and seven days of wound healing (Cottle and Kolattukudy 1982b; Hammerschmidt 1985; Lulai and Corsini 1998). Accordingly, total peroxidase activity, presumed to be involved in polymerization of the aromatic domain, and its anchorage to the cell wall, increased gradually in the suberizing cells four days post-wounding and remained high thereafter (Borchert 1978).

The suberin phenolic domain provides esterification sites for the aliphatic components; hence, initiation of aliphatics deposition in the periderm tissue was detected upon completion of suberin phenolic depositions at the first, outer layer of the phellem (Lulai and Corsini 1998). With the use of cycloheximide and actinomycin D, it was demonstrated that RNA and protein synthesis of ω-hydroxyacid dehydrogenase, the enzyme involved in the formation of suberin aliphatic components, occurs between 72 and 96 h post-wounding (Agrawal and Kolattukudy 1977). Measurement of the accumulation of the aliphatic components indicated a slower increase up to five days post-wounding followed by a dramatic increase up to day ten of suberization (Agrawal and Kolattukudy 1977; Lulai and Corsini 1998). The temporal incorporation of $[1-^{14}C]$oleic acid and $[1-^{14}C]$acetate into aliphatic components of suberin supported this time course (Dean and Kolattukudy 1977).

Thus, in potato, deposition of the suberin components initiates one to two days post-wounding, and formation of fully protective tissue is complete about ten days later.

The above time course of suberin accumulation up to the formation of a fully protective barrier may vary with potato genotype (Lulai and Orr 1995). Such variation is also found in woody plants, e.g. suberin accumulation in peach bark wounds was found to be highly cultivar-dependent (Biggs and Miles 1988).

6.5.2 Physiological Factors that Determine the Rate of Wound Healing

The above-described time course for the development of a suberized periderm barrier in the potato-tuber-disc system was determined under controlled conditions; however, physiological as well as environmental conditions influence the capacity to heal wounds. The effect of temperature and relative humidity on periderm formation has been extensively studied in the context of the industrial/agricultural practice of curing skinning injuries of newly harvested potato tubers. Histological observations indicated that wound-induced periderm formation occurs most rapidly at 25°C; wound healing was delayed at lower temperatures (10°C–15°C), while temperatures as high as 35°C prevented periderm formation (Morris et al. 1989; Thomas 1982). Similarly, in wound-healing peach bark, phellogen regeneration was strongly influenced by temperature and occurred more rapidly under moderate climatic conditions (Biggs and Miles 1988). Nevertheless, high humidity (98%) in combination with the optimal temperature (25°C) hastened wound-induced periderm formation in potato tubers (Morris et al. 1989).

The effect of temperature on wound-induced suberin formation was determined in the potato-tuber-disc system by monitoring the development of water-loss resistance. The optimum temperature for the formation of aliphatic suberin monomers and the development of resistance to water-vapor conduction was 26.4°C, whereas alkane synthesis was optimal at 18.6°C (Dean 1989). Lower temperatures reduced suberin monomer production. The difference in optimal temperatures for the synthesis of suberin aliphatic and wax components may derive from the temperature sensitivity of fatty-acid-chain-elongation reactions: C_{21} alkane dominates the alkanes found at 19.5°C whereas at 29.4°C, the three homologs C_{21}, C_{23} and C_{25}, are found in equal proportions (Dean 1989). Temperature dependence of periderm development and suberin biosynthesis implies an effect of seasonal variation on the plant's ability to recover from injury.

Wound depth affects wound-healing rate: shallow tuber wounds begin to heal faster than deeper ones (Lulai and Orr 1995). The age of the wounded tissue also affects the healing process. Kumar and Knowles (2003) showed that aging potato tubers lose the capacity to heal wounds during storage, together with their ability to sustain superoxide production on the wound surface. They suggested that high activity of superoxide dismutase (SOD) in the aging tissues limits the accumulation of superoxide radicals, thereby reducing the wounded tissue's ability to oxidize 1-aminocyclopropane-1-carboxylic acid (ACC) to wound-induced ethylene. The lower availability of superoxide radicals may further affect the rate of the protective layer's evolution as they are required for polymerization of the suberin aromatic domain (Bernards and Razem 2001; Razem and Bernards 2003). In addition, older tubers exhibit a reduced ability to increase PAL activity in response to wounding, which potentially limits the availability of suberin aromatic precursors and further contributes to age-induced loss in wound-healing ability and pathogen resistance (Kumar and Knowles 2003).

6.5.3 Hormonal Involvement

Suberization, as measured by resistance of the tissue surface to water-vapor diffusion, was inhibited by millimolar concentrations of IAA, severely inhibited by micromolar concentrations of cytokinin, but stimulated by abscisic acid (ABA; Lulai and Orr 1995; Soliday et al. 1978). The involvement of ABA in wound healing was first observed when the hormone was released into the washing medium within one day of cutting the potato discs; addition of ABA to the incubation medium of the washed tuber discs reversed the washing-induced inhibition of suberization (Soliday et al. 1978). It was therefore proposed that during the early phase of wound healing, ABA plays a role in triggering a chain of biochemical processes which eventually result in the formation of a suberization-inducing factor, responsible for the induction of enzymes involved in suberin biosynthesis. The effect of ABA on suberization was further studied in potato-tuber tissue culture by measuring the deposition of suberin components. ABA treatment resulted in an increase in both aromatic and aliphatic polymeric components of suberin, as well as hydrocarbons and fatty alcohols characteristic of waxes associated with potato suberin, and certain enzymes postulated to be involved in suberization (Cottle and Kolattukudy 1982).

As ethylene is a well-known wound-response hormone (Abeles et al. 1992; O'Donnell et al. 1996), its role in wound-induced suberization of potato tuber was examined over a 9-day wound-healing period, using a variety of inhibitors of ethylene biosynthesis and action (Lulai and Suttle 2004). Ethylene evolution was stimulated by wounding and reached a maximum two to three days post-wounding, after which it gradually declined. However, the hormone was not required for wound-induced accumulation of suberin polyphenolics in the sealing layer (i.e., suberization of existing cells at the wound surface) during the first two to four days of wound healing, or for the subsequent suberization of phellem cells (between four and nine days; Lulai and Suttle 2004). Although the effects of ABA, IAA and ethylene were examined with respect to the suberization processes, their involvement in the evolution of wounding periderm, i.e. the initiation of phellogen initials, regulation of cell division, and determination of phellem and phelloderm fate, remained unclear.

The time course of periderm formation and its suberization may differ *in planta*, and may depend on various physiological and environmental stimuli; nevertheless, the rate and extent of suberin accumulation are more important than the actual number of new phellem cells or the thickness of the new suberized layers. This means that heavily suberized periderm composed of relatively few cell layers may be a more effective barrier to pathogen ingression than a thicker periderm (i.e., one with more cell layers) that is relatively less suberized (Biggs 1989; Biggs and Miles 1988).

6.6 Perspectives

Wound-periderm formation and suberization of cell walls at the wounding site are fundamental features of the plant's response to disruption of its natural outer protective membranes. The rate of establishment of suberized periderm following injury is an important factor in the plant's resistance to indirect damage following the wounding, such as water loss and pathogen invasion. However, the regulation, developmental stages and biosynthetic pathway of periderm development and suberization are still not completely understood. Moreover, herbivore feeding differs from plain injury as the former secrete saliva which may interact with chemical factors involved in the healing process. The effect of the plant-herbivore interaction and the resultant signaling with respect to suberized periderm development has not been considered experimentally, and requires further clarification.

References

Abeles F, Morgan P, Saltveit M (1992) Ethylene in plant biology. Academic Press, London
Agrawal VP, Kolattukudy PE (1977) Biochemistry of suberization. Omega-hydroxyacid oxidation in enzyme preparations from suberizing potato tuber disks. Plant Physiol 59:667–672
Aldea M, Hamilton JG, Resti JP, Zangerl AR, Berenbaum MR, DeLucia EH (2005) Indirect effects of insect herbivory on leaf gas exchange in soybean. Plant Cell Environ 28:402–411
Arrieta-Baez D, Stark RE (2006) Using trifluoroacetic acid to augment studies of potato suberin molecular structure. J Agric Food Chem 54:9636–9641
Artschwager E (1924) Studies on the potato tuber. J Agric Res 27:809–835
Artschwager E (1927) Wound periderm formation in the potato as affected by temperature and humidity. J Agr Res 35:995–1000
Bakker JC (1988) Russeting (cuticle cracking) in glasshouse tomatoes in relation to fruit growth. J Hort Sci 63:459–463
Bakker ME, Baas P (1993) Cell walls in oil and mucilage cells. Acta Bot Neer 42:133–139
Beckman CH (2000) Phenolic-storing cells: keys to programmed cell death and periderm formation in wilt disease resistance and in general defence responses in plants? Physiol Mol Plant Pathol 57:101–110
Beisson F, Li Y, Bonaventure G, Pollard M, Ohlrogge JB (2007) The acyltransferase GPAT5 is required for the synthesis of suberin in seed coat and root of *Arabidopsis*. Plant Cell 19:351–368
Bernards MA (2002) Demystifying suberin. Can J Bot 80:227–240
Bernards MA, Lewis NG (1992) Alkyl ferulates in wound healing potato tubers. Phytochemistry 31:3409–3412
Bernards MA, Lewis NG (1998) The macromolecular aromatic domain in suberized tissue: a changing paradigm. Phytochemistry 47:915–933
Bernards MA, Lopez ML, Zajicek J (1995) Hydroxycinnamic acid derived polymers constitute the polyaromatic domain of suberin. J Biol Chem 270:7382–7386
Bernards MA, Razem FA (2001) The poly(phenolic) domain of potato suberin: a non-lignin cell wall bio-polymer. Phytochemistry 57:1115–1122

Bernards MA, Susag LM, Bedgar DL, Anterola AM, Lewis NG (2000) Induced phenylpropanoid metabolism during suberization and lignification: a comparative analysis. J Plant Physiol 157:601–607
Biggs AR (1989) Temporal changes in the infection court after wounding of peach bark and their association with cultivar variation in infection by *Leucostoma persoonii*. Phytopathology 79:627–630
Biggs AR, Miles NW (1988) Association of suberin formation in uninoculated wounds with susceptibility to *Leucostoma cincta* and *L. persoonii* in various peach cultivars. Phytopathology 78:1070–1074
Bloch R (1941) Wound healing in higher plants. Bot Rev 7:110–146
Borchert R (1978) Time course and spatial distribution of phenylalanine ammonia-lyase and peroxidase activity in wounded potato tuber tissue. Plant Physiol 62:789–793
Bostock RM, Stermer BA (1989) Perspective on wound healing in resistance to pathogens. Annu Rev Phytopathol 27:342–371
Cochrane MP, Paterson L, Gould E (2000) Changes in chalazal cell walls and in the peroxidase enzymes of the crease region during grain development in barley. J Exp Bot 51:507–520
Cottle W, Kolattukudy PE (1982) Abscisic acid stimulation of suberization: induction of enzymes and deposition of polymeric components and associated waxes in tissue cultures of potato tuber. Plant Physiol 70:775–780
Cottle W, Kolattukudy PE (1982b) Biosynthesis, deposition, and partial characterization of potato suberin phenolics. Plant Physiol 69:393–399
Dean BB (1989) Deposition of aliphatic suberin monomers and associated alkanes during aging of *Solanum tuberosum* L. tuber tissue at different temperatures. Plant Physiol 89:1021–1023
Dean BB, Kolattukudy PE (1976) Synthesis of suberin during wound-healing in jade leaves, tomato fruit, and bean pods. Plant Physiol 58:411–416
Dean BB, Kolattukudy PE (1977) Biochemistry of suberization. Incorporation of [1 $-^{14}$ C]oleic acid and [1 $-^{14}$ C]acetate into aliphatic components of suberin in potato tuber disks (*Solanum tuberosum*). Plant Physiol 59:48–54
Esau K (1965) The periderm. In: Plant anatomy, 2nd edn. Wiley, New York
Espelie KE, Sadek NZ, Kolattukudy PE (1980) Composition of suberin-associated waxes from the subterranean storage organs of 7 plants—parsnip, carrot, rutabaga, turnip, red beet, sweet potato and potato. Planta 148:468–476
Evert RF, Russin WA, Bosabalidis AM (1996) Anatomical and ultrastructural changes associated with sink-to-source transition in developing maize leaves. Int J Plant Sci 157:247–261
Fabbri AA, Fanelli C, Reverberi M, Ricelli A, Camera E, Urbanelli S, Rossini A, Picardo M, Altamura MM (2000) Early physiological and cytological events induced by wounding in potato tuber. J Exp Bot 51:1267–1275
Franceschi VR, Krokene P, Christiansen E, Krekling T (2005) Anatomical and chemical defenses of conifer bark against bark beetles and other pests. New Phytol 167:353–375
Gandini A, Pascoal C, Silvestre AJD. (2006) Suberin: a promising renewable resource for novel macromolecular materials. Prog Pol Sci 31:878–892
Graça J, Pereira H (2000) Suberin structure in potato periderm: glycerol, long-chain monomers, and glyceryl and feruloyl dimers. J Agric Food Chem 48:5476–5483
Graça J, Santos S (2007) Suberin: a biopolyester of plants' skin. Mol Biosci 7:128–135
Hammerschmidt R (1984) Rapid deposition of lignin in potato tuber tissue as a response to fungi non-pathogenic on potato. Physiol Plant Pathol 24:33–42
Hammerschmidt R (1985) Determination of natural and wound-induced potato tuber suberin phenolics by thioglycolic acid derivatization and cupric oxide oxidation. Potato Res 28:123–127
Hawkins S, Boudet A (1996) Wound-induced lignin and suberin deposition in a woody angiosperm (*Eucalyptus gunnii* Hook): histochemistry of early changes in young plants. Protoplasma 191:96–104

Held BM, Wang H, John I, Wurtele ES, Colbert JT (1993) An mRNA putatively coding for an O-methyltransferase accumulates preferentially in maize roots and is located predominantly in the region of the endodermis. Plant Physiol 102:1001–1008

Ichihara Y, Fukuda K, Suzuki K (2000) The effect of periderm formation in the cortex of *Pinus thunbergii* on early invasion by the pinewood nematode. Forest Pathol 30:141–148

Keren-Keiserman A, Tanami Z, Shoseyov O, Ginzberg I (2004) Peroxidase activity associated with suberization processes of the muskmelon (*Cucumis melo*) rind. Physiol Plant Pathol 121:141–148

Kobayashi M, Nakagawa H, Asaka T, Matoh T (1999) Borate-rhamnogalacturonan II bonding reinforced by Ca^{2+} retains pectic polysaccharides in higher-plant cell walls. Plant Physiol 119:199–203

Kolattukudy PE (1977) Lipid polymers and associated phenols, their chemistry, biosynthesis and role in pathogenesis. Recent Adv Phytochem 77:185–246

Kolattukudy PE (1980) Biopolyester membranes of plants: cutin and suberin. Science 208:990–1000

Kolattukudy PE (1984) Biochemistry and function of cutin and suberin. Can J Bot 62:2918–2933

Kolattukudy PE (2001) Polyesters in higher plants. Adv Biochem Eng Biotechnol 71:2–49

Kolattukudy P, Agrawal VP (1974) Structure and composition of aliphatic constituents of potato tuber skin (suberin). Lipids 9:682–691

Kolattukudy PE, Dean BB (1974) Structure, gas-chromatographic measurement, and function of suberin synthesized by potato tuber tissue slices. Plant Physiol 54:116–121

Kumar GXM, Knowles NR (2003) Wound-induced superoxide production and PAL activity decline with potato tuber age and wound healing ability. Physiol Plant 117:108–117

Lapierre C, Pollet B, Negrel J (1996) The phenolic domain of potato suberin: structural comparison with lignins. Phytochemistry 42:949–953

Lofty S, Negrel J, Javelle F (1994) Formation of ω-feruloyloxypalmitic acid by an enzyme from wound-healing potato tuber discs. Phytochemistry 35:1419–1424

Lulai EC, Corsini DL (1998) Differential deposition of suberin phenolic and aliphatic domains and their roles in resistance to infection during potato tuber (*Solanum tuberosum* L.) wound-healing. Physiol Mol Plant Pathol 53 209–222

Lulai EC, Freeman TP (2001) The importance of phellogen cells and their structural characteristics in susceptibility and resistance to excoriation in immature and mature potato tuber (*Solanum tuberosum* L.) periderm. Ann Bot 88:555–561

Lulai EC, Morgan WC (1992) Histochemical probing of potato periderm with neutral red: a sensitive cytofluorochrome for the hydrophobic domain of suberin. Biotech Histochem 67:185–195

Lulai EC, Orr PH (1995) Porometric measurements indicate wound severity and tuber maturity affect the early stages of wound-healing. Am Potato J 72:225–241

Lulai EC, Suttle JC (2004) The involvement of ethylene in wound-induced suberization of potato tuber (*Solanum tuberosum* L.): a critical assessment. Postharvest Biol Technol 34:105–112

Major IT, Constabel CP (2006) Molecular analysis of poplar defense against herbivory: comparison of wound- and insect elicitor-induced gene expression. New Phytol 172:617–635

Matzke K, Riederer M (1991) A comparative study into the chemical constitution of cutins and suberins from *Picea abies* (L.) Karst, *Quercus robur* L., and *Fagus sylvatica* L. Planta 185:233–245

Morelli JK, Shewmaker CK, Vayda ME (1994) Biphasic stimulation of translational activity correlates with induction of translation elongation factor 1 subunit α upon wounding in potato tubers. Plant Physiol 106:897–903

Morelli JK, Zhou W, Yu J, Lu C, Vayda ME (1998) Actin depolymerization affects stress-induced translational activity of potato tuber tissue. Plant Physiol 116:1227–1237

Morris SC, Forbessmith MR, Scriven FM (1989) Determination of optimum conditions for suberization, wound periderm formation, cellular desiccation and pathogen resistance in wounded *Solanum tuberosum* tubers. Physiol Mol Plant Pathol 35:177–190

Negrel J, Javelle F, Paynot M (1993) Wound-induced tyramine hydroxycinnamoyl transferase in potato (*Solanum tuberosum*) tuber discs. J Plant Physiol 142:518–524

Negrel J, Lotfy S, Javelle F (1995) Modulation of the activity of two hydroxycinnamoyl transferases in wound-healing potato tuber discs in response to pectinase or absisic acid. J Plant Physiol 146:318–322

Negrel J, Pollet B, Lapierre C (1996) Ether-linked ferulic acid amides in natural and wound periderms of potato tuber. Phytochemistry 43:1195–1199

O'Donnell PJ, Calvert C, Atzorn R, Wasternack C, Leyser HMO, Bowles DJ (1996) Ethylene as a signal mediating the wound response of tomato plants. Science 274:1914–1917

Parker CC, Parker ML, Smith AC, Waldron KW (2001) Pectin distribution at the surface of potato parenchyma cells in relation to cell–cell adhesion. J Agric Food Chem 49:4364–4371

Quiroga M, Guerrero C, Botella MA, Barcelo A, Amaga I, Medina MI, Alfonso FJ, de Forchetto SM, Tigier M, Valpnesta V (2000) A tomato peroxidase involved in the synthesis of lignin and suberin. Plant Physiol 122:1119–1127

Ralph SG, Yueh H, Friedmann M, Aeschliman D, Zeznik JA, Nelson CC, Butterfield YSN, Kirkpatrick R, Liu J, Jones SJM, Marra MA, Douglas CJ, Ritland K, Bohlmann J (2006) Conifer defence against insects: microarray gene expression profiling of Sitka spruce (*Picea sitchensis*) induced by mechanical wounding or feeding by spruce budworms (*Choristoneura occidentalis*) or white pine weevils (*Pissodes strobi*) reveals large-scale changes of the host transcriptome. Plant Cell Environ 29:1545–1570

Razem FA, Bernards MA (2003) Reactive oxygen species production in association with suberization: evidence for an NADPH-dependent oxidase. J Exp Bot 54:935–941

Reeve RM, Hautala E, Weaver ML (1969) Anatomy and compositional variation within potatoes. 1. Developmental histology of the tuber. Am Potato J 46:361–373

Reymond P, Weber H, Damond M, Farmer EE (2000) Differential gene expression in response to mechanical wounding and insect feeding in *Arabidopsis*. Plant Cell 12:707–720

Rittinger PA, Biggs AR, Peirson DR (1987) Histochemistry of lignin and suberin deposition in boundary layers formed after wounding in various plant species and organs. Can J Bot 65:1886–1892

Roberts E, Kolattukudy PE (1989) Molecular cloning, nucleotide sequence, and abscisic acid induction of a suberization-associated highly anionic peroxidase. Mol Gen Genet 217:223–232

Roberts E, Kutchan T, Kolattukudy PE (1988) Cloning and sequencing of cDNA for a highly anionic peroxidase from potato and the induction of its mRNA in suberizing potato tubers and tomato fruits. Plant Mol Biol 11:15–26

Sabba RP, Lulai EC (2002) Histological analysis of the maturation of native and wound periderm in potato (*Solanum tuberosum* L.) tuber. Ann Bot 90:1–10

Sabba RP, Lulai EC (2004) Immunocytological comparison of native and wound periderm maturation in potato tuber. Am J Potato Res 81:119–124

Sabba RP, Lulai EC (2005) Immunocytological analysis of potato tuber periderm and changes in pectin and extensin epitopes associated with periderm maturation. J Am Hort Sci 130:936–942

Schmidt A, Grimm R, Schmodt J, Scheel D, Strack D, Rosahl S (1999) Cloning and expression of a potato cDNA encoding hydroxycinnamoyl-CoA: tyramine N-(hydroxycinnamoyl)-transferase. J Biol Chem 274:4273–4280

Schönherr J (1976) Water permeability of isolated cuticular membranes: the effect of cuticular waxes on diffusion of water. Planta 131:159–164

Schreiber L, Franke R, Hartmann K (2005) Wax and suberin development of native and wound periderm of potato (*Solanum tuberosum* L.) and its relation to periderm transpiration. Planta 220:520–530

Schreiber L, Hartmann K, Skrabs M, Zeier J (1999) Apoplastic barriers in roots: chemical composition of endodermal and hypodermal cell walls. J Exp Bot 50:1267–1280

Schreiber L, Skrabs M, Hartmann K, Becker D, Cassagne C, Lessire R (2000) Biochemical and molecular characterization of corn (*Zea mays* L.) root elongases. Biochem Soc Trans 28:647–649

Simons RK, Chu MC (1978) Periderm morphology of mature Golden Delicious apple with special reference to russeting. Sci Hort 8:333–340

Soliday CL, Dean BB, Kolattukudy PE (1978) Suberization: inhibition by washing and stimulation by abscisic acid in potato disks and tissue culture. Plant Physiol 61:170–174

Soliday CL, Kolattukudy PE, Davis RW (1979) Chemical and ultrastructural evidence that waxes associated with the suberin polymer constitute the major diffusion barrier to water-vapor in potato tuber (*Solanum tuberosum* L). Planta 146:607–614

Thomas P (1982) Wound-induced suberization and periderm development in potato tubers as affected by temperature and gamma irradiation. Potato Res 25:155–164

Thomson N, Evert RF, Kelman A (1995) Wound healing in whole potato tubers: a cytochemical, fluorescence, and ultrastructural analysis of cut and bruise wounds. Can J Bot 73:1436–1450

Vogt E, Schonherr J, Schmidt HW (1983) Water permeability of periderm membranes isolated enzymatically from potato tubers (*Solanum tuberosum* L). Planta 158:294–301

Yan B, Stark RE (2000) Biosynthesis, molecular structure, and domain architecture of potato suberin: a ^{13}C NMR study using isotopically labeled precursors. J Agric Food Chem 48:3298–3304

Zeier J, Ruel K, Ryser U, Schreiber L (1999) Chemical analysis and immunolocalisation of lignin and suberin in endodermal and hypodermal/rhizodermal cell walls of developing maize (*Zea mays* L.) primary roots. Planta 209:1–12

Zimmermann HM, Hartmann K, Schreiber L, Steudle E (2000) Chemical composition of apoplastic transport barriers in relation to radial hydraulic conductivity of corn roots (*Zea mays* L.). Planta 210:302–311

Chapter 7
Traumatic Resin Ducts and Polyphenolic Parenchyma Cells in Conifers

Paal Krokene, Nina Elisabeth Nagy and Trygve Krekling

Conifers integrate multiple constitutive and inducible defenses into a coordinated, multitiered defense strategy. Constitutive defenses, established before an attack, represent a fixed cost and function as an insurance against inevitable attacks. Inducible defenses, mobilized in response to an attack, represent a variable resistance that is turned on when it is needed. Polyphenolic parenchyma cells (PP cells) that are specialized for synthesis and storage of phenolic compounds are abundant in the phloem of all conifers. In addition to being a prominent constitutive defense component, PP cells are also involved in a range of inducible defense responses, including activation of existing PP cells, production of new PP cells, and wound periderm formation. Their abundance and varied defensive roles make the PP cells the single most important cell type in conifer defense. Another important defense are traumatic resin ducts which are induced in many conifers after various biotic or abiotic challenges. Traumatic resin ducts are primarily formed in the xylem where they appear in tangential rows, but inducible resin ducts are also formed in the phloem of some conifers. Activation of PP cells and formation of traumatic resin ducts take place through the octadecanoid pathway, involving jasmonate and ethylene signaling.

7.1 Introduction

The conifers include about 630 species of woody gymnosperms that are widely distributed, from Arctic and alpine timber lines to tropical forests (Farjon 2001). Even though the conifers are extremely species-poor compared to the enormous diversity of angiosperms, extant conifers are in many ways remarkably successful plants. Many species are the dominant plants over huge areas, and are major forest trees with immense ecological and economical values. Conifers grow taller, stouter, and older than any other trees, with record holders such as coast redwood *Sequoia sempervirens* (115.6 m tall), Montezuma cypress *Taxodium mucronatum*

P. Krokene
Norwegian Forest and Landscape Institute, N-1432 Ås, Norway
e-mail: Paal.Krokene@skogoglandskap.no

A. Schaller (ed.), *Induced Plant Resistance to Herbivory*,
© Springer Science+Business Media B.V. 2008

(11.4 m diameter) and Great Basin bristlecone pine *Pinus longaeva* (4844 years). Taxonomically, the conifers constitute the division Pinophyta, which includes a single class (Pinopsida) with a single extant order (Pinales or Coniferales). The most species-rich and geographically widespread of the seven to eight recognized families of the Pinales is the Pinaceae with about 225 species, including important Northern Hemisphere forest trees in the genera *Pinus*, *Picea*, *Larix* and *Abies*. Other important conifer families are the Cupressaceae (~135 species), the Podocarpaceae (~185 species) and the Araucariaceae (41 species), all of which include many tropical and Southern Hemisphere species (Farjon 2001).

A key to the conifer's success are their effective defensive strategies that have enabled them to fend of attackers over a phylogenetic history of more than 200 million years. Long-lived and apparent plants such as conifers are bound to be found by natural enemies during ecological time, and thus need a wide range of defenses that are effective against both specialist and generalist enemies (Feeny 1976). This is achieved by integrating multiple constitutive and inducible defenses into a coordinated, multitiered defense strategy (Franceschi et al. 2005). Constitutive defenses, such as cork bark and preformed resin and phenols, act as a first line of defense that inhibits or slows down an initial attack, and if this is not sufficient, inducible defenses are mobilized to kill or compartmentalize the attackers. The temporal and spatial integration of the constitutive and inducible defenses probably contribute to reducing the total costs of defense. Two types of inducible defenses can be distinguished in conifers based on their level of organizational complexity. At the simplest level are defense responses that are a result of changed metabolism of existing cells, such as lignification of cell walls, induction of resin production in preformed resin ducts, and activation of the hypersensitive response. At a more complex organizational level are induced defense responses that involve changes in cell division and differentiation, leading to formation of e.g. traumatic resin ducts and a wound periderm.

The constitutive and inducible defenses in conifers may be both mechanical and chemical in nature. Mechanical defenses consist of structural elements that provide toughness or thickness to tissues, and thereby interfere with penetration and degradation by attackers. Prominent mechanical defenses in conifer stems include the outer defense barrier (the periderm and cork bark), tough elements of the phloem (e.g., sclerenchyma and calcium oxalate crystals), and tissue impregnation with polymers (e.g., lignins and suberin) (Wainhouse et al. 1990; Kartuch et al. 1991; Franceschi et al. 2000, 2005; Hudgins et al. 2003b). Chemical defenses include substances with toxic or inhibitory effects, such as allelochemicals, proteins, and enzymes. Chemical defenses in conifers often consist of pools of stored chemicals (e.g., phenolics or terpenoids) that are released after attack (Phillips and Croteau 1999; Schmidt et al. 2005). There is an obvious overlap between different ways of classifying conifer defenses, as the same defense traits can have both chemical and mechanical properties, and be both constitutively expressed and inducible. Resin for example, a major defense component in many conifers, is produced constitutively and stored in preformed resin ducts, but its production can also be induced in preformed resin ducts (Ruel et al. 1998; Lombardero et al. 2000; Hudgins and

Franceschi 2004), or in inducible traumatic resin ducts (Bannan 1936, Hudgins et al. 2003a, 2004). Furthermore, resin is chemically toxic to insects and fungi, as well as having mechanical defense properties by being sticky and exuded under pressure when attacking organisms are penetrating the bark (e.g., Raffa et al. 1985; Cook and Hain 1988).

In this chapter we describe two important anatomically and chemically based defense components of conifer bark and sapwood. Parenchyma cells with characteristic polyphenolic content occur in the bark of all conifers and are the most abundant living cell type of the secondary phloem. These polyphenolic parenchyma cells (PP cells) are specialized for synthesis and storage of phenolic compounds, and due to their ubiquitous occurrence and varied defensive roles, they may be regarded as the single most important cell type in conifer defense (Franceschi et al. 2005). PP cells are produced during normal phloem development, but undergo extensive changes in response to attack, and are thus a major site of both constitutive and induced defenses (Franceschi et al. 1998, 2000). Another major defense component in conifers are various resin containing structures that may be constitutively present or induced in the bark and sapwood of many species. We will focus on traumatic resin ducts (TRDs) which, as their name implies, are induced after various biotic or abiotic challenges. TRDs in the strict sense are found in the xylem, where they appear as tangential rows, but inducible resin ducts are also formed in the phloem of some conifers (Hudgins et al. 2004). Induction of PP cells and TRDs has been associated with the phenomenon of systemic induced resistance (Bonello and Blodgett 2003) that has been described in only a few members of the Pinaceae so far (Reglinsky et al. 1989; Christiansen et al. 1999b; Krokene et al. 1999, 2000, 2003; Bonello et al. 2001, Bonello and Blodgett 2003), but that is likely to be widespread in conifers.

7.2 Polyphenolic Parenchyma Cells

7.2.1 Anatomy and Development of PP Cells

Polyphenolic parenchyma cells seem to play a prominent role in the defense of all conifer families that have been studied (Franceschi et al. 1998, 2000; Hudgins et al. 2003a, 2004) (Fig. 7.1). Considering their abundance and importance in conifer bark, surprisingly little was published on their structure and function until 1998, when Franceschi and co-workers published the first in a series of papers on PP cell anatomy, development, and roles in constitutive and inducible defenses (Franceschi et al. 1998, 2000, 2005; Krekling et al. 2000, 2004; Krokene et al. 2003; Hudgins et al. 2003a, 2004; Nagy et al. 2004). Norway spruce has served as a model species for most of this research, but studies on other conifers in the Cupressaceae, Podocarpaceae, Araucariaceae and Taxaceae have demonstrated the general importance of PP cells in conifer defense (Hudgins et al. 2003a, 2004).

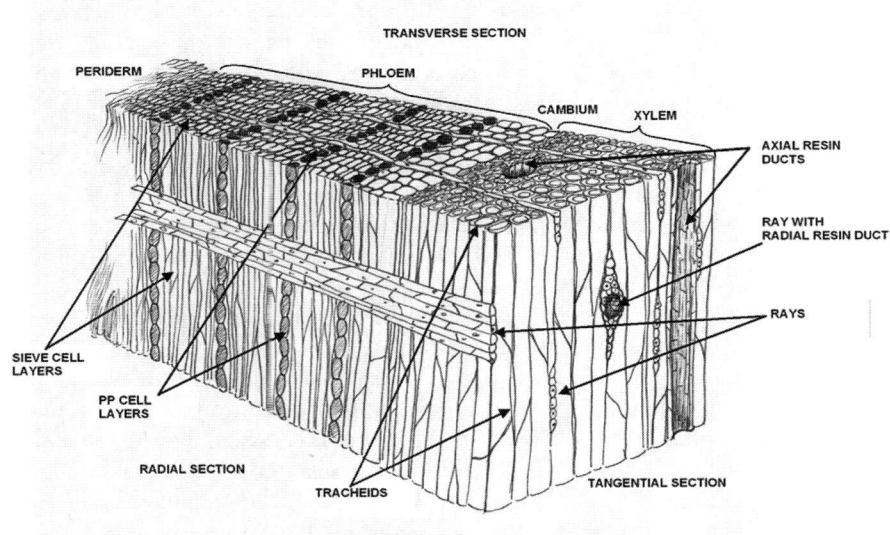

Fig. 7.1 Three-dimensional line drawing showing the basic bark and sapwood anatomy of conifer stems, using Norway spruce as an example. Characterizing features are the multiple concentric layers of defensive structures in the bark, including the periderm and multiple layers of polyphenolic parenchyma cells, and the sapwood with axial and radial resin ducts

Microscopy examination of conifer bark reveals that conifer defenses are laid out in a concentric, multilayered design around the vascular cambium (Franceschi et al. 2005). This is exemplified in the Pinaceae where the PP cells are organized in characteristic concentric rows (one cell wide) in the secondary phloem, separated by 9–12 rows of sieve cells, and associated albuminous cells interspersed with a few scattered brachysclereids or stone cells (Franceschi et al. 1998; Krekling et al. 2000; Figs. 7.1 and 7.2A). In temperate conifers, the PP cell cylinders form annual rings in the phloem, since a new row is laid down at the beginning of every growth season (Alfieri and Evert 1973; Krekling et al. 2000; Fig. 7.3). Running perpendicular to the tangentially oriented PP cell layers, the radial rays span from the secondary phloem into the xylem. Close to the cambium they are usually uniseriate, consisting of ray parenchyma only, but they frequently become multiseriate in the older parts of the phloem with a central radial resin duct (Krekling et al. 2000; Fig. 7.3A).

In the non-Pinaceae, PP cells and rays are organized similarly, but the intervening blocks of sieve cells are only a few cell rows thick (Fig. 7.4B), and normally intersected by an almost complete tangential row of sclerenchymatic fiber cells (Fig. 7.2B). The fiber cells have lignified secondary wall thickenings and serve as structural elements as well as mechanical defense (Hudgins et al. 2003a, 2004). The highly ordered arrangement of densely spaced fiber rows and PP cells probably

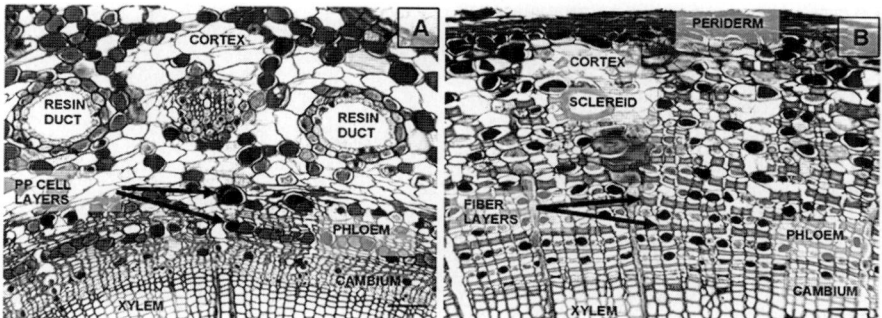

Fig. 7.2 Cross-sections showing the concentric organization of tissues in the bark (cortex and secondary phloem) and in the xylem of two different conifers. (**A**) A young stem of balsam fir (*Abies balsamea*), representing the Pinaceae-type phloem anatomy, with two annual layers of polyphenolic parenchyma (PP) cells, separated by a thicker layer of sieve cells. Large axial resin ducts are found in the cortex. (**B**) Young stem of bald cypress (*Taxodium distichum*) showing typical non-Pinaceae phloem anatomy, with alternating rows of fiber cells and PP cells, separated by thin layers of sieve cells. Both PP cells and fiber cells are laid down in annual layers, with one row of cells per year. Bars, 100 μm

constitutes a formidable physical barrier to organisms that attempt to penetrate the phloem (Hudgins et al. 2004). The fiber cells develop from precursor cells that are distinct from PP cells and normally require two to three years to become fully lignified (Franceschi et al. 2005).

The importance of PP cells in conifer defense is indicated by the facts that they are produced early in both primary and secondary growth, and that considerable resources are spent on their development and maintenance. PP cells are produced by the interfascicular cambium during the earliest stages of primary growth in Norway spruce, and by the second year of growth, regular PP cell layers are produced (Krekling et al. 2000; Fig. 7.2A). In seasonal climates, a new annual layer of PP cells forms soon after the cambium resumes activity after winter dormancy at the beginning of each growth season, from an undifferentiated cell layer that was produced towards the end of the previous growth season (Krekling et al. 2000). Interestingly, a new PP cell layer is also produced after other major disruptions of cambial activity in Norway spruce, such as insect or pathogen attack (Krekling et al. 2000, 2004; Krokene et al. 2003).

Maturation of PP cells is a slow process that may take five years or more in Norway spruce (Krekling et al. 2000). During this process the cells increase in size and gradually become more rounded, the cell walls thicken, and the phenolic bodies become more extensive. Additional PP cells may develop from undifferentiated axial parenchyma cells in between the regular annual layers during the first five to eight years after formation (Krekling et al. 2000). As a consequence of the gradual increase in PP cell size during maturation, the surrounding sieve cells are progressively crushed and compacted and become non-functional (Figs. 7.3A and C). The rows of PP cells, however, stay alive for many years; in 100-year-old

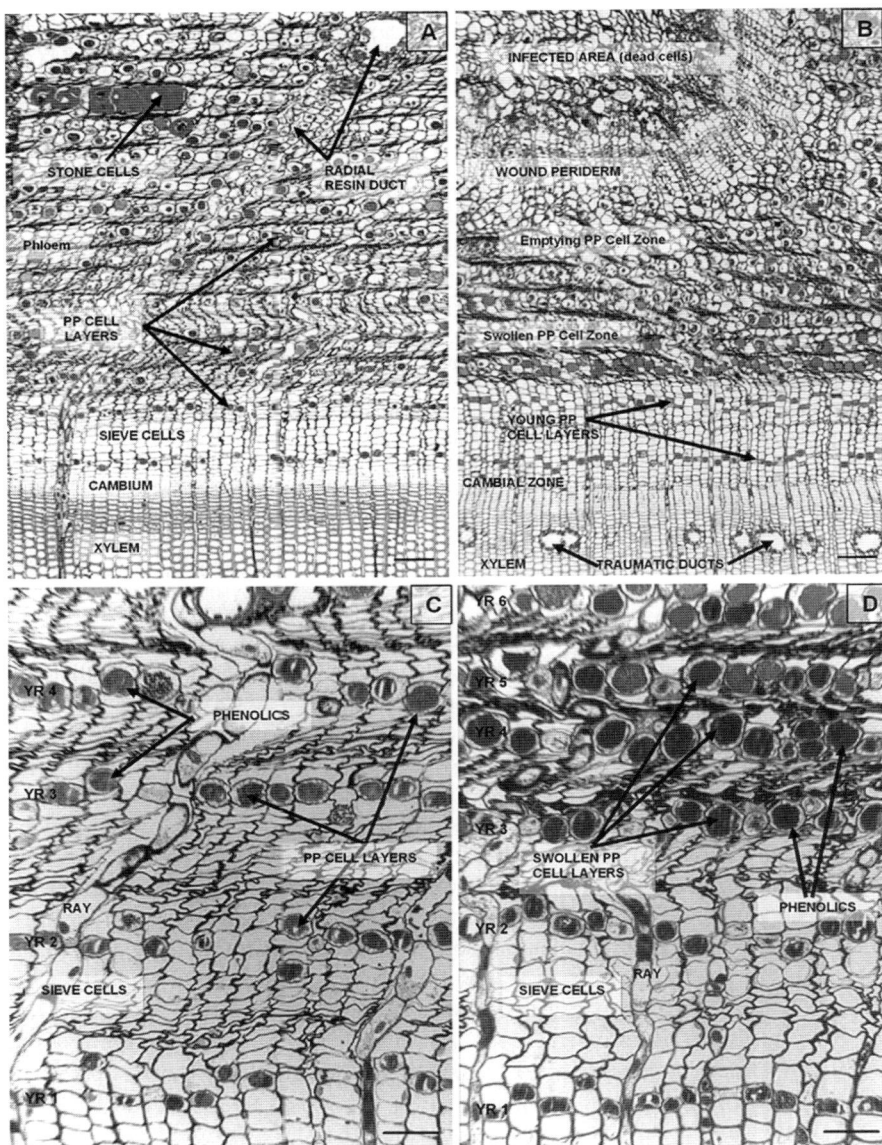

Fig. 7.3 Cross-sections of the secondary phloem of Norway spruce (*Picea abies*), showing the anatomy and organization of polyphenolic parenchyma (PP) cells. (**A**) Cross section from control tree showing the normal, constitutive tissue with concentric layers of PP cells separated by sieve cells. A large radial resin duct inside a multiseriate ray is traversing the phloem and the xylem. Clusters of heavily lignified stone cells are randomly distributed in the older parts of the phloem. These sclerenchyma cells are derived from PP cells as the phloem ages. Bar, 100 μm. (**B**) Cross section showing induction of wound periderm formation and PP cells in the bark after inoculation with the blue-stain fungus *Ceratocysis polonica*. The wound periderm consists of suberized and lignified cells. Below the wound periderm close to the cambium the PP cells are enlarged due to

Norway spruce trees PP cell rows formed more than 70 year ago contained living cells (Krekling et al. 2000). PP cells in older phloem layers may undergo cell division, and this enables the tree to maintain intact layers of PP cells as the stem diameter increases during growth (Krekling et al. 2000). In addition, new PP cells are recruited in the older layers of the secondary phloem by differentiation of existing residual parenchyma cells or ray cell derivatives (Fahn 1990; Krekling et al. 2000).

The phenolic material characterizing the PP cells is stored inside a large vacuole that may fill most of the cell volume (the phenolic body). The phenolic nature of the PP cells is inferred from their bright yellow fluorescence when exposed to blue light (450–490 nm; Fig. 7.4A), strong staining (and quenching of fluorescence) by osmium tetroxide and the periodic acid-Schiff procedure (Fig. 7.3), and the presence of phenylalanine ammonia lyase (PAL), a key enzyme in phenolic synthesis (Parham and Kaustinen 1976; Franceschi et al. 1998). The vacuolar content of the PP cells is highly dynamic and changes in appearance over the year, as well as after pathogen attack or mechanical wounding (Fig. 7.3D). In Norway spruce the density of the phenolic body increases during autumn and appears dense-staining and granular by mid-winter. During spring and summer the density decreases again and reverts to a homogenous, light-staining appearance by mid-summer (Krekling et al. 2000). Attempts have been made to classify PP cells into four types based on the appearance of the vacuolar content (Franceschi et al. 1998), but any relationship between vacuolar morphology and resistance remains unclear.

In addition to being a major storage site for phenolic compounds, PP cells also accumulate and store carbohydrates, in the form of starch grains, and lipids (Murmanis and Evert 1967; Alfari and Evert 1973; Krekling et al. 2000; Fig. 7.4C). Starch reserves build up through the summer months in Norway spruce, are completely absent in the winter, and begin to accumulate again in the spring (Krekling et al. 2000). In addition to serving as energy reserves, the starch and lipid of the PP cells can also be used for rapid synthesis of defensive chemicals in response to attack.

Since PP cells and ray cells make up the bulk of the living cells in the phloem they are probably involved in radial and circumferential transmission of signals that activate defense responses (Franceschi et al. 2000; Krekling et al. 2004). Consistent with this notion, there are numerous symplasmic cell connections (plasmodesmata) facilitating transport of signal molecules and nutrients between PP cells, between ray cells and adjacent PP cells, and between ray cells (Krekling et al. 2000).

PP cell are also involved in production of important mechanical defenses that are found in the secondary phloem of all conifers (Franceschi et al. 2005).

Fig. 7.3 (continued) mobilization of their polyphenolic vacuolar content, whereas PP cells in the immediate vicinity of the wound periderm have emptied their content. Traumatic resin duct formation has been induced in the xylem. Bar, 100 μm. (**C**) Larger magnification of normal, non-activated annual layers of PP cells in the bark. Four annual rows of PP cells are shown, separated by 10–12 layers of sieve cells. Bar, 50 μm. (**D**) Larger magnification of bark three weeks after inoculation with *Ceratocystis polonica*, showing PP cells that have mobilized their phenolic content and increased in size. The older layers of sieve cells are compressed into dense areas of cell walls by the swelling PP cells. Bar, 50 μm

Fig. 7.4 Large magnification cross-sections of polyphenolic parenchyma (PP) cells. The PP cells are named after their phenolic vacuolar inclusions that cover most of the cell volume. (**A**) The polyphenolic material is emitting a characteristic bright, yellowish fluorescence when illuminated with blue light (450–490 nm). (**B**) PP cells in Pacific yew (*Taxus brevifolia*) viewed in the SEM microscope in a back-scattered mode, demonstrating how the cell walls are encrusted with extra-cellular calcium-oxalate crystals. Small extra-cellular crystals are characteristic of non-Pinaceae conifers, in contrast to the Pinacea species that have larger, intra-cellular crystals. (**C**) The PP cells store starch grains in their cytoplasm, clearly visible after staining with the carbohydrate specific Periodic acids Schiff's procedure. Bars, 25 μm

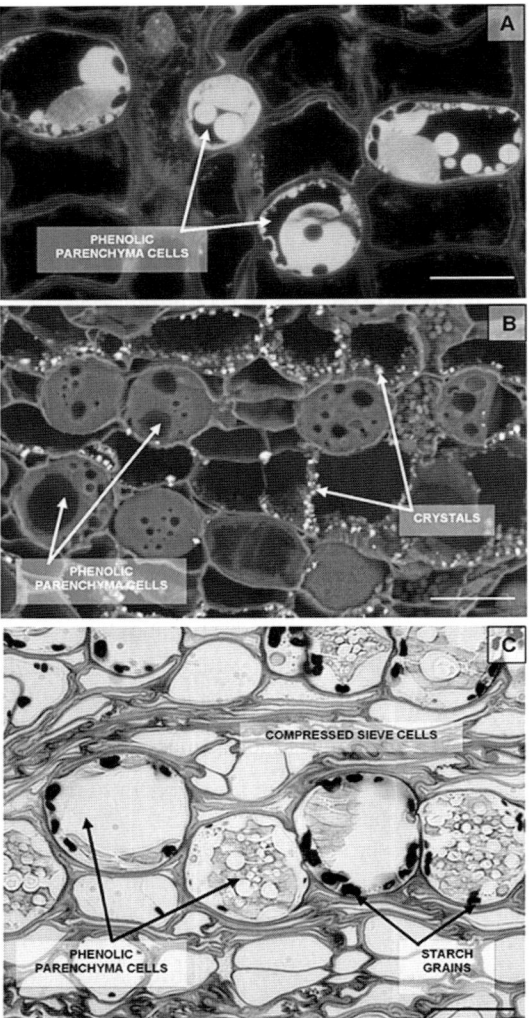

The distribution of stone cells (sclereids) in the secondary phloem of Norway spruce suggests that these cells are derived from PP cells (Franceschi et al. 2005). Stone cells are massive, irregularly shaped cells with thick lignified cell walls that occur as single cells or in small clusters in all Pinaceae plus *Araucaria* (Franceschi et al. 2005). Due to their physical toughness, stone cells can be a deterrent to bark boring organisms (Wainhouse et al. 1990, 1997; Hudgins et al. 2004). In the Pinaceae modified PP cells are producing calcium oxalate crystals that are likely to interfere with bark penetration as well (Hudgins et al. 2003b; Franceschi et al. 2005; Fig. 7.4B). Unlike normal PP cells, these modified cells are dead at maturity and have suberized cell walls. Crystals of calcium oxalate may also occur in the vacuoles of some of the PP cells along with the phenolic material (Franceschi et al. 1998).

7.2.2 Inducible Responses in PP Cells

In addition to being a prominent constitutive component in the secondary phloem, PP cells are also involved in a range of inducible defense responses, including activation of existing cells, production of extra PP cells, and wound periderm formation. Activation or swelling of existing PP cells begins a few days after an attack and results in a four-fold increase in the PP cell volume and a change in the appearance of the phenolic content (Franceschi et al. 1998, 2000, 2005; Krokene et al. 2003; Fig. 7.3). The dramatic increase in PP cell size leads to extensive compression of the surrounding sieve cells, and at least in the Pinaceae, which produce many layers of sieve cells per year, the induced phloem is transformed into dense blocks of cell walls separated by layers of swollen PP cells (Figs. 7.3B and D). In addition to serving as a physical barrier to penetration by invading organisms this combined cell wall/PP cell barrier appears to be reinforced by phenolics released from the induced PP cells and deposited in the surrounding sieve cell walls (Franceschi et al. 2000).

New or extra PP cells may be induced to develop from undifferentiated parenchyma cells in between the regular annual PP cell layers. This is not part of the normal developmental program of the secondary phloem and is only seen after severe challenges such as massive fungal infection (Krekling et al. 2000, 2004; Krokene et al. 2003). Production of extra PP cells may also be induced by treatment with methyl jasmonate (Krokene et al. 2007), an elicitor of plant defense responses that induces similar responses as fungal pathogens and insects in conifers (Franceschi et al. 2002; Martin et al. 2002; Hudgins et al. 2004; Krokene et al. 2008). With time, extra PP cells appear several centimeters away from the point of induction and can be quite abundant, covering 50% of the phloem circumference in Norway spruce 15 weeks after a massive fungal infection (Krokene et al. 2003). Because differentiation of extra PP cells is a relatively slow process compared to the activation of existing ones, it probably enhances long-term resistance rather than resistance to an on-going attack. Production of extra PP cells seems to be involved in the phenomenon of systemic induced resistance in Norway spruce (Krokene et al. 2003).

PP cells are also instrumental in the process of wound periderm formation (Franceschi et al. 2000). This is an essential part of conifer defenses, as it serves to wall off damaged regions and re-establish a continuous surface barrier (see also Ginzberg this volume). Wound periderms are produced by activation of existing PP cells, which begin to divide to form a wound phellogen or cork cambium (Fig. 7.3B). Similar to the normal periderm this meristem produces cork tissue (phellem) outwards and phelloderm inwards, and thus regenerates a continuous surface barrier (Franceschi et al. 2005). Damaged tissues outside this new periderm are isolated from nutrient supplies through the secondary phloem and will eventually be shed as new phloem is produced by the vascular cambium.

7.2.3 Evidence for Defensive Roles of PP Cells

The anatomical organization, chemical nature and inducible responses of PP cells suggest that they are important in resistance against attacking organisms. A defensive role for PP cells is supported by inoculation experiments with bark beetle-associated bluestain fungi, showing that tree resistance developed over a time frame that paralleled the time required to activate PP cells (Krokene et al. 2003). Trees pretreated with a sublethal fungal infection three to nine weeks before a massive artificial infection showed strong activation of PP cells, and were much more resistant to the subsequent challenge than untreated trees (Krokene et al. 1999, 2003). Trees pretreated one week before mass-infection showed no activation of PP cells and were as susceptible as untreated control trees (Krokene et al. 2003). In this experiment, the constitutive PP cell barrier was circumvented, as fungus was artificially inoculated into wounds going all the way to the vascular cambium. When fungal inoculations are made into shallow bark wounds, ending in the middle of the PP cell layers, fungal growth is completely restricted, confirming a role in defense also for the constitutive, concentric layers of PP cells (Franceschi et al. 2000). Many bluestain fungi are successful conifer pathogens because they circumvent bark defenses by hitching a ride with scolytid bark beetles tunneling through the bark to the vascular cambium (e.g., Paine et al. 1997; Franceschi et al. 2000, 2005).

The exact chemical nature of the phenolic material inside the PP cells is not known. The conclusion that the material is polyphenolic in nature is based on microscopical evidence, such as structural similarity to polyphenolics in other plant cell types, autofluorescence, staining properties, and localization of PAL to the plasma membrane and cytoplasm of PP cells by immunolocalization (Franceschi et al. 1998). Other important enzymes involved in phenol synthesis, such as chalcon and stilbene synthases, have also been found to increase in concentration in conifer bark after fungal infection, emphasizing the role of phenolics in bark defenses (Brignolas et al. 1995a; Richard et al. 2000; Nagy et al. 2004). There are also numerous direct chemical analyses of conifer phloem showing high concentrations of soluble phenolics, such as flavonoids and stilbenes (Lindberg et al. 1992; Lieutier et al. 1991, 1996, 2003; Brignolas et al. 1995a, b, 1998; Dixon and Paiva 1995; Bois and Lieutier 1997; Evensen et al. 2000; Viiri et al. 2001; Zeneli et al. 2006). However, the soluble phenolics that lend themselves to routine chemical analysis do not appear to be crucial for defense, as they were only weakly inducible by fungal inoculation, mechanical wounding, and bark beetle attack (Erbilgin et al. 2006; Zeneli et al. 2006). PP cell activation may thus be related to changes in more complex phenolics, such as high molecular weight condensed tannins or cell wall-bound phenolics, but this remains to be tested.

7.3 Traumatic Resin Ducts

7.3.1 Resin-Based Defenses in Conifers

Conifer resin is composed largely of terpenes, made up chiefly of monoterpenes (C_{10}) and diterpenes (C_{20}), with small amounts of sesquiterpenes (C_{15}) and other compounds (Gershenzon and Croteau 1991; Bohlmann this volume). Resin is generally considered to be important in conifer defense because of its physical properties and its toxic and repellent effects on many herbivores and pathogens (Reid et al. 1967; Shrimpton and Whitney 1968; Bordasch and Berryman 1977; Raffa et al. 1985; Gijzen et al. 1993; Bohlmann this volume). Conifer resin is synthesized and accumulated in specialized secretory structures of different anatomical complexity, ranging from scattered individual resin cells and sac-like resin cavities lined with epithelial cells, to complex networks of interconnected radial and axial resin ducts. These resin structures may be present constitutively or they may be inducible, and they are found in many different tissues, including xylem, phloem, cortex and needles.

Resin producing structures are found in many conifer lineages, but they are not uniformly important. In fact, most conifers in the families Araucariaceae, Podocarpaceae, Cupressaceae, Cephalotaxaceae, and Taxaceae (the non-Pinaceae families) lack preformed resin structures in phloem and xylem (Bannan 1936; Fahn 1979; Wu and Hu 1997), and inducible resin structures only appear in some genera (Hudgins et al. 2004; Franceschi et al. 2005; Table 7.1). Instead, the non-Pinaceae species seem to rely mainly on massive constitutive bark defenses in the form of PP cells and lignified fiber cells (see Section 7.2.1). It is a striking and almost paradoxical fact that the Pinaceae conifers that are relying on resin-based defenses are more susceptible to tree-killing bark beetles (family Scolytidae) as compared to species in the non-Pinaceae group which rarely suffer from bark beetle attacks (Hudgins et al. 2004; Franceschi et al. 2005). The specialization of tree-killing bark beetles on resin-producing conifers includes the use of resin as a precursor for production of population-aggregating pheromones (e.g., Vanderwel and Oehlschlager 1987). Effective aggregation pheromones are probably essential for tree-killing, since small herbivores like bark beetles must attack in large numbers in order to overwhelm and kill large host trees (Franceschi et al. 2005). Bark beetles are thus using one of the tree's primary chemical defenses to their advantage, illustrating how co-evolution between insect herbivores and host trees can turn a defense into a weakness. Still, since most conifers, both Pinaceae and non-Pinaceae species, usually are well-defended against bark beetles, these contrasting strategies may also serve as an example of how completely different defense strategies can be effective against the same pest (Franceschi et al. 2005).

Constitutive resin structures are present in the cortex and phloem of most Pinaceae and a few non-Pinaceae conifers, usually in the form of resin ducts or resin cavities (Fig. 7.5; Table 7.1). In addition, many Pinaceae species have scattered constitutive resin ducts in normal xylem (Table 7.1). The axial resin ducts in the cortex (the

Table 7.1 Different types of resin structures in conifer stems. The genera are grouped by family, starting with the Pinaceae (showing all 11 genera), continuing with Araucariaceae (1 of 3 genera), Podocarpaceae (1 of 18–19 genera), Cupressaceae (7 of 33 genera), Cephalotaxaceae (1 of 3 genera), and Taxaceae (2 of 3 genera)

Genus	Species	Cortical ducts	Phloem Constitutive	Inducible	Xylem Constitutive	Inducible	Reference
Abies	>40	yes	cavities	no?	no	cavities	Hudgins et al. (2004) and Wu and Hu (1997)
Cedrus	2–4	yes	cavities	no?	no	axial	Hudgins et al. (2004) and Wu and Hu (1997)
Pseudolarix	1	no	cavities	no?	no	cavities	Penhallow (1907), Bannan 1936 and Wu and Hu (1997)
Tsuga	9	yes	cavities	no?	no	cavities	Hudgins et al. (2004) and Wu and Hu (1997)
Nothotsuga	1	yes	radial	no?	no	axial	Wu and Hu (1997)
Keteleeria	3	yes	radial	no?	axial	axial?	Wu and Hu (1997)
Larix	14	no	radial	no?	network[1]	axial	Hudgins et al. (2003a) and Wu and Hu (1997)
Pseudotsuga	5	yes	radial	no?	network	axial	Hudgins et al. (2004) and Wu and Hu (1997)
Cathaya	1	yes	radial	no?	network	axial?	Wu and Hu (1997)
Picea	35	yes	radial	no?	network	axial	Hudgins et al. (2004) and Wu and Hu (1997)
Pinus	115	yes	radial	no?	network	(axial)[2]	Hudgins et al. (2004) and Wu and Hu (1997)
Araucaria	19		(no)[3]	axial	no	no	Hudgins et al. (2004)
Agathis	21		no	no?	no	no?	Penhallow (1907)
Wollemia	1		no?	no?	no	no?	Heady et al. (2002)

7 Induced Anatomical Defenses in Conifers

Table 7.1 (continued)

Genus	Species	Cortical ducts	Phloem Constitutive	Phloem Inducible	Xylem Constitutive	Xylem Inducible	Reference
Podocarpus + 17–18 genera	105 >65		no	no	no	no	Hudgins et al. (2004)
Thuja	5		cells	no?	cells?	no?	Hudgins et al. (2003b)
Chamaecyparis	6		no	axial	no	no	Yamada et al. (2002)
Sequoia	1		radial?	no	no	axial	Hudgins et al. (2004)
Sequoiadendron	1		radial?	no	no	axial	Hudgins et al. (2004)
Metasequoia	1		radial?	no	no	axial	Hudgins et al. (2004)
Cryptomeria	1		no	axial	no	no	Hudgins et al. (2004)
Cupressus + 26 genera	>16 >117		(no)[3]	axial	no	no	Hudgins et al. (2004)
Torreya +2 genera	5–6 16		no	no	no	no	Penhallow (1907)
Taxus +2 genera	9 2		no	no	no	no	Hudgins et al. (2003a)

[1] Network = combination of radial and axial resin ducts
[2] Weak induction of resin ducts that are restricted to the immediate vicinity of the point of induction
[3] Non-functional bundles of epithelial cells are produced constitutively and activated in response to damage/attack

outer part of the young stem bark) may be very large compared to other resin ducts (Figs. 7.2A and 7.5A). They are formed in a circumferential ring as the cortex is produced during primary stem development, and may be present for at least 25 years in Norway spruce (Christiansen et al. 1999a). The cortex is an important defense barrier in the early stages of stem growth, but its defensive roles are gradually taken over by the secondary phloem as this develops more extensive layers (Franceschi et al. 2005). The constitutive resin ducts in the phloem are always radial, and reside within multiseriate radial rays surrounded by ray parenchyma cells. The cross-section of these ducts increases in size in the older parts of the bark and can be quite extensive (Christiansen et al. 1999a; Fig. 7.3A). Radial resin ducts occur primarily in the phloem, but some run continuously from the phloem to the sapwood, where they are interconnected with the axial resin ducts (Fig. 7.5E; Christiansen et al. 1999a; Nagy et al. 2000). A complex constitutive resin duct network with interconnected radial and axial ducts in phloem and xylem are found only in some Pinaceae genera (*Pinus, Picea, Larix, Pseudotsuga, Keteleeria, Cathaya*; Table 7.1), and is most highly developed in the genus *Pinus* which seems to rely heavily on constitutive resin (Bannan 1936; Vité 1961; Johnson and Croteau 1987).

Resin ducts are lined with plastid-enriched epithelial cells that produce and secrete resin into an extracellular lumen, where it is stored under pressure (Charon et al. 1987; Gershenzon and Croteau 1990; Nagy et al. 2000; Fig. 7.6A). The simpler resin cells accumulate resin internally under pressure and may expand into quite large structures (Fig. 7.5B). Resin ducts form schizogenously as the epithelial cells pull apart during resin duct formation (Nagy et al. 2000). The epithelial cells lining radial and axial resin ducts are thin-walled and long-lived, in contrast to the epithelial cells of the resin cavities, which are short-lived and gradually become lignified during development (Bannan 1936; Fahn 1979).

7.3.2 Anatomy and Development of Traumatic Resin Ducts

In addition to constitutive resin structures many conifers develop inducible resin structures in response to biotic and abiotic challenges, such as insect attack, pathogen infection, frost damage, or physical wounding (e.g., Bannan 1936; Alfaro 1995; Kytö et al. 1996; Hudgins et al. 2003a, 2004). In the Pinaceae, inducible resin structures usually occur as so-called pathological or traumatic axial resin ducts in the xylem, but axial resin ducts are also induced in the phloem of a few non-Pinaceae genera (Table 7.1). The nature of the vascular cambium implies that inducible resin ducts must be axial and not radial. The cambium can relatively quickly be re-programmed to produce functional axial resin ducts that may be several meters long (e.g., Christiansen et al. 1999b; Nagy et al. 2000), whereas it would require years to produce radial ducts of any extent.

TRDs form in tangential bands in the xylem of many Pinaceae species (Figs. 7.3B and 7.5D). These bands are often surrounded by small tracheids with thickened cell walls and are visible to the naked eye as a punctuated line within the annual ring

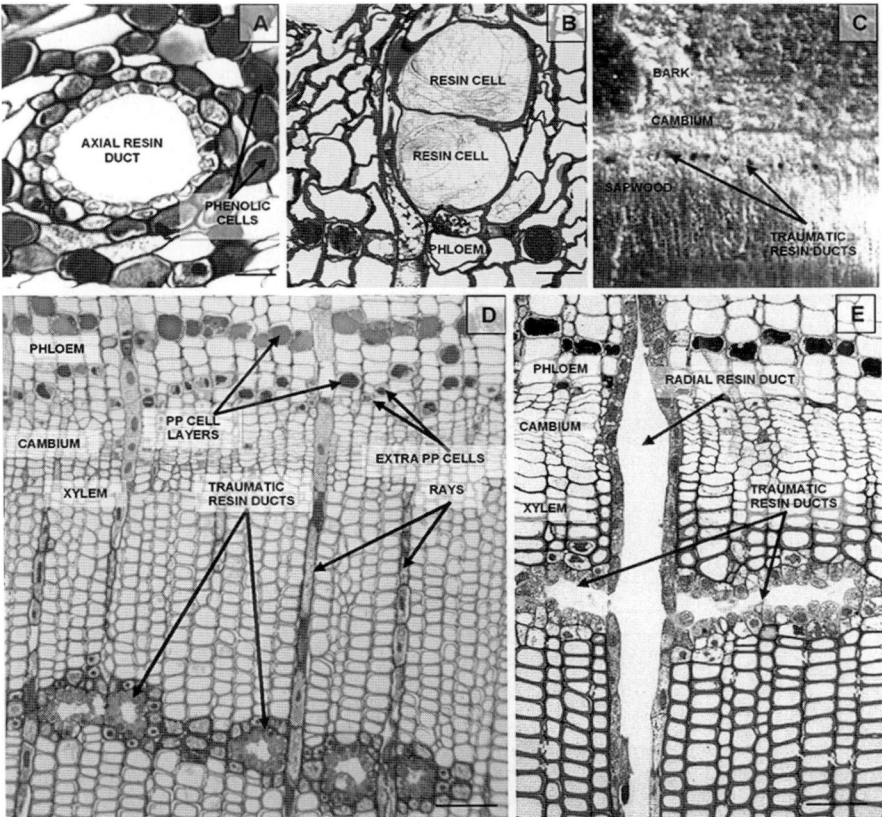

Fig. 7.5 Cross-sections of resin-containing structures in conifers. (**A**) An axially oriented resin duct in the outer cortical region of the phloem of balsam fir (*Abies balsamea*). Bar, 50 μm. The circular cortical resin ducts are lined with small epithelial cells which are surrounded by sheet cells containing phenolic material. (**B**) In grand fir (*Abies grandis*) resin is produced and stored in individual resin cells in the secondary phloem. Bar, 50 μm. (**C**) Low magnification of Norway spruce (*Picea abies*) bark and sapwood, showing traumatic resin ducts within the sapwood 3 weeks after inoculation with the blue-stain fungus *Ceratocystis polonica*. The traumatic resin ducts are visible as small, tangentially oriented channels close to the cambium. (**D**) Closer look at traumatic resin ducts in the xylem of Norway spruce after fungal inoculation. The traumatic resin ducts are characteristically associated with radial rays. (**E**) Cross-section demonstrating luminal continuity between several coalescing traumatic resin ducts and a radial resin duct traversing the phloem and xylem. Bars, 50 μm

(false annual rings; Fig. 7.5C). The formation of functional TRDs with secretory activity is quite rapid, and requires two to four weeks in Norway spruce (Nagy et al. 2000). TRDs develop in the same way as normal axial resin ducts, by schizogenesis between incipient epithelial cells when these are still close to the cambial zone (Fig. 7.6B). Sometimes large transverse resin ducts can be produced by tangential anastomosis of neighboring ducts (Gerry 1942; Fig. 7.5E). If the inducing

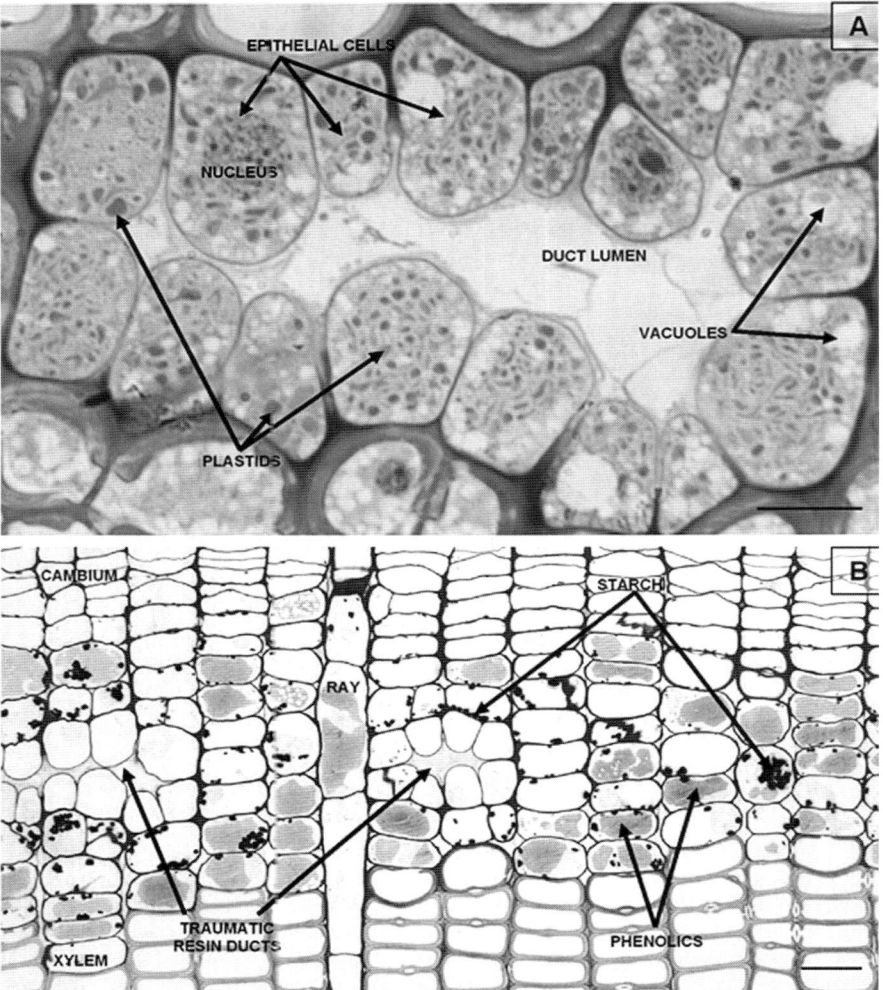

Fig. 7.6 Cross-sections of the epithelial cells of developing traumatic resin ducts. (**A**) In Norway spruce, the epithelial cells lining the resin duct are characterized by having dense cytoplasm, with many plastids and vacuoles, and conspicuously enlarged nuclei. The numerous plastids are the primary site for resin synthesis. Bar, 10 μm. (**B**) Early in their development the traumatic resin ducts are often surrounded by living parenchyma cells full of starch grains and phenolic material. Bar, 25 μm

stimulus is strong, an almost complete tangential ring of TRDs can be produced, and with time, the ducts may extend several meters from the point of induction in spruce (Christiansen et al. 1999b). In other genera, such as *Abies*, *Tsuga*, *Cedrus* and *Pseudolarix*, TRDs mainly form in the vicinity of wounds (Bannan 1936; Wu and Hu 1997).

The TRD system may represent a considerable volume, and contribute significantly to a tree's overall resin production. After massive induction, TRDs are much more abundant than normal xylem ducts, and may extend over 50%–100% of the xylem circumference in Norway spruce (Franceschi et al. 2002; Krokene et al. 2003). TRDs are also much larger than constitutive xylem ducts; in Norway spruce for example, they are often twice as wide and represent a volume that is four times greater per unit duct length. TRDs form an interconnected network of resin-filled cavities tangentially in the new sapwood, as well as radially with luminal continuity between adjacent TRDs and radial constitutive resin ducts (Fig. 7.5E). This organization facilitates resin flow both axially along the trunk, as well as radially towards the bark surface (Nagy et al. 2000; Krokene et al. 2003).

TRDs appear to be derived from xylem mother cells (Werker and Fahn 1969; Nagy et al. 2000). The strict distribution pattern of TRDs in tangential bands implies a combination of predisposition and a tangential spread of a stimulus resulting in coordinated induction and differentiation of the xylem mother cells. With increasing distance from the point of induction, TRDs are gradually reduced in size and number, and are found closer and closer to the cambium (Krekling et al. 2004). There thus appears to be a time-lag in the production and differentiation of the TRDs away from the stimulus, which would be consistent with a signal propagating a developmental wave traveling in the axial direction at about 2.5 cm per day (Krekling et al. 2004). TRD formation appears to be activated through the octadecanoid pathway, involving jasmonate and ethylene signaling (Franceschi et al. 2002; Martin et al. 2002; Hudgins et al. 2003a, 2004, 2006; Hudgins and Franceschi 2004; Miller et al. 2005; Ralph et al. 2007).

7.3.3 Defensive Roles of TRDs

TRD formation in Norway spruce is part of a complex defense response that is rapid and coordinated, resulting in an extensive resin-secreting network. The result is massive accumulation of resin and enhanced resin flow at the site of attack (Phillips and Croteau 1999; Nagy et al. 2000). Artificial inoculation or wounding, leading to TRD formation and increased synthesis of resin, can thus be used as a technique to improve and prime a tree's resistance against insects and pathogens. For example, Norway spruce trees that were pretreated with a sublethal fungal infection developed abundant TRDs which rendered the trees much more resistant to a subsequent massive infection after three to nine weeks (Krokene et al. 1999, 2003). The ability of a tree to undergo traumatic resinosis can be used to screen for resistance (Krokene et al. 2008). For example, resistant clones of Norway spruce and white spruce (*Picea glauca*) produce more TRDs more quickly as compared to susceptible clones (Tomlin et al. 1998; Nagy et al. 2000; Krekling et al. 2004). In white spruce, large inner and dense outer resin canals were correlated with resistance against the white pine weevil, *Pissodes strobi* (Peck) (Alfaro et al. 2004).

There is ample evidence that conifer resin is effective in defense against insects. In several conifers, the concentration of monoterpenes is enhanced in tissues attacked by bark beetles and their associated blue-stain fungi (Russell and Berryman 1976; Cates and Alexander 1982; Raffa and Berryman 1982, 1983; Paine et al. 1987; Lewinsohn et al. 1991). Tomlin et al. (2000) found that simulated white pine weevil damage enhanced the monoterpene to resin acid ratio in stem regions distant from the wounded site in resistant trees, but only slightly in susceptible trees. Monoterpene-enriched resin, which is less viscous than constitutive resin, probably flows more readily into oviposition cavities and larval mines, thus being more effective in inhibiting/killing immature weevils. Resin may also inhibit invading insects and mycelial growth of fungal pathogens due to its toxic properties. For example, Klepzig et al. (1995) found that monoterpenes in red pine resin inhibit fungal germination and mycelial growth, as well as beetle tunneling. Alpha-pinene in particular, the most abundant monoterpene in red pine phloem, exhibited strong biological activity towards insects and fungi (Klepzig et al. 1995).

Resin diterpenes are considered to be insecticidal and fungicidal, and are important in wound sealing. An open wound is a potential infection pathway for attacking insects and pathogenic fungi, and effective wound sealing is decisive in preventing invasions. The diterpene abietic acid is important for oxidative polymerization of resin and dehydroabietic acid for resin viscosity and crystallization (Phillips and Croteau 1999). Resistant white spruce trees appear to produce more volatile terpenes and diterpene resin acids at the area of wounding than susceptible trees (Tomlin et al. 2000).

The induction of functional TRDs is usually too slow to be effective against rapid mass-attacks by bark beetles, which may be completed in a few days (Paine et al. 1997). However, TRD formation may contribute to defense when bark beetle attacks are slowed down by e.g. unfavorable weather conditions (Franceschi et al. 2000), and may also be important in protecting the trees against opportunistic wound colonizers after e.g. mechanical damage to the bark.

7.4 Perspectives

Invaders attempting to colonize healthy conifers have to struggle their way through multiple layers of periderm, PP cells and/or fiber cells. However, the ray cells running axially through the secondary phloem and xylem represent a potential weak point in this defense barrier. Since the rays are the only tissue type in the stem that is oriented radially they form a major direct link between phloem and sapwood. This potential loophole must therefore be expected to be well-defended, and immunolabeling studies in Norway spruce have indeed demonstrated that rays are a major site for PAL activity (Franceschi et al. 1998), and probably other defensive elements as well. Anatomical studies in Scots pine suggest that pathogen infection induces changes in ray cells, including increase in cell size and a more active cytoplasm that may be related to increased cellular activity and resin production (Nagy et al. 2006). However, as far as we know these preliminary studies are the only ones looking at inducible defenses in ray cells, and there are no detailed studies of ray cell structure and function. The rays are not only involved in defending the bark and outer sapwood, but are probably also instrumental in heartwood formation and defense against e.g. wood rotting fungi, since they are the only route for translocation of substances between the living phloem and the inner sapwood (Shain 1967; Kwon et al. 2001; Franceschi et al. 2005). With their radial orientation, the rays are also expected to play important roles in signaling and transport of chemicals between phloem and xylem.

Chemical and molecular characterization of the central cell types in conifer defense, such as radial rays, PP cells, and resin duct epithelial cells, is strongly needed. The data available so far represent whole tissues (e.g., the whole secondary phloem), and the results are thus an average of information of different tissue and cell types. Recent technological and analytical developments have made it possible to dissect specific cell types and characterize their content using both chemical and molecular techniques. Laser microdissection enables effective isolation of single cells and is widely used in the biomedical field (Emmert-Buck et al. 1996; Day et al. 2005). More recently, laser microdissection has been adapted for use with many plant species, including *Arabidopsis*, rice, maize and others (Ohtsu et al. 2007). The only example from conifers so far is the dissection and chemical analysis of stone cells in Norway spruce (Li et al. 2007). Using laser microdissection in combination with sensitive instrumental analysis such as mass spectroscopy and cryogenic NMR-spectroscopy, it will be possible to chemically characterize pure cell populations. This represents a new approach for the cell-specific localization of low-molecular weight compounds in plants, and has been termed micro-phytochemistry (Li et al. 2007). In addition to metabolites, proteins, DNA, and RNA can be analyzed from microdissected samples. This powerful tool can thus give us detailed information about the chemical and molecular nature of all the important cell types in conifer defense.

References

Alfaro RI (1995) An induced defense reaction in white spruce to attack by the white-pine weevil, *Pissodes strobe*. Can J For Res 25:1725–1730

Alfaro RI, VanAkker L, Jaquish B, King J (2004) Weevil resistance of progeny derived from putatively resistant and susceptible interior spruce parents. For Ecol Man 202:369–377

Alfieri FJ, Evert RF (1973) Structure and seasonal development of secondary phloem in Pinaceae. Bot Gaz 134:17–25

Bannan MW (1936) Vertical resin ducts in the secondary wood of the abietineae. New Phytol. 35:11–46

Bois E, Lieutier F (1997) Phenolic response of Scots pine clones to inoculation with *Leptographium wingfieldii*, a fungus associated with *Tomicus piniperda*. Plant Physiol Biochem 35:819–825

Bonello P, Gordon TR, Storer AJ (2001) Systemic induced resistance in Monterey pine. For Pathol 31:99–106

Bonello P, Blodgett JT (2003) *Pinus nigra–Sphaeropsis sapinea* as a model pathosystem to investigate local and systemic effects of fungal infection of pines. Physiol Mol Plant Pathol 63:249–261

Bordasch RP, Berryman AA (1977) Host resistance to the fir engraver beetle, *Scolytus ventralis* (Coleoptera: Scolytidae). 2. Repellency of *Abies grandis* resins and some monoterpenes. Can Entomol 109:95–100

Brignolas F, Lacroix B, Lieutier F, Sauvard D, Drouet A, Claudot AC, Yart A, Berryman AA, Christiansen E (1995a) Induced responses in phenolic metabolism in two Norway spruce clones after wounding and inoculation with *Ophiostoma polonicum*, a bark beetle-associated fungus. Plant Physiol 109:821–827

Brignolas F, Lieutier F, Sauvard D, Yart A, Drouet A, Claudot AC (1995b) Changes in soluble-phenol content of Norway-spruce (*Picea abies*) phloem in response to wounding and inoculation with *Ophiostoma polonicum*. Eur J For Pathol 25:253–265

Brignolas F, Lieutier F, Sauvard D, Christiansen E, Berryman AA (1998) Phenolic predictors for Norway spruce resistance to the bark beetle *Ips typographus* (Coleoptera: Scolytidae) and an associated fungus, *Ceratocystis polonica*. Can J For Res 28:720–728

Cates RG, Alexander H (1982) Host resistance and susceptibility. In: Mitton JB, Sturgeon KB (eds) Bark beetles in North American conifers. A system for the study of evolutionary biology. University of Texas Press, Austin, pp 212–263

Charon J, Launay J, Carde J-F (1987) Spatial organization and volume density of leucoplasts in pine secretory cells. Protoplasma 138:45–53

Christiansen E, Franceschi VR, Nagy NE, Krekling T, Berryman AA, Krokene P, Solheim H (1999a) Traumatic resin duct formation in Norway spruce after wounding or infection with a bark beetle-associated blue-stain fungus, *Ceratocystis polonica*. In: Lieutier F, Mattson WJ, Wagner MR (eds) Physiology and genetics of tree-phytophage interactions. Les Colloques de l'INRA, INRA Editions, Versailles, pp 79–89

Christiansen E, Krokene P, Berryman AA, Franceschi VR, Krekling T, Lieutier F, Lönneborg A, Solheim H (1999b) Mechanical injury and fungal infection induce acquired resistance in Norway spruce. Tree Physiol 19:399–403

Cook SP, Hain FP (1988) Toxicity of host monoterpenes to *Dendroctonus frontalis* and *Ips calligraphus* (Coleoptera: Scolytidae). J Entomol Sci 23:287–292

Day RC, Grossniklaus U, Macknight RC (2005) Be more specific! Laser-assisted microdissection of plant cells. Trends Plant Sci 10:397–406

Dixon RA, Paiva NL (1995) Stress-induced phenylpropanoid meatabolism. Plant Cell 7:1085–1097

Emmert-Buck MR, Bonner RF, Smith PD, Chuaqui RF, Zhuang ZP, Goldstein SR, Weiss RA, Liotta LA (1996) Laser capture microdissection. Science 274:998–1001

Erbilgin N, Kroken, P, Christiansen E, Zeneli G, Gershenzon J (2006) Exogenous application of methyl jasmonate elicits defenses in Norway spruce (*Picea abies*) and reduces host colonization by the bark beetle *Ips typographus*. Oecologia 148:426–436

Evensen PC, Solheim H, Høiland K, Stenersen J (2000) Induced resistance of Norway spruce, variation of phenolic compounds and their effects on fungal pathogens. For Pathol 30:97–108

Fahn A (1990) Plant anatomy. Pergamon Press, Oxford

Fahn A (1979) Resin ducts of the coniferae. In: Secretory tissues in plants. Academic Press, London

Farjon A (2001) World checklist and bibliography of conifers, 2nd edn. Royal Botanical Gardens, Kew, Richmond

Feeny P (1976) Plant apparency and chemical defense. In: Wallace JW, Mansell RT (eds) Biochemichal interaction between plants and insects. Plenum Press, New York, pp 1–40

Franceschi VR, Krekling T, Berryman AA, Christiansen E (1998) Specialized phloem parenchyma cells in Norway spruce (Pinaceae) bark are an important site of defense reactions. Am J Bot 85:601–615

Franceschi VR, Krokene P, Krekling T, Christiansen E (2000) Phloem parenchyma cells are involved in local and distant defense responses to fungal inoculation or bark beetle attack in Norway spruce (Pinaceae). Am J Bot 87:314–326

Franceschi VR, Krekling T, Christiansen E (2002) Application of methyl jasmonate on *Picea abies* (Pinaceae) stems induces defense related responses in phloem and xylem. Am J Bot 89:578–586

Franceschi VR, Krokene P, Christiansen E, Krekling T (2005) Anatomical and chemical defences of conifer bark against bark beetles and other pests (Tansley Review). New Phytol 167:353–376
Gershenzon J, Croteau RB (1991). Terpenoids. In: Rosenthal GA, Berenbaum MR (eds) Herbivores: their interactions with secondary plant metabolites. Vol I, the chemical participants. Academic Press, San Diego, pp 165–219
Gijzen M, Lewinsohn E, Savage TJ, Croteau RB (1993) Conifer monoterpenes – biochemistry and bark beetle chemical ecology. ACS Symp. Ser. 525:8–22
Heady RD, Banks JG, Evans PD (2002) Wood anatomy of wollemi pine (*Wollemia nobilis*, Araucariaceae). IAWA Journal 23:339–357
Hudgins JW, Christiansen E, Franceschi VR (2003a) Methyl jasmonate induced changes mimicking anatomical defenses in diverse members of the Pinaceae. Tree Physiol 23:361–371
Hudgins JW, Krekling T, Franceschi VR (2003b) Distribution of calcium oxalate crystals in the secondary phloem of conifers: a constitutive defense mechanism? New Phytol 159:677–690
Hudgins JW, Christiansen E, Franceschi VR (2004) Induction of anatomically based defense responses in stems of diverse conifers by methyl jasmonate: a phylogenetic perspective. Tree Physiol 24:251–264
Hudgins JW, Franceschi VR (2004) Methyl jasmonate-induced ethylene production is responsible for conifer phloem defense responses and reprogramming of stem cambial zone for traumatic resin duct formation. Plant Physiol 13:2134–2149
Johnson MA, Croteau R (1987) Biochemistry of conifer resistance to bark beetles and their fungal symbionts. In: Fuller G, Nes WD (eds) Ecology and metabolism of plant lipids. American Chemical Society, Washington DC, pp 76–92
Kartuch B, Karuch R, Weilgong P (1991) Site-specific differences in calcium oxalate content of the secondary phloem of spruce (*Picea abies* Karst). Flora 185:377–384
Klepzig KD, Kruger EL, Smalley EB, Raffa KF (1995) Effects of biotic and abiotic stress on induced accumulation of terpenes and phenolics in red pines inoculated with bark beetle-vectored fungus. J Chem Ecol 21:601–626
Krekling T, Franceschi VR, Berryman AA, Christiansen E (2000) The structure and development of polyphenolic parenchyma cells in Norway spruce (*Picea abies*) bark. Flora 195:354–369
Krekling T, Franceschi VR, Krokene P, Solheim H (2004) Differential anatomical responses of Norway spruce tissues to sterile and fungus-infected wounding. Trees 18:1–9
Krokene P, Christiansen E, Solheim H, Berryman AA, Franceschi VR (1999) Induced resistance to pathogenic fungi in Norway spruce. Plant Physiol 121:565–570
Krokene P, Nagy NE, Solheim H (2008) Methyl jasmonate and oxalic acid treatment of Norway spruce: anatomically based defence responses and increased resistance against fungal infection. Tree Physiol 28:29–35
Krokene P, Solheim H, Langström B (2000) Fungal infection and mechanical wounding induce disease resistance in Scots pine. Eur J Plant Pathol 106:537–541
Krokene P, Solheim H, Christiansen E (2003) Inducible anatomical defense responses in Norway spruce stems and their possible role in induced resistance. Tree Physiol 23:191–197
Kwon M, Davin LB, Lewis NG (2001) In situ hybridization and immunolocalization of lignan reductases in woody tissue: implications for heartwood formation and other forms for vascular tissue preservation. Phytochemistry 57:899–914
Kytö M, Niemelä P, Annila E (1996) Vitality and bark beetle resistance of fertilized Norway spruce. For Ecol Manag 84:149–157
Lewinsohn E, Gijzen M, Croteau R (1991) Defense-mechanisms of conifers – differences in constitutive and wound-induced monoterpene biosynthesis among species. Plant Physiol 96:44–49
Li SH, Schneider B, Gershenzon J (2007). Microchemical analysis of laser-microdissected stone cells of Norway spruce by cryogenic nuclear magnetic resonance spectroscopy. Planta 225:771–779
Lieutier F, Yart A, Jay-Allemand C, Delorme L (1991) Preliminary investigations on phenolics as a response of Scots pine phloem to attacks by bark beetles and associated fungi. Eur J For Pathol 21:354–364

Lieutier F, Sauvard D, Brignolas F, Picron V, Yart A, Bastien C, Jay-Allemand C (1996) Changes in phenolic metabolites of Scots-pine phloem induced by *Ophiostoma brunneo-ciliatum*, a bark-beetle-associated fungus. Eur J For Pathol 26:145–158

Lieutier F, Brignolas F, Sauvard D, Yart A, Galet C, Brunet M, van de Sype H (2003) Intra- and inter-provenance variability in phloem phenols of *Picea abies* and relationship to a bark beetle-associated fungus. Tree Physiol 23:247–256

Lindberg M, Lundgren L, Graf R, Johansson M (1992) Stilbenes and resin acids in relation to the penetration of *Heterobasidion annosum* through the bark of *Picea abies*. Eur J For Pathol 22:95–106

Lombardero MJ, Ayres MP, Lorio PL, Ruel JJ (2000) Environmental effects on constitutive and inducible resin defences of *Pinus taeda*. Ecol Lett 3:329–339

Martin D, Tholl D, Gershenzon J, Bohlmann J (2002) Methyl jasmonate induces traumatic resin ducts, terpenoid resin biosynthesis, and terpenoid accumulation in developing xylem of Norway spruce stems. Plant Physiol 129:1003–1018

Miller B, Madilao LL, Ralph S, Bohlmann J (2005) Insect-induced conifer defense. White pine weevil and methyl jasmonate induce traumatic resinosis, de novo formed volatile emissions, and accumulation of terpenoid synthase and putative octadecanoid pathway transcripts in Sitka spruce. Plant Physiol 137:369–382

Murmanis L, Evert RF (1967) Parenchyma cells of secondary phloem in *Pinus strobus*. Planta 73:301–318

Nagy NE, Fossdal CG, Krokene P, Krekling T, Lönneborg A, Solheim H (2004) Induced responses to pathogen infection in Norway spruce phloem: changes in polyphenolic parenchyma cells, chalcone synthase transcript levels and peroxidase activity. Tree Physiol 24:505–515

Nagy NE, Franceschi VR, Solheim H, Krekling T, Christiansen E (2000) Wound-induced traumatic resin duct development in stems of Norway spruce (Pinaceae): anatomy and cytochemical traits. Am J Bot 87:302–313

Nagy NE, Krokene P, Solheim H (2006) Anatomical-based defense responses of Scots pine (*Pinus sylvestris*) stems to two fungal pathogens. Tree Physiol 26:159–167

Ohtsu K, Takahashi H, Schnable PS, Nakazono M (2007) Cell type-specific gene expression profiling in plants by using a combination of laser microdissection and high-throughput technologies. Plant Cell Physiol 48:3–7

Paine TD, Blanche CA, Nebeker TE, Stephen FM (1987) Composition of loblolly pine resin defenses: comparison of monoterpenes from induced lesion and sapwood resin. Can J For Res 17:1202–1206

Paine TD, Raffa KF, Harrington TC (1997) Interactions among scolytid bark beetles, their associated fungi, and live host conifers. Ann Rev Entomol 42:179–206

Parham RA, Kaustinen HM (1976) Differential staining of tannin in sections of epoxy-embedded plant cells. Stain Technol 51:237–240

Penhallow DP (1907) A manual of the North American gymnosperms: exclusive of the Cycadales but together with certain exotic species. Athenaeum Press, Boston, MA. pp 404

Phillips MA, Croteau RB (1999) Resin-based defense in conifers. Trends Plant Sci 4:184–190

Raffa KF, Berryman AA (1982) Accumulation of monoterpenes and associated volatiles following inoculation of grand fir with a fungus transmitted by the fir engraver *Scolytus ventralis* (Coleoptera: Scolytidae). Can Entomol 114:797–810

Raffa KF, Berryman AA (1983) Physiological aspects of lodgepole pine wound responses to a fungal symbiont of the mountain pine beetle, *Dendroctonus ponderosae* (Coleoptera: Scolytidae). Can Entomol 115:723–734

Raffa KF, Berryman AA, Simasko J, Teal W, Wong BL (1985) Effects of grand fir monoterpenes on the fir engraver, *Scolytus ventralis* (Coleoptera: Scolytidae), and its symbiotic fungus. Environ Entomol 14:552–556

Ralph SG, Hudgins JW, Jancsik S, Franceschi VR, Bohlmann J (2007) Aminocyclopropane carboxylic acid synthase is a regulated step in ethylene-dependent induced conifer defense. Full-length cDNA cloning of a multigene family, differential constitutive, and wound- and insect-induced expression, and cellular and subcellular localization in spruce and Douglas fir. Plant Physiol 143:410–424

Reglinski T, Stavely FJL, Taylor JT (1998) Induction of phenylalanine ammonia lyase activity and control of *Sphaeropsis sapinea* infection in *Pinus radiata* by 5-chlorosalicylic acid. Eur J For Pathol 28:153–158

Reid RW, Whitney HS, Watson JA (1967) Reactions of lodgepole pine to attack by *Dendroctonus ponderosae* Hopkins and blue stain fungus. Can J Bot 45:1115–1126

Richard S, Lapointe G, Rutledge G, Séguin A (2000) Induction of chalcone synthase expression in white spruce by wounding and jasmonate. Plant Cell Physiol 41:982–987

Ruel JJ, Ayres MP, Lorio PL (1998) Loblolly pine responds to mechanical wounding with increased resin flow. Can J For Res 28:596–602

Russell CE, Berryman AA (1976) Host resistance to the fir engraver beetle. 1. Monoterpene composition of *Abies grandis* pitch blisters and fungus-infected wounds. Can J Bot 54:14–18

Schmidt A, Zeneli G, Hietala AM, Fossdal CG, Krokene P, Christiansen E, Gershenzon J (2005) Induced chemical defenses in conifers: biochemical and molecular approaches to studying their function. In: Romeo J (ed) Chemical ecology and phytochemistry of forests and forest ecosytems. Elsevier, Amsterdam, pp 1–28

Shain L (1967) Resistance of sapwood in stems of loblolly pine to infection by *Fomes annosus*. Phytopathology 57:1034–1045

Shrimpton DM, Whitney HS (1968) Inhibition of growth of blue stain fungi by wood extractives. Can J Bot 46:757–761

Tomlin ES, Alfaro RI, Borden JH, He FL (1998) Histological response of resistant and susceptible white spruce to simulated white pine weevil damage. Tree Physiol 18: 21–28

Tomlin ES, Antonejevic E, Alfaro RI, Borden JH (2000) Changes in volatile terpene and diterpene resin acid composition of resistant and susceptible white spruce leaders exposed to simulated white pine weevil damage. Tree Physiol 20:1087–1095

Vanderwel D, Oehlschlager AC (1987) Biosynthesis of pheromones and endocrine regulation of pheromone production in Coleoptera. In: Prestwich GD, Blomquist GJ (eds) Pheromone biochemistry. Academic Press, New York, pp 175–215

Vité JP (1961) The influence of water supply on oleoresin exudation pressure and resistance to bark beetle attack in *Pinus ponderosa*. Contributions from the Boyce Thompson Institute 21:37–66

Viiri H, Annila E, Kitunen V, Niemelä P (2001) Induced responses in stilbenes and terpenes in fertilized Norway spruce after inoculation with blue-stain fungus, *Ceratocystis polonica*. Trees 15:112–122

Wainhouse D, Cross DJ, Howell RS (1990) The role of lignin as a defense against the spruce bark beetle *Dendroctonus micans* – effect on larvae and adults. Oecologia 85:257–265

Wainhouse D, Rose DR, Peace AJ (1997) The influence of preformed defences on the dynamic wound response in spruce bark. Funct Ecol 11:564–572

Werker E, Fahn A (1969) Resin ducts of *Pinus halepensis* Mill.—their structure, development and pattern of arrangement. Bot J Linn Soc 62:379–411

Wu H, Hu ZH (1997) Comparative anatomy of resin ducts of the Pinaceae. Trees Struct Funct 11:135–143

Zeneli G, Krokene P, Christiansen E, Krekling T, Gershenzon J (2006) Methyl jasmonate treatment of large Norway spruce (*Picea abies*) trees increases the accumulation of terpenoid resin components and protects against infection by *Ceratocystis polonica*, a bark beetle-associated fungus. Tree Physiol 26:977–988

Part B
Production of Secondary Metabolites

Chapter 8
Insect-Induced Terpenoid Defenses in Spruce

Jörg Bohlmann

Conifers produce an array of hundreds of different terpenoids as part of their complex chemical, physical, and ecological defenses against insect pests and pathogens. Terpenoid chemicals exist both as constitutive and as massively induced defenses in conifers. It is thought that the diversity of terpenoid chemicals serves as a multilayered chemical shield in long-lived conifer trees that provides a lasting protection against the much faster evolving insect pests and potential pathogens. The formation of terpenoid defenses in conifers involves the activity of two pathways, the methylerythritol phosphate pathway and the mevalonate pathway, which lead to the five-carbon precursors of terpenoid biosynthesis. The many monoterpenoids, sesquiterpenoids, and diterpene resin acids, which are present in oleoresin mixtures, are then formed by families of enzymes belonging to the classes of prenyl transferases, terpenoid synthases, and cytochrome P450 dependent monooxygenases. The genes for almost all the enzymatic steps in terpenoid oleoresin biosynthesis have been identified in species of spruce, which have thus been established as a conifer reference system to study constitutive and induced terpenoid defenses using biochemical, molecular genetic, and genomic and proteomic approaches. At the histological and cellular levels, oleoresin terpenoids are produced and accumulate constitutively in large quantities in specialized anatomical structures that are found in most organs and tissues. In many conifer species, biosynthesis and accumulation of terpenoids is further enhanced as part of the induced defense in response to insect attack or fungal infection. In this chapter, I will discuss selected aspects of terpenoid defenses in conifers against insects and pathogens.

8.1 Introduction

Conifers (i.e., members of the pine family; Pinaceae) include some of the longest-living and tallest organisms on earth. Their tall physical structure and long lifespan

---

J. Bohlmann
Michael Smith Laboratories, University of British Columbia, Vancouver, B.C., Canada
e-mail: bohlmann@msl.ubc.ca

make them prominent targets in space and time for many potential herbivores and pathogens. The successful defense and resistance of conifers against most herbivores and pathogens can be explained, at least in part, by the formation of a diverse array of monoterpenoid, sesquiterpenoid, and diterpene resin acid defense chemicals. These compounds accumulate in large amounts in form of preformed or induced oleoresin mixtures. Some terpenoids, in particular monoterpenes and sesquiterpenes, can also be actively emitted as volatile organic compounds with functions as semiochemicals from conifer needles. Conifer oleoresin is stored in specialized anatomical structures such as resin canals, resin blisters, or resin cells in stems, roots, or needles. The developmental programs for the constitutive formation of these specialized, resin-accumulating structures are not understood; except, it is now well established that the formation of so-called traumatic resin ducts is induced in the cambium zone of conifer stems in response to insect or fungal attack, by mechanical wounding, or by chemical elicitation with methyl jasmomate (MeJA) or ethylene treatment (Krokene et al. this volume). In recent years, several reviews have been published on the topic of terpenoid oleoresin defenses in conifers (e.g., Bohlmann and Croteau 1999; Phillips and Croteau 1999; Trapp and Croteau 2001a; Keeling and Bohlmann 2006a, b).

The families of terpenoid synthases (TPS) and cytochrome P450 dependent monooxygenases (P450) play a central role in the formation of terpenoid chemical diversity and for phenotypic plasticity in conifer defense (Bohlmann et al. 1998a; Martin et al. 2004; Ro et al. 2005). Both, the TPS and P450 are encoded in multi-gene families in conifers (Martin et al. 2004; Hamberger and Bohlmann 2006). These gene families are thought to contribute, on the genomic, molecular, and bio-chemical levels, much to the diversity and plasticity of constitutive and induced terpenoid defenses in the long-lived conifers. In contrast to the TPS and P450, we know relatively little about the early steps of terpenoid biosynthesis in conifers with only a few studies published on prenyltransferases (PT; Tholl et al. 2001; Burke and Croteau 2002; Martin et al. 2002; Schmidt and Gershenzon 2007) or earlier steps in conifer isoprenoid biosynthesis (Phillips et al. 2007). However, a recent large-scale spruce genomics project has identified genes (i.e., expressed sequences tags, ESTs; and full-length cDNAs, FL-cDNAs) for almost every single step in conifer monoter-penoid, sesquiterpenoid, and diterpenoid biosynthesis (Ralph et al. 2006; Ralph and Bohlmann, in preparation).

In building on pioneering work by Clarence (Bud) Ryan and coworkers, the use of MeJA as an elicitor to induce plant defenses enabled us and others to develop a detailed characterization of inducible terpenoid defenses in several conifer species such as Norway spruce (*Picea abies*), Sitka spruce (*P. sitchensis*), White spruce (*P. glauca*) and Douglas fir (*Pseudotsuga menziesii*; e.g., Franceschi et al. 2002; Martin et al. 2002, 2003a; Fäldt et al. 2003; Hudgins et al. 2003; Huber et al. 2005a, b; Miller et al. 2005; Ralph et al. 2006; Erbilgin et al. 2006; Zeneli et al. 2006; Phillips et al. 2007). By treating conifer seedlings, mature trees, or even cell cultures with MeJA, it became possible to accurately measure the quantitative and qualitative changes of traumatic resin terpenoids, detect the dynamics of enzyme activities and transcript levels induced in the traumatic terpenoid defense response, and analyze the induced and active emission of terpenoid volatiles. In addition, the non-invasive treatment with MeJA, as opposed to mechanical wounding or insect attack, allowed for the analysis of temporal and spatial patterns of cell differentiation during traumatic resin duct formation in the cambium zone of spruce stems. Application of MeJA provided a powerful tool to mimic at least some of the effects of insect feeding in spruce, as has been shown in the defense response of Sitka spruce to white pine weevil (*Pissodes strobi;* Miller et al. 2005). Beyond the use of MeJA as an exogenous elicitor, both ethylene and octadecanoid signaling appear to be involved in the endogenous signaling of the induced traumatic resin response in spruce and Douglas fir (Hudgins and Franceschi 2004; Miller et al. 2005; Hudgins et al. 2006; Ralph et al. 2007).

The MeJA- or insect-induced accumulation of terpenoids in traumatic resin ducts in stems and roots, as well as the induced emission of terpenoid volatiles from needles, is controlled at least in part by up-regulation of terpenoid biosynthesis (Martin et al. 2002, 2003a; Huber et al. 2005b; Miller et al. 2005; Keeling and Bohlmann, 2006a). In this process, transcript levels and/or enzyme activities for several steps in terpenoid formation are up-regulated in the methylerythritol phosphate (MEP) pathway, in the subsequent prenyltransferase steps, and in the late steps catalyzed by TPS and P450s. The analysis of more than 200,000 spruce ESTs and 6,464 high-quality finished FLcDNAs along with transcriptome and proteome analyses have identified many other insect- and MeJA-regulated processes in the induced chemical defense and resistance of Sitka spruce against white pine weevil and spruce budworms (*Choristoneura occidentalis;* Ralph et al. 2006, 2007b; Lippert et al. 2007; Ralph and Bohlmann, in preparation).

8.2 Biochemistry of Terpenoid Biosynthesis in Conifer Defense

The progress in research on the biochemistry and molecular biology of terpenoids in conifer defense has recently been reviewed (Martin and Bohlmann 2005; Keeling and Bohlmann 2006a, b; Phillips et al. 2006) and a number of earlier reviews on this topic are available (e.g., Bohlmann and Croteau 1999; Phillips and Croteau 1999; Trapp and Croteau 2001a). Terpenoids represent the largest group of known plant secondary metabolites. The three large classes of monoterpenoids, sesquiterpenoids, and diterpene resin acids are the most abundant terpenoid defense chemicals in conifers. In particular, the monoterpenoids and diterpene resin acids constitute much of the conifer oleoresin volume. In addition, sesquiterpenoids contribute substantially to the structural diversity of chemicals in the oleoresin mixture. The biosynthesis of monoterpenoids, sesquiterpenoids, and diterpene resin acids begins with the formation of the five carbon building blocks, isopentenyl diphosphate (IDP) and its isomer, dimethylallyl diphosphate (DMAPP). In plants, two pathways exist for the formation of these precursors, the mevalonic acid (MEV) pathway and the MEP pathway. Based on large-scale EST and FLcDNA sequencing, all but one gene for the MEV and MEP pathways have been cloned from Sitka spruce and/or white spruce (Ralph et al. 2006; Ralph and Bohlmann, in preparation). These genomics efforts in Sitka spruce and white spruce, by extension of gene discovery to other conifer species, led to the identification and functional characterization of a small family of differentially MeJA- and fungal-induced genes for the first step of the MEP pathway, deoxyxylulose phosphate synthase (DXPS), in Norway spruce (Phillips et al. 2007). Based on our studies with Norway spruce DXPS in cell suspension cultures, it appears that the different isoforms of DXPS may have specific functions in regulation of substrate flux in primary isoprenoid metabolism and secondary terpenoid defense metabolism. Following the formation of IDP and DMAPP, a group of PTs catalyze 1-4 condensation reactions coupling IPP with an allylic prenyl diphosphate. Specifically, geranyl diphosphate (GDP) synthase forms the C_{10} precursor of monoterpenoids, farnesyl diphosphate (FDP) synthase forms the C_{15} precursor of sesquiterpenoids, and geranylgeranyl diphosphate (GGDP) synthase produces the C_{20} precursor for diterpenoids. A few conifer PTs have been characterized in grand fir (*Abies grandis;* Tholl et al. 2001; Burke and Croteau 2002) and more recently in Norway spruce (Schmidt and Gershenzon 2007) and a large number of distinct PTs have been identified with ESTs and FLcDNAs in Sitka spruce and white spruce (Ralph and Bohlmann, in preparation). Although it is thought that PTs may control flux of pathway intermediates at branch points of monoterpenoid, sesquiterpenoid, and diterpenoid biosynthesis in the induced defense of conifers, except for a few studies showing up-regulation of some PT activities and transcript levels, this aspect of the regulation of terpenoid biosynthesis is not well understood and will require much further characterization (Martin et al. 2002; Schmidt and Gershenzon 2007; Ralph and Bohlmann, in preparation).

TPS utilize the three prenyl diphosphates GDP, FDP, and GGDP as substrates in the formation of the hundreds of structurally diverse monoterpenoids, sesquiterpenoids, and diterpenoids. The TPS exist in large gene families and are arguably the best characterized genes and enzymes of chemical defenses in conifers (Bohlmann et al. 1998a; Martin et al. 2004; Keeling and Bohlmann 2006a). Much of the work on TPS in induced conifer defense is based on the early and pioneering studies by Rodney Croteau and coworkers at Washington State University. The TPS employ an electrophilic reaction mechanism, assisted by divalent metal ion cofactors (Davis and Croteau 2000). The prenyl diphosphate substrates are ionized or protonated by TPS to produce reactive carbocation intermediates. The carbocation intermediates can then rearrange within the spatial constraints of the TPS active site and are eventually quenched to yield the many different cyclic and acyclic terpenoid products (Starks et al. 1997; Cane 1999; Wise and Croteau 1999; Christianson 2006). While some TPS form only a single product, the majority of TPS characterized to date, including many TPS identified in conifers, produce arrays of multiple products from a single substrate (Steele et al. 1998; Fäldt et al. 2003; Martin et al. 2004). Based on their substrate specificities, the single- and multi-product TPS are grouped into classes of mono-TPS, sesqui-TPS and di-TPS, which are responsible for the formation of the many simple (acyclic or single ring structure) and more intricate (two or more ring structures) terpene skeletons of conifer mono-, sesqui-, and diterpenoids. Many of the known conifer TPS appear to exert tight control over the stereochemistry of products formed and usually one enantiomer dominates any given TPS product profile. To date, approximately 50 different conifer TPS have been cloned and many of them have been functionally characterized in Norway spruce, Sitka spruce and white spruce (Byun McKay et al. 2003, 2006; Fäldt et al. 2003; Martin et al. 2004; Keeling et al. 2006a; Ralph and Bohlmann, in preparation). In addition, TPS have been cloned and characterized from grand fir (e.g., Stofer Vogel et al. 1996; Bohlmann et al. 1997, 1998a, 1999; Steele et al. 1998), loblolly pine (Phillips et al. 2003; Ro and Bohlmann 2006) and Douglas fir (Huber et al. 2005a).

While the majority of terpenoid products formed by the conifer mono-TPS and sesqui-TPS accumulate in the oleoresin without any further apparent biochemical modifications, the diterpene olefins produced by the di-TPS may be subject to oxidations catalyzed by P450s (Keeling and Bohlmann 2006b). To date only a single P450 in the formation of diterpene resin acids has been cloned and functionally characterized (Ro et al. 2005). The loblolly pine PtAO P450 enzyme, abietadienol/abietadienal oxidase, is a multi-substrate and multi-functional diterpene oxidase in the CYP720 family of plant P450 enzymes. PtAO catalyzes at least two of the three consecutive oxidation steps in the formation of conifer diterpene resin acids and the enzyme efficiently uses several different diterpene alcohols and diterpene aldehydes as substrates. Through the combination of accepting multiple diterpenoid substrates and by catalyzing at least two consecutive oxidations, this P450 enzyme leads to the formation of several different diterpene resin acids found in conifer oleoresin. The ability of PtAO to catalyze consecutive oxidation steps is similar to the activity of angiosperm P450s in the biosynthesis of the diterpene gibberellin phytohormones (Ro et al. 2005; Keeling and Bohlmann 2006b). A further analysis of the nearly half a million conifer ESTs from spruce and loblolly pine revealed a large number of new P450s in the CYP85 clan that may be involved in terpenoid oxidation in conifers (Hamberger and Bohlmann 2006). While conifer P450s putatively involved in terpenoid phytohormone formation, such as gibberellic acid biosynthesis, appear to be expressed as single-copy genes in Sitka spruce and loblolly pine, we found substantial expansion of conifer-specific P450 subfamilies that are likely associated with terpenoid secondary metabolism and defense. Since conifer P450 enzymes are extremely difficult to study as native enzymes in crude or cell-free extracts from bark, needle, or wood tissues, it was necessary to develop an efficient recombinant expression and assay system for functional biochemical characterization of candidate full-length P450 cDNAs. The system developed for the characterization of the PtAO enzyme involved a combination of yeast expression of P450 candidate cDNAs together with expression of cytochrome P450 reductase, in vivo feeding experiments with diterpene substrates, in vitro enzyme assays, and ultimately the development of an engineered yeast strain that produces conifer diterpene resin acids de novo based on the simultaneous expression of GGDP synthase, diterpene synthase, P450, and P450 reductase (Ro et al. 2005).

Based on all current information, the biosynthesis of terpenoids in conifers involves several subcellular compartments (Keeling and Bohlmann 2006a). Enzymes of the MEV pathway are thought to be localized to the cytosol and endoplasmic reticulum, and the MEP pathway is localized in plastids. While GDP and GGDP synthases are most likely localized to plastids, FDP synthase is thought to reside in the cytosol and possibly in mitochondria. Based on the presence or absence of transit peptides, mono-TPS and di-TPS are also present in plastids, while sesqui-TPS appear to be cytosolic. Finally, the PtAO P450 is associated with the endoplasmic reticulum (Ro et al. 2005; Ro and Bohlmann 2006). It has been speculated that enzymes of terpenoid oleoresin formation are predominantly present in specialized epithelial cells lining the surface of resin ducts or resin blisters (Keeling and Bohlmann 2006b). However, testing this idea requires further validation by immuno-localization or cell-specific transcriptome and proteome analysis. Nevertheless, the involvement of several subcellular compartments in oleoresin terpenoid biosynthesis, the possible association of terpenoid biosynthesis with specialized cells, as well as the massive sequestration and accumulation of oleoresin terpenoids in the extracellular space of resin ducts or resin blister mandates efficient and specialized transport systems for intermediates and end-products of terpenoid biosynthesis. Such transport systems would be essential to deliver the large amounts of lipophilic terpenoids against steep concentration gradients and across cell membranes and the cell wall into the extracellular storage sites of specialized anatomical structures. The combined genomic and proteomic analysis of specialized cells types based on laser-assisted microdissection along with yeast expression systems developed for the de novo formation of conifer terpenoids in vivo (Ro et al. 2005) are now being explored for the discovery of candidate transport systems in conifer oleoresin biosynthesis and terpenoid accumulation.

8.3 Evolution of the Conifer TPS and P450 Gene Families for Terpenoid Defense

The cloning and characterization of several dozen different TPS from several conifer species, including Norway spruce, Sitka spruce, loblolly pine, grand fir, and Douglas fir, enabled phylogenetic reconstruction of the gymnosperm TPS family, thereby shedding some light on the evolution of the larger TPS family in plants in general. Based on overall protein sequence relatedness, gene structure similarities, and catalytic mechanisms, all plant TPS are believed to have arisen from a common ancestor, which may have been closely related to the known conifer di-TPS and to the recently characterized bifunctional *ent*-kaurene synthase from the moss *Physcomitrella patens* (Bohlmann et al. 1998a; Trapp et al. 2001b; Martin et al. 2004; Hayashi et al. 2006). It is possible that such an ancestral TPS was involved in the biosynthesis of the precursors of gibberellic acid, copalyl diphosphate and *ent*-kaurene. The evolution of the large family of plant TPS enzymes as we know them today occurred apparently to some large extend independently in the separate angiosperm and gymnosperm lineages involving numerous events of gene duplication and subsequent functional specialization, i.e. neo-functionalization and sub-functionalization (Martin et al. 2004; Keeling and Bohlmann 2006a; Keeling et al. 2008).

Based on amino acid sequence similarity, the TPS family can be divided into seven subfamilies designated TPS-a through TPS-g (Bohlmann et al. 1998a; Martin et al. 2004). The conifer mono-TPS, sesqui-TPS and di-TPS cluster separately from the angiposperm TPS into the TPS-d subfamily, which is further divided into groups TPS-d1 (mostly mono-TPS), TPS-d2 (mostly sesqui-TPS) and TPS-d3 (mostly di-TPS; all containing an ancestral 200-amino acid motif). Both phylogenetic analyses and analysis of gene structure position the conifer di-TPS closest to the putative ancestor of plant TPS (Bohlmann et al. 1998a; Trapp and Croteau 2001b; Martin et al. 2004).

TPS of conifer secondary metabolism are believed to have evolved from TPS involved in primary metabolism. In the evolution of conifer TPS in terpenoid secondary metabolism a massive expansion of gene families and radiation of a diversity of biochemical functions has occurred (Martin et al. 2004). The same may be the case for the P450s (Hamberger and Bohlmann 2006). In contrast to the TPS and P450 of terpenoid secondary metabolism with their adaptive functions in conifer defense and resistance against insects and pathogens, to the best of current knowledge, the TPS and P450s in primary gibberellic acid phytohormone formation seem to be conserved and have undergone very little if any radiation in conifers.

8.4 Biosynthesis and Accumulation of Terpenoid Defenses Requires Specialized Cells

Accumulation of large amounts of hydrophobic terpenoids in constitutive and induced (i.e., traumatic) oleoresin requires specialized anatomical structures in conifers stems, roots, foliage, and cones. Unless sequestered into extracellular spaces of such specialized structures, the hydrophobic terpenoids would interfere with biochemical processes, integrity of membranes and cell structure of the terpenoid producing cells. Sequestration and accumulation of terpenoid oleoresin can be achieved with simple, short-lived resin blisters or with complex, long-lived resin duct systems commonly found in conifer species that produce large quantities of oleoresin. As part of their ability to massively increase the biosynthesis and accumulation of terpenoid defenses, conifers also develop additional traumatic resin ducts (TRD) when challenged by insect attack or fungal inoculation or in response to simulated insect attack induced by treatment with MeJA or ethylene (e.g., Alfaro 1995; Nagy et al. 2000; Franceschi et al. 2002, 2005; Martin et al. 2002; Byun McKay et al. 2003; Hudgins et al. 2003; Krekling et al. 2004; Hudgins and Franceschi 2004; Huber et al. 2005b; Krokene et al. this volume). The de novo formation of TRD occurs within the cambium zone and outermost layers of developing xylem due to a transient change in cambial activity that initiates resin duct epithelial cells in lieu of wood-forming tracheids. Lumenal continuity between TRD, radial ducts, and resin ducts in the bark establishes a three-dimensional resin duct reticulum enabling enhanced biosynthesis, accumulation and flow of resin. The induced formation of TRD is associated with increased biosynthesis and accumulation of terpenoid oleoresin and involves induced gene expression and enzyme activities of several mono-TPS, sesqui-TPS, and di-TPS (Martin et al. 2002; Byun McKay et al. 2003, 2006; Fäldt et al. 2003; Huber et al. 2005b; Miller et al. 2005) and also involves increased levels of transcripts for octadecanoid and ethylene formation (Miller et al. 2005; Hudgins et al. 2006; Ralph et al. 2007a). Contact of TRD with ray parenchyma cells may enable signaling of induced terpenoid defenses between xylem, cambium, and bark tissues (Nagy et al. 2000; Franceschi et al. 2005; Hudgins et al. 2006; Ralph et al. 2007a). Given the abundance of relevant gene probes available from spruce EST and FLcDNAs projects (Ralph et al. 2006; Ralph and Bohlmann, in preparation) combined with immuno-localization, in-situ hybridization, and cell- or tissue specific micro-dissection techniques, it is now possible to test the cell-specific localization of transcripts and proteins for constitutive and induced terpenoid formation and its defense signaling in conifers (Hudgins et al. 2006; Keeling and Bohlmann 2006b; Ralph et al. 2007a).

8.5 Molecular Biology of Insect-and MeJA-Induced Terpenoid Defenses in Conifers

The use of MeJA to induce chemical and anatomical defenses in conifers has been of critical importance in the characterization of insect-induced terpenoid defense responses in species of spruce (Franceschi et al. 2002; Martin et al. 2002, 2003a; Fäldt et al. 2003; Miller et al. 2005; Erbilgin et al. 2006; Zeneli et al. 2006). Similar to the effect of real insect attack or fungal inoculations, exogenous treatment of conifers with MeJA induces the development of TRD, terpenoid accumulation, as well as terpenoid volatile emissions. Treatment of trees with MeJA provided a means by which to induce conifer defense responses without mechanically injuring the tree. This is especially vital to the quantification of induced terpenoid defenses where any injury may reduce the biological capacity of the bark, cambium, or xylem tissue to respond, but would also enhance the loss of terpenoids through the wound site. Instead, treatment with MeJA allowed for the characterization of the induced enzyme activities and gene expression of terpenoid biosynthesis in stems and foliage of spruces, and enabled the detailed quantitative and qualitative analysis of terpenoids in control and induced trees. This approach also enabled the characterization of active emission of terpenoid volatiles from the needles of Sitka spruce and Norway spruce (Martin et al. 2002, 2003a; Miller et al. 2005).

The cloning and functional characterization of families of TPS and P450 in species of spruce and other conifers provided an initial set of valuable probes for gene expression analysis of MeJA- or insect-induced terpenoid defenses (Fäldt et al. 2003; Byun McKay et al. 2003, 2006; Miller et al. 2005; Ro et al. 2005). TPS of all biochemical classes were induced by weevil-feeding or MeJA-treatment in stems of Sitka spruce, with some of the strongest responses observed for the mono-TPS and di-TPS. In loblolly pine, we also found a strong MeJA-induced increase of transcript levels of the PtAO P450 gene along with increased transcript levels of a diterpene synthase that produces the substrate precursor for PtAO (Ro et al. 2005; Ro and Bohlmann 2006). TPS gene expression and TPS enzyme activities are also elevated in foliage of Norway spruce and Sitka spruce following MeJA elicitation or weevil attack (Martin et al. 2003a; Miller et al. 2005). Induced gene expression of TPS in Norway spruce and Sitka spruce is associated with increased TPS enzyme activities and with the elevated accumulation of oleoresin terpenoids in stems or the release of terpenoid volatiles from needles (Martin et al. 2002; Miller et al. 2005). In recent work with Sitka spruce, we have established comprehensive maps of weevil-induced gene expression profiles for additional steps of terpenoid biosynthesis, indicating that TPS, P450, PTs, and a few genes in the MEP and MEV pathways are the strongest up-regulated transcripts in the weevil-induced terpenoid defense response in Sitka spruce (Ralph and Bohlmann, in preparation). In addition, microarray analyses on platforms with ~21,800 spotted spruce cDNAs identified thousands of transcript species as differentially expressed in response to insect attack in Sitka spruce (Ralph et al. 2006; Ralph and Bohlmann, in preparation).

8.6 Effects of Conifer Terpenoids on Insects and Insect-Associated Pathogens

Terpenoids in form of oleoresins and volatile emissions are a critical component of conifer interactions with other organisms such as defense and resistance against insects and pathogens (Langenheim 2003; Raffa et al. 2005). In general, terpenoid chemicals can protect conifers by providing mechanical barriers in form of 'pitch' or crystallized resin at wound sites, through toxicity against insects and insect-associated pathogens, or by interrupting essential processes in insect biology. In addition, in indirect defense systems, conifer terpenoids may function as signals for predators and parasites of the attacking herbivore.

Specific functions of individual terpenoid compounds in conifer defense have been difficult to establish experimentally (Keeling and Bohlmann 2006a). While the complexity of hundreds of terpenoid chemicals that form the constitutive and inducible oleoresin blends is likely to provide a major advantage for the sustainability of a conifer defense system that is both stable and flexible, the same chemical complexity also poses a substantial challenge for researchers to pinpoint specific defense activities conclusively to individual terpenoid compounds. It is also possible that the mixture of terpenoids may be more effective than any individual compound alone. The lack of appropriate mutants for most conifers and the slow process of genetic transformation of conifers in general make it extremely difficult to manipulate complete terpenoid profiles or individual terpenoid compounds for in vivo elucidation of their defense functions with intact trees. Therefore, most tests of conifer terpenoids for their effects on insects or insect-associated pathogens have relied on exposure of insects or pathogens to isolated compounds or blends of these compounds. In such experiments, for example, the sesquiterpenoid compounds juvabione, farnesol, and farnesal have been shown to interrupt development and maturation of insects by interfering with insect endocrine systems (Schmialek 1963; Slama et al. 1965).

The tree-killing activity of certain bark beetles involves an association of these insects with a community of fungi and bacteria some of which are known to be pathogenic and contribute directly to tree mortality. It is therefore important to establish the effect of terpenoid defenses with these pathogens as well. For example, the diterpene resin acids abietic acid and isopimaric acid strongly inhibit spore germination of the blue-stain fungus *Ophiostoma ips*, a conifer pathogen that is symbiotically associated with the pine engraver *Ips pini* (Kopper et al. 2005). The same study also showed that abietic acid inhibits mycelial growth of *Ophiostoma ips*.

Several possible approaches can be considered for future research to test the role of individual terpenoids or groups of terpenoids in conifer defense. Targeted manipulation of terpenoid biosynthesis or genetic association studies are likely to provide the most informative approaches. Both approaches require funding commitments of many years, and genetic association studies also require access to substantial biological resources in form of established breeding programs and provenance trials. Fortunately, relevant biological materials are available to study the role of constitutive and traumatic terpenoids in the resistance of white spruce and Sitka spruce against the white pine weevil (Alfaro et al. 2002, 2004; King et al. 1997, 2004). In Sitka spruce, lines with extremely divergent profiles for individual terpenoids have been identified in metabolite profiling studies and the chemical profiles have been associated with resistance against the white pine weevil (Roberts, Keeling, and Bohlmann, unpublished results). In studies with white spruce, it has been established that insect-induced formation of traumatic resin ducts is positively correlated with resistance to white pine weevil (Alfaro et al. 1996, 2002; Tomlin et al. 1998). As far as targeted manipulation of terpenoids in conifers is concerned, as a proof of principle, we have recently transformed a sesquiterpenoid synthase, E-α-bisabolene synthase, under the control of a wound- and insect-inducible promoter in white spruce (Godard et al. 2007). The use of an insect-inducible promoter will allow us to manipulate and test terpenoid defenses in a more realistic fashion than what can be accomplished with constitutive over-expression (Godard et al. 2007).

Although terpenoids have important roles in the protection of conifers against insects, they are also vital for the successful attack against conifers by certain bark beetles and their associated tree-killing fungi. The large topic of conifer terpenoids and terpenoid pheromones as signals in bark beetle biology, as well as the topic of de novo formation of terpenoid pheromones in bark beetles have recently been addressed in several excellent reviews (e.g., Seybold and Tittiger 2003; Seybold et al. 2006; Raffa et al. 2005). It is well established that terpenoid volatiles emitted from conifer host trees function as semiochemicals in the identification of host and non-host conifer species as well as in the identification of individual susceptible trees within a given species. Once a tree has been attacked by a bark beetle, enzymes of the insect and the insect-associated microorganisms are required for detoxification of host terpenoid chemicals. A recent analysis of ESTs of the fungal pathogen *Ophiostoma clavigerum*, a blue-stain fungus associated with the mountain pine beetle (*Dendroctonus ponderosa*), revealed an overrepresentation of transcripts encoding for P450 and transport proteins in fungi exposed to terpenoids present in the conifer host tree (Diguistini et al. 2007). Beyond simple detoxification of terpenoid defenses, oxidation of monoterpenoids by bark beetles or by their associated fungi results in the formation of pheromones for the coordination of mass attack and mating of hundreds or thousands of insects on individual trees. Recent biochemical and molecular studies showed that bark beetle P450 enzymes play a central role in the conversion of conifer host defense chemicals into insect terpenoid pheromones (Huber et al. 2007; Sandstrom et al. 2006). A mass-attack by bark beetles that is coordinated by terpenoid pheromones allows the insect to overwhelm and exhaust the defense system of the much larger host tree and thus permits the insects and fungal pathogens to kill not only individual conifer trees but in epidemic situations leads to the destructions of conifer forest over landscape areas of millions of hectares. While much research has focused on the role of host monoterpenes as precursors for bark beetle sex and aggregation pheromones, it is now well established, based on precursor feeding experiments, enzyme characterization and gene discovery, that some species of coniferophagus bark beetles are able to produce monoterpenoid pheromones de novo (e.g., Gilg et al. 2005; Huber et al. 2007; Keeling et al. 2004, 2006; Martin et al. 2003b; Sandstrom et al. 2006; Seybold and Tittiger 2003; Seybold et al. 1995, 2000, 2006).

In addition to the prominent role of terpenoids in the direct defense of conifers, indirect defense systems mediated by conifer volatiles have also been established in several conifers. For example, the egg deposition by the sawfly *Diprion pini* elicits emissions of terpenoid and other volatiles from Scots pine (*P. sylvestris*) which can result in the attraction of a parasitic wasp (Mumm and Hilker 2006). Over the millions of years of co-evolution of conifers with insects and insect-associated microorganisms, terpenoids may have played a substantial role in shaping the many interactions of conifers with insects and pathogens.

Acknowledgments Research on conifer defenses in my laboratory has been generously supported with grants from the Natural Sciences and Engineering Research Council of Canada (NSERC), the Human Frontiers Science Program (HFSP), Genome Canada, Genome British Columbia, and the Province of British Columbia. I wish to thank the members of my group, both past and present, for their many excellent contributions to our program and our collaborators in the BC Ministry of Forests and Range and the Canadian Forest Service for continuous support of the research in my group. Salary support has been provided, in part, by a UBC Distinguished University Scholar Award and an NSERC E.W.R. Steacie Memorial Fellowship. My research on conifer defense has been inspired by a very fine group of Washington State University eminent scholars: Dr. Rodney Croteau, who pioneered biochemical research on conifer terpenoids; Dr. Vince Franceschi, who introduced me to the fascinating system of traumatic resin cells; and last but not least Dr. Clarence (Bud) Ryan whose pioneering work on induced plant defense against insects provided a foundation for my research on induced defenses in conifers.

References

Alfaro RI (1995) An induced defense reaction in white spruce to attack by the white-pine weevil, *Pissodes strobi*. Can J For Res 25:1725–1730

Alfaro RI, Kiss GK, Yanchuk A (1996) Variation in the induced resin response of white spruce, *Picea glauca*, to attack by *Pissodes strobi*. Can J For Res 26:967–972

Alfaro RI, Borden JH, King JN, Tomlin ES, McIntosh RL, Bohlmann J (2002) Mechanisms of resistance in conifers against shoot infesting insects. In: Wagner MR, Clancy KM, Lieutier F,

Paine TD (eds) Mechanisms and deployment of resistance in trees to insects. Kluwer Academic Press, Dordrecht, pp 101–126

Alfaro RI, van Akker L, Jaquish B, King JN (2004) Weevil resistance of progeny derived from putatively resistant and susceptible interior spruce parents. For Ecol Manage 202:369–377

Bohlmann J, Steele CL, Croteau R (1997) Monoterpene synthases from grand fir (*Abies grandis*). cDNA isolation, characterization, and functional expression of myrcene synthase, (−)-(*4S*)-limonene synthase, and (−)-(*1S,5S*)-pinene synthase. J Biol Chem 272:21784–21792

Bohlmann J, Meyer-Gauen G, Croteau R (1998a) Plant terpenoid synthases: Molecular biology and phylogenetic analysis. Proc Natl Acad Sci USA 95:4126–4133

Bohlmann J, Crock J, Jetter R., Croteau R (1998b) Terpenoid-based defenses in conifers: cDNA cloning, characterization, and functional expression of wound-inducible (*E*)-α-bisabolene synthase from grand fir (*Abies grandis*). Proc Natl Acad Sci USA 95:6756–6761

Bohlmann J, Croteau R (1999) Diversity and variability of terpenoid defenses in conifers: molecular genetics, biochemistry and evolution of the terpene synthase gene family in grand fir (*Abies grandis*). In: Chadwick DJ, Goode JA (eds) Insect plant interactions and induced plant fefense. John Wiley, West Sussex, pp 132–146

Bohlmann J, Phillips M, Ramachandiran V, Katoh S, Croteau R (1999) cDNA cloning, characterization, and functional expression of four new monoterpene synthase members of the *Tpsd* gene family from grand fir (*Abies grandis*). Arch Biochem Biophys 368:232–243

Burke C, Croteau R (2002) Geranyl diphosphate synthase from *Abies grandis*: cDNA isolation, functional expression, and characterization. Arch Biochem Biophys 405:130–136

Byun McKay SA, Hunter W, Goddard KA, Wang S, Martin DM, Bohlmann J, Plant AL (2003) Insect attack and wounding induce traumatic resin duct development and gene expression of (−)-pinene synthase in Sitka spruce. Plant Physiol 133:368–378

Byun McKay A, Godard KA, Toudefallah M, Martin DM, Alfaro R, King J, Bohlmann J, Plant AL (2006) Wound-induced terpene synthase gene expression in Sitka spruce that exhibit resistance or susceptibility to attack by the white pine weevil. Plant Physiol 140:1009–1021

Cane DE (1999) Sesquiterpene biosynthesis: cyclization mechanisms. In: Cane DE (ed) Comprehensive natural products chemistry: isoprenoids, including carotenoids and steroids, vol 2. Pergamon Press, Oxford, pp 155–200

Christianson DW (2006) Structural biology and chemistry of the terpenoid cyclases. Chem Rev 106:3412–3442

Davis EM, Croteau R (2000) Cyclization enzymes in the biosynthesis of monoterpenes, sesquiterpenes, and diterpenes. Top Curr Chem 209:53–95

DiGuistini S, Ralph S, Lim Y, Holt R, Jones S, Bohlmann J, Breuil C (2007) Generation and annotation of lodgepole pine and oleoresin-induced expressed sequences from the blue-stain fungus *Ophiostoma clavigerum*, a Mountain Pine Beetle-associated pathogen. FEMS Microbiol Lett 267:151–158

Erbilgin N, Krokene P, Christiansen E, Zeneli G, Gershenzon J (2006) Exogenous application of methyl jasmonate elicits defenses in Norway spruce (*Picea abies*) and reduces host colonization by the bark beetle *Ips typographus*. Oecologia 148:426–436

Fäldt J, Martin D, Miller B, Rawat S, Bohlmann J (2003) Traumatic resin defense in Norway spruce (*Picea abies*): Methyl jasmonate-induced terpene synthase gene expression, and cDNA cloning and functional characterization of (+)-3-carene synthase. Plant Mol Biol 51:119–133

Franceschi VR, Krekling T, Christiansen E (2002) Application of methyl jasmonate on *Picea abies* (Pinaceae) stems induces defense-related responses in phloem and xylem. Am J Bot 89:578–586

Franceschi VR, Krokene P, Christiansen E, Krekling T (2005) Anatomical and chemical defenses of conifer bark against bark beetles and other pests. New Phytol 167:353–376

Gilg AB, Bearfield JC, Tittiger C, Welch WH, Blomquist GJ (2005) Isolation and functional expression of the first animal geranyl diphosphate synthase and its role in bark beetle pheromone biosynthesis. Proc Natl Acad Sci, USA 102:9760–9765

Godard KA, Byun-McKay A, Levasseur C, Plant A, Séguin A, Bohlmann J (2007) Testing of a heterologous, wound- and insect-inducible promoter for functional genomics studies in conifer defense. Plant Cell Rep 26:2083–2090 (doi:10.1007/s00299-007-0417-5)

Hamberger B, Bohlmann J (2006) Cytochrome P450 monooxygenases in conifer genomes: Discovery of members of the terpenoid oxygenase superfamily in spruce and pine. Biochem Soc Transactions 34:1209–1214

Hayashi KI, Kawaide H, Notomi M, Sagiki Y, Matsuo A, Nozaki H (2006) Identification and functional analysis of bifunctional *ent*-kaurene synthase from the moss *Physcomitrella patens*. FEBS Lett 580:6175–6181

Huber DPW, Philippe RN, Godard KA, Bohlmann J (2005a) Characterization of four terpene synthase cDNAs from methyl jasmonate-induced Douglas-fir, *Pseudotsuga menziesii*. Phytochemistry 66:1427–1439

Huber DPW, Phillippe RN, Madilao L, Sturrock RN, Bohlmann J (2005b) Changes in anatomy and terpene chemistry in roots of Douglas-fir seedlings following treatment with methyl jasmonate. Tree Physiol 25:1075–1083

Huber DPW, Erickson ML, Leutenegger CM, Bohlmann J, Seybold SJ (2007) Isolation and extreme sex-specific expression of cytochrome P450 genes in the bark beetle, *Ips paraconfusus*, following feeding on the phloem of host ponderosa pine, *Pinus ponderosa*. Insect Mol Biol 16:335–349

Hudgins JW, Christiansen E, Franceschi VR (2003) Methyl jasmonate induces changes mimicking anatomical defenses in diverse members of the Pinaceae. Tree Physiol 23:361–371

Hudgins JW, Franceschi VR (2004) Methyl jasmonate-induced ethylene production is responsible for conifer phloem defense responses and reprogramming of stem cambial zone for traumatic resin duct formation. Plant Physiol 135:2134–2149

Hudgins JW, Ralph SG, Franceschi VR, Bohlmann J (2006) Ethylene in induced conifer defense: cDNA cloning, protein expression, and cellular and subcellular localization of 1-aminocyclopropane-1-carboxylate oxidase in resin duct and phenolic parenchyma cells. Planta 224:865–877

Keeling CI, Blomquist GJ, Tittiger C (2004) Coordinated gene expression for pheromone biosynthesis in the pine engraver beetle, *Ips pini* (Coleoptera: Scolytidae). Naturwissenschaften 91:324–328

Keeling CI, Bohlmann J (2006a) Genes, enzymes and chemicals of terpenoid diversity in the constitutive and induced defence of conifers against insects and pathogens. New Phytol 170:657–675

Keeling CI, Bohlmann J (2006b) Diterpene resin acids in conifers. Phytochemistry 67:2415–2423

Keeling CI, Bearfield JC, Young S, Blomquist GJ, Tittiger C (2006) Effects of juvenile hormone on gene expression in the pheromone-producing midgut of the pine engraver beetle, *Ips pini*. Insect Mol Biol 15:207–216

Keeling CI, Weißhaar S, Lin R, Bohlmann J (2008) Functional plasticity of paralogous diterpene synthases involved in conifer defense. Proc Natl Acad Sci USA, in press

King JN, Yanchuk AD, Kiss GK, Alfaro RI (1997) Genetic and phenotypic relationships between weevil (*Pissodes strobi*) resistance and height growth in spruce populations of British Columbia. Can J For Res 27:732–739

King JN, Alfaro RI, Cartwright C (2004) Genetic resistance of Sitka spruce (*Picea sitchensis*) populations to the white pine weevil (*Pissodes strobi*): distribution of resistance. Forestry 4:269–278

Kopper BJ, Illman BL, Kersten PJ, Klepzig KD, Raffa KF (2005) Effects of diterpene acids on components of a conifer bark beetle-fungal interaction: tolerance by *Ips pini* and sensitivity by its associate *Ophiostoma ips*. Environ Entomol 34:486–493

Krekling T, Franceschi VR, Krokene P, Solheim H (2004) Differential anatomical response of Norway spruce stem tissues to sterile and fungus infected inoculations. Trees-Struct Funct 18:1–9

Langenheim JH (2003) Plant resins: chemistry, evolution, ecology, and ethnobotany. Timber Press Inc, Portland, OR, p 586

Lippert D, Chowrira S, Ralph SG, Zhuang J, Aeschliman D, Ritland C, Ritland K, Bohlmann J (2007) Conifer defense against insects: proteome analysis of Sitka spruce (*Picea sitchensis*) bark induced by mechanical wounding or feeding by white pine weevils (*Pissodes strobi*). Proteomics 7:248–270

Martin D, Tholl D, Gershenzon J, Bohlmann J (2002) Methyl jasmonate induces traumatic resin ducts, terpenoid resin biosynthesis, and terpenoid accumulation in developing xylem of Norway spruce stems. Plant Physiol 129:1003–1018

Martin DM, Gershenzon J, Bohlmann J (2003a) Induction of volatile terpene biosynthesis and diurnal emission by methyl jasmonate in foliage of Norway spruce (*Picea abies*). Plant Physiol 132:1586–1599

Martin D, Bohlmann J, Gershenzon J, Francke W, Seybold SJ (2003b) A novel sex-specific and inducible monoterpene synthase activity associated with a pine bark beetle, the pine engraver, *Ips pini*. Naturwissenschaften 90:173–179

Martin DM, Fäldt J, Bohlmann J (2004) Functional characterization of nine Norway spruce *TPS* genes and evolution of gymnosperm terpene synthases of the *TPS-d* subfamily. Plant Physiol 135:1908–1927

Martin D, Bohlmann J (2005) Molecular biochemistry and genomics of terpenoid defenses in conifers. Rec Adv Phytochem 39:29–56

Miller B, Madilao LL, Ralph S, Bohlmann J (2005) Insect-induced conifer defense. White pine weevil and methyl jasmonate induce traumatic resinosis, de novo formed volatile emissions, and accumulation of terpenoid synthase and octadecanoid pathway transcripts in Sitka spruce. Plant Physiol 137:369–382

Mumm R, Hilker M (2006) Direct and indirect chemical defense of pine against folivorous insects. Trends Plant Sci 11:351–358

Nagy NE, Franceschi VR, Solheim H, Krekling T, Christiansen E (2000) Wound-induced traumatic resin duct development in stems of Norway spruce (Pinaceae): anatomy and cytochemical traits. Am J Bot 87:302–313

Phillips MA, Croteau RB (1999) Resin-based defenses in conifers. Trends Plant Sci 4:184–190

Phillips MA, Wildung MR, Williams DC, Hyatt DC, Croteau R (2003) cDNA isolation, functional expression, and characterization of (+)-α-pinene synthase and (−)-α-pinene synthase from loblolly pine (*Pinus taeda*): Stereocontrol in pinene biosynthesis. Arch Biochem Biophys 411:267–276

Phillips MA, Bohlmann J, Gershenzon J (2006) Molecular regulation of induced terpenoid biosynthesis in conifers. Phytochem Rev 5:179–189

Phillips MA, Walter MH, Ralph SG, Dabrowska P, Luck K, Urós EM, Boland W, Strack D, Rodríguez-Conceptión M, Bohlmann J, Gershenzon J (2007) Functional identification and differentential expression of 1-deoxy-D-xylulose-5-phosphate synthase in induced terpenoid resin formation of Norway spruce (*Picea abies*). Plant Mol Biol 65:243–257 (doi:10.1007/s11103-007-9212-5)

Raffa KF, Aukema BH, Erbilgin N, Klepzig KD, Wallin KF (2005) Interactions among conifer terpenoids and bark beetles across multiple levels of scale: An attempt to understand links between population patterns and physiological processes. Rec Adv Phytochem 32: 79–118

Ralph SG, Yueh H, Friedmann M, Aeschliman D, Zeznik JA, Nelson CC, Butterfield YSN, Kirkpatrick R, Liu J, Jones SJM, Marra MA, Douglas CJ, Ritland K, Bohlmann J (2006) Conifer defense against insects: Microarray gene expression profiling of Sitka spruce (*Picea sitchensis*) induced by mechanical wounding or feeding by spruce budworms (*Choristoneura occidentalis*) or white pine weevils (*Pissodes strobi*) reveals large-scale changes of the host transcriptome. Plant Cell Environ 29:1545–1570

Ralph SG, Hudgins JW, Jancsik S, Franceschi VR, Bohlmann J (2007a) Aminocyclopropane carboxylic acid synthase is a regulated step in ethylene-dependent induced conifer defense. Full-length cDNA cloning of a multigene family, differential constitutive, and wound- and insect-induced expression, and cellular and subcellular localization in spruce and Douglas fir. Plant Physiol 143:410–424

Ralph SG, Jancsik S, Bohlmann J (2007b) Dirigent proteins in conifer defense II: Extended gene discovery, phylogeny, and constitutive and stress-induced gene expression in spruce (*Picea* spp.). Phytochemistry 68:1974–1990

Ro DK, Arimura G, Lau SYW, Piers E, Bohlmann J (2005) Loblolly pine abietadienol/abietadienal oxidase *PtAO* is a multi-functional, multi-substrate cytochrome P450 monooxygenase. Proc Natl Acad Sci, USA 102:8060–8065

Ro DK, Bohlmann J (2006) Diterpene resin acid biosynthesis in loblolly pine (*Pinus taeda*): Functional characterization of abietadiene/levopimaradiene synthase (*PtTPS-LAS*) cDNA and subcellular targeting of PtTPS-LAS and abietadienol/abietadienal oxidase (PtAO, CYP720B1). Phytochemistry 67:1572–1578

Sandstrom P, Welch WH, Blomquist GJ, Tittiger C (2006) Functional expression of a bark beetle cytochrome P450 that hydroxylates myrcene to ipsdienol. Insect Biochem Mol Biol 36:835–845

Schmialek P (1963) Compounds with juvenile hormone action. Z Naturforsch 18:516–519.

Schmidt A, Gershenzon J (2007) Cloning and characterization of isopentenyl diphosphate synthases with farnesyl diphosphate and geranylgeranyl diphosphate synthase activity from Norway spruce (*Picea abies*) and their relation to induced oleoresin formation. Phytochemistry 68:2649–2659 (doi:10.1016/j.phytochem.2007.05.037).

Seybold SJ, Quilici DR, Tillman JA, Vanderwel D, Wood DL, Blomquist GJ (1995) De novo biosynthesis of the aggregation pheromone components ipsenol and ipsdienol by the pine bark beetles, *Ips paraconfusus* Lanier and *Ips pini* (Say) (Coleoptera: Scolytidae). Proc Natl Acad Sci USA 92:8393–8397

Seybold SJ, Bohlmann J, Raffa KF (2000) The biosynthesis of coniferophagous bark beetle pheromones and conifer isoprenoids: Evolutionary perspective and synthesis. Can Entomol 132:697–753

Seybold SJ, Tittiger C (2003) Biochemistry and molecular biology of de novo isoprenoid pheromone production in the Scolytidae. Annu Rev Entomol 48:425–453

Seybold SJ, Huber DPW, Lee LC, Bohlmann J (2006) Pine monoterpenes and pine bark beetles: a marriage of convenience for defense and chemical communication. Phytochem Rev 5:143–178

Slama K, Williams CM (1965) Juvenile hormone activity for the bug *Pyrrhocoris apterus*. Proc Natl Acad Sci USA 54:411–414

Starks CM, Back KW, Chappell J, Noel JP (1997) Structural basis for cyclic terpene biosynthesis by tobacco 5-*epi*-aristolochene synthase. Science 277:1815–1820

Steele CL, Crock J, Bohlmann J, Croteau R (1998) Sesquiterpene synthases from grand fir (*Abies grandis*): Comparison of constitutive and wound-induced activities, and cDNA isolation, characterization, and bacterial expression of δ-selinene synthase and γ-humulene synthase. J Biol Chem 273:2078–2089

Stofer Vogel B, Wildung MR, Vogel G, Croteau R (1996) Abietadiene synthase from grand fir (*Abies grandis*) – cDNA isolation, characterization, and bacterial expression of a bifunctional diterpene cyclase involved in resin acid biosynthesis. J Biol Chem 271:23262–23268

Tomlin ES, Alfaro RI, Borden JH, He FL (1998) Histological response of resistant and susceptible white spruce to simulated white pine weevil damage. Tree Physiol 18:21–28

Tholl D, Croteau R, Gershenzon J (2001) Partial purifications and characterization of the short-chain prenyltransferases, geranyl diphosphate synthase and farnesyl diphosphate synthase, from *Abies grandis* (grand fir). Arch Biochem Biophys 386:233–242

Trapp SC, Croteau R (2001a) Defensive resin biosynthesis in conifers. Annu Rev Plant Physiol Plant Mol Biol 52:689–724

Trapp SC, Croteau RB (2001b) Genomic organization of plant terpene synthases and molecular evolutionary implications. Genetics 158:811–832

Wise ML, Croteau R (1999) Monoterpene biosynthesis. In: Cane DE (ed) Comprehensive natural products chemistry: isoprenoids, including carotenoids and steroids, vol 2. Pergamon Press, Oxford, pp 97–154

Zeneli G, Krokene P, Christiansen E, Krekling T, Gershenzon J (2006) Methyl jasmomate treatment of mature Norway spruce (*Picea abies*) trees increases the accumulation of terpenoid resin components and protects against infection by *Ceratocystis polonica*, a bark beetle-associated fungus. Tree Physiol 26:977–988

Chapter 9
Phenylpropanoid Metabolism Induced by Wounding and Insect Herbivory

Mark A. Bernards and Lars Båstrup-Spohr

Wounding of plant tissue, whether by biotic or abiotic means, results in a massive re-arrangement of metabolism, with the primary objectives of isolating the affected tissue and limiting the extent of further damage. Within this context, newly synthesized phenylpropanoid-derived compounds are known to provide anti-feedant, anti-microbial, and cytotoxic effects, as well as building blocks for structural macromolecules (e.g., lignin, suberin) and condensed tannins. The literature describing hydroxycinnamates (and their conjugates), flavonoids and condensed tannins vis a vis their roles in plant/herbivore interactions is incomplete. Nevertheless, it is clear that the biosynthesis of phenylalanine, a necessary precursor, is induced by herbivore damage, and there is a coordinate induction of transcripts/enzymes associated with hydroxycinnamic acid, flavonoid, and condensed tannin formation. Recent genome scale studies have provided evidence for the involvement of many genes in the response to herbivory, including those of phenylpropanoid metabolism. The next steps will certainly involve the functional characterization of these genes. Forward progress is dependent on expanding analyses to more plant-insect interactions, as well as monitoring plant reactions to herbivory (or wounding) over relevant time frames. Defining the biochemistry of the phenylpropanoid metabolism of specific plant-insect interactions, both at the transcript and enzyme levels, will lead to the discovery of new pathways, or variations on known pathways. In defining plant/herbivore-specific variations in phenylpropanoid metabolism we will discover the underlying strategies used by plants to defend themselves and gain insights into the role(s) of phenylpropanoids in these processes.

9.1 Introduction

The overall success of plants defending themselves against herbivores likely depends on a combination of both pre-formed defenses (phytochemicals, physical barriers, etc.) and those induced by the chewing, boring, or sucking of insects.

M.A. Bernards
Department of Biology, The University of Western Ontario, London, ON, Canada, N6A 5B7
e-mail: bernards@uwo.ca

This chapter summarizes the field of wound- and insect herbivore-induced phenylpropanoid metabolism, with an emphasis on the types of molecules involved, and the enzymes and pathways associated with their biosynthesis. The chapter begins with an overview of the types of phenolic compounds known to accumulate in response to wounding or herbivory, and some of their putative roles in defense against herbivorous insects, followed by a summary of the literature describing the enzymology of phenylpropanoid metabolism induced by wounding and/or herbivory. It is important to note up front that the literature on herbivore-induced phenylpropanoid enzymes at the protein/enzyme activity level is sparse. Few contributions describing the enzymology of wound- and/or herbivory-induced phenylpropanoid metabolism have been made since the summary chapter written by Constable (1999). Instead, much of the recent literature has focused on the induction of specific genes at the transcript level and the use of genome wide screening of herbivory and wound-induced genes using available microarrays. With the latter, there is much reliance on database annotation for the identification of genes, with only a small subset of these having been verified by functional analysis, so far. As we look ahead to what needs to be accomplished in the future to further our understanding of the role of phenolics in the overall interaction between plants and their insect pests, functional characterization of the dozens of genes putatively involved in induced phenylpropanoid metabolism will be high priority.

Since there are chapters elsewhere in this book that specifically address volatile aromatics (Chapters 19–21), wound periderm (Chapter 6), and polyphenol oxidase (Chapter 12), these topics will not be repeated herein. Similarly, there is an entire section in this book devoted to signals in defense; as such the topic of wound-induced signaling (e.g., via the jasmonate pathway) will not be covered.

9.2 Phenolic Compounds and Plant-Herbivore Interactions

9.2.1 Roles for Phenolics in Plant-Herbivore Interactions

Phenolic compounds are aromatic compounds bearing one or more hydroxyl group on an aromatic ring. Phenylpropanoids, strictly speaking, are phenolic compounds derived from phenylalanine. And while there are >8000 known phenolic compounds, roughly categorized into 14 compound classes based on the number of carbons and their arrangement (e.g., Strack 1997), only a small fraction of them have been implicated in plant-herbivore interactions. Amongst these are the benzoic acids, hydroxycinnamic acids (and their conjugates), furanocoumarins, coumarins, stilbenes, flavonoids (especially flavonols), hydrolysable tannins, condensed tannins (catechin polymers; Fig. 9.1), and lignin (reviewed in Constable 1999). Wounding and herbivory also induce the formation of other poly(phenolic)-based cell wall modifications, especially suberin (Ginzberg this volume).

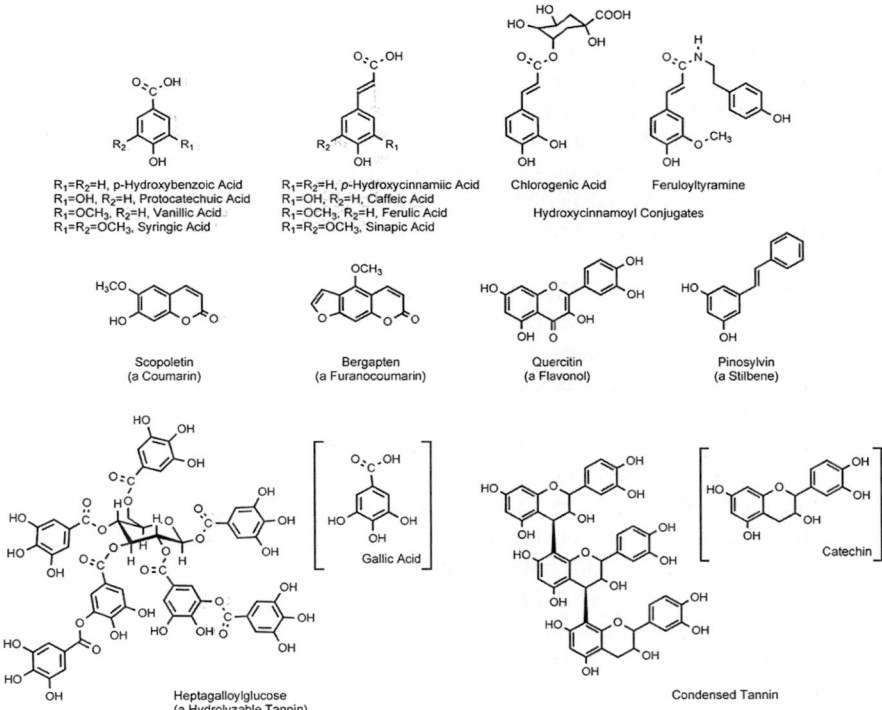

Fig. 9.1 Representative phenolics associated with wounding and herbivory. Examples from the main classes of phenolic compounds known to be associated with either the response of plants to mechanical wounding and/or herbivory are shown. For the tannins, repeating phenolic units are shown in square brackets. See text for details

Different phenolic compounds play different roles in the plant-herbivore interaction. For example, hydroxycinnamic acids may act as cell wall cross-links that fortify and protect plant cell walls against chewing damage (Santiago et al. 2005, 2006). Tannins, which accumulate in many plant species, in response to herbivory, especially within trees where they are most common (Bernays et al. 1989; Peters and Constabel 2002), exert their biological effect through protein binding. They have been shown to affect a wide variety of herbivores from insects to mammals (for a review see Bernays et al. 1989). In insects, tannins can work by producing lesions in the midgut of these animals. Insects that normally feed on tannins have been found to have a relatively thick protective peritrophic membrane lining the midgut epithelium. This is thought to be evolutionary evidence that tannins play a role in plant-insect interactions. Tannins are mainly found in woody species and therefore primarily exert their effect on species that forage on trees (Bernays et al. 1989; Peters and Constabel 2002).

Condensed and non-condensed tannins contribute to the overall oxidative potential of cells (e.g., Barbehenn et al. 2005, 2006), as do other flavonoids (reviewed in Treutter 2005). Paradoxically, one of the hallmark features of phenylpropanoids in general is their anti-oxidative properties (i.e., they are readily oxidized to relatively stable quinone intermediates), but in the context of resistance to herbivores this translates into a class of compounds that, once oxidized can interfere with proteins in the digestive tract of chewing insects and mammals. This can result in an overall decline in nutritional quality of plant material and/or cause oxidative stress in herbivores (Bi and Felton 1997; Barbehenn et al. 2005, 2006). The effect of defense compounds on herbivores also varies according to the compound. For example, defense compounds can influence herbivore settling, feeding, oviposition, growth fecundity and/or fertility (Walling 2000).

Since several types of phenolic compounds (e.g., hydroxycinnamates, hydroxycinnamoyl conjugates, flavonoids) possess anti-microbial properties (e.g., Malmberg 1984; Ding et al. 2000), their accumulation in tissue damaged by wounding (including chewing damage) could play a dual role: direct defense against herbivores and protection against opportunistic pathogens. For example, in the wound response of potato tubers, where the predominant role of induced phenylpropanoid metabolism appears to be associated with the formation of the poly (phenolic) domain of suberin, other phenolics (e.g., hydroxycinnamoyl amides, chlorogenic acid) that are not only involved in suberin formation also accumulate (Cottle and Kolattukudy 1982; Malmberg 1984; Razem and Bernards 2002). The suberin phenolics play a role in forming a physical barrier, while the soluble phenolics possess anti-microbial properties and presumably protect against opportunistic pathogens.

9.2.2 Constitutive vs. Induced Accumulation of Phenolics

Phenolics are widely distributed in plants, and accumulate during normal growth and development. Consequently, phenolics are constitutively present in plants prior to insect or mammalian herbivory-induced damage (i.e., wounding). Indeed, roles for phenolics as pre-formed (constitutive) defenses against herbivory are well documented (e.g., Ding et al. 2000; Mutikainen et al. 2000; Treutter 2005; Santiago et al. 2005, 2006) and include physical barriers such as cell wall bound phenolics, lignins, suberin, and cuticle-associated phenolics as well as stored compounds that have a deterring (anti-feedant) or directly toxic (insecticidal) effect on herbivores (Walling 2000). Induced defenses, on the other hand, are characterized by the synthesis of new compounds or structural barriers after herbivore-induced damage has occurred, and includes both proteins and secondary metabolites. Induced defenses can work both locally, near the point of attack and/or systemically, i.e., throughout the plant.

Since producing defense related compounds is metabolically expensive, constitutive and induced defense mechanisms can be seen as two different strategies for avoiding herbivores. Constitutive defenses are always present, but potentially energetically costly. For example, tannins accumulate as constitutive defense in poplar leaves (Arnold et al. 2004), but the extent of accumulation is dependent on the sink strength of the developing leaves and associated with the import of carbohydrates from older source leaves. If other sinks, such as the roots, are removed by girdling and thus decreasing competition for the carbohydrates, an increase in leaf tannin concentration occurs. Increased phenolic accumulation in association with high sink strength, does not appear to be universal, however, since the opposite phenomenon has been found in the gymnosperm species *Pinus sylvestris,* where the production of phenolics was negatively correlated with sink strength (Honkanen et al. 1999).

Similarly, phenolics can also be induced in poplar leaves as a response to grazing by gypsy moth larvae, mechanical wounding and treatment with jasmonic acids (Arnold et al. 2004). In this study phenolic responses in leaves, induced by herbivores, were found in the first six to eight days post emergence, when the whole canopy was treated with jasmonic acid. During this period leaves are characterized by sink strength indicated by the activity of the enzyme cell wall invertase that facilitates phloem unloading to the cells. Arnold et al. (2004) found that when carbohydrate source leaves were removed, the induced response to jasmonic acid was significantly reduced. These results indicate that even a local defense response in developing foliage depends to some degree on long distance carbohydrate transport from source tissue. Thus, the role of tannins in plant defense, and maybe other phenylpropanoids as well, is not only dependent on the plant's genetic basis to produce them, but also on what stage a particular leaf is in and if there is a source of carbohydrates to supply the production.

Induced defenses are only invoked after tissue damage has occurred. While this potentially provides defense at a lower energetic cost it also comes with some loss of biomass. Thus plants need to balance the allocation of carbon and nitrogen resources between these two types of defenses and vegetative and reproductive growth to insure survival in the long term (Walling 2000). However, it is not always easy to distinguish between constitutive and induced defense mechanisms since some defense-related compounds are constitutively synthesized and stored, and also synthesized de novo as a response to herbivore damage (Ding et al. 2000; Gatehouse 2002).

To add to this complexity, induced metabolic pathways differ when plants are exposed to different types of herbivory, and even species of herbivores. Generally insects that use a piercing/sucking mode of feeding elicit a response similar to that of fungal and bacterial plant pathogens, while chewing insects produce a response similar to that of mechanical wounding (e.g., Walling 2000). In the case of chewing insects, it has been reported that the damage made by insects is not equivalent to mechanical wounding alone (Hartley and Firn 1989) since, for example, insect regurgitant can increase the response to chewing herbivory over that induced by mechanical wounding alone (Korth and Dixon 1997). Further to this, some plants react specifically to one species of herbivore or even a specific life stage within a species (van der Ven et al. 2000), which may correspond to the content of the saliva in the insect species (Walling 2000). However, the general response of plants to wounding and herbivore damage, vis a vis induced phenylpropanoid metabolism and the accumulation of phenolic compounds, is essentially the same: i.e., the same types of compounds are induced to accumulate, and thereby the same enzymes involved. Notwithstanding the subtle details defining specific plant-herbivore interactions, the wound response literature, which is more complete than that of herbivore response literature, can provide a model for the phenylpropanoid metabolic processes involved in the response of plants to herbivore damage.

9.3 Induced Biosynthesis of Phenolic Compounds

Historically, phenylpropanoid metabolism has been divided into two main stages, 'general phenylpropanoid metabolism' (i.e., the biosynthesis of hydroxycinnamic acids) and pathway specific metabolism (i.e., hydroxycinnamic acids as precursors to specific classes of phenylpropanoids such as flavonoids or coumarins; e.g., Strack 1997). More recently it has been shown that the interconversion between hydroxycinnamic acids does not necessarily lie within a defined general pathway (reviewed in Dixon et al. 2001; Humphreys and Chapple 2002), and in fact branches into compound class-specific pathways as early as cinnamate (or cinnamoyl-CoA), or after its hydroxylation to form *p*-coumarate. Consequently, 'general phenylpropanoid metabolism' only consists of two common steps: the deamination of phenylalanine by phenylalanine ammonia-lyase (PAL) and the 4-hydroxylation of cinnamate by cinnamate-4-hydroxylase (C4H).

The main classes of soluble phenolic compounds involved in the response of plants to wounding and herbivory are hydroxycinnamic acids (and their conjugates and derivatives) and tannins (via flavonoids), and these will be considered separately below. However, phenylpropanoids are derived from phenylalanine, which also has to be synthesized de novo as part of the overall wound/herbivory response, and it is important to consider this as an early stage of induced phenylpropanoid metabolism.

9.3.1 The Shikimate Pathway

All three aromatic amino acids, phenylalanine, tyrosine, and tryptophan are synthesized via the shikimate pathway. With respect to induced phenolic metabolism, the synthesis of phenylalanine (Fig. 9.2; Table 9.1) is of particular importance since a significant flux of carbon must go through this intermediate. The first enzyme in the shikimate pathway is 3-deoxy-D-*arabino*-heptulosonate-7-phosphate (DAHP) synthase, which catalyses the condensation of one molecule each of phospho*enol*pyruvate and erythrose-4-phosphate; both of these starting materials are derived from primary metabolism. DAHP synthase has been shown to be induced, at the transcript, protein, and enzyme activity levels, in wounded potato tubers and tomato fruits (Dyer et al. 1989). More recently, DAHP synthase gene up-regulation in Sitka spruce bark following mechanical wounding and weevil feeding, as well as in buds following spruce bud worm larvae feeding has been demonstrated at the transcript level, using a microarray gene profiling approach (Ralph et al. 2006). In addition, shikimate kinase was also shown to be coordinately up-regulated with DAHP synthase in the same tissues (Ralph et al. 2006). Similarly, wounding has been

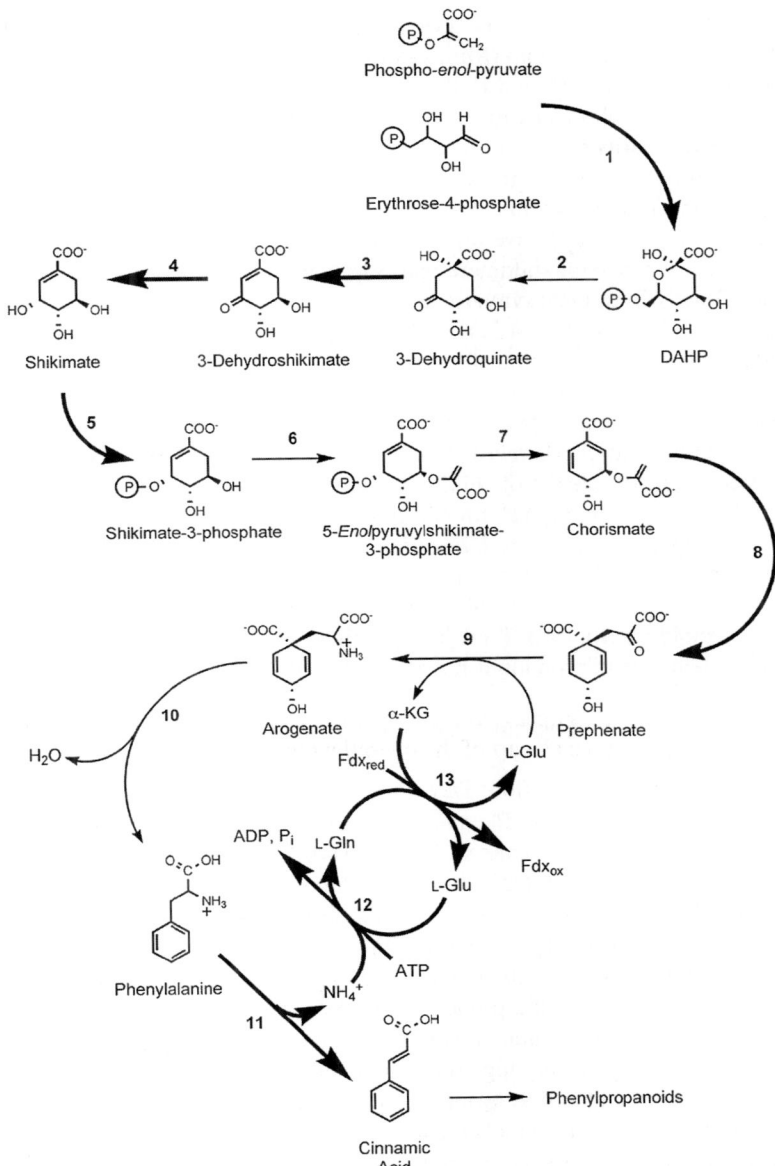

Fig. 9.2 The shikimate pathway and N-recycling in plants. The shikimic acid pathway, from phospho-*enol*-pyruvate and erythrose-4-phosphate to phenylalanine is shown. Cofactors and co-substrates are not shown for clarity. Refer to Table 9.1. Bold arrows indicate steps for which wounding and/or herbivory have been shown to result in induced gene expression and/or enzyme activity. Enzymes are: (1) 3-deoxy-D-*arabino*-heptulosonate-7-phosphate (DAHP) synthase; (2) 3-dehydroquinate synthase; (3) 3-dehydroquinate dehydratase; (4) shikimate dehydrogenase; (5) shikimate kinase; (6) 5-*enol*pyruvylshikimate-3-phosphate (EPSP) synthase;

shown to induce enzyme activity levels of dehydroquinate dehydratase/shikimate dehydrogenase in sweet potato (Minamikawa et al. 1966) and chorismate mutase in potato tubers (Kuroki and Conn 1988). These studies provide a critical link between primary carbon metabolism and the de novo synthesis of phenolics in response to wounding and herbivory. No other enzymes of the shikimate pathway have been shown to be directly involved in the response of plants to wounding or herbivory, either at the transcript or protein levels. However, a recent genomic-scale exploration of gene expression during active lignin formation in poplar (Ehlting et al. 2005) has demonstrated that several shikimate pathway genes are co-coordinately regulated, including 3-dehydroquinate synthase, 3-dehydroquinate dehydratrase/shikimate dehydrogenase, shikimate kinase, chorismate synthase, chorismate mutase, putative aspartate: prephenate aminotransferases and arogenate dehydratase. Significantly, neither arogenate dehydrogenase (which forms tyrosine from arogenate), nor any of the genes associated with tryptophan biosynthesis (which branches off of the main shikimate pathway at chorismate) were up-regulated during lignification, suggesting a process-specific subset of shikimate pathway genes to be involved. The synchronized upregulation of genes for shikimate pathway enzymes and PAL was also observed in tomato cell cultures in response to fungal elicitors (Görlach et al. 1995). Thus, while not strictly wound- or herbivory-based studies, these data illustrate the coordinate regulation of the biosynthetic pathway providing the precursor for phenylpropanoid metabolism. Presumably, this level of coordinate regulation is in operation when phenylpropanoid metabolism is induced by wounding or herbivore damage.

Herbivory and wounding have been clearly linked to the induction of PAL at the transcript, protein and enzyme levels (e.g., Bi and Felton 1995; Hartley and Firn 1989; Bernards et al. 2000; Devoto et al. 2005; Qiuju et al. 2005; Saltveit et al. 2005; Ralph et al. 2006). Phenylalanine ammonia-lyase serves as the gateway enzyme into phenylpropanoid metabolism, and while it is considered in more detail in Section 9.3.2, it is mentioned here since metabolism of phenylalanine represents an exit point from the shikimate pathway and provides a strong metabolic draw on it. The deamination of phenylalanine by PAL yields two products: cinnamic acid, the metabolic fate of which will be discussed in greater detail in Section 9.3.2, and ammonia. As noted above, the production of defense compounds is metabolically expensive, and the 'cost' of inducing phenylpropanoid metabolism for chemical defense could potentially be very high if one mole of ammonia were lost for each mole of phenylpropanoid metabolite formed. However, plants have evolved an efficient N-recycling mechanism in which the ammonia liberated from phenylalanine by PAL is reused in the production of more phenylalanine (Fig. 9.2). This recycling

Fig. 9.2 (continued) (7) chorismate synthase; (8) chorismate mutase; (9) aspartate:prephenate aminotransferase; (10) arogenate dehydratase; (11) phenylalanine ammonia-lyase; (12) glutamine synthetase; (13) ferredoxin-dependent glutamine:2-oxoglutarate aminotransferase. α-KG, α-ketoglutarate (2-oxoglutarate); Fdx_{red}, reduced ferredoxin; Fdx_{ox}, oxidized ferredoxin. (Adapted in part from Razal et al. 1996; Strack 1997; Croteau et al. 2000)

Table 9.1 Enzymes of the shikimate pathways leading to phenylalanine and their corresponding genes in *Arabidopsis*

	Enzyme name	E.C number[1]	Reaction catalyzed	Representative *Arabidopsis* genes[2]
1[3]	3-Deoxy-D-arabino-heptulosonate-7-phosphate Synthase (DAHPS)	2.5.1.54	phospho*enol*pyruvate + erythrose 4-phosphate + H_2O → 3-Deoxy-D-*arabino*-heptulosonate-7-phosphate + P_i	At1g22410, At4g33510, At4g39980
2	3-Dehydroquinate Synthase	4.2.3.4	3-Deoxy-D-*arabino*-heptulosonate-7-phosphate → 3-dehydroquinate + P_i	At5g66120
3/4	3-Dehydroquinate Dehydratase/ Shikimate Dehydrogenase	4.2.1.10/ 1.1.1.25	3-dehydroquinate → 3-dehydroshikimate + H_2O 3-dehydroshikimate + $NADP^+$ → shikimate + $NADPH, H^+$	At3g06350
5	Shikimate Kinase (SK)	2.7.1.71	shikimate + ATP → shikimate-3-phosphate + ADP	At1g06890, At1g60530, At2g14050, At2g16790, At2g21940, At2g35500, At3g50950, At3g26900, At4g39540, At5g22370, At5g45210, At5g47050
6	5-*enol*pyruvylshikimate 3-phosphate Synthase (EPSPS)	2.5.1.19	shikimate-3-phosphate + phospho*enol*pyruvate → 5-*enol*pyruvylshikimate-3-phosphate + P_i	At1g48860, At2g45300
7	Chorismate Synthase (CS)	4.2.3.5	5-*enol*pyruvylshikimate-3-phosphate → chorismate + P_i	At1g48850
8	Chorismate Mutase (CM)	5.4.99.5	chorismate → prephenate	At1g69370, At3g29200, At5g10870
9	Aspartate-Prephenate Aminotransferase	2.6.1.78	prephenate + aspartate → arogenate + oxaloacetate	Unknown in *Arabidopsis*
10	Arogenate Dehydratase (ADT)	4.2.1.91	arogenate → phenylalanine + CO_2 + H_2O	At1g08250, At1g11790, At2g27820, At3g07630, At3g44720, At5g22630
11	Phenylalanine Ammonia-lyase (PAL)	4.3.1.5	phenylalanine → cinnamate + NH_4^+	At2g37040, At3g10340, At3g53260, At5g04230
12	Glutamine Synthetase (GS)	6.3.1.2	glutamate + NH_4^+ + ATP → glutamine + ADP + P_i	At1g48470, At1g66200, At3g17820, At3g53180, At5g16570, At5g35630, At5g37600
13	Ferredoxin-dependent glutamate synthase	1.4.7.1	glutamine + 2-oxoglutarate + Fdx_{red} → 2 glutamate + Fdx_{ox}	At2g41220

[1] Enzyme Nomenclature (web version) www.chem.qmul.ac.uk/iubmb/enzyme/
[2] Gene identifiers are from Ehlting et al. 2005, Ralph et al. 2006, and The *Arabidopsis* Information Resource (TAIR) (www.arabidopsis.org). Not all genes listed have been functionally characterized; nor are the lists considered complete.
[3] Enzyme number refer to the corresponding reaction in Fig. 9.2.

mechanism, which involves the glutamine synthetase/ferredoxin-dependent glutamine 2-oxoglutarate amino transferase (GS/GOGAT) ammonia assimilation system, has been demonstrated in wound healing potato tubers (Razal et al. 1996), sweet potato tubers (Singh et al. 1998), and in lignifying cell cultures of *Pinus taeda* (van Heerden et al. 1996). Thus, the heavy demand for phenylalanine brought about by wound-induced phenylpropanoid metabolism is met by the induction of the shikimate pathway and an efficient N-recycling system. However, with the exception of demonstrated GS up-regulation post wounding and/or herbivore damage in Sitka spruce (Ralph et al. 2006), these aspects of induced phenylpropanoid metabolism have not been demonstrated in plant-herbivore interactions. While it is logical to assume that the shikimate and N-recycling pathways are more broadly associated in plant-herbivore interactions resulting in de novo synthesis of phenolics, confirmation of their participation is lacking.

9.3.2 Induced General Phenylpropanoid Metabolism and the Origin of Hydroxycinnamates

The induction of phenylpropanoid metabolism is implied in the accumulation of newly formed phenolic compounds in several plants in response to herbivore damage, e.g., *p*-coumaric and ferulic acids in wheat exposed to *Sitodipolis mosellana* (Ding et al. 2000) and corn exposed to *Sesamina noonagrioides* (Santiago et al. 2005, 2006), tannins in many tree species (Bernays et al. 1989; Peters and Constabel 2002; Arnold et al. 2004), and phenolics in *Medicago truncatula* damaged by *Spodoptera littoralis* (Leitner et al. 2005) There are relatively few reports in which the activity of phenylpropanoid biosynthesis enzymes is measured directly.

Phenylalanine ammonia-lyase (PAL) is the gateway enzyme providing C_6–C_3 carbon skeletons for the formation of essentially all phenylpropanoids and their derivatives (Fig. 9.3, Table 9.2). The conversion of Phe to cinnamate + NH_4^+ is considered to be physiologically irreversible, and thus is a good indicator of carbon commitment to phenylpropanoid metabolism. Phenylalanine ammonia-lyase is represented by a small gene family (reviewed in Dixon and Paiva 1995; Dixon et al. 2001). For example, at least four genes have been annotated as PAL in the *Arabidopsis* database, while ten uniquely expressed sequence tags (ESTs) are described for Sitka spruce (Ralph et al. 2006). Not surprisingly, this step in the phenylpropanoid metabolic pathway has received a great deal of attention in the literature, albeit largely as a function of biotic and abiotic stress induction; e.g., response to pathogen attack, exposure to UV light, wounding etc. (Strack 1997; Croteau et al. 2000). From a herbivory induction standpoint, increased PAL activity has been directly measured in a few insect/plant systems, e.g., *Helicoverpa zea* damaged soybean leaves (Bi and Felton 1995); *Apocheima pilosaria* damaged birch (Hartley and Firn 1989); gypsy moth larvae damaged poplar leaves (Arnold et al. 2004). At the transcript level, PAL has been shown to be induced by insect damage of *Arabidopsis* (Reymond et al. 2004) and Sitka spruce (Ralph et al. 2006). There are many more

9 Wound-Induced Phenylpropanoid Metabolism

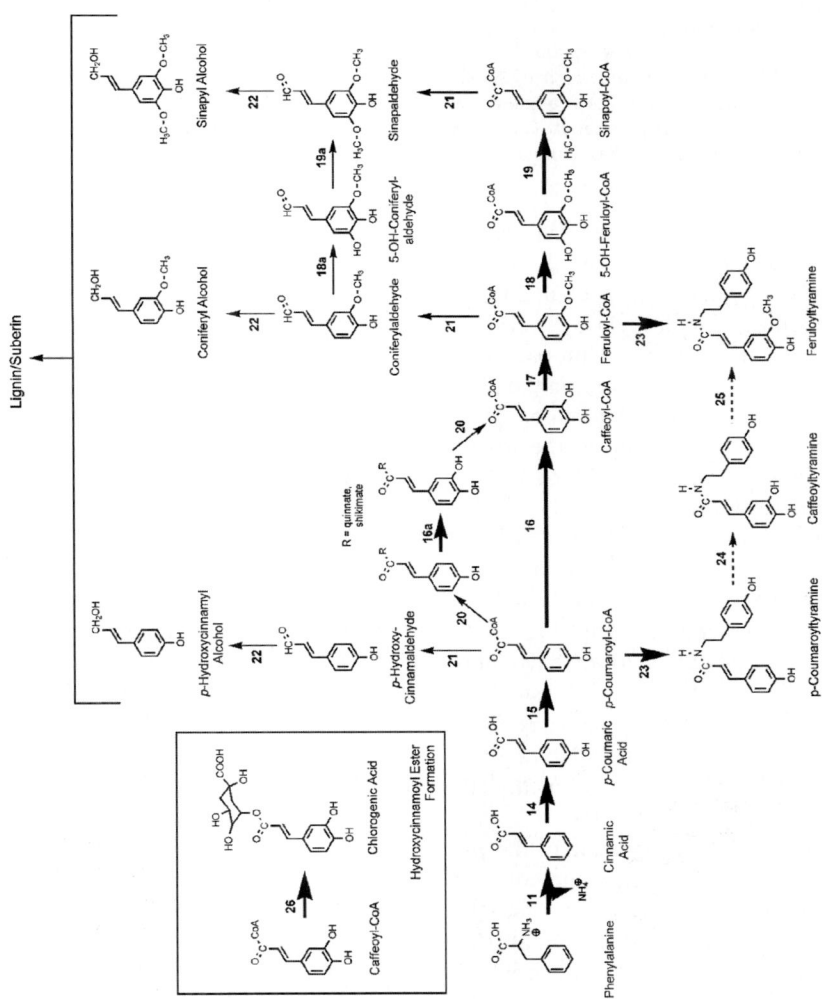

Fig. 9.3 (continued)

Fig. 9.3 (continued) Biosynthesis of hydroxycinnamic acids and their conjugates and derivatives. The biosynthesis of hydroxycinnamates, hydroxycinnamoyl conjugates, and derivatives of hydroxycinnamates is depicted. Cofactors and co-substrates are not shown for clarity. Refer to Table 9.2. Bold arrows indicate steps for which wounding and/or herbivory have been shown to result in induced gene expression and/or enzyme activity, while dashed arrows are hypothetical steps. Enzymes are: (11) phenylalanine ammonia-lyase; (14) cinnamate-4-hydroxylase; (15) 4-coumaroyl-CoA ligase; (16/16a) coumarate-3-hydroxylase; (17) caffeoyl-CoA O-methyltransferase; (18/18a) ferulate-5-hydroxylase; (19/19a) caffeic acid/5-hydroxyferulic acid O-methyltransferase; (20) hydroxycinnamoyl CoA:shikimate/quinate hydroxycinnamoyltransferase; (21) cinnamoyl-CoA reductase; (22) cinnamyl alcohol dehydrogenase; (23) hydroxycinnamoyl CoA:tyramine hydroxycinnamoyltransferase; (24) *p*-coumaroyltyramine-3-hydroxylase; (25) caffeoyltyramine O-methyltransferase; (26) caffeoyl CoA:quinate caffeoyltransferase. (Adapted in part from Strack 1997; Croteau et al. 2000; Humphreys and Chapple 2002)

examples of wound-induced PAL, both at the transcript expression level (e.g., Reymond et al. 2000; Devoto et al. 2005; Ralph et al. 2006) and enzyme level (e.g., Bernards et al. 2000; Qiuju et al. 2005; Saltveit et al. 2005).

Cinnamate-4-hydroxylase (C4H) catalyses the first of potentially many hydroxylation reactions in phenylpropanoid metabolism, generating the first 'phenolic' compound, *p*-coumaric acid, in the phenylpropanoid metabolic pathway (Fig. 9.3). It is a cytochrome P-450 monooxygenase of the CYP73A subfamily. Cinnamate-4-hydroxylase is induced at the transcript level in wound- and insect-damaged Sitka spruce (Ralph et al. 2006) and lignifying poplar stems (Ehlting et al. 2005). Similarly, transcription of C4H in jerusalem artichoke (Batard et al. 1997) and *Camptotheca acuminata* leaves (Kim et al. 2005) in response to wounding, has been demonstrated. The formation of *p*-coumaric acid represents a critical step in phenylpropanoid metabolism since this intermediate can be utilized as the precursor to a myriad of end products (Figs. 9.3 and 9.4). For example, *p*-coumaroyl-CoA is the direct precursor to all flavonoids (Fig. 9.4), and ultimately the condensed tannins (Section 9.3.3) as well as the monolignols (via *p*-coumaroylquinnate or *p*-coumaroylshikimate intermediates) and potentially, other hydroxycinnamoyl conjugates (e.g., hydroxycinnamoyl tyramines; Fig. 9.3). It is also the common substrate for further hydroxylation and methylation to generate characteristic aromatic substitution patterns of phenolics.

Historically, hydroxylation of the *p*-coumaroyl sub-structure has been depicted as occurring at the level of the coenzyme-A thioester; however, in the context of monolignol formation, it has recently been shown that this reaction proceeds after ester transfer to shikimate or quinate (Schoch et al. 2001). Regardless, coenzyme-A thioester formation, via 4-coumaroyl-CoA ligase (4CL), is an important activation step. As with PAL, 4CL is represented by a gene family in plants, with a range of 4CL-like genes identified via sequence similarity analysis, and observation of their coordinate regulation during lignification (Ehlting et al. 2005). Amongst these, at least fifteen 4CL/4CL-like genes are induced by wounding, weevil damage and/or budworm damage in Sitka spruce (Ralph et al. 2006). Similarly, 4CLs are induced by wounding at the transcript level in several other plant species, including parsley and tobacco carrying one of the parsley genes (Ellard-Ivey and Douglas 1996), poplar

Table 9.2 Enzymes of phenylpropanoid metabolism leading to hydroxycinnamic acids and their conjugates, and their corresponding genes in *Arabidopsis*

Enzyme name	E.C number[1]	Reaction catalyzed	Representative *Arabidopsis* genes[2]
14[3] Cinnamate-4-Hydroxylase (C4H) (CYP73A)	1.14.13.11	cinnamate + NADPH,H$^+$ + O$_2$ → p-coumarate + NADP$^+$ + H$_2$O	At2g30490, At2g30160, At2g30600, At2g30580,
15 4-Coumaroyl-CoA Ligase (4CL)	6.2.1.12	p-coumarate + coenzyme A + ATP → p-coumaroyl-CoA + AMP + PP$_i$	At1g51680, At1g65060, At3g21230, At3g21240
16/16a Coumarate-3-Hydroxylase (C3H) (CYP98A)	1.14.13.36	p-coumaroyl-CoA[quinate/shikimate] + NADPH,H$^+$ + O$_2$ → caffeoyl-CoA[quinate/shikimate] + NADP$^+$ + H$_2$O	At1g74540, At1g74550, At2g40890
17 Caffeoyl-CoA O-Methyltransferase (CoAOMT)	2.1.1.104	caffeoyl-CoA + S-adenosyl-L-methionine → feruloyl-CoA + S-adenosyl-L-homocysteine	At1g24735, At1g67980, At1g67990, At3g61990, At4g26220, At4g34050
18/18a Ferulate-5-Hydroxylase (F5H) (CYP84A)	1.14.13.XX	ferulate + NADPH,H$^+$ + O$_2$ → 5-hydroxyferulate + NADP$^+$ + H$_2$O	At4g36220, At5g04330
19/19a Caffeic Acid/5-Hydroxyferulic acid O-Methyltransferase (COMT)	2.1.1.68	caffeate/5-hydroxyferulate+S-adenosyl-L-methionine → ferulate/sinapate+S-adenosyl-L-homocysteine	At5g54160
20 Hydroxycinnamoyl CoA:Shikimate/ Quinate hydroxycinnamoyltransferase (CST/QST)	2.3.1.133	p-coumaroyl-CoA + quinate/shikimate → p-coumaroyl-quinate/shikimate + coenzyme A	At5g48930
21 Cinnamoyl-CoA Reductase (CCR)	1.2.1.44	hydroxycinnamoyl-CoA + NADPH,H$^+$ → hydroxycinnamaldehyde + NADP$^+$ + coenzyme A	At1g15950, At1g80820
22 Cinnamyl Alcohol Dehydrogenase (CAD)	1.1.1.195	hydroxycinnamaldehyde + NADPH,H$^+$ → hydroxycinnamyl alcohol + NADP$^+$	At1g72680, At3g19450, At4g34230

Table 9.2 (continued)

Enzyme name		E.C number[1]	Reaction catalyzed	Representative Arabidopsis genes[2]
23	Hydroxycinnamoyl CoA:Tyramine hydroxycinnamoyltransferase	2.3.1.XX	hydroxycinnamoyl-CoA + tyramine → hydroxycinnamoyltyramine + coenzyme A	Unknown in Arabidopsis
24	p-Coumaroyltyramine-3-hydroxylase (CT3H)	1.14.13.XX	p-coumaroyltyramine + NADPH,H$^+$ + O$_2$ → caffeoyltyramine + NADP$^+$ + H$_2$O	Unknown in Arabidopsis
25	Caffeoyltyramine O-Methyltransferase (CTOMT)	2.1.1.XX	caffeoyltyramine + S-adenosyl-L-methionine → feruloyltyramine+S-adenosyl-L-homocysteine	Unknown in Arabidopsis
26	Caffeoyl CoA:Quinate caffeoyltransferase (CQC)	2.3.1.99	caffeoyl-CoA + quinate → caffeoylquinate + coenzyme A	Unknown in Arabidopsis

[1] Enzyme Nomenclature (web version) www.chem.qmul.ac.uk/iubmb/enzyme/
[2] Gene identifiers are from Ehlting et al. 2005, Ralph et al. 2006, and The *Arabidopsis* Information Resource (TAIR) (www.arabidopsis.org). Not all genes listed have been functionally characterized; nor are the lists considered complete.
[3] Enzyme number refer to the corresponding reaction in Fig. 9.3.

9 Wound-Induced Phenylpropanoid Metabolism

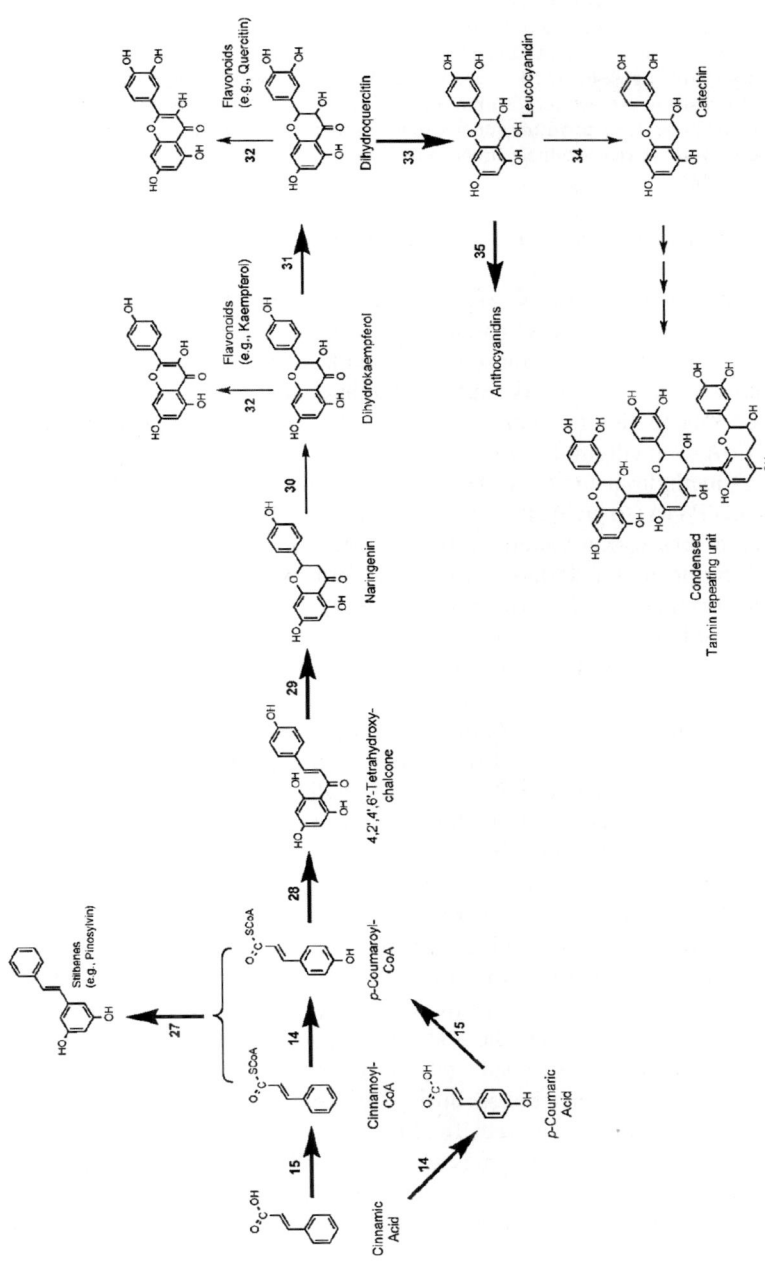

Fig. 9.4 (continued)

Fig. 9.4 (continued) Biosynthesis of flavonoids, stilbenes and condensed tannins. The biosynthesis of flavonoids, stilbenes, and condensed tannins is depicted. Cofactors and co-substrates are not shown for clarity. Refer to Table 9.3. Bold arrows indicate steps for which wounding and/or herbivory have been shown to result in induced gene expression and/or enzyme activity. Enzymes are: (14) cinnamate-4-hydroxylase; (15) 4-coumaroyl-CoA ligase; (27) stilbene synthase; (28) chalcone synthase; (29) chalcone isomerase; (30) flavonone 3-hydroxylase; (31) flavonoid 3′-hydroxylase; (32) flavonol synthase; (33) dihydroflavonol 4-reductase; (34) leucoanthocyanidin reductase; (35) anthocyanidin synthase (leucoanthocyanidin dioxygenase). (Adapted in part from Strack 1997; Croteau et al. 2000)

(Peters and Constabel 2002), and at the enzyme level in potato tubers (Bernards et al. 2000).

Coumarate-3-hydroxylase (C3H; CYP98A) catalyses the hydroxylation of *p*-coumarate to form the caffeoyl (*o*-dihydroxy) ring substitution common in many phenolics. As with 4CL, C3H transcripts are induced by wounding, weevil damage, and/or budworm damage in Sitka spruce (Ralph et al. 2006), and are coordinately upregulated during lignification (Ehlting et al. 2005). Subsequent methylation (caffeoyl-CoA-*O*-methyltransferase; CCoAOMT and/or caffeic acid/5- hydroxyferulic acid *O*-methyltransferase; COMT) and 5-hydroxylation (ferulate-5-hydroxylase; F5H; CYP84A) to yield guaiacyl- and syringyl-substituted hydroxycinnamates follow similar expression patterns after wounding and insect damage (Reymond et al. 2004; Ralph et al. 2006), or during development (Ehlting et al. 2005). Consequently, the formation of phenylpropanoids with typical ring substitution patterns is readily achieved in response to mechanical wounding and chewing insect damage to plant tissue, in a coordinate fashion. However, with the exception of some plant-insect interactions (e.g., Ding et al. 2000; Santiago et al. 2005, 2006), hydroxycinnamic acids themselves do not generally accumulate in response to insect damage or wounding, but rather act as precursors to the other phenylpropanoid derivatives that do. And it is in this regard that species-specific end products, and the unique metabolism leading to them, differentiates between responses of different plants to environmental stimuli, including insect damage.

The formation of physical barriers such as lignin or suberin in response to wounding is well documented (e.g., Bernards and Lewis 1998; Ginzberg this volume). One of the critical steps committing carbon skeletons to the formation of monolignols (essential building blocks for both lignin and suberin) is catalyzed by the enzyme cinnamoyl-CoA oxidoreductase (CCR), which reduces hydroxycinnamoyl-CoA derivatives of hydroxycinnamic acids to their corresponding aldehydes (Fig. 9.3, Table 9.2). And while CCR is coordinately up-regulated with other genes of the early steps of phenylpropanoid metabolism during lignification (Ehlting et al. 2005), it appears to be down-regulated upon weevil and/or budworm herbivory (Ralph et al. 2006). This apparent contradiction represents an important example where wounding and herbivory-induced tissue damage are not equivalent. Instead, the hydroxycinnamates generated in Sitka spruce by the up-regulation of genes of the early steps in the phenylpropanoid pathway appear to be channeled into flavonoids (see below).

Mechanical wounding results in the accumulation of myriad hydroxycinnamoyl conjugates such as chlorogenic acid (Cottle and Kolattukudy 1982; Razem and

Bernards 2002) and hydroxycinnamoyl amides (Malmberg 1984; Razem and Bernards 2002) in potatoes and tomatoes (Pearce et al. 1998). With respect to hydroxycinnamoyltyramines, a hydroxycinnamoyl CoA:tyramine hydroxycinnamoyltransferase (HTH) enzyme has been shown to be induced by wounding (Negrel et al. 1995), which utilizes either p-coumaroyl-CoA or feruloyl-CoA as hydroxycinnamoyl donor (Fig. 9.3). Interestingly, even though feruloyltyramine has been shown to be incorporated into the phenolic domain of wound suberin (Negrel et al. 1996), it is the p-coumaroyl derivative that accumulates when poly(phenolic) domain formation is inhibited (Razem and Bernards 2002). Subsequently, a unique coumarate-3-hydroxylase (C3H) that utilizes p-coumaroyltyramine as substrate in wound-healing potato tubers has been measured in vitro (Pabani and Bernards, unpublished results). This result serves to emphasize the central role of p-coumaroyl-CoA in hydroxycinnamic acid metabolism as well as the power of using wound-induced (i.e., either mechanical of herbivory associated) tissues as a source of discovering novel biochemistry.

9.3.3 Flavonoids and Tannins

Flavonoids are compounds with a C_{15} aglycone skeleton and form the building blocks of the condensed tannins (Fig. 9.4, Table 9.3). The first compound in the pathway, naringenin chalcone (4,2′,4′,6′-tetrahydroxychalcone), is formed by the condensation of p-coumaroyl-CoA with three molecules of malonyl-CoA, catalyzed by the enzyme chalcone synthase (CHS). Chalcone synthase is considered to be the rate limiting enzyme in the flavonoid pathway (Strack 1997), and has been shown to be induced at both the protein and transcript level by wounding of Norway spruce (Brignolas et al. 1995) and white spruce (Richard et al. 2000), respectively, as well as at the transcript level by wounding and weevil and/or budworm herbivory of Sitka spruce (Ralph et al. 2006). Interestingly, stilbene synthase, which catalyzes the same condensation reaction as CHS, but results in a different folding and cyclyzation sequence to generate stilbenoids (e.g., pinosylvin; Fig. 9.4), is also induced at the protein level by wounding in Norway spruce (Brignolas et al. 1995) and at transcript level by wounding and weevil and/or budworm herbivory of Sitka spruce (Ralph et al. 2006). Presumably the combined synthesis of stilbenes and condensed tannins (from CHS-derived flavonoids) represents a concerted response to both deter insects and prevent opportunistic pathogens from establishing themselves in wounded tissues.

Naringenin chalcone is converted into naringenin via a stereospecifc ring closure and isomerization step catalyzed by the enzyme chalcone isomerase (CHI). Chalcone isomerase transcription is induced by wounding and weevil and/or budworm herbivory of Sitka spruce (Ralph et al. 2006). The metabolic fate of naringenin is an important branchpoint in flavonoid biosynthesis, giving rise to the flavonols (e.g., kaempferol, quercitin), dihydroflavonols (e.g., dihydrokaempferol, dihydroquercitin), isoflavonoids (e.g., genistein) and flavones (e.g., apigenin). Collectively, these compounds are referred to as flavonoids. However, while flavonoids themselves have been reported to have deterring effects on insect herbivores (e.g.,

Table 9.3 Selected enzymes of flavonoids and stilbene synthesis, and their corresponding genes in *Arabidopsis*

Enzyme name	E.C number[1]	Reaction catalyzed	Representative *Arabidopsis* genes[2]
27[3] Stilbene Synthase (STS)	2.3.1.146	cinnamoyl-CoA + 3 malonyl-CoA → pinosylvin + 4 coenzyme A + 4CO_2	Unknown in *Arabidopsis*
28 Chalcone Synthase (CHS)	2.3.1.74	p-coumaroyl-CoA + 3 malonyl-CoA → naringenin chalcone + 4 coenzyme A + 3CO_2	At1g02050, At4g00040, At4g34850, At5g13930
29 Chalcone Isomerase CHI	5.5.1.6	naringenin chalcone → naringenin	At1g53520, At5g05270, At3g55120, At5g66220
30 Flavanone 3-Hydroxylase (FHT)	1.14.11.9	naringenin + 2-oxoglutarate + O_2 → dihydrokaempferol + succinate + CO_2 + H_2O	At1g78550, At3g19000, At3g51240, At4g10490, At4g16330, At5g24530
31 Flavonoid 3′-Hydroxylase (F3H)	1.14.13.21	dihydrokaempferol + NADPH,H^+ + O_2 → dihydroquercitin + $NADP^+$ + H_2O	At5g07990
32 Flavonol Synthase (FLS)	1.14.11.23	dihydrokaempferol/dihydroquercitin + 2-oxoglutarate + O_2 → kaempferol/quercitin + succinate + CO_2 + H_2O	At1g49390, At2g44800, At3g19010, At3g50210, At5g08640, At5g43935, At5g63580, At5g63590, At5g63595, At5g63600
33 Dihydroflavonol 4-Reductase (DFR)	1.1.1.219	quercitin + NADPH,H^+ → leucocyanidin + $NADP^+$	At5g42800
34 Leucoanthocyanidin Reductase (LAR)	1.17.1.3	leucocyanidin + NADPH,H^+ → catechin + $NADP^+$ + H_2O	At1g61720
35 Anthocyanidin Synthase (ANS) (Leucoanthocyanidin Dioxygenase)	1.14.11.19	leuco(antho)cyanidin + 2-oxoglutarate + O_2 → anthocyanidin + succinate + CO_2 + 2H_2O	At2g38240, At4g22880

[1] Enzyme Nomenclature (web version) www.chem.qmul.ac.uk/iubmb/enzyme/
[2] Gene identifiers are from Ehlting et al. 2005, Ralph et al. 2006, and The *Arabidopsis* Information Resource (TAIR) (www.arabidopsis.org). Not all genes listed have been functionally characterized; nor are the lists considered complete.
[3] Enzyme number refer to the corresponding reaction in Fig. 9.4.

Treutter 2005), they are not as prominent as the condensed tannins with respect to plant-insect interactions.

Condensed tannins, also called proanthocyanindins, are formed from the covalent linking of flavan-4-ols (e.g., catechin) and/or flavan-3,4-diols (Fig. 9.4). In Sitka spruce, transcripts for several key steps in the formation of (+)-catechin (the major condensed tannin monomer) are induced several fold by wounding and weevil and/or budworm herbivory (Ralph et al. 2006). These include flavonoid 3'-hydroxylase (a CYP75B1 enzyme that hydroxylates dihydrokaempferol to yield dihydroquercitin) and dihydroflavonol 4-reductase. This latter gene, which codes for the enzyme that catalyzes the reduction of the 4'-ketone of the dihydroflavonol ring structure to yield the flavan-3,4-diol leucocyanidin, represents a key step in the formation of condensed tannins. In an earlier report, Peters and Constabel (2002) showed that (in addition to the accumulation of condensed tannins) herbivory induced the expression of this enzyme. Dihydroflavonol 4-reductase catalyzes the penultimate step in the pathway to condensed tannins; the last known step, which converts leuco(antho)cyanidin to (+)-catechin is catalyzed by leucoanthocyanidin reductase. There are no reports of this enzyme being induced by either wounding or herbivory. Similarly, the key step converting naringenin to dihydrokaempferol, catalyzed by flavonone 3-hydroxylase (a 2-oxoglutarate-dependent dioxygenase) does not appear to be induced. It may be that there is sufficient constitutive activity of these key enzymes, the latter of which is also important in the formation of other flavonoids, to support wound/herbivory-induced accumulation. Interestingly, in the Sitka spruce response to wounding, weevil, and budworm damage, the anthocyanidin synthase (leucoanthocyanidin dioxygenase) gene is also induced (Ralph et al. 2006). The enzyme product of this gene catalyses the first step in the formation of anthocyanidins from leucocyanidin, and therefore represents an alternative fate for the carbon skeletons otherwise destined for tannin biosynthesis.

The formation of condensed tannins from (+)-catechin and other flavan-4-ols and flavan-3,4-diols remains uncharacterized. It is interesting to note, however, that a number of laccase/diphenol oxidase genes are also induced by wounding and herbivory in Sitka spruce (Ralph et al. 2006); however, whether they are involved in tannin formation remains unknown. It seems clear from the work of Peters and Constabel (2002) and more recently by Ralph et al. (2006) that there is a concerted, coordinate induction of many genes associated with condensed tannin formation, in response to wounding and herbivory, at least in tree species that normally form condensed tannins as defense compounds.

9.4 Concluding Remarks and Perspectives

In the preceding sections, the induction of specific subclasses of phenylpropanoid metabolites (hydroxycinnamates and their conjugates; flavonoids; condensed tannins) were described. This treatment of induced phenylpropanoid metabolism, at the transcript and/or enzyme level, is by no means exhaustive. Indeed, a number of other plant/herbivore interactions involving other types of phenylpropanoids have been described, including furanocoumarins in wild parsnip and parsely (e.g., Zangerl 1990, 1999; Lois 1991; Olson and Roseland 1991; Berenbaum and Zangerl 1996), benzoic acids in *Nicotiana* species (Spencer and Towers 1991; Wildermuth 2006), and hydrolysable tannins in red oak (Schultz and Baldwin 1982). These represent other plant/herbivore specific phenomena that could also provide insights into how plants respond to this type of environmental stress. Nevertheless, the literature describing hydroxycinnamates (and their conjugates), flavonoids, and condensed tannins vis a vis their roles in plant/herbivore interactions may serve as model, providing some general points. First, it is clear that the biosynthesis of phenylalanine, a necessary precursor to most phenylpropanoids, is induced by herbivore damage, presumably along with elements of a nitrogen recycling system that minimizes the energetic impact of using this amino acid as a precursor. Second, there is a coordinate induction of transcripts/enzymes associated with hydroxycinnamic acid formation. These too are essential as precursors to the more specific phenylpropanoids that define different plant-herbivore interactions. Third, there is a coordinate induction of several genes in the flavonoid pathway, leading to the accumulation of flavonoids as well as providing precursors for the condensed tannins.

Clearly the literature describing phenylpropanoid metabolism induced by herbivory is incomplete. Recent microarray (genome scale) studies (Reymond et al. 2004; Ehlting et al. 2005; Ralph et al. 2006) have provided evidence for the involvement of many genes in the response of plants to herbivory, including confirmation of the involvement of many expected ones. The next step will certainly involve the functional characterization of these genes, including their selective knock-down and cloning. Forward progress is dependent on expanding analyses to more plant and insect species, as well as monitoring plant reactions to herbivory (or wounding) over relevant time frames. In addition, much needs to be done to advance from our current state of knowledge, i.e., of which genes are transcribed in response to herbivory, to knowledge of how the gene products are assembled into functional metabolic pathways (e.g., post-transcriptional and post-translational modifications, activators, inhibitors, etc.).

The study of wound and/or herbivory induced phenylpropanoid metabolism also provides the opportunity to discover novel biochemistry. Defining the phenylpropanoid metabolism of specific plant-insect interactions, both at the transcript and enzyme levels, will lead to the discovery of new pathways, or at least variations on known pathways. It is in defining plant/herbivore-specific variations in phenylpropanoid metabolism that we will discover the underlying strategies used by plants to defend themselves, as well as gain insights into the role(s) of phenylpropanoids in these processes.

Acknowledgments The authors are indebted to Dr. S.E. Kohalmi for thoughtful reading of the manuscript. Work conducted in MAB's lab was supported by an NSERC Discovery Grant.

References

Arnold T, Appel H, Patel V, Stochum E, Kavalie A, Schultz J (2004) Carbohydrate translocation determines the phenolic content of *Populus* foliage: a test of the sink-source model of plant defense. New Phytol 164:157–164

Barbehenn RV, Cheek S, Gasperut A, Lister E, Maben R (2005) Phenolic compounds in red oak and sugar maple leaves have prooxidant activities in the midguts of *Malacosoma disstria* and *Orgyia leucostigma* caterpillars. J Chem Ecol 31:969–988

Barbehenn RV, Jones CP, Karonen M, Salminen J-P (2006) Tannin composition affects the oxidative activities of tree leaves. J Chem Ecol 32:2235–2251

Batard Y, Schalk M, Pierrel M-A, Zimmerlin A, Durst F, Werck-Reichart D (1997) Regulation of the cinnamate 4-hydroxylase (CYP73A1) in jerusalem artichoke tubers in response to wounding and chemical treatments. Plant Physiol 113:951–959

Berenbaum MR, Zangerl AR (1996) Phytochemical diversity: adaptation or random variation? In: Romeo JT, Saunders JA, Barbosa P (eds) Phytochemical diversity and redundancy in ecological interactions. Plenum Press, New York, pp 1–24

Bernards MA, Lewis NG (1998) The macromolecular aromatic domain in suberized tissue: a changing paradigm. Phytochemistry 47:915–933

Bernards MA, Susag LM, Bedgar DB, Anterola AM, Lewis NG (2000) Induced phenylpropanoid metabolism during suberization and lignification: a comparative analysis. J Plant Physiol 157:601–607

Bernays EA, Driver GC, Bilgener M (1989) Herbivores and plant tannins. Adv Ecol Res 19: 263–302

Bi JL, Felton GW (1995) Foliar oxidative stress and insect herbivory: primary compounds, secondary metabolites, and reactive oxygen species as components of induced resistance. J Chem Ecol 21:1511–1530

Bi JL, Felton GW (1997) Antinutritive and oxidative components as mechanisms of induced resistance in cotton. J Chem Ecol 23:97–117

Brignolas F, Lacroix B, Lieutier F, Sauvard D, Drouet A, Claudot, A-C, Yart A, Berryman AA, Christiansen E (1995) Induced responses in phenolic metabolism in two Norway spruce clones after wounding and inoculations with *Ophiostoma polonicum*, a bark beetle-associated fungus. Plant Physiol 109:821–827

Constable CP (1999) A survey of herbivore-induced defense proteins and phytochemicals. In: Agrawal AA, Tuzun S, Bent E (eds) Induced plant defenses against pathogens and herbivores: biochemistry, ecology and agriculture. The American Phytopathological Society, St. Paul, pp 137–166

Cottle W, Kolattukudy PE (1982) Biosynthesis, deposition, and partial characterization of potato suberin phenolics. Plant Physiol 69:393–399

Croteau R, Kutchan TM, Lewis NG (2000) Natural products (Secondary Metabolites). In: Buchanan B, Gruissem R, Jones R (eds) Biochemistry & molecular biology of plants. American Society of Plant Biologists, Rockville, pp 1250–1318

Devoto A, Magusin A, Chang H-S, Chilcott C, Zhu T, Turner JG (2005) Expression profiling reveals *COI1* to be a key regulator of genes involved in wound- and methyl jasmonate-induced secondary metabolism, defense, and hormone interactions. Plant Mol Biol 58:497–513

Ding H, Lamp RL, Ames N (2000) Inducible production of phenolic acids in wheat and antibiotic resistance to *Sitodiplosis mosellana*. J Chem Ecol 26:969–984

Dixon RA, Chen F, Guo D, Parnathi K (2001) The biosynthesis of monolignols: a 'metabolic grid', or independent pathways to guaiacyl and syringyl units? Phytochemistry 57: 1069–1084

Dixon RA, Paiva NL (1995) Stress-induced phenylpropanoid metabolism. Plant Cell 7:1085–1097

Dyer WE, Henstrand JM, Handa AK, Herrmann KM (1989) Wounding induces the first enzyme of the shikimate pathway in Solanaceae. Proc Natl Acad Sci USA 86:7370–7373

Ehlting J, Mattheus N, Aeschliman DS, Li E, Hamberger B, Cullis IF, Zhuang J, Kaneda M, Mansfield SD, Samuels L, Ritland K, Ellis BE, Bohlmann J, Douglas CJ (2005) Global transcript profiling of primary stems from *Arabidopsis thaliana* identifies candidate genes for missing links in lignin biosynthesis and transcriptional regulators of fiber differentiation. Plant J 42:618–640

Ellard-Ivey M, Douglas CJ (1996) Role of jasmonates in the elicitor- and wound-inducible expression of defense genes in parsley and transgenic tobacco. Plant Physiol 112: 183–192

Gatehouse JA (2002) Plant resistance towards insect herbivores: a dynamic interaction. New Phytol 156:145–169

Görlach J, Raesecke HR, Rensch D, Regenass M, Roy P, Zla M, Keel C, Boller T, Amrhein N, Schmid J (1995) Temporally distinct accumulation of transcripts encoding enzymes of the prechorismate pathway in elicitor-treated, cultured tomato cells. Proc Natl Acad Sci USA 92:3166–3170

Hartley SE, Firn RD (1989) Phenolic biosynthesis, leaf damage, and insect herbivory in birch (*Betula pendula*). J Chem Ecol 15(1):275–283

Honkanen T, Haukioja E, Kituren V (1999) Responses of *Pinus sylvestris* branches to simulated herbivory are modified by tree sink/source dynamics and by external resources. Functional Ecol 13:126–140

Humphreys JM, Chapple C (2002) Rewriting the lignin road map. Current Opin Plant Biol 5: 224–229

Kim D-G, Kim T-J, Lee S-H, Lee I (2005) Effect of wounding and chemical treatments on expression of the gene encoding cinnamate-4-hydroxylase in *Campotheca acuminata* leaves. J Plant Biol 48:298–303

Korth KL, Dixon RA (1997) Evidence for chewing insect-specific molecular events distinct from general wound response in leaves. Plant Physiol 115:1299–1305

Kuroki G, Conn EE (1988) Increased chorismate mutase levels as a response to wounding in *Solanum tuberosum* L. tubers. Plant Physiol 86: 895–898

Leitner M, Boland W, Mithöfer A (2005) Direct and indirect defenses induced by piercing-sucking and chewing herbivores in *Medicago truncatula*. New Phytol 167:597–606

Lois R, Hahlbrock K (1991) Differential wound activation of members of the phenylalanine ammonia-lyase and 4-coumarate: CoA ligase gene families in various organs of parsley plants. Z Naturforschung 47c:90–94

Malmberg A (1984) *N*-Feruloylputrescine in infected potato tubers. Acta Chem Scand B 38: 153–155

Minamikawa T, Kojima M, Uritari I (1966) Dehydroquinate hydro-lyase and shikimate: NADP oxidoreductase in sliced roots of sweet potato. Arch Biochem Biophys 117:194–195

Mutikainen P, Walls M, Ovaska J, Keinänen M, Julkunen-Tiitto R, Vapaavouri E (2000) Herbivore resistance in *Betula pendula*: Effect of fertilization, defoliation and plant genotype. Ecology 81:49–65

Negrel J, Lotfy S, Javelle F (1995) Modulation of the activity of two hydroxycinnamoyl transferases in wound-healing potato tuber discs in response to pectinase or abscisic acid. J Plant Physiol 146:318–322

Negrel J, Pollet B, Lapierre C (1996) Ether-linked ferulic acid amides in natural and wound periderms of potato tuber. Phytochemistry 43:1195–1199

Olson MM, Roseland CR (1991) Induction of the coumarins scopoletin and ayapin in sunflower by insect-feeding stress and effects of coumarins on the feeding of sunflower beetle (Coleoptera: Chrysomelidae). Environ Entomol 20:1166–1172

Pearce G, Marchand PA, Griswold J, Lewis NG, Ryan CA (1998) Accumulation of feruloyltyramine and *p*-coumaroyltyramine in tomato leaves in response to wounding. Phytochemistry 47(4):659–664

Peters D, Constabel CP (2002) Molecular analysis of herbivore-induced condensed tannin synthesis: cloning and expression of dihydroflavonol reductase from trembling aspen (*Populus tremulus*). Plant J 32:701–712

Qiuju Q, Xueyan S, Liang P, Xiwu G (2005) Induction of phenylalanine ammonia-lyase and lipoxygenase in cotton seedlings by mechanical wounding and aphid infestation. Prog Nat Sci 15(5):419–423

Ralph SG, Yueh H, Friedmann M, Aeschliman D, Zeznik JA, Nelson CC, Butterfield YSN, Kirkpatrick R, Liu J, Jones SJM, Marra MA, Douglas CJ, Ritland K, Bohlmann J (2006) Conifer defence against insects: microarray gene expression profiling of Sitka spruce (*Picea sitchensis*) induced by mechanical wounding or feeding by spruce budworms (*Choristoneura occidentalis*) or white pine weevils (*Pissodes strobi*) reveals large-scale changes of the host transcriptome. Plant Cell Environ 29:1545–1570

Razal RA, Ellis S, Singh S, Lewis NG, Towers GHN (1996) Nitrogen recycling in phenylpropanoid metabolism. Phytochemistry 41:31–36

Razem FA, Bernards MA (2002) Hydrogen peroxide is required for poly(phenolic) domain formation during wound-induced suberization. J Agric Food Chem 50:1009–1015

Reymond P, Bodenhausen N, van Poecke RMP, Krishnamurthy V, Dicke M, Farmer EE (2004) A conserved transcript pattern in response to a specialist and a generalist herbivore. Plant Cell 16:3132–3147

Reymond P, Weber H, Damond M, Farmer EE (2000) Differential gene expression in response to mechanical wounding and insect feeding in *Arabidopsis*. Plant Cell 12: 707–719

Richard S, Lapointe G, Rutledge RG, Séguin A (2000) Induction of chalcone synthase expression in white spruce by wounding and jasmonate. Plant Cell Physiol 41(8): 982–987

Saltveit ME, Choi Y-J, Tomás-Berberán FA (2005) Involvement of components of the phospholipid-signaling pathway in wound-induced phenylpropanoid metabolism in lettuce (*Lactuca Sativa*) leaf tissue. Physiol Plant 125:345–355

Santiago R, Butron A, Arnason JT, Reid LM, Souto XC, Malvar RA (2006) Putative role of cell wall phenylpropanoids in *Sesamia noonagrioides* (Lepidoptera: Noctuidae) resistance. J Agric Food Chem 54:2274–2279

Santiago R, Malvar RA, Baamonde MD, Revilla P, Souto XC (2005) Free phenols in maize pith and their relationship with resistance to *Sesamia nonagrioides* (Lepidoptera: Noctuidae) attack. J Econ Entomol 98 (4):1349–1356

Schoch G, Goepfert S, Morant M, Hehn A, Meyer D, Ullmann P, Werck-Reichhart D (2001) CYP98A3 from *Arabidopsis thaliana* is a 3'-hydroxylase of phenolic esters, a missing link in the phenylpropanoid pathway. J Biol Chem 276:36566–36574

Schultz JC, Baldwin IT (1982) Oak leaf quality declines in response to defoliation by gypsy moth larvae. Science 217:149–151

Singh S, Lewis NG, Towers GHN (1998) Nitrogen recycling during phenylpropanoid metabolism in sweet potato tubers. J Plant Physiol 153:316–323

Spencer PA, Towers GHN (1991) Restricted occurrence of acetophenone signal compounds. Phytochemistry 30:2933–2937

Strack D (1997) Phenolic metabolism. In: Dey PM, Harborne JB (eds) Plant biochemistry. Academic Press, New York, pp 387–416

Treutter D (2005) Significance of flavonoids in plant resistance and enhancement of their biosythesis. Plant Biol 7:581–591

van der Ven WTG, LeVesque CS, Perring TM, Walling LL (2000) Local and systemic changes in squash gene expression in response to silverleaf whitefly feeding. Plant Cell 12:1409–1423

van Heerden PS, Towers GHN, Lewis NG (1996) Nitrogen metabolism in lignifying *Pinus taeda* cell cultures. J Biol Chem 271:12350–12355

Walling LL (2000) The myriad plant responses to herbivores. J Plant Growth Regul 19:195–216

Wildermuth MC (2006) Variations on a theme: synthesis and modification of plant benzoic acids. Current Opin Plant Biol 9:288–296

Zangerl AR (1990) Furanocoumarin induction in wild parsnip: evidence for an induced defense against herbivores. Ecology 71:1933–1940

Zangerl AR (1999) Locally-induced responses in plants: the ecology and evolution of restrained defense. In: Agrawal AA, Tuzun S, Bent E (eds) Induced plant defenses against pathogens and herbivores: biochemistry, ecology and agriculture. The American Phytopathological Society, St. Paul, pp 231–249

Chapter 10
Defense by Pyrrolizidine Alkaloids: Developed by Plants and Recruited by Insects

Thomas Hartmann and Dietrich Ober

Pyrrolizidine alkaloids are constitutively expressed toxic plant secondary compounds with sporadic occurrence in distantly related angiosperm families. Physiological and ecological aspects of their role in defense against herbivores are discussed. In *Senecio* species senecionine *N*-oxide is synthesized as backbone structure in roots, distributed through the phloem all over the plant and diversified by peripheral reactions yielding the species-specific alkaloid profiles. Except structural diversification the alkaloids do not exhibit any turnover or degradation. They are, however, spatially mobile, slowly allocated and accumulated at strategic important sites of defense, e.g. inflorescences and epidermal tissues. Based on the molecular evolution of the first pathway-specific enzyme, evidence is presented that the pathways of pyrrolizidine alkaloids evolved independently in the various angiosperm taxa. Pyrrolizidine alkaloids are the best studied example for a plant defense system recruited by herbivorous insects. Various adapted insects sequester the alkaloids and utilize them for their own protection against predators and parasitoids. Adapted arctiids (Lepidoptera) are chosen to illustrate the specific integration of a plant defense system into an insect's biology. This integration concerns sensory recognition and efficient exploitation of plant alkaloid sources as well as maintenance and allocation of the alkaloids in a non-toxic and metabolically safe state. These biochemical adaptations assure the unique role of the alkaloids in the insect's behavior, for instance, male-to-female transfer of the alkaloids that guarantees an efficient protection of the eggs by coating with alkaloids from both parents.

10.1 Introduction

Pyrrolizidine alkaloids (PAs) are a class of constitutively expressed plant defense compounds against herbivores (Hartmann and Witte 1995; Hartmann and Ober 2000). Why should a constitutive defense be considered in a book dedicated to

---

T. Hartmann
Institute of Pharmaceutical Biology, Technical University of Braunschweig, D-38106 Braunschweig, Germany
e-mail: t.hartmann@tu-bs.de

induced plant defenses? There are good reasons to do so. Firstly, for many chemical defenses the borders between constitutive and induced defenses are floating. Terpenoids (see Bohlmann this volume), phenolics (see Bernards and Båstrup-Spohr this volume), and even alkaloids such as nicotine (see Steppuhn and Baldwin this volume) are generally expressed constitutively at a certain level which under herbivore pressure can be boosted by induced synthesis. Secondly, in comparison to preformed defenses, induced defenses are less expensive due to lower metabolic costs and allow the focused accumulation of high levels of defense compounds. Constitutive defenses are likely to be more ancient, and provided the basis for the evolution of efficiently regulated inducible defenses. The survival of an herbivorous insect depends greatly on its ability to cope with the plant anti-herbivore endowments. For an herbivore switching to a new potential food-plant it should be easier to become adapted to a preformed defense than to an inducible one. Some herbivores that successfully adapted to a constitutive plant defense even adopted the plant defense system for their own benefit. The mechanistic background of such insect–plant adaptations provides fascinating insights into the abilities of insects to integrate plant-derived defense devices into their own biology.

In the first part we review the present knowledge of plant PAs as defense compounds that evolved probably independently in different angiosperm taxa. The second part deals with specialized herbivorous insects that independently acquired the ability to sequester plant PAs. It will be shown that the acquisition of plant PAs may greatly affect the biology of the sequestering insect herbivore.

10.2 Plant Pyrrolizidine Alkaloids

10.2.1 Occurrence and Structural Types

The occurrence of PAs is restricted to the angiosperms. They are found scattered and often sporadically in distantly related families. Well known sources are the tribes Senecioneae and Eupatorieae of the Asteraceae, and many genera of the Boraginaceae (Hartmann and Witte 1995). In the Fabaceae, PAs are only found in the tribe Crotalariae where they substitute the quinolizidine alkaloids (Wink and Mohamed 2003). Within the Apocynaceae and Orchidaceae, few genera are known to contain PAs (Hartmann and Witte 1995) while in the Convolvulaceae PAs are only found in a few related species (Jenett-Siems et al. 2005).

PAs encompass a rich diversity of about 400 compounds. Despite their diversity they all have one feature in common, they are ester alkaloids composed of a necine base esterified to one or more necic acids (Fig. 10.1). All necine bases are derivatives of the unique bicyclic aminoalcohol, 1-hydroxymethylpyrrolizidine. The necic acids represent a variety of often rather complex branched-chain aliphatic and less abundant aromatic acids. The majority of known PAs can be attributed either to the senecionine type (Fig. 10.1B) or the lycopsamine type (Fig. 10.1C). The former are found mainly in the Asteraceae, tribe Senecioneae, the latter type in three plant

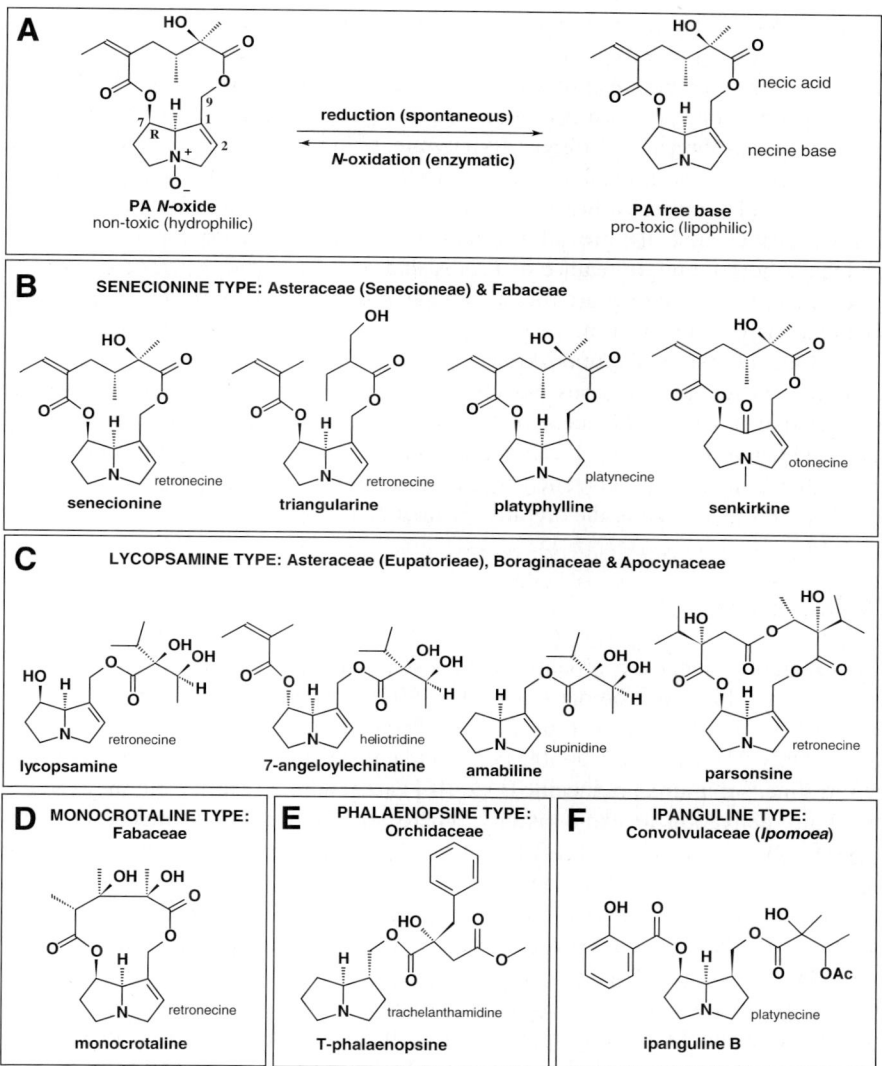

Fig. 10.1 Major structural types of naturally occurring plant pyrrolizidine alkaloids (PAs). **A**: In many plants 1,2-unsaturated PAs are present as polar, non-toxic *N*-oxides which are easily reduced producing the respective pro-toxic free base. **B**: The senecionine type represents the major class of macrocyclic PAs. **C**: the lycopsamine type represents the major class of monoesters and open-chain diesters. **D**: The monocrotaline type occurs only in Fabaceae. **E** and **F**: The phalaenopsine and ipangularine types represent 1,2-saturated PAs

taxa, the Asteraceae, tribe Eupatorieae, the Boraginaceae, and the Apocynaceae. PAs of the senecionine type represent typically macrocyclic diesters occasionally accompanied by related open-chain diesters such as triangularine (Fig. 10.1B). PAs of the lycopsamine type are open-chain monoesters or diesters containing at least one unit of 2-isopropyl-2,3-dihydroxybutyrate, the unique C_7-necic acid, which in nature occurs almost exclusively in PAs. In the Apocynaceae, these PAs are occasionally found as macrocyclic triesters such as parsonsine (Fig. 10.1C). In both types of PAs, retronecine is the prevailing necine base but occasionally replaced by other necines. Another unique feature of PAs is that at least in the Asteraceae, Boraginaceae, and Fabaceae they are found as polar N-oxides which, however, are easily spontaneously reduced to the respective free bases (Hartmann and Toppel 1987). Only otonecine esters like senkirkine (Fig. 10.1B) which cannot be N-oxidized are genuinely present as free bases. Species of the Apocynaceae appear to contain PAs only as free bases. Orchidaceae and Convolvulaceae contain only PAs with 1,2-saturated necine bases (Fig. 10.1E and F). These PAs are either found as free bases like in the Convolvulaceae (Jenett-Siems et al. 1998), or as mixtures of free bases and their N-oxides like in the orchids (Frölich et al. 2006).

10.2.2 Toxicity

PA-containing plants are probably the most common poisonous plants affecting livestock, wildlife, and humans (Cheeke 1998; Prakash et al. 1999; Stegelmeier et al. 1999). Toxicity induced by PAs has been studied most intensively in vertebrates. Only PAs with a 1,2-double bond in the necine base moiety, e.g. retronecine, heliotridine, supinidine, otonecine (Fig. 10.1) are potentially toxic, but they require metabolic activation to exert toxicity. There are three principle metabolic pathways for the metabolism of 1,2-unsaturated PAs in vertebrates (Fu et al. 2004; Fig. 10.2): (i) hydrolysis of the ester bonds yielding the free necine base; (ii) N-oxidation of the free base, and (iii) oxidation of the necine base moiety to the corresponding dehydropyrrolizidine (pyrrolic ester) derivative. Hydrolysis and N-oxidation are considered as detoxification pathways whereas the formation of a dehydropyrrolizidine creates reactive intermediates that easily form adducts with biological nucleophiles resulting in severe cell toxicity (Fig. 10.2). In vertebrates, this bioactivation is mediated by hepatic cytochrome P450 monooxygenases, particularly CYP3A and CYP2B isoforms (Huan et al. 1998a). Multisubstrate flavine monooxygenases (Huan et al. 1998b) and to a lesser extent P450 enzymes (Williams et al. 1989a, b) are involved in the detoxification of PAs to the corresponding N-oxides.

In herbivorous insects, bioactivation of PAs comparable to that of vertebrates exists, since insects possess a rich range of P450 enzymes involved in xenobiotic metabolism (Brattsten 1992; Glendinning and Slansky 1995; Scott et al. 1998; Glendinning 2002). Mutagenic effects of PAs have been demonstrated in *Drosophila* (Frei et al. 1992), and acute toxicity in larval development of *Philosamia ricini* (Saturniidae; Narberhaus et al. 2005).

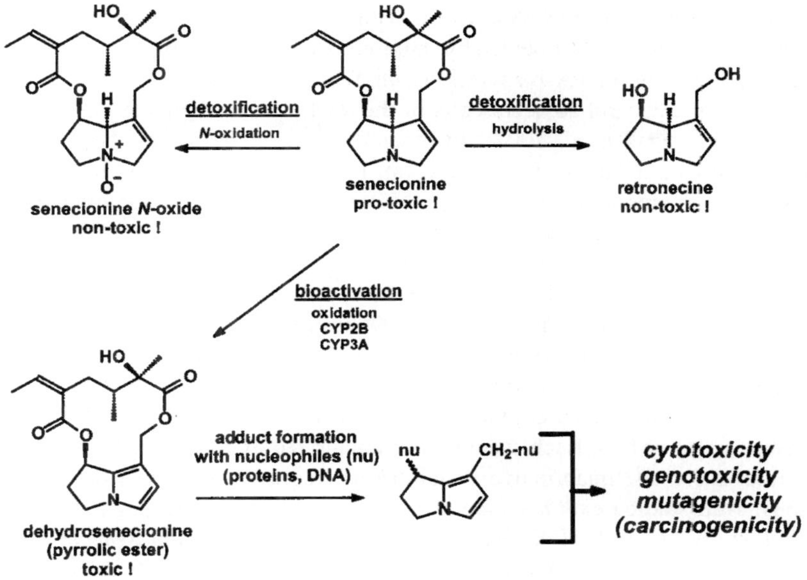

Fig. 10.2 Detoxification of pro-toxic 1,2-unsaturated pyrrolizidine alkaloids in vertebrates and their conversion into toxic pyrrolic intermediates by cytochrom P450-mediated bioactivation

Many plants store their PAs as non-toxic N-oxides (see Section 5.2.1). However, any herbivorous vertebrate or insect feeding on these plants absorb the PAs as pro-toxic free base since in the reducing gut milieu any ingested N-oxide is easily converted into the respective free base (Mattocks 1986; Lindigkeit et al. 1997). Thus, for a non-adapted herbivore, both forms are pro-toxic. PAs that lack the 1,2-double-bond are not bioactivated (Fu et al. 2004) and therefore in the above discussed sense are non-toxic.

10.2.3 Features of a Constitutive Chemical Defense: Physiological Aspects

The basic requirements for a constitutive defense are simple: The defense compound(s) should accumulate stably in the target tissues at concentrations needed to prevent damage by an antagonist, e.g. an insect herbivore. PAs fulfill these requirements in general: they are accumulated in sufficient quantities and are metabolically stable, i.e. they do not show turnover or degradation (Hartmann and Dierich 1998; Frölich et al. 2007). However, PAs are spatially mobile and carried from their sites of synthesis to their sites of accumulation (Hartmann et al. 1989) where they are concentrated within cellular vacuoles (Ehmke et al. 1988). These metabolic characteristics have been established for *Senecio* species which synthesize senecionine N-oxide as PA backbone structure in the roots. Senecionine N-oxide is distributed through the phloem all over the plant, and is chemically diversified (see below) to produce the species-specific PA profiles. Within roots, PA biosynthesis is restricted to specialized cells, as documented by immunolocalization of homospermidine synthase a key enzyme of PA biosynthesis. In *Senecio,* these cells comprise groups of endodermis cells and adjacent cells of the cortex parenchyma directly opposite to the phloem (Moll et al. 2002). In other PA-producing species the sites of PA synthesis may be different. In *Eupatorium,* PA biosynthesis occurs in all cells of the root cortex parenchyma (Anke et al. 2004), in *Phalaenopsis* (Orchidaceae), PA synthesis is restricted to the tips of the aerial roots (Frölich et al. 2006). In the Boraginaceae the site of synthesis may vary from species to species, e.g. only roots (*Symphytum officinale*), only shoots (*Heliotropium indicum*), or shoots and roots (*Cynoglossum officinale*, van Dam et al. 1995b; Frölich et al. 2007). This indicates that for a constitutive defense compound not the site of synthesis *per se* but the site of accumulation is of ultimate importance.

Like for other secondary compounds, chemical diversity and variability are typical features also of PAs. Each species is characterized by its unique PA profile, and great qualitative and quantitative variability is frequently observed between populations. Well studied examples are populations of *Senecio jacobaea* and *Senecio erucifolius* (Witte et al. 1992; Vrieling and de Boer 1999), and *Senecio vulgaris* and *Senecio vernalis* (von Borstel et al. 1989). Structural diversification has been studied using five *Senecio* species which differ in their PA profiles (Hartmann and Dierich 1998). All species synthesize senecionine *N*-oxide as backbone structure which is diversified by simple one or two-step transformations (Fig. 10.3). These reactions affect mainly the necic acid moiety; they include dehydrogenations, various position-specific hydroxylations and epoxidations, and *O*-acetylations. These 'peripheral reactions' proceed slowly and often differ in efficiency between plant organs. Therefore, the PA profile of an individual plant is the result of a dynamic equilibrium of a number of interfering processes such as the rate of supply with de-novo-synthesized senecionine *N*-oxide, specificity and efficiency of the transformations, the velocity of allocation and, finally, the tissue- and cell-specific storage. Any genetic alteration affecting the efficiency of one of the 'peripheral reactions' would drastically modify the PA profile without affecting its overall quantity. Total

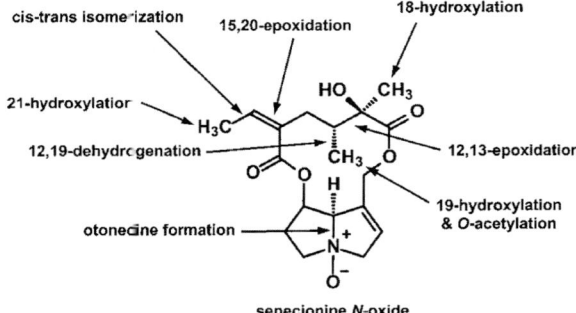

Fig. 10.3 Diversification of senecionine *N*-oxide by peripheral one-step or two-step reactions producing the various species-specific PA-profiles in *Senecio* species

plant PA quantity is determined exclusively by the rate of root-specific formation of senecionine *N*-oxide. This mechanistic scenario explaining the great variation of PA profiles between populations is supported by other studies. A genetic analysis of *S. jacobaea* populations (Vrieling et al. 1993) revealed that the large phenotypic variation in both total PA content and the qualitative PA profiles is largely due to genetic variation. A comparative analysis of the PA profiles of 24 species of the *Senecio* section *Jacobaea* based on a maximum parsimony cladogram inferred from DNA sequence and morphological data (Pelser et al. 2005) resulted in a largely incidental distribution of 26 biogenetically related PAs within the whole clade. Differences in the PA profiles cannot be explained by gain and loss of PA-specific genes during evolution, but rather by a transient switch-off and switch-on of gene expression. Apparently the species are able to randomly shuffle patterns of senecionine-derived derivatives from the 'genetic pool' of the clade, and thus generating PA profiles of changing diversity and variability.

10.2.4 Features of a Constitutive Defense System: Ecological Aspects

As most plant toxins, PAs are bitter tasting for humans and strong feeding deterrents for vertebrates and invertebrates (Boppré 1986). However, direct evidence for their defense function in plants is still sparse. In contrast to an inducible defense where cause (attack) and effect (formation of a defense compound) are obvious, a constitutive defense can only be judged by the absence ore presence of potential herbivores. Therefore, evidence supporting a function of PAs in plant defense is mostly indirect. One line of indirect evidence is tissue-specific accumulation of PAs. Flowers are often the organs showing the highest PA concentrations. This has been demonstrated for *Senecio* (Hartmann and Zimmer 1986), *Heliotropium* (Boraginaceae; Frölich et al. 2007), and *Phalaenopsis* (Orchidaceae; Frölich et al. 2006). In flowering specimens, the amounts of PAs associated with the inflorescences account for 50%–90% of total PAs, indicating an efficient protection of the most valuable reproductive plant organs. In rosette leaves of *C. officinale,* the youngest leaves contain up to 190 times higher PA levels than older leaves (van Dam et al. 1994). Generalist herbivorous insects only feed on the PA-poor old leaves whereas a specialist herbivore, adapted to PAs, prefers the protein-rich young leaves (van Dam et al. 1995a). In vegetative stems of herbal species like *Senecio,* PAs accumulate preferentially in the peripheral cell layers (Hartmann et al. 1989), and they are even found in trace amounts on the leaf-surface (Vrieling and Derridj 2003). Again, the functionally valuable young tissues as well as peripheral tissues which provide the first contact for an herbivore are efficiently protected. Another line of indirect evidence indicating a defense role of PAs is the chemical diversity and great variability of the PA profiles. The effect of pure PAs on five generalist insect herbivores revealed that structurally related PAs differ in their deterrent effects and can even act synergistically if applied in mixtures (Macel et al. 2005). This result was taken as argument to emphasize the importance of generalist herbivores on the evolution and maintenance of PA diversity. Further evidence is derived from the observation that the geographical distribution of certain *S. jacobaea* chemotypes is apparently not related to common ancestry but due to similar selection pressure, e.g. herbivores, in a certain region (Macel et al. 2004).

10.2.5 Molecular Evolution of the PA Pathways in Angiosperms

All plant PAs share the carbon skeleton of 2-hydroxymethylpyrrolizidine in their necine base moiety (Fig 10.1). Detailed tracer studies carried out in the late 1970s and 1980s identified the symmetrical polyamine homospermidine as biosynthetic precursor of the necine base moiety (Spenser 1985; Robins 1989). The formation of homospermidine is catalyzed by homospermidine synthase (EC 2.5.1.45, HSS) which was first isolated and characterized from *Senecio* root cultures (Böttcher et al. 1993, 1994). The enzyme was purified, cloned and functionally expressed in *Escherichia coli* (Ober and Hartmann 1999). It is encoded by a gene that originated by gene duplication and subsequent diversification from an ancestral gene for deoxyhypusine synthase (EC 2.5.1.46, DHS; Ober and Hartmann 2000). DHS catalyzes the first of two steps in the posttranslational activation of the eukaryotic initiation factor 5A (eIF5A; Caraglia et al. 2001). DHS is a highly conserved enzyme, ubiquitously present in all Eukarya and Archaea. In the activation process of eIF5A, it catalyzes the transfer of the aminobutyl moiety of spermidine to a specific lysine residue of the eIFA precursor protein (Fig. 10.4A). As a side-reaction, DHS is able to catalyze the formation of homospermidine from putrescine (Fig. 10.4A) and thus precisely the main-reaction of HSS. A comparison of the enzymatic and molecular properties of the two enzymes revealed that HSS retained all properties of DHS except the ability to bind the eIF5A precursor protein (Fig. 10.4B; Ober et al. 2003). The evolution of HSS is a conclusive example for evolution by change of function: the duplicate of a ubiquitous gene (DHS) functionally involved in the activation of a regulatory protein was recruited and integrated as HSS into a completely different functional environment, i.e. the biosynthesis of low molecular defense compounds (Fig. 10.4C; Ober and Hartmann 2000; Ober 2005).

The confirmed ancestry of HSS from ubiquitously occurring DHS offered the opportunity to address the intriguing question whether the PA pathways within the angiosperms are of monophyletic or polyphyletic origin. Analyzing the pairs of cDNA sequences encoding HSS and DHS from various angiosperm species revealed at least four independent duplication events of the gene encoding DHS resulting in each case in a new gene encoding HSS (Reimann et al. 2005; Fig. 10.5). These duplications occurred early in the evolution of the monocots and of the Boraginaceae, respectively, and even twice within the family of the Asteraceae (i.e., once within the Senecioneae and the Eupatorieae, respectively). Further duplication events are likely in those families with PA-producing species that were not included in this study. These data show that the scattered occurrence of PAs within the angiosperms is a result of convergent evolution. This observation is remarkable against the background that the PA structures produced are often identical. PAs of the lycopsamine type, for example, occur within the tribe Eupatorieae (Asteraceae) and within the

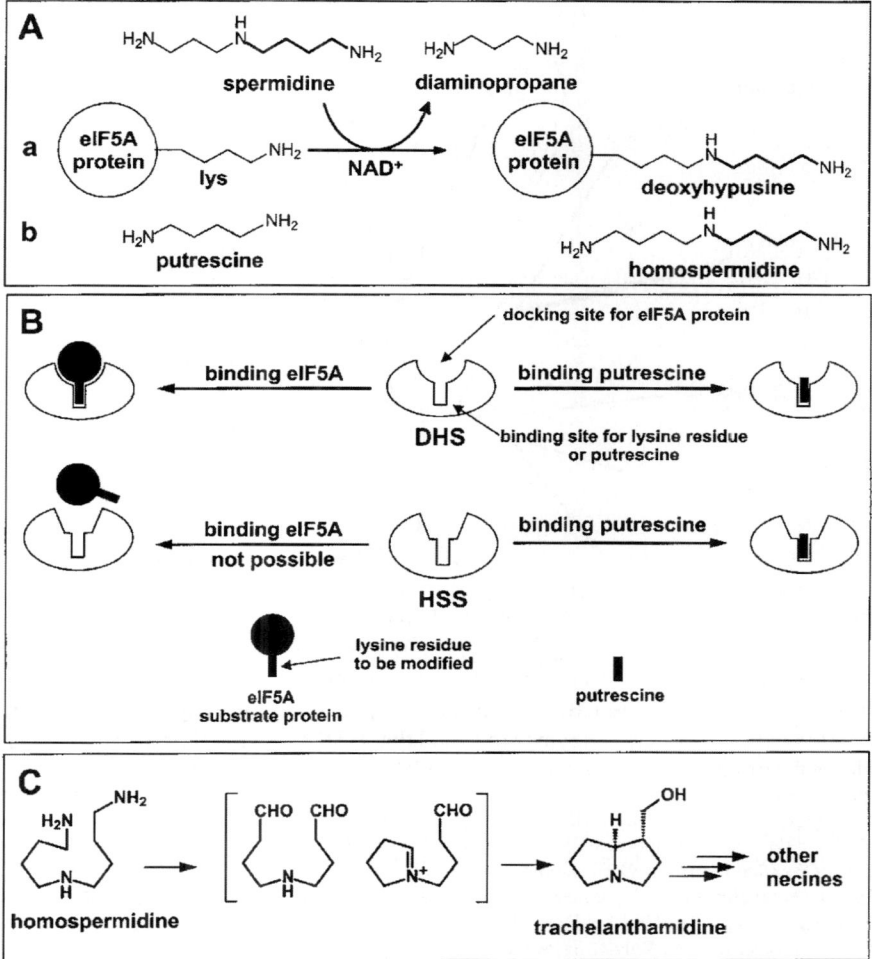

Fig. 10.4 Origin of homospermidine synthase (HSS) by duplication of the gene encoding deoxyhypusine synthase (DHS). **A**: DHS catalyzes the aminobutylation of a specific lysine residue of the eIF5A precursor protein (a) and, as side reaction, converts putrescine to homospermidine (b). **B**: HSS is a DHS that lost the ability to bind the eIF5A precursor protein. Its reactivity and kinetics with putrescine remain unaltered. **C**: HSS is only known from PA containing plants, it was recruited and integrated into PA biosynthesis and catalyzes the formation of homospermidine, the unique carbon skeleton of the necine bases

families of the Boraginaceae and the Apocynaceae. Despite the independent origin of HSS within these families, the PAs share a number of identical structural features (Fig. 10.1C), such as retronecine as necine base, and the unique C_7-necid acid, 2-isopropyl-2,3-dihydroxybutyrate. Further enzymes of PA biosynthesis will have to be analyzed to get an impression of how the whole pathway was recruited in separate angiosperm lineages.

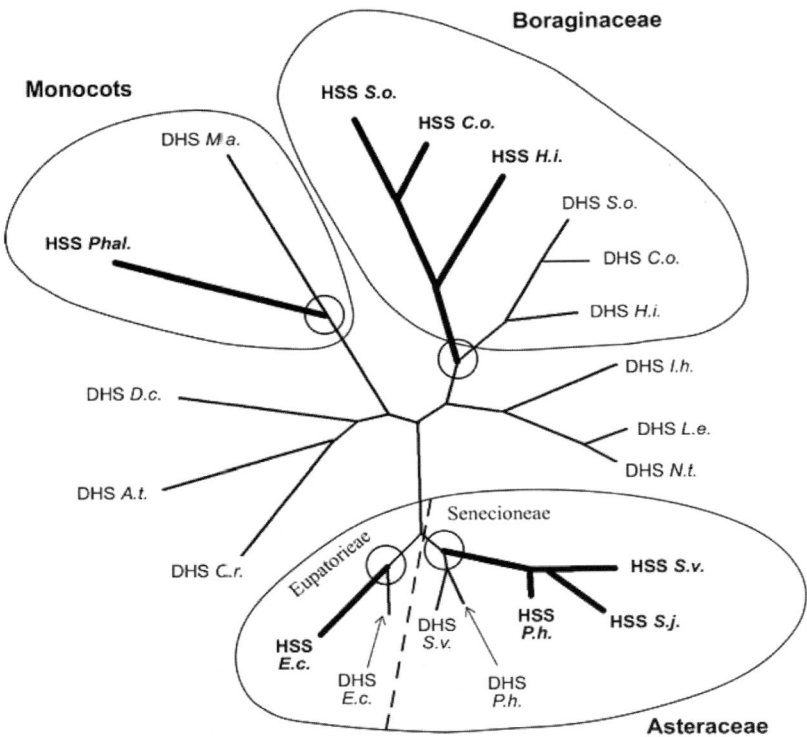

Fig. 10.5 Phylogenetic tree of HSS and DHS encoding cDNA sequences of various angiosperm species. Branches of HSS encoding sequences are shown in bold, originating in each case at the duplication event of the DHS-encoding gene (labeled by circles). Modified from Reimann et al. (2004). Species abbreviations: *A.t.*, *Arabidopsis thaliana*; *C.o.*, *Cynoglossum officinale*; *C.r.*, *Crotalaria retusa*; *D.c.*, *Dianthus caryophyllus*; *E.c.*, *Eupatorium cannabinum*; *H.i.*, *Heliotropium indicum*; *I.h.*, *Ipomoea hederifolia*; *L.e.*, *Lycopersicon esculentum*; *M.a.*, *Musa acuminata*; *N.t.*, *Nicotiana tabacum*; *P.h.*, *Petasites hybridus*; *Phal*, *Phalaenopsis* spec.; *S.j.*, *Senecio jacobaea*; *S.v.*, *Senecio vernalis*; *S.o.*, *Symphytum officinale*

10.3 Utilization of Plant Derived Pyrrolizidine Alkaloids by Insects

10.3.1 Occurence and Distribution of PA Sequestration Among Insects

Almost 40 years ago, Miriam Rothschild and her colleagues showed that the larvae of the European cinnabar moth, *Tyria jacobaeae* (Arctiidae) sequester PAs from their host plant *S. jacobaea* and maintain the alkaloids through all life stages (Aplin et al. 1968). In the following decades, many further examples of PA-sequestering herbivorous insects became known (Hartmann and Witte 1995; Hartmann 1999; Hartmann and Ober 2000). PA sequestration is common within the tiger moth family (Arctiidae; Conner and Weller 2004). About 70 species are known that either sequester PAs with their larval food, like *T. jacobaeae,* or obtain the alkaloids as adults from dead or withered parts of PA-containing plants. In the latter case, adult moths regurgitate on the plant surface and then reimbibe the fluid containing the dissolved alkaloids (Boppré 1986; Schneider 1987). This behavior in which the plant is only used as 'drug source' is called pharmacophagy (Boppré 1984). Pharmacophagy is also common among PA-sequestering butterflies including danaine and ithomiine nymphalid butterflies (Boppré 1990). Pioneering work on PA-sequestration by danaid and ithomiine butterflies has been performed in Australia (Edgar 1982) and Brasil (Brown 1984, 1987; Trigo et al. 1996).

Other examples for PA sequestration are found among beetle species. Chrysomelid leaf beetles (Coleoptera) are known for their diverse endogenous chemical defenses. Some beetles possess numerous exocrine glands, from which scores of diminutive droplets of defensive secretions are released when the beetle is attacked. Leaf beetles generally synthesize their own defense compounds, but in some cases, they recruit plant defenses. Species of the palaearctic genus *Oreina* feeding on PA-containing plants like *Adenostyles alliariae* (Asteraceae) sequester PAs from their host plants, and concentrate them up to 0.3 mol l^{-1} in their defense secretions (Pasteels et al. 1988; Hartmann et al. 1997). The same adaptation evolved independently in species of the neotropical genus *Platyphora* feeding on species containing PAs of the lycopsamine type (Pasteels et al. 2001). Further examples of Coleoptera that sequester plant PAs are flea beetles of the genus *Longitarsus* feeding on PA-containing species of the Asteraceae and Boraginaceae (Haberer and Dobler 1999; Dobler et al. 2000), and the Brasilian soldier beetle *Chauliognathus falax* (Cantharidae) feeding on *Senecio* inflorescences (Klitzke and Trigo 2000). Within the Orthoptera, a single PA-sequestering species was described, the polyphagous grasshopper *Zonocerus variegatus* (Bernays et al. 1977). Within the Hemiptera, PA-sequestering species are found in three unrelated taxa, including the largid bug *Largus rufipennis* sucking on *Senecio* stems (Klitzke and Trigo 2000), and two phloem feeders, i.e. the aphid *Aphis jacobaeae* feeding on various *Senecio* species (Witte et al. 1990), and the Mexican scale insect *Ceroplastes albolineatus* feeding on *Pittocaulon* (ex *Senecio*) *praecox* (Marín Loaiza et al. 2007).

All quoted examples for PA sequestration must have evolved independently indicating a significant benefit of PA acquisition for the insects. In most cases, PA sequestration was found to be accompanied by highly specific biochemical adaptations. Leaf beetles, for instance, possess different but specific adaptations to prevent the accumulation of detrimental concentrations of PAs as pro-toxic free bases in metabolically active tissues. They have most efficient membrane carriers which catalyze the transfer of PAs into their defensive glands, and concentrate them in their defense secretions (Hartmann et al. 1999, 2001). In the next chapter, PA-sequestering arctiid moths are selected to exemplify the amazingly sophisticated integration of plant-derived PAs into the moth's biology.

10.3.2 PA Sequestration by Arctiid Moths

10.3.2.1 Behavioral and Chemo-Ecological Aspects: The *Utetheisa ornatrix* Story

The role of plant-derived PAs in the performance of an insect has most completely been elucidated in *U. ornatrix* (Eisner et al. 2002). Larvae of this moth obtain their PAs from *Crotalaria* spp. and retain the alkaloids through metamorphosis. At mating, the male advertises his PA load to the female through the PA-derived male courtship pheromone, hydroxydanaidal. The pheromone is emitted from a pair of androconial scent-brushes (coremata) which the male everts during close-range precopulatory interactions with the female. Females are able to measure the pheromone concentration and differentiate between males that contain unequal pheromone quantities and accept eventually males having higher levels (Conner et al. 1990). During insemination the male transmits a portion of his PAs to the female. These alkaloids are transmitted together with the female's own load to the eggs during oviposition (Dussourd et al. 1988; Iyengar et al. 2001).

PAs protect adults against predation by spiders (Eisner and Eisner 1991; Gonzalez et al. 1999). The PA endowment was also shown to protect the eggs, as the most endangered life stage of the moth, against predation by coccinelid beetles (Dussourd et al. 1988), ants (Hare and Eisner 1993), and chrysopid larvae (Eisner et al. 2000), as well as parasitization by a parasitoid wasp, *Trichogramma ostriniae* (Bezzerides et al. 2004).

10.3.2.2 Physiological and Biochemical Aspects

Arctiids that sequester PAs from their host plants must have acquired at least two novel capabilities: they need to recognize PAs and they must be able to prevent self-poisoning. It is known that PAs are feeding stimulants for PA-adapted arctiids (Boppré 1986; Schneider 1987), but only recently have arctiid caterpillars been shown to possess single sensory neurons in their mouthparts that respond specifically and sensitively (threshold of response $<10^{-9}$ M) to all major structural types of PAs (Fig 10.1), but not to other alkaloids (Bernays et al. 2002a, b). Arctiids absorb PAs as pro-toxic free bases since any N-oxide ingested with the food is reduced in the gut. The absorbed free bases are immediately converted into the respective non-toxic PA N-oxides. This N-oxidation is catalyzed by a flavine-dependent monooxygenase (senecionine N-oxygenase; EC 1.14.13.101) localized as soluble enzyme in the hemolymph. The enzyme specifically converts any pro-toxic free base into its non-toxic N-oxide (Lindigkeit et al. 1997). Its acquisition in ancestral arctiids appears to be the prerequisite for PA sequestration (Naumann et al. 2002). Sensory recognition and specific detoxification allow adapted monophagous arctiids like *U. ornatrix* and *T. jacobaeae* to perceive and maintain PAs from their host plant. However, polyphagous species like *Estigmene acrea* or *Grammia geneura* for which more than 65 larval host plant species have been recorded need additional adaptations. Both species exploit PA-containing plants such as *Senecio* or *Crotalaria* in their natural environment, e.g. Arizona grass-lands, as 'drug sources' (Hartmann et al. 2005a, b). Since they cannot survive on PA containing plants as sole food source, extensive feeding on a PA plant causes a transient loss in the response of the PA-sensitive neurons allowing the caterpillar to leave its alkaloid source and search for better quality food (Bernays et al. 2003). This mechanism apparently allows the caterpillar to find a suitable balance between the intake of nutrients and 'defense drugs' (Singer et al. 2004a, b). Moreover, the phagostimulatory response to all kinds of PAs requires the ability of polyphagous caterpillar to cope with any naturally occurring plant PA source. There are at least three ways how they handle PAs (Fig. 10.6): (1) Macrocyclic PAs and PA of the lycopsamine type (Fig. 10.1) are sequestered and maintained unaltered through all life-stages (Fig. 10.6 IIA); (2) PAs that cannot be detoxified by N-oxidation like the otonecine derivatives (e.g., senkirkine; Fig. 10.1) are excluded from sequestration (Fig. 10.6 IIC), and (3) PAs that are sequestered by larvae but need to be trans-esterified before transmission to the subsequent life stages (Fig. 10.6 IIB). The latter case includes mainly PA monoesters and open-chain diesters that do not belong to the lycopsamine type. These PAs are hydrolyzed in the larvae and the resulting necine bases are re-esterified with necic acids of insect origin. Arctiid-specific retronecine esters were frequently observed, i.e. callimorphine (Edgar et al. 1980; Ehmke et al. 1990) and creatonotine (Hartmann et al. 1990; Schulz et al. 1993). These compounds represent two classes of arctiid-specific PAs, the callimorphines and the creatonotines (Fig 10.6 I; Hartmann et al. 2005a, b; Beuerle et al. 2007). Arctiids are able to re-esterify all kinds of necine bases derived from plant PAs and are thus able to salvage necine bases and maintain them as insect-made esters through all life stages. These amazing metabolic adaptations allow polyphagous arctiids to utilize almost the entire diversity of naturally occurring plant PAs efficiently. This also accounts for the formation of the male courtship pheromone (hydroxydanaidal) that signals the female the male's PA load. Its biosynthesis proceeds via a retronecine O^7-monoester (Schulz et al. 1993). Probably a certain proportion of all naturally occurring retronecine esters are partially hydrolyzed and the resulting retronecine is converted into the pheromone via the creatonotines as common intermediates (Fig. 10.6 IID; Hartmann et al. 2003, 2005b).

10.4 Perspectives

The evidence discussed here favors the hypothesis that PAs represent efficient defense compounds that may protect organisms on two trophic levels: the plant level where they evolved probably under the pressure of herbivory and the insects level where certain species evolved adaptations that allowed them to adopt these plant defense compounds and utilize them for their own protection against predators and parasitoids.

In both cases selected mechanisms are required, either, as in plants, to synthesize, maintain and apply these compounds as a chemical defense barrier, or, as in adapted

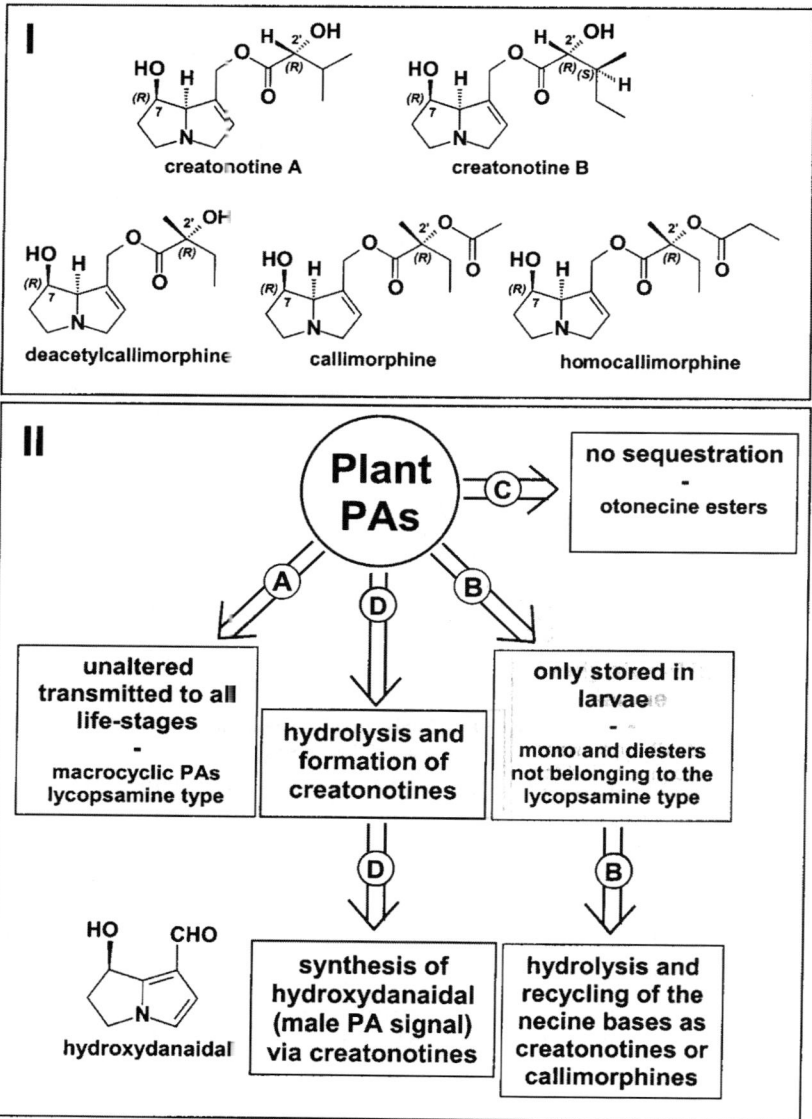

Fig. 10.6 (**I**) The creatonotines and callimorphines are two classes of arctiid-specific pyrrolizidine alkaloids made of a necine base of plant origin and necic acids provided by the insect. (**II**) Polyphagous arctiid larvae are able to cope with all structural types of plant PAs. Depending on the structure, PAs are either utilized unaltered (route A), or they are hydrolyzed and the necine bases are salvaged as creatonotines or callimorphines (route B), or they are excluded from sequestration like the otonecine esters (route C). Species which produce the male courtship pheromone, hydroxydanaidal, synthesize this 'PA signal' via creatonotines as common intermediates (route D)

insects, to recognize, ingest, safely maintain, and employ the adopted compounds in their species-specific behavior and defense strategies. To unravel the evolutionary origin of the acquired mechanisms involved in these processes on both trophic levels is one of the most challenging future goals. The independent evolutionary origins of the biosynthetic pathways in plants, and the multiple independent adaptations of various insects to PA sequestration and utilization pose further exciting questions.

Acknowledgments The authors thank the Deutsche Forschungsgemeinschaft (DFG) for continues financial support.

References

Anke S, Niemuller D, Moll S, Hansch R, Ober D (2004) Polyphyletic origin of pyrrolizidine alkaloids within the Asteraceae. Evidence from differential tissue expression of homospermidine synthase. Plant Physiol 136:4037–4047
Aplin RT, Benn MH, Rothschild M (1968) Poisonos alkaloids in the body tissues of the cinnabar moth (*Callimorpha jacobaeae* L.). Nature 219:747–748
Bernays EA, Chapman RF, Hartmann T (2002a) A highly sensitive taste receptor cell for pyrrolizidine alkaloids in the lateral galeal sensillum of a polyphagous caterpillar, *Estigmene acrea*. J Comp Physiol A 188:715–723
Bernays EA, Chapman RF, Hartmann T (2002b) A taste receptor neurone dedicated to the perception of pyrrolizidine alkaloids in the medial galeal sensillum of two polyphagous arctiid caterpillars. Physiol Entomol 27:1–10
Bernays EA, Edgar JA, Rothschild M (1977) Pyrrolizidine alkaloids sequestered and stored by the aposematic grasshopper, *Zonocerus variegatus*. J Zool, 182:85–87
Bernays EA, Rodrigues D, Chapman RF, Singer MS, Hartmann T (2003) Loss of gustatory responses to pyrrolizidine alkaloids after their extensive ingestion in the polyphagous caterpillar *Estigmene acrea*. J Exp Biol 206:487–4496
Beuerle T, Theuring C, Klewer N, Schulz S, Hartmann T (2007) Absolute configuration of the creatonotines and callimorphines, two classes of arctiid-specific pyrrolizidine alkaloids. Insect Biochem Mol Biol 37:80–89
Bezzerides A, Yong TH, Bezzerides J, Husseini J, Ladau J, Eisner M, Eisner T (2004) Plant-derived pyrrolizidine alkaloid protects eggs of a moth (*Utetheisa ornatrix*) against a parasitoid wasp (*Trichogramma ostriniae*). Proc Natl Acad Sci USA 101:9029–9032
Boppré M (1984) Redefining 'pharamcophagy'. J Chem Ecol 10:1151–1154
Boppré M (1986) Insects pharmacophagously utilizing defensive plant chemicals (pyrrolizidine alkaloids). Naturwissenschaften 73:17–26
Boppré M (1990) Lepidoptera and pyrrolizidine alkaloids: exemplification of complexity in chemical ecology. J Chem Ecol 16:165–185
Böttcher F, Adolph RD, Hartmann T (1993) Homospermidine synthase, the first pathway-specific enzyme in pyrrolizidine alkaloid biosynthesis. Phytochemistry 32:679–689
Böttcher F, Ober D, Hartmann T (1994) Biosynthesis of pyrrolizidine alkaloids: putrescine and spermidine are essential substrates of enzymatic homospermidine formation. Can J Chem 72:80–85
Brattsten LB (1992) Metabolic defenses against plant allelochemicals. In: Rosenthal GA, Berenbaum MR (eds) Herbivores: their interactions with secondary metabolites, vol 2. Academic Press, San Diego, pp 175–242
Brown KSJ (1984) Adult-obtained pyrrolizidine alkaloids defend ithomiine butterflies against a spider predator. Nature 309:707–709

Brown KSJ (1987) Chemistry at the solanaceae/ithomiinae interface. Ann Missouri Bot Gard 74: 359–397

Caraglia M, Marra M, Giuberti G, D'Alessandro AM, Budillon A, del Prete S, Lentini A, Beninati S, Abbruzzese A (2001) The role of eukaryotic initiation factor 5A in the control of cell proliferation and apoptosis. Amino Acids 20:91–104

Cheeke PR (ed) (1998) Natural toxicants in feeds, forages, and poisonous plants. Interstate, Danville

Conner WE, Roach B, Benedict E, Meinwald J, Eisner T (1990) Courtship pheromone production and body size as correlates of larval diet in males of the arctiid moth *Utetheisa ornatrix*. J Chem Ecol 16:543–552

Conner WE, Weller, SJ (2004) A quest for alkaloids: the curious relationship between tiger moths and plants containing pyrrolizidine alkaloids. In: Cardé RT, Millar JG (eds) Advances in insect chemical ecology. University Press, Cambridge, pp 248–282

Dobler S, Haberer W, Witte L, Hartmann T (2000) Selective sequestration of pyrrolizidine alkaloids from diverse host plants by *Longitarsus* flea beetles. J Chem Ecol 26:1281–1298

Dussourd DE, Ubik K, Harvis C, Resch J, Meinwald J, Eisner T (1988) Biparental defensive endowment of eggs with acquired plant alkaloid in the moth *Utetheisa ornatrix*. Proc Natl Acad Sci USA 85:5992–5996

Edgar JA (1982) Pyrrolizidine alkaloids sequestered by Salomon Island Danainae butterflies. The feeding preferences of the Danainae and Ithomiinae. J Zool 196:385–399

Edgar JA, Culvenor CCJ, Cockrum PA, Smith LW (1980) Callimorphine: identification and symthesis of the cinnabar moth 'metabolite'. Tetrahedron Lett 21:1383–1384

Ehmke A, Von Borstel K, Hartmann T (1988) Alkaloid N-oxides as transport and vacuolar storage compounds of pyrrolizidine alkaloids in *Senecio vulgaris* L. Planta 176:83–90

Ehmke A, Witte L, Biller A, Hartmann T (1990) Sequestration, N-oxidation and transformation of plant pyrrolizidine alkaloids by the arctiid moth *Tyria jacobaeae* L. Z. Naturforsch C 45:1185–1192

Eisner T, Eisner M (1991) Unpalatability of the pyrrolizidine alkaloid-containing moth *Utetheisa ornatrix* and its larva to wolf spiders. Psyche Cambridge 98:111–118

Eisner T, Eisner M, Rossini C, Iyengar VK, Roach BL, Benedikt E, Meinwald J (2000) Chemical defense against predation in an insect egg. Proc Natl Acad Sci USA 97:1634–1639

Eisner T, Rossini C, Gonzalez A, Iyengar VK, Siegler MVS, Smedley SR (2002) Paternal investment in egg defence. In: Hilker M, Meiners T (eds) Chemoecology of insect eggs and egg deposition. Blackwell, Oxford, pp 91–116

Frei H, Lüthy J, Bräuchli J, Zweifel U, Wurgler FE, Schlatter C (1992) Structure/activity relationships of the genotoxic potencies of sixteen pyrrolizidine alkaloids assayed for the induction of somatic mutation and recombination in wing cells of *Drosophila melanogaster*. Chem Biol Interact 83:1–22

Frölich C, Hartmann T, Ober D (2006) Tissue distribution and biosynthesis of 1,2-saturated pyrrolizidine alkaloids in *Phalaenopsis* hybrids (Orchidaceae). Phytochemistry 67:1493–1502

Frölich C, Ober D, Hartmann T (2007) Tissue distribution, core biosynthesis and diversification of pyrrolizidine alkaloids of the lycopsamine type in three Boraginaceae species. Phytochemistry 68:1026–1037

Fu PP, Xia Q, Lin G, Chou MW (2004) Pyrrolizidine alkaloids – genotoxicity, metabolism enzymes, metabolic activation, and mechanisms. Drug Metab Rev 36:1–55

Glendinning JI (2002) How do herbivorous insects cope with noxious secondary plant compounds in their diet? Entomol Exp Appl 104:15–25

Glendinning JI, Slansky F (1995) Consumption of a toxic food by caterpillars increases with dietary exposure: support for a role of induced detoxification enzymes. J Comp Physiol A 176:337–345

Gonzalez A, Rossini C, Eisner M, Eisner T (1999) Sexually transmitted chemical defense in a moth (*Utetheisa ornatrix*). Proc Natl Acad Sci USA 96:5570–5574

Haberer W, Dobler S (1999) Quantitative analysis of pyrrolizidine alkaloids sequestered from diverse host plants in *Longitarsus* flea beetles (Coleoptera, Chrysomelidae). Chemoecology 9:169–179

Hare JF, Eisner T (1993) Pyrrolizidine alkaloid deters ant predators of *Utetheisa ornatrix* eggs: effects of alkaloid concentration, oxidation state, and prior exposure of ants to alkaloid-laden prey. Oecologia 96:9–18

Hartmann T (1999) Chemical ecology of pyrrolizidine alkaloids. Planta 207:483–495

Hartmann T, Biller A, Witte L, Ernst L, Boppre M (1990) Transformation of plant pyrrolizidine alkaloids into novel insect alkaloids by arctiid moths (Lepidoptera). Biochem Syst Ecol 18:549–554

Hartmann T, Dierich B (1998) Chemical diversity and variation of pyrrolizidine alkaloids of the senecionine type: biological need or coincidence? Planta 206:443–451

Hartmann T, Ehmke A, Eilert U, von Borstel, K, Theuring C (1989) Sites of synthesis, translocation and accumulation of pyrrolizidine alkaloid *N*-oxides in *Senecio vulgaris* L. Planta 177:98–107

Hartmann T, Ober D (2000) Biosynthesis and metabolism of pyrrolizidine alkaloids on plants and specialized insect herbivores. Top Curr Chem 209:207–243

Hartmann T, Theuring C, Bernays EA (2003) Are insect-synthesized retronecine esters (creatonotines) the precursors of the male courtship pheromone in the arctiid moth *Estigmene acrea*? J Chem Ecol 29:2603–2608

Hartmann T, Theuring C, Beuerle T, Bernays EA, Singer MS (2005a) Acquisition, transformation and maintenance of plant pyrrolizidine alkaloids by the polyphagous arctiid *Grammia geneura*. Insect Biochem Mol Biol 35:1083–99

Hartmann T, Theuring C, Beuerle T, Klewer N, Schulz S, Singer, MS, Bernays EA (2005b) Specific recognition, detoxification and metabolism of pyrrolizidine alkaloids by the polyphagous arctiid *Estigmene acrea*. Insect Biochem Mol Biol 35:391–411

Hartmann T, Theuring C, Schmidt J, Rahier M, Pasteels JM (1999) Biochemical strategy of sequestration of pyrrolizidine alkaloids by adults and larvae of chrysomelid leaf beetles. J Insect Physiol 45:1085–1095

Hartmann T, Theuring C, Witte L, Pasteels JM (2001) Sequestration, metabolism and partial synthesis of tertiary pyrrolizidine alkaloids by the neotropical leaf-beetle *Platyphora boucardi*. Insect Biochem Mol Biol 31:1041–1056

Hartmann T, Toppel G (1987) Senecionine *N*-oxide, the primary product of pyrrolizidine alkaloid biosynthesis in root cultures of *Senecio vulgaris*. Phytochemistry 26:1639–1644

Hartmann T, Witte L (1995) Pyrrolizidine alkaloids: chemical, biological and chemoecological aspects. In: Pelletier SW (ed) Alkaloids: chemical and biological perspectives, vol 9. Pergamon Press, Oxford, pp 155–233

Hartmann T, Witte L, Ehmke A, Theuring C, Rowell-Rahier M, Pasteels JM (1997) Selective sequestration and metabolism of plant derived pyrrolizidine alkaloids by chrysomelid leaf beetles. Phytochemistry 45:489–497

Hartmann T, Zimmer M (1986) Organ-specific distribution and accumulation of pyrrolizidine alkaloids during the life history of two annual *Senecio* species. J Plant Physiol 122:67–80

Huan JY, Miranda CL, Buhler DR, Cheeke PR (1998a) The roles of CYP3A and CYP2B isoforms in hepatic bioactivation and detoxification of the pyrrolizidine alkaloid senecionine in sheep and hamsters. Toxicol Appl Pharmacol 151:229–235

Huan JY, Miranda CL, Buhler DR, Cheeke PR (1998b) Species differences in the hepatic microsomal enzyme metabolism of the pyrrolizidine alkaloids. Toxicol Lett 99:127–137

Iyengar VK, Rossini C, Eisner T (2001) Precopulatory assessment of male quality in an arctiid moth (*Utetheisa ornatrix*): hydroxydanaidal is the only criterion of choice. Behav Ecol Sociobiol 49:283–288

Jenett-Siems K, Ott SC, Schimming T, Siems K, Muller F, Hilker M, Witte L, Hartmann T, Austin DF, Eich E (2005) Ipangulines and minalobines, chemotaxonomic markers of the infrageneric *Ipomoea* taxon subgenus Quamoclit, section Mina. Phytochemistry 66:223–231

Jenett-Siems K, Schimming T, Kaloga M, Eich E, Siems K, Gupta MP, Witte L, Hartmann T (1998) Pyrrolizidine alkaloids of *Ipomoea hederifolia* and related species. Phytochemistry 47:1551–1560

Klitzke CF, Trigo JR (2000) New records of pyrrolizidine alkaloid-feeding insects. Hemiptera and Coleoptera on *Senecio brasiliensis*. Biochem Syst Ecol 28:313–318

Lindigkeit R, Biller A, Buch M, Schiebel HM, Boppré M, Hartmann T (1997) The two faces of pyrrolizidine alkaloids: the role of the tertiary amine and its N-oxide in chemical defense of insects with acquired plant alkaloids. Eur J Biochem 245:626–636

Macel M, Bruinsma M, Dijkstra SM, Ooijendijk T, Niemeyer HM, Klinkhamer PGL (2005) Differences in effects of pyrrolizidine alkaloids on five generalist insect herbivore species. J Chem Ecol 31:1493–1508

Macel M, Vrieling K, Klinkhamer PG (2004) Variation in pyrrolizidine alkaloid patterns of *Senecio jacobaea*. Phytochemistry 65:865–873

Marín Loaiza JC, Céspedes CL, Beuerle T, Theuring C, Hartmann T (2007) *Ceroplastes albolineatus*, the first scale insect shown to sequester pyrrolizidine alkaloids from its host-plant *Pittocaulon praecox*. Chemoecology 17:109–115

Mattocks AR (ed) (1986) Chemistry and toxicology of pyrrolizidine alkaloids. Academic Press, London

Moll S, Anke S, Kahmann U, Hänsch R, Hartmann T, Ober D (2002) Cell specific expression of homospermidine synthase, the entry enzyme of the pyrrolizidine alkaloids in *Senecio vernalis* in comparison to its ancestor deoxyhypusine synthase. Plant Physiol 130:47–57

Narberhaus I, Zintgraf V, Dobler S (2005) Pyrrolizidine alkaloids on three trophic levels – evidence for toxic and deterrent effects on phytophages and predators. Chemoecology 15:121–125

Naumann C, Hartmann T, Ober D (2002) Evolutionary recruitment of a flavin-dependent monooxygenase for the detoxification of host plant-acquired pyrrolizidine alkaloids in the alkaloid-defended arctiid moth *Tyria jacobaeae*. Proc Natl Acad Sci USA 99:6085–6090

Ober D (2005) Seeing double – gene duplication and diversification in plant secondary metabolism. Trends Plant Sci 10:444–449

Ober D, Harms R, Witte L, Hartmann T (2003) Molecular evolution by change of function: alkaloid-specific homospermidine synthase retained all properties of deoxyhypusine synthase except binding the eIF5A precursor protein. J Biol Chem 278:12805–12815

Ober D, Hartmann T (1999) Homospermidine synthase, the first pathway-specific enzyme of pyrrolizidine alkaloid biosynthesis, evolved from deoxyhypusine synthase. Proc Natl Acad Sci USA 96:14777–14782

Ober D, Hartmann T (2000) Phylogenetic origin of a secondary pathway: the case of pyrrolizidine alkaloids. Plant Mol Biol 44:445–450

Pasteels JM, Rowell-Rahier M, Randoux T, Braekman JC, Daloze D (1988) Pyrrolizidine alkaloids of probable host-plant origin in the pronotal and elytral secretion of the leaf beetle *Oreina cacaliae*. Entomol Exp Appl 49:55–88

Pasteels JM, Termonia A, Windsor D, Witte L, Theuring C, Hartmann T (2001) Pyrrolizidine alkaloids and pentacyclic triterpene saponins in the defensive secretions of *Platyphora* leaf beetles. Chemoecology 11 113–120

Pelser PB, de Vos H, Theuring C, Beuerle T, Vrieling K, Hartmann T (2005) Frequent gain and loss of pyrrolizidine alkaloids in the evolution of *Senecio* section *Jacobaea* (Asteraceae). Phytochemistry 66:1285–1295

Prakash AS, Pereira TN, Reilly PEB, Seawright AA (1999) Pyrrolizidine alkaloids in human diet. Mutat Res 443:53–67

Reimann A, Nurhayati N, Backenköhler A, Ober D (2004) Repeated evolution of the pyrrolizidine alkaloid-mediated defense system in separate angiosperm lineages. Plant Cell 16:2772–2784

Robins DJ (1989) Biosynthesis of pyrrolizidine alkaloids. Chem Soc Rev 18:375–408

Schneider D (1987) The strange fate of pyrrolizidine alkaloids. In: Chapman RF, Bernays EA, Stoffolano JG (eds) Perspectives in chemoreception and behavior. Springer, New York, pp 123–142

Schulz S, Francke W, Boppré M, Eisner T, Meinwald J (1993) Insect pheromone biosynthesis: stereochemical pathway of hydroxydanaidal production from alkaloidal precursors in *Creatonotos transiens* (Lepidoptera, Arctiidae). Proc Natl Acad Sci USA 90:6834–6838

Scott JG, Liu N, Wen Z (1998) Insect cytochromes P450: diversity, insecticide resistance and tolerance to plant toxins. Comp Biochem Physiol C Pharmacol Toxicol Endocrinol 121: 147–155

Singer MS, Carrière Y, Theuring C, Hartmann T (2004a) Disentangeling food quality from resistance against parasitoids: diet choice by a generalist caterpillar. Am Nat 164:424–429

Singer MS, Rodrigues D, Stireman JOI, Carrière Y (2004b) Roles of food quality and enemy-free space in host use by a generalist insect herbivore. Ecology 85:2747–2753

Spenser ID (1985) Stereochemical aspects of the biosynthetic routes leading to the pyrrolizidine and quinolizidine alkaloids. Pure Appl Chem 57:453–470

Stegelmeier BL, Edgar JA, Colegate SM, Gardner DR, Schoch TK, Coulombe RA, Molyneux RJ (1999) Pyrrolizidine alkaloid plants, metabolism and toxicity. J Nat Tox 8:95–116

Trigo JR, Brown KS, Henriques SA, Barata LES (1996) Qualitative patterns of pyrrolizidine alkaloids in ithomiinae butterflies. Biochem Syst Ecol 24:181–188

van Dam NM, Verporte R, van der Mejden E (1994) Extreme differences in pyrrolizidine alkaloid levels between leaves of *Cynoglossum officinale*. Phytochemistry 37:1013–1016

van Dam NM, Vuister LWM, Bergshoff C, de Vos H, van der Meijden E (1995a) The 'raison d'être' of pyrrolizidine alkaloids in *Cynoglossum officinale*: deterrent effects against generalist herbivores. J Chem Ecol 21:507–523

van Dam NM, Witte L, Theuring C, Hartmann T (1995b) Distribution, biosynthesis and turnover of pyrrolizidine alkaloids in *Cynoglossum officinale*. Phytochemistry 39:287–292

von Borstel K, Witte L, Hartmann T (1989) Pyrrolizidine alkaloid patterns in populations of *Senecio vulgaris, Senecio vernalis* and their hybrids. Phytochemistry 28:1635–1638

Vrieling K, de Boer NJ (1999) Host-plant choice and larval growth in the cinnabar moth: do pyrrolizidine alkaloids play a role. Entomol Exp Appl 91:251–257

Vrieling K, de Vos H, van Wijk CAM (1993) Genetic analysis of the concentrations of pyrrolizidine alkaloids in *Senecio jacobaea*. Phytochemistry 32:1141–1144

Vrieling K, Derridj S (2003) Pyrrolizidine alkaloids in and on the leaf surface of *Senecio jacobaea* L. Phytochemistry 64:1223–1228

Williams DE, Reed RL, Kedzierski B, Dannan GA, Guengerich FP, Buhler DR (1989a) Bioactivation and detoxication of the pyrrolizidine alkaloid senecionine by cytochrome P-450 enzymes in rat liver. Drug Metab Dispos 17:387–392

Williams DE, Reed RL, Kedzierski B, Ziegler DM, Buhler DR (1989b) The role of flavin-containing monooxygenase in the N-oxidation of the pyrrolizidine alkaloid senecionine. Drug Metab Dispos 17:380–386

Wink M, Mohamed GIA (2003) Evolution of chemical defense traits in the Leguminosae: mapping of distribution patterns of secondary metabolites on a molecular phylogeny inferred from nucleotide sequences of the *rbc*L gene. Biochem Syst Ecol 31:897–917

Witte L, Ehmke A, Hartmann T (1990) Interspecific flow of pyrrolizidine alkaloids; from plants via aphids to ladybirds. Naturwissenschaften 77:540–543

Witte L, Ernst L, Adam H, Hartmann T (1992) Chemotypes of two pyrrolizidine alkaloid-containing *Senecio* spp. Phytochemistry 31:559–566

Part C
Anti-nutritional Enzymes and Proteins

Chapter 11
Plant Protease Inhibitors: Functional Evolution for Defense

Maarten A. Jongsma and Jules Beekwilder

In this chapter we review the ways in which plants and plant predators have evolved creative solutions in the battle for scarce essential amino acids. We show that on both the plant and insect side many different gene families of inhibitors and proteases are involved with intricate specialized expression patterns among the different members. They provide an evolutionary and ecological genetic fingerprint of their functional role. Some of the hypotheses we put forward are currently only supported by circumstantial evidence, but provide attractive alleys which functional evolution may have gone. At first glance, this topic of plant-insect/pathogen interactions appears to be only a clash of two protein molecules. However, due to the opportunistic nature of evolution, this battle may have taken on many new appearances. The availability of fully sequenced genomes on both the plant and insect side, and the excellent tools to specifically knock down and overexpress specific genes provide exactly the instruments needed to allow a whole new set of discoveries in this area.

11.1 Introduction

The research in the field of plant protease inhibitors (PIs) has a 60 year long history dating back to the years just after the second world war when Kunitz (1946), Bowman (1946) and Birk et al. (1963) first purified and characterized the protease inhibitors in soybeans. Then, in the following decades, new gene families were added to the initial spectrum, to the extent that in plants we are now aware of 13 different protease inhibitor gene families known to target all main families of proteases (Rawlings et al. 2006), and often aimed at inhibiting the proteases of herbivores or plant pathogens.

The antinutritional effects of proteinase inhibitors in animal and human nutrition were already known early on, but their role in plant defense became much

---

M.A. Jongsma
Plant Research International B.V., Wageningen University and Research Center, 6700 AA Wageningen, The Netherlands
e-mail: maarten.jongsma@wur.nl

more evident when Green and Ryan discovered that also leaves of plants induced high levels of protease inhibitors in a specific response to mechanical and insect damage (Green and Ryan 1972). The advent of genetic modification allowed the overexpression of protease inhibitors which subsequently provided the first *in planta* demonstrations of the role of protease inhibitors in protecting plants against insects (Hilder et al. 1987; Abdeen et al. 2005), but also against microbes (Qu et al. 2003), and viruses (Gutierrez-Campos et al. 1999). Natural mutants of trypsin inhibitors in wild tobacco plants recently provided the first true evidence that the inhibitors exert a certain level of control even against their natural 'adapted' pests (Zavala et al. 2004).

Despite the dramatically raised plant PI levels in defense against herbivorous insects, the effects on herbivore mortality or development are often relatively minor or even absent. Yet, even small effects on development may still be ecologically relevant, and affect the next generation with severely reduced fecundity (De Leo and Gallerani 2002). However, the observation that some insects circumvent the effects of the inhibitors by responding with the induction of PI-insensitive proteases created the awareness that here an intriguing battle of inhibitor and protease molecules was being staged (Jongsma et al. 1995; Jongsma and Bolter 1997). It was subsequently shown that proteases had acquired mutations rendering them sterically insensitive to the inhibitors (Bown et al. 1997; Volpicella et al. 2003; Bayes et al. 2005). Other insects were very effective in degrading the inhibitors with specific proteases (Michaud 1997; Giri et al. 1998). But despite these insights, we think that we have actually advanced only a little bit on the track of unraveling the mutual plays of trick and deceit that have evolved around this form of plant defense.

This chapter essentially deals with a description of how these interactions have been understood so far and what may be in for the future. Despite being an 'old' field of research, tantalizing results lie ahead if the subject matter is placed in its ecological, physiological, cytological, and finally molecular context with plant predators and pathogens.

11.2 An Array of Gene Families with a Diversity of Roles

11.2.1 The Arabidopsis and Rice Genomes

So far, the rich resource of fully sequenced plant genomes has not been exploited to its full extend with respect to the specific area of protease inhibitors. For the purpose of this review we therefore made an inventory of the complete genetic repertoire of protease inhibitors of two plant species: the fully annotated genomes of *Arabidopsis thaliana* (thale cress, a crucifer), and *Oryza sativa* (rice). With *Arabidopsis* being a dicotyledonous species, and rice being a monocot, these plant species are only distantly related in the plant kingdom. Conveniently, the MEROPS database (Rawlings et al. 2006) has interrogated these genomes for the presence of sequences homologous to known protease inhibitor sequences.

Across all organisms, the inhibitors in MEROPS have been classified into 68 families, referred to as I01, I02, etc. Thirteen of these families are present in plants. In a specific plant species like *Arabidopsis* the genome contains 81 putative inhibitor genes, representing seven of those families (Fig. 11.1A), while the genome of *Oryza* contains 121 putative inhibitor genes, representing eight inhibitor families (Fig. 11.1B). There is considerable overlap in the inhibitor families represented in both plants. Six families I03 (Kunitz-type inhibitors), I04 (serpins), I13 (eglin C-homologues), I25 (cystatins), I29 (cathepsin prodomains) and I51 (CPY-inhibitor homologues) occur in both plants. One could consider those families as general plant inhibitors, likely occurring in most plant species.

In addition to general protease inhibitor families, many plant families have inhibitors that are uniquely found only in certain groups of plants. In *Oryza* this includes members of the families I06 (trypsin/alpha-amylase inhibitors), which are typical for monocots, and I13 (Bowman–Birk inhibitors), which apart from monocots also occur in Leguminosae, potato and sunflower, but not in *Arabidopsis*. *Arabidopsis* on the other hand expresses members of the I18 family (mustard trypsin inhibitor homologues), which is unique for Cruciferae. There are more of those species-specific protease inhibitors. For example, only solanaceous species are known to have inhibitors of family I20 (potato inhibitor 2) and I37 (potato metallocarboxypeptidase inhibitor), and only Cucurbitaceae are known to have family I07 (squash inhibitors), while family I67 (bromein) is uniquely found in pineapple.

The overview in Fig. 11.1 is based on sequence data only and is likely to overestimate the actual number of encoded inhibitors. Not every sequence that was annotated as such has actually been demonstrated to encode an active protease inhibitor. For instance, the I51 CPY inhibitor homologues are related to the yeast

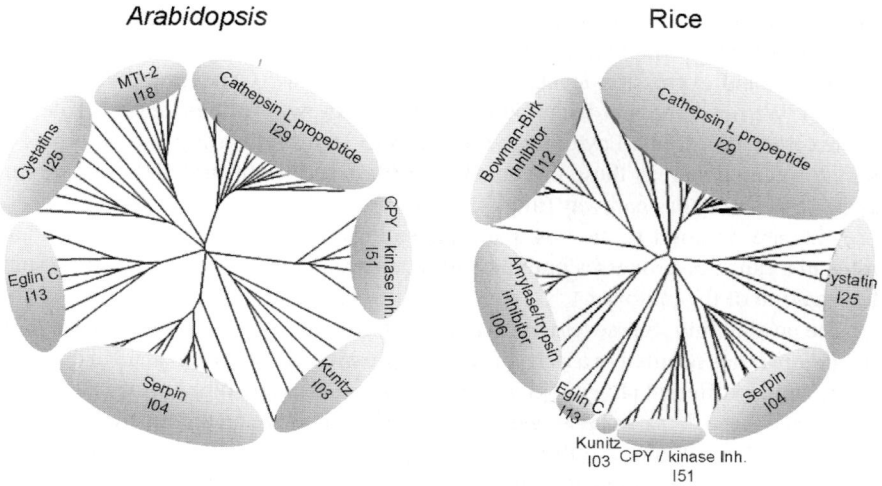

Fig. 11.1 Phylogenetic tree based on the mature protein sequences of all the protease inhibitors of *Arabidopsis* and rice in the MEROPS database (Rawlings et al. 2006)

carboxypeptidase Y inhibitor, but also to a collection of known kinase inhibitors. So far, protease inhibitor activity has not been demonstrated for any of these plant proteins. Also noteworthy in this respect is the largest family of inhibitors in both plants, the I29 cathepsin-prodomains. I29s can occur either as free inhibitors, or as prodomain of a cathepsin. In all *Arabidopsis* and most of the *Oryza* I29 members, the inhibitor module in the protein is followed by a cathepsin domain, indicating that these proteins merely function as pro-domain to prevent premature activation of the cathepsins. In *Oryza*, a small group of four sequences (indicated in Fig. 11.1B) is formed by I29 members that are not part of a cathepsin sequence, and may thus function as inhibitors of other proteases. However, this potential function has not yet been assessed.

Most inhibitors in *Oryza* and *Arabidopsis* target serine proteases, such as trypsins and chymotrypsins, which may relate to the fact that many herbivores, such as lepidopteran insects but also mammals, use these proteases for protein digestion. Only family I25 (cystatins) targets cysteine proteases, while family I03, the Kunitz inhibitors, also comprises members specific for cysteine and aspartic proteases.

11.2.2 Roles in Defense, but Also in Other Fields

A direct relationship to defense against arthropods, nematodes, fungi, and/or bacteria has been established for most inhibitor families, such as I12 (Bowman–Birk), I03 (Kunitz), I06 (trypsin/α-amylase), I25 (cystatins), I13 (eglin C) and I18 (mustard trypsin inhibitor). An exception may be found in the serpin family (I04). They have been demonstrated to possess inhibitor activity in plants, but their functional role is still unclear. Serpins do accompany storage proteins in seeds of monocots, where they could protect these from proteolysis or still act in defense (Hejgaard 2005). In *Arabidopsis*, however, none of its members are induced by jasmonate or salicylate treatments and expression levels are generally very low (Genevestigator; data not shown). We observed that when potato cystatins, known for their defense role, are overexpressed in tomato cytoplasm this tends to lead to dwarfing and chlorosis of tomato plants whereas these effects are not observed when the inhibitors are targeted to the secretory pathway (unpublished results). This does emphasize that these proteins are not inert to the host plant physiology and may play subtle secondary roles in a range of processes unrelated to defense and dependent on the subcellular localization of the inhibitor.

The availability for several years now of full genome DNA microarrays of *Arabidopsis* has resulted in a rich source of expression data for each gene in different tissues under different physiological conditions and stresses. Without any wetlab experiments, these data can be mined and used to hypothesize on the functions of individual members of gene families (Zimmermann et al. 2004). As an example, we did this exercise for the unique cruciferous MTI-2 gene family with six gene members which was also studied by Clauss and Mitchell-Olds (2004). Table 11.1 summarizes the presently available data for tissue-specific gene expression in *Arabidopsis*. It is

11 Plant Protease Inhibitors

Table 11.1 Tissue-specific constitutive expression without inducer of all six members of the MTI-2 gene family in *Arabidopsis*. The average fluorescence signals for these tissues are shown

Anatomy	Acc. nr.	At2g43510	At2g43520	At2g43530	At2g43535	At1g47540	At2g43550
	Potential inducer	*P. syringae* senescence		jasmonate wounding	darkness		jasmonate
		ATTI1	ATTI2	ATTI3	ATTI4	ATTI7	ATTI6
	# of Chips	Mean	Mean	Mean	Mean	Mean	Mean
0 callus	6	**5604**	**190**	**440**	**2030**	**113**	**39**
1 cell suspension	87	**5766**	**255**	**307**	**5771**	**70**	**298**
2 seedling	493	**1832**	**697**	**2123**	**5944**	**91**	**1879**
21 cotyledons	9	763	618	3879	1127	42	4049
22 hypocotyl	6	453	413	6104	3540	82	3552
23 radicle	9	121	173	357	7690	84	672
3 inflorescence	232	**1939**	**3428**	**4203**	**827**	**6685**	**2710**
31 flower	85	2058	925	5116	513	62	3842
311 carpel	15	1314	563	7154	827	47	3844
3111 ovary	4	1613	376	5021	479	49	3334
3112 stigma	3	2363	148	2340	170	69	1868
312 petal	6	991	75	1566	156	50	4727
313 sepal	6	13776	337	2229	236	47	2098
314 stamen	15	575	154	1056	191	101	1253
3141 pollen	2	452	86	59	396	157	12
315 pedicel	3	489	652	2716	752	69	3546
32 silique	19	1542	2004	4746	738	9370	1089
33 seed	53	2661	9864	801	1624	25735	176
34 stem	28	1212	533	3827	489	56	1961

Table 11.1 (continued)

Anatomy	Acc. nr.	At2g43510	At2g43520	At2g43530	At2g43535	At1g47540	At2g43550
	Potential inducer	P. syringae senescence		jasmonate wounding	darkness		jasmonate
		ATTI1	ATTI2	ATTI3	ATTI4	ATTI7	ATTI6
	# of Chips	Mean	Mean	Mean	Mean	Mean	Mean
35 node	3	260	536	4362	454	39	4957
36 shoot apex	25	1858	3680	7164	807	41	3873
37 cauline leaf	3	1636	178	2129	193	38	392
4 rosette	**710**	**3473**	**1499**	**6035**	**925**	**55**	**4228**
41 juvenile leaf	87	2036	782	3846	722	69	2377
42 adult leaf	243	6346	2449	8560	1007	69	5533
43 petiole	12	559	467	4303	548	27	3986
44 senescent leaf	3	3454	102	995	79	49	131
45 hypocotyl	12	2703	3858	3470	2097	83	415
451 xylem	3	3441	4618	1676	587	71	116
452 cork	3	812	225	1249	383	90	571
5 roots	**236**	**1967**	**109**	**229**	**4793**	**72**	**408**
52 lateral root	4	611	303	1300	7892	101	548
53 root tip	4	148	162	281	2351	147	89
54 elongation zone	7	190	116	131	4685	129	157
55 root hair zone	4	296	114	126	1008	92	112
56 endodermis	3	483	127	187	1037	170	49
57 endodermis + cortex	3	155	108	300	974	104	306
58 epid atrichoblasts	3	498	156	326	9012	136	88
59 lateral root cap	3	838	162	594	1579	131	174
60 stele	3	244	137	163	1576	190	94

Inducer refers to circumstances which will activate the gene on top of the developmentally regulated expression in different tissues provided in the table. Results were obtained from the Gene Atlas of Genevestigator (Zimmermann et al. 2004) at https://www.genevestigator.ethz.ch/. Most data possessed standard errors of 5%–20% of the mean. ATTI numbers refer to Clauss and Mitchell-Olds (2004).

remarkable to see that the apparent functional redundancy of these genes (they all encode trypsin inhibitors) at least partly finds its explanation in the fact that their tissue specificity differs widely between for example roots (mainly At2g43535), leaves (mainly At2g43530; At2g43550), or seeds (mainly At1g47540; At2g43520). Furthermore, they are also regulated very differently under different stresses and physiological conditions. Remarkably, only At2g43530 and At2g43550 are clearly responsive to jasmonates whereas At2g43510 is specifically responsive to microbial infection by *Pseudomonas syringae* (not shown).

11.3 Structure–Function Relationships

11.3.1 Specificity for Proteases: The Different Models

As mentioned in Section 11.1, plants have evolved different classes of molecules to inhibit different types of proteases often occurring outside their own system as part of the herbivore gut or the microbial extracellular fluid. The different classes of inhibitors are distinguished by the structure of their polypeptide backbone. The mechanism of inhibition, however, is mostly the same for all classes, and involves an interaction of inhibitor residues with the active site of the target protease in a canonical (substrate-like) manner through an exposed reactive site loop (Bode and Huber 2000). For serine proteinase inhibitors, most residues interacting with the proteinase are located on a single loop, to which the P1 residue is central. The P1 residue of protease substrates is the one on the amino-side of the hydrolyzed bond, and is often crucial for substrate recognition. The surrounding residues (P2, P3 etc. at its N-terminus, and P1', P2' etc. at its C-terminus) play secondary roles in the interaction of proteases and their substrates (Berger and Schechter 1970). Both in inhibitors and in substrates, the P1 residue fits into the S1 substrate binding site of the proteinase. Unlike normal peptide substrates, the inhibitor residues around P1 interact with the enzyme through complementary polar and hydrophobic interactions, and are held in position by strong bonds with the inhibitor scaffold. This prevents immediate dissociation of the complex, thus keeping the enzyme inactive (Bode and Huber 2000).

The P1 residue is usually the primary determinant of inhibitor specificity. In the case of trypsin inhibitors, the P1 residue is usually a positively charged residue (Arg or Lys), while in the case of chymotrypsin inhibitors, the P1 is hydrophobic (Leu, Ile, Trp or Phe). The relative importance of the P1 residue varies between inhibitors. For instance, the mustard trypsin inhibitor MTI-2, with an Arg at position P1, is a very potent trypsin inhibitor, but is unable to inhibit chymotrypsin. When the MTI-2 P1 is changed into Leu, the inhibitor becomes active against chymotrypsin, and looses most of its activity towards trypsin (Volpicella et al. 2001). Thus, the specificity of MTI-2 is clearly governed by the P1 residue. For PI-2, on the other hand, inhibitor variants with Arg at the P1 position inhibit both trypsin and chymotrypsin (Beekwilder et al. 2000). This indicates that, in the case of PI-2, contacts between enzyme and inhibitor outside the P1 position play an essential role in recognition. In general, however, the nature of the P1 residue is quite predictive of the enzyme class which is inhibited.

The contact residues outside the P1 play a role in the recognition between enzyme and substrate. This is especially so for regulatory proteinases which specifically activate pro-proteins. Thrombin for instance has a canyon-like active site groove, build up from two relatively large loops around residues 60 and 150. As a result, only appropriately shaped substrates (such as fibrinogen) and inhibitors have access to its active site (Bode and Huber 2000). Most proteinases contain a substrate groove, but it is usually much wider, and thereby much less selective than in the case of thrombin. This is particularly true for trypsin-like proteinases with a wide substrate range, such as those involved in digestion of dietary protein.

Still, the inhibitor residues around P1 within a gene family are hypervariable relative to the rest of the protein. By using arrays of fluorogenic substrates, the importance of residues adjacent to the P1 residue for cleavage efficiency can be explored (Gosalia et al. 2005). Such exercises reveal that related proteinases can develop different affinities for substrates (and inhibitors) by interacting with the amino-acids around the primary cleavage site. Laskowski and coworkers analyzed the enzyme-binding loop of inhibitors from the ovomucoid trypsin inhibitor family in over 100 bird species, and found that the residues known to be in contact with the target enzymes are particularly variable between these species (Laskowski et al. 1987). These differences likely indicate adaptations of the inhibitor family towards different target proteinases, which, in turn, exhibit amino-acid changes at the binding interface.

More profound differences exist between different families of proteinase inhibitors. These differences include the number of contacts they make to different loops of the proteinase. In Table 11.2 we have summarized the contacts found in crystal structures of three classes of plant inhibitors with trypsin-like proteinases. While all inhibitors make contacts through their P1 residue, the other residues make

Table 11.2 Interaction between proteinase residues and different inhibitors. The first column shows the regions of the proteinase, exposed to the inhibitor (Bode and Huber 2000). The second column indicates to which substrate-binding pockets the region contributes. The third, fourth, and fifth columns show the residues within the proteinase loops which are in contact to the inhibitor, according to Greenblatt et al. (1989), Song and Suh (1998), and Koepke et al. (2000). Residues are numbered according to the chymotrypsin nomenclature

Region	Pocket	Contacts to SKTI	Contacts to PI-2	Contacts to BBI
40 loop	S1' S2' S3'	40–42	38–42	41
	S1	57	57	57
60 loop	S1' S2	60	–	–
90 loop	S2	94, 96, 97, 99, 102	–	97, 99
150 loop	S2'	151	–	–
170 loop	S2, S3, S4	–	169,171	–
	S1	189–195	190–195	189–195
217 loop	S3, S4, S5	214–217, 219, 220	214–217, 219	213, 215–217, 219

the difference. For instance, PI-2 has interactions with the loop around trypsin residue 170 (involved in the S2, S3, and S4 sites of the proteinase), while SKTI and BBI lack these interactions. Conversely, SKTI makes a number of contacts to the loop around trypsin residue 150, while these are missing in BBI and PI-2. BBI has very limited contact to loops composing the S1', S2' etc. pockets, touching only trypsin residue 41, while SKTI and especially PI-2 have many residues that contact the S1', S2' etc pockets. In Fig. 11.2 exemplary proteinase-inhibitor contacts are shown emphasizing the fact that a single proteinase makes many different contacts to different types of inhibitors. This indicates that different proteinase inhibitor families are complementary with regard to their contact sites to the proteinases. The observed complementarity may be crucial for effective defense against herbivores with different digestive proteinases, and may explain the existing variety of inhibitor families.

11.3.2 Multimeric and Multidomain Proteins: If You Can't Beat Them, Join or Exclude Them

One additional fascinating property of plant protease inhibitors is their tendency to occur as multimeric or multidomain proteins often targeting proteases from different classes and even amylases. This issue was discussed by us before in relation to the observation of the Terra group that some of the digestive enzymes of *Spodoptera frugiperda* occur as multimeric complexes (Ferreira et al. 1994; Jongsma and Bolter 1997). One hypothesis was that if both the inhibitors and the

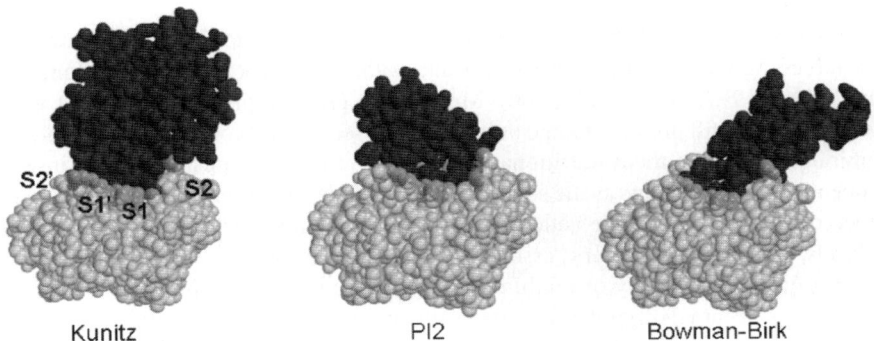

Fig. 11.2 Proteinase – inhibitor complexes. The three complexes depict porcine trypsin in complex with SKTI (*left*), PI-2 (*middle*) and BBI (*right*) (Greenblatt et al. 1989; Koepke et al. 2000; Song and Suh 1998). The enzyme is in light color, and the inhibitors are in dark. The enzymes from the original X-ray structures (porcine pancreas trypsin in case of SKTI, *Streptomyces griseus* proteinase B in case of PI-2, and bovine pancreas trypsinogen in case of BBI) have all been replaced by porcine pancreas trypsin by molecular replacement, using the catalytic triad residues as fixed coordinates, and the orientation in this view is for all enzymes the same. Enzyme sites S2', S1', S1 and S2 are indicated in the Kunitz structure

(multimeric) proteases/amylases had two or more contact sides to each other this would lead to polymerization in one or more dimensions. Such a polymer of concatenated protein would be less effective in accessing protein substrates, and, with fewer degrees of freedom, better prevent the dissociation of the inhibitor-enzyme complex. In this way protease-inhibitor complexes with poor dissociation constants could still remain effective in controlling the enzyme activity. Multidomain or multimeric associations of inhibitors could, thus, be an evolutionary advantageous trait to the plant. It was not clear, however, how multimeric associations of proteases could be advantageous to the insect. This was discussed in a second study on *Heliothis virescens* (Brito et al. 2001) where the authors found the multimeric forms of trypsin to be specifically induced in response to PIs and, furthermore, to be less sensitive to PIs. This led them to hypothesize that the multimeric association might render them sterically less sensitive to the inhibitors. If the enzymes would be complexed not in juxtaposition but with the active sites close together this would certainly be a real possibility as previously observed for human tryptase (Pereira et al. 1998). Thus, there are ways for both plants and insects of associating proteases or inhibitors to potentially increase or decrease inhibition. More detailed research is needed to demonstrate the reality and success of these strategies both on the plant and insect side. Surely, however, these structural features provide an extra dimension to the study and understanding of this form of plant defense.

11.4 The Insects Fight Back

11.4.1 Evolution of Steric Occlusion

Plants have evolved different inhibitor classes as a presumably successful strategy of defending themselves against insect and pathogen attack (Ryan 1990; Jongsma and Bolter 1997; Ferry et al. 2006). Most plants produce protease inhibitors upon wounding, and fill protein storage organs such as seeds, fruits or tubers with protease inhibitors, to make them nutritionally unsuitable for herbivores. However, insects depend on plant proteins as the source of amino acids and strong selection pressures have operated to avoid the action of the inhibitors. In the previous paragraph we already discussed multimeric associations as one way which possibly evolved to sterically avoid the entry of inhibitors in the active sites of the protease. This possibility is still largely hypothetical, however, and requires experimental verification. A second well-documented strategy has evolved mainly in generalist insects such as *Helicoverpa* and *Spodoptera* species, which must adapt to many different configurations of inhibitors. The successful colonization of many different plant species by these insects is based on their ability to switch to a different set of proteases upon encountering protease inhibitors in their diet. Characteristic of this induced set of proteases is their insensitivity to a broad range of inhibitors. This special property was first noted a decade ago (Jongsma et al. 1995), and has since then been found in many insect pests (Bown et al. 1997, 2004; Mazumdar-Leighton and

Broadway 2001a; Volpicella et al. 2003; Gruden et al. 2004; Liu et al. 2004; Moon et al. 2004; Bayes et al. 2005; Chougule et al. 2005).

What causes the resistance of these proteases to inhibition? Several different lines of research have been deployed to understand the induced proteinases. By analyzing gene expression in the gut, sequences of proteases have been identified that are induced upon ingestion of protease inhibitors (Bown et al. 1997; Mazumdar-Leighton and Broadway 2001b; Gruden et al. 2004). However, in the absence of a suitable recombinant expression system, it has so far not been possible to analyze the reactivity of different inhibitors to individual proteases. Proteinases have been purified from herbivore guts to identify the dominant proteinases involved in this switch, and to analyze their interaction with plant inhibitors in inhibition assays and by protein modeling (Volpicella et al. 2003, 2006; Gruden et al. 2004). Proteinases in the *Helicoverpa* midgut, induced by ingestion of SKTI, were compared to enzymes isolated from uninduced midguts. Both classes of enzymes were equally active on proteinaceous substrates, but differed extremely in their sensitivity to inhibitors from four different classes. The dominant trypsin isolated from SKTI-induced guts was completely insensitive to >1000 fold molar excess of any of the four tested inhibitors, each of them with different structural folds (Volpicella et al. 2003). On the other hand, also enzymes with differential sensitivity to different inhibitor classes were observed. For instance, a chymotrypsin induced by SKTI was more susceptible to PI-2 than to other inhibitors (Volpicella et al. 2006). Other groups have observed differences in effectiveness of chickpea-derived and potato-derived inhibitors to semi-purified *Helicoverpa armigera* proteinases (Telang et al. 2005). These data show that a panel of different proteinases can be deployed by adapted pest insects, each with different inhibitor sensitivity.

What makes a proteinase insensitive to an inhibitor? The use of synthetic substrates revealed potentially relevant differences between SKTI-induced and constitutive enzymes. Benzoyl-Arg-pNA, which carries a benzoyl group at the P2 position, was digested ten times less efficiently by SKTI-induced trypsin as compared to uninduced trypsin, while substrates with Phe or Arg in the P2 position did not discriminate the two proteinases (Volpicella et al. 2003). This suggests that an important structural feature exists at the S2 site of induced trypsin, which correlates with its resistance to inhibitors. However, the nature of that structural feature cannot be further explored in the absence of a successful recombinant expression system and crystal structures of the proteins.

Unequivocal insight in the structural basis of the resistance of insect proteases to plant inhibitors was obtained for the first time from the crystal structure of *Helicoverpa zea* carboxypeptidase B (Bayes et al. 2005, 2006). This enzyme, referred to as CPBHz, resists inhibition by potato carboxypeptidase inhibitor (PCI), which inhibits all other known enzymes from this class, including *Helicoverpa* carboxypeptidase A (CPAHa). The CPBHz crystal structure clearly showed that two surface loops of the enzyme assumed an unusual position in comparison to related inhibited enzymes such as CPAHa. In high-affinity complexes, these regions stabilize the interaction between the carboxypeptidase and the inhibitor. This interaction is initially formed by insertion of the flexible C-terminal tail of PCI, which presents the P1 and P1' residue to the active site pocket (Rees and Lipscomb 1982; Molina et al. 1994). However, due to the displacement of these loops on the surface of the enzyme, the inhibitor binds to a more distant position relative to the active site. As a consequence, the P1 and P1' residues are unable to penetrate the active site deep enough to achieve a strong interaction, and the interaction is easily lost.

The evolutionary mechanism by which the proteases evolve towards insensitivity for the inhibitors can be deduced in part from a comparison of the mutated sequence with its closest PI-sensitive homologue. Displacement of protease loops is clearly involved, as is apparent from the CPBHz structure. In one of the two loops, displacement is accompanied by insertion of a residue relative to inhibitor-sensitive enzymes, while in the other loop, a single residue is deleted. In both loops, a the observed number of mutations was higher as compared to other parts of the molecule (Bayes et al. 2005). Interestingly, one could predict that the plant could re-establish inhibition of CPBHz by inserting an additional residue in the substrate-like tail of PCI, thus allowing its P1 residue to reach the active site pocket, but such an inhibitor with an extended tail has not yet been found or evolved.

11.4.2 Inactivation with Specific Proteases

Although generally stable against many proteases, some insects have evolved specific proteases capable of effectively degrading plant protease inhibitors. We observed this phenomenon recently with *Plutella xylostella* degrading MTI-2 when expressed in *Arabidopsis* (Yang et al. submitted). Others found degradation of dietary protease inhibitors by insects like black vine weevil, *Phaedon cochleria* and *H. armigera* (Michaud 1997; Girard et al. 1998; Giri et al. 1998). In our opinion, many cases of specific inactivation may have escaped detection due to the experimental set up chosen by most authors, in which comparing short and prolonged preincubation of the inhibitors with the gut proteases is not standard procedure.

11.5 Perspectives

11.5.1 Regulation is the Key: Did Insects and Plants Both Think of It...?

The existence of PI-insensitive proteases in the insect gut and their specific upregulation in response to proteinase inhibitors in the diet provides a clear benefit to the insect and is likely monitored by some kind of feedback mechanism which has not yet been elucidated. However, the parallel observation of sensitive enzymes being downregulated could be beneficial to both insects and plants alike. The insects don't waste their energy on useless (inhibited) proteases, but the plants similarly, if they would govern this process, could spend less energy on synthesizing an effective concentration of inhibitors. Indeed if plants could successfully antagonize the regulatory feedback system operational in insect guts, for PI-sensitive and -insensitive proteases alike, they would have found a second way independent of direct inhibition to achieve the same net result: lower protease levels in the gut, leading to slower development and higher mortality. Hijacking the regulation of proteases in the insect gut is, therefore, a holy grail for plants, and preventing them from achieving this goal is the lifeline of the insects: an attractive second stage of evolutionary battle between proteases and inhibitors.

However, has the script been written? How likely is it, that plants have evolved such signals for the specific purpose of downregulating proteases in insects? Currently, there are a few peptide hormones known for receptor-mediated up- or downregulation of serine proteases. In mammals the monitor peptide, member of the PSTI gene family, and the luminal cholecystokinin releasing factor LCRF are known to upregulate pancreatic protease secretion by stimulating cholecystokinin release in the blood (Iwai et al. 1988; Spannagel et al. 1996). Excess proteases subsequently degrade the monitor/LCRF peptides and stop the secretion. In mosquitoes after a blood meal the oostatic hormone or trypsin modulating oostatic factor (TMOF) was the only peptide found so far to actively signal the downregulation of trypsin secretion into the gut (Borovsky et al. 1990).

Remarkably the TMOF peptide, which consists of only ten amino acids, is also active in the lepidopteran herbivore *H. virescens*. TMOF actively downregulates proteinase secretion when injected into the hemolymph of the larvae (Nauen et al. 2001), but surprisingly also when it is expressed in plants fused to other peptides, and when it reaches the larvae through the gut after eating the plant leaves. Expressed as a potentially cleavable six-domain repeat, or replacing systemin in prosystemin, or in C-terminal fusion to the coat protein of TMV, TMOF reduced larval development and decreased both gut trypsin and chymotrypsin activity (Tortiglione et al. 2002, 2003; Borovsky et al. 2006).

The TMOF peptide presumably signals through the interaction with receptors present on cells lining the gut. Such receptors are also potential targets for plant peptides or proteins to upset gut physiology. This brings us to the last unexplored area of plant protease inhibitors: the function(s) of the N- and C-terminal peptides which extend from most of the basic inhibitor domains that have been characterized in plants. The extensions appear in significant variation, both in length and sequence, within most inhibitor gene families. As an example, Table 11.3 shows the peptides extending from the mature forms of the Bowman–Birk inhibitors found in soybean and peanut. These sequences represent specific insertions and/or deletions. The extensions are not likely to be involved in the subcellular targeting of inhibitors, since only the mature proteins are shown and the targeting peptides are already removed. Therefore, these peptide domains may serve a direct role in plant defense. Potential functions include the modification of the pI of the protein molecule to favor the interaction with the protease, or else, the binding to cellular receptors lining the herbivore gut to modulate or antagonize the regulatory processes of the insect. Understanding the regulation of protease expression in insects and the potential role of plant protease inhibitors to interfere with this regulation promises to be a rewarding research area.

Table 11.3 Comparison of the N- and C-terminal peptides of the Bowman–Birk inhibitors of soybean and peanut extending outside the basic inhibitory domain for which no function has been assigned

Source	Code[1]	N-terminal sequence[3]	Inhibitor[2]	C-terminal sequence[3]
Soybean	MER21210	SDQSSSYDDDEYSKPCC-	-inhibitor-	-CKSRDD
Soybean	MER21212	SDHSSSDDESSKPCC-	-inhibitor-	-CKSSDEDDD
Soybean	MER19956	SDHHQHSNDDDSSKPCC-	-inhibitor-	-CKPSEDDKENY
Soybean	MER19957	SDHHQHSNDDESSKPCC-	-inhibitor-	-CKPSQDDKENY
Peanut	MER18102	EASSSSDDNVCC-	-inhibitor-	-CRS
Peanut	MER18107	AASDCC-	-inhibitor-	-CA

[1] Codes refer to the MEROPS protease inhibitor database: http://merops.sanger.ac.uk/
[2] This represents the conserved two domains involved in protease inhibition
[3] The sequences represent the mature stored forms after removal of the signal peptide and other targeting peptides

11.5.2 Creative Evolution

In this chapter we reviewed the ways in which plants and plant predators have evolved creative solutions in the battle for scarce essential amino acids. We showed that on both the plant and insect side many different gene families are involved with intricate specialized expression patterns among the different members. They provide an evolutionary and ecological genetic fingerprint of their functional role, and, if anything, this chapter highlighted the little that is actually known about the intricate fabric of functional relationships that are still out there to be discovered and admired.

Some of the hypotheses we put forward are currently only supported by circumstantial evidence, but provide attractive alleys which functional evolution may have gone. At first glance, this topic of plant–insect/pathogen interactions appears to be only a clash of two protein molecules. However, due to the opportunistic nature of evolution, this battle may have taken on many new appearances. We hope our account of these possibilities provides inspiration to researchers to investigate them. The availability of fully sequenced genomes on both the plant and insect side, and the excellent tools to specifically knock down and overexpress specific genes provide exactly the instruments needed to allow a whole new set of discoveries in this area.

References

Abdeen A, Virgos A, Olivella E, Villanueva J, Aviles X, Gabarra R, Prat S (2005) Multiple insect resistance in transgenic tomato plants over-expressing two families of plant proteinase inhibitors. Plant Mol Biol 57:189–202

Bayes A, Comellas-Bigler M, de la Vega MR, Maskos K, Bode W, Aviles FX, Jongsma MA, Beekwilder J, Vendrell J (2005) Structural basis of the resistance of an insect carboxypeptidase to plant protease inhibitors. Proc Natl Acad Sci USA 102:16602–16607

Bayes A, de la Vega MR, Vendrell J, Aviles FX, Jongsma MA, Beekwilder J (2006) Response of the digestive system of *Helicoverpa zea* to ingestion of potato carboxypeptidase inhibitor and characterization of an uninhibited carboxypeptidase B. Insect Biochem Mol Biol 36:654–664

Beekwilder J, Schipper B, Bakker P, Bosch D, Jongsma M (2000) Characterization of potato proteinase inhibitor II reactive site mutants. Eur J Biochem 267:1975–1984

Berger A, Schechter I (1970) Mapping the active site of papain with the aid of peptide substrates and inhibitors. Philos Trans R Soc Biol Sci 257:249–264

Birk Y, Gertler A, Khalef S (1963) A pure trypsin inhibitor from soya beans. Biochem J 87:281–284

Bode W, Huber R (2000) Structural basis of the endoproteinase–protein inhibitor interaction. Biochim Biophys Acta 1477:241–252

Borovsky D, Carlson DA, Griffin PR, Shabanowitz J, Hunt DF (1990) Mosquito oostatic factor: a novel decapeptide modulating trypsin-like enzyme biosynthesis in the midgut. FASEB J 4:3015–3020

Borovsky D, Rabindran S, Dawson WO, Powell CA, Iannotti DA, Morris TJ, Shabanowitz J, Hunt DF, DeBondt HL, DeLoof A (2006) Expression of Aedes trypsin-modulating oostatic factor on the virion of TMV: a potential larvicide. Proc Natl Acad Sci USA 103:18963–18968

Bowman DE (1946) Differentiation of soy bean antitryptic factors. Proc Soc Exp Biol Med 63:547–550

Bown DP, Wilkinson HS, Gatehouse JA (1997) Differentially regulated inhibitor-sensitive and insensitive protease genes from the phytophagous insect pest, *Helicoverpa armigera*, are members of complex multigene families. Insect Biochem Mol Biol 27:625–638

Bown DP, Wilkinson HS, Gatehouse JA (2004) Regulation of expression of genes encoding digestive proteases in the gut of a polyphagous lepidopteran larva in response to dietary protease inhibitors. Physiol Entomol 29:278–290

Brito LO, Lopes AR, Parra JRP, Terra WR, Silva MC (2001) Adaptation of tobacco budworm *Heliothis virescens* to proteinase inhibitors may be mediated by the synthesis of new proteinases. Comp Biochem Physiol Biochem Mol Biol 128:365–375

Chougule NP, Giri AP, Sainani MN, Gupta VS (2005) Gene expression patterns of *Helicoverpa armigera* gut proteases. Insect Biochem Mol Biol 35:355–367

Clauss MJ, Mitchell-Olds T (2004) Functional divergence in tandemly duplicated *Arabidopsis thaliana* trypsin inhibitor genes. Genetics 166:1419–1436

De Leo F, Gallerani R (2002) The mustard trypsin inhibitor 2 affects the fertility of *Spodoptera littoralis* larvae fed on transgenic plants. Insect Biochem Mol Biol 32:489–496

Ferreira C, Capella AN, Sitnik R, Terra WR (1994) Properties of the digestive enzymes and the permeability of the peritrophic membrane of *Spodoptera frugiperda* (Lepidoptera) larvae. Comp Biochem Physiol Physiol 107:631–640

Ferry N, Edwards MG, Gatehouse J, Capell T, Christou P, Gatehouse AMR (2006) Transgenic plants for insect pest control: a forward looking scientific perspective. Transgenic Res 15:13–19

Girard C, Le Metayer M, Bonade-Bottino M, Pham-Delegue MH, Jouanin L (1998) High level of resistance to proteinase inhibitors may be conferred by proteolytic cleavage in beetle larvae. Insect Biochem Mol Biol 28:229–237

Giri AP, Harsulkar AM, Deshpande VV, Sainani MN, Gupta VS, Ranjekar PK (1998) Chickpea defensive proteinase inhibitors can be inactivated by podborer gut proteinases. Plant Physiol 116:393–401

Gosalia DN, Salisbury CM, Ellman JA, Diamond SL (2005) High throughput substrate specificity profiling of serine and cysteine proteases using solution-phase fluorogenic peptide microarrays. Mol Cell Prot 4:626–636

Green TR, Ryan CA (1972) Wound-induced proteinase inhibitor in plant leaves – possible defense mechanism against insects. Science 175:776–777

Greenblatt HM, Ryan CA, James MN (1989) Structure of the complex of *Streptomyces griseus* proteinase B and polypeptide chymotrypsin inhibitor-1 from russet burbank potato tubers at 2.1 Å resolution. J Mol Biol 205:201–228

Gruden K, Kuipers AGJ, Guncar G, Slapar N, Strukelj B, Jongsma MA (2004) Molecular basis of colorado potato beetle adaptation to potato plant defence at the level of digestive cysteine proteinases. Insect Biochem Mol Biol 34:365–375

Gutierrez-Campos R, Torres-Acosta JA, Saucedo-Arias LJ, Gomez-Lim MA (1999) The use of cysteine proteinase inhibitors to engineer resistance against potyviruses in transgenic tobacco plants. Nature Biotech 17:1223–1226

Hejgaard J (2005) Inhibitory plant serpins with a sequence of three glutamine residues in the reactive center. Biol Chem 386:1319–1323

Hilder VA, Gatehouse AMR, Sheerman SE, Barker RF, Boulter D (1987) A novel mechanism of insect resistance engineered into tobacco. Nature 330:160–163

Iwai K, Fushiki T, Fukuoka S (1988) Pancreatic-enzyme secretion mediated by novel peptide: monitor peptide hypothesis. Pancreas 3:720–728

Jongsma MA, Bakker PL, Peters J, Bosch D, Stiekema WJ (1995) Adaptation of *Spodoptera exigua* larvae to plant proteinase-inhibitors by induction of gut proteinase activity insensitive to inhibition. Proc Natl Acad Sci USA 92:8041–8045

Jongsma MA, Bolter C (1997) The adaptation of insects to plant protease inhibitors. J Insect Physiol 43:885–895

Koepke J, Ermler U, Warkentin E, Wenzl G, Flecker P (2000) Crystal structure of cancer chemopreventive Bowman–Birk inhibitor in ternary complex with bovine trypsin at 2.3 Å resolution. Structural basis of janus-faced serine protease inhibitor specificity. J Mol Biol 298:477–491

Kunitz M (1946) Crystalline soybean trypsin inhibitor. J Gen Physiol 29:149–154

Laskowski M, Kato I, Ardelt W, Cook J, Denton A, Empie MW, Kohr WJ, Park SJ, Parks K, Schatzley BL, Schoenberger OL, Tashiro M, Vichot G, Whatley HE, Wieczorek A, Wieczorek M (1987) Ovomucoid 3rd domains from 100 avian species – isolation, sequences, and hypervariability of enzyme-inhibitor contact residues. Biochemistry 26:202–221

Liu YL, Salzman RA, Pankiw T, Zhu-Salzman K (2004) Transcriptional regulation in southern corn rootworm larvae challenged by soyacystatin N. Insect Biochem Mol Biol 34:1069–1077

Mazumdar-Leighton S, Broadway RM (2001a) Identification of six chymotrypsin cDNAs from larval midguts of *Helicoverpa zea* and *Agrotis ipsilon* feeding on the soybean (Kunitz) trypsin inhibitor. Insect Biochem Mol Biol 31:633–644

Mazumdar-Leighton S, Broadway RM (2001b) Transcriptional induction of diverse midgut trypsins in larval *Agrotis ipsilon* and *Helicoverpa zea* feeding on the soybean trypsin inhibitor. Insect Biochem Mol Biol 31:645–657

Michaud D (1997) Avoiding protease-mediated resistance in herbivorous pests. Trends Biotechnol 15:4–6

Molina MA, Marino C, Oliva B, Aviles FX, Querol E (1994) C-tail valine is a key residue for stabilization of complex between potato inhibitor and carboxypeptidase-A. J Biol Chem 269:21467–21472

Moon J, Salzman RA, Ahn JE, Koiwa H, Zhu-Salzman K (2004) Transcriptional regulation in cowpea bruchid guts during adaptation to a plant defence protease inhibitor. Insect Mol Biol 13:283–291

Nauen R, Sorge D, Sterner A, Borovsky D (2001) Tmof-like factor controls the biosynthesis of serine proteases in the larval gut of *Heliothis virescens*. Arch Insect Biochem Physiol 47:169–180

Pereira PJB, Bergner A, Macedo-Ribeiro S, Huber R, Matschiner G, Fritz H, Sommerhoff CP, Bode W (1998) Human β-tryptase is a ring-like tetramer with active sites facing a central pore. Nature 392:306–311

Qu LJ, Chen J, Liu MH, Pan NS, Okamoto H, Lin ZZ, Li CY, Li DH, Wang JL, Zhu GF, Zhao X, Chen X, Gu HG, Chen ZL (2003) Molecular cloning and functional analysis of a novel type of Bowman–Birk inhibitor gene family in rice. Plant Physiol 133:560–570

Rawlings ND, Morton FR, Barrett AJ (2006) MEROPS: the peptidase database. Nucl Acids Res 34:D270–D272

Rees DC, Lipscomb WN (1982) Refined crystal-structure of the potato inhibitor complex of carboxypeptidase-A at 2.5-Å resolution. J Mol Biol 160:475–498

Ryan CA (1990) Protease inhibitors in plants – genes for improving defenses against insects and pathogens. Ann Rev Phytopathol 28:425–449

Song HK, Suh SW (1998) Kunitz-type soybean trypsin inhibitor revisited: refined structure of its complex with porcine trypsin reveals an insight into the interaction between a homologous inhibitor from *Erythrina caffra* and tissue-type plasminogen activator. J Mol Biol 275:347–363

Spannagel AW, Green GM, Guan DF, Liddle RA, Faull K, Reeve JR (1996) Purification and characterization of a luminal cholecystokinin-releasing factor from rat intestinal secretion. Proc Natl Acad Sci USA 93:4415–4420

Telang MA, Giri AP, Sainani MN, Gupta VS (2005) Characterization of two midgut proteinases of *Helicoverpa armigera* and their interaction with proteinase inhibitors. J Insect Physiol 51:513–522

Tortiglione C, Fanti P, Pennacchio F, Malva C, Breuer M, De Loof A, Monti LM, Tremblay E, Rao R (2002) The expression in tobacco plants of *Aedes aegypti* trypsin modulating oostatic factor (aea-tmof) alters growth and development of the tobacco budworm, *Heliothis virescens*. Mol Breed 9:159–169

Tortiglione C, Fogliano V, Ferracane R, Fanti P, Pennacchio F, Monti LM, Rao R (2003) An insect peptide engineered into the tomato prosystemin gene is released in transgenic tobacco plants and exerts biological activity. Plant Mol Biol 53:891–902

Volpicella M, Ceci LR, Cordewener J, America T, Gallerani R, Bode W, Jongsma MA, Beekwilder J (2003) Properties of purified gut trypsin from *Helicoverpa zea*, adapted to proteinase inhibitors. Eur J Biochem 270:10–19

Volpicella M, Ceci LR, Gallerani R, Jongsma MA, Beekwilder J (2001) Functional expression on bacteriophage of the mustard trypsin inhibitor MTI-2. Biochem Biophys Res Commun 280:813–817

Volpicella M, Cordewener J, Jongsma MA, Gallerani R, Ceci LR, Beekwilder J (2006) Identification and characterization of digestive serine proteases from inhibitor-resistant *Helicoverpa zea* larval midgut. J Chrom B Analyt Technol Biomed Life Sci 833:26–32

Yang L, Fang ZY, Dicke M, van Loon JJA, Jongsma MA (2008) The diamondback moth, *Plutella xylostella* specifically inactivates Mustard Trypsin Inhibitor 2(MTI2) to overcome host plant defence (submitted)

Zavala JA, Patankar AG, Gase K, Hui DQ, Baldwin IT (2004) Manipulation of endogenous trypsin proteinase inhibitor production in *Nicotiana attenuata* demonstrates their function as antiherbivore defenses. Plant Physiol 134:1181–1190

Zimmermann P, Hirsch-Hoffmann M, Hennig L, Gruissem W (2004) Genevestigator. *Arabidopsis* microarray database and analysis toolbox. Plant Physiol 136:2621–2632

Chapter 12
Defensive Roles of Polyphenol Oxidase in Plants

C. Peter Constabel and Raymond Barbehenn

Plant polyphenol oxidases (PPOs) are widely distributed and well-studied oxidative enzymes, and their effects on discoloration in damaged and diseased plant tissues have been known for many years. The discovery in C.A. Ryan's laboratory in the mid-1990s that tomato PPO is induced by the herbivore defense signals systemin and jasmonate, together with seminal work on PPO's possible effects on herbivorous insects by G. Felton and S. Duffey has motivated many studies of PPO in the context of plant-herbivore defense. The cloning and characterization of PPO cDNAs from multiple plant species now allows for direct testing of defensive functions of PPO using transgenic plants. These have shown that PPO can contribute to insect herbivore and pathogen resistance, although how this occurs is only now being investigated more closely. Here we review progress in the functional analysis of PPO in plant defense against pests, and describe recent results that address the mechanisms of PPO as an anti-herbivore protein. We suggest that assumptions of how PPO functions as an anti-nutritive defense against lepidopterans needs to be re-examined in light of the near anoxic conditions in lepidopteran midguts. Ultimately, the efficacy of PPO should be directly tested in a greater variety of plant-insect interactions. In addition, the identification of the endogenous PPO substrates will help to define defensive and potentially other roles for PPO in plants.

12.1 Introduction

Polyphenol oxidases (PPOs) are ubiquitous copper-containing enzymes which use molecular oxygen to oxidize common *ortho*-diphenolic compounds such as caffeic acid and catechol to their respective quinones. PPO-generated quinones are highly reactive and may cross-link or alkylate proteins, leading to the commonly observed brown pigments in damaged plant tissues and plant extracts. The conspicuous pigments are generally undesirable in food products, and the role of PPO in browning

---

C.P. Constabel
Centre for Forest Biology and Department of Biology, University of Victoria, Victoria, BC, Canada V8W 3N5,
e-mail: cpc@uvic.ca

has prompted numerous studies on PPO in food and beverages. In parallel, the potential roles for PPO in plant defense against pests have motivated many studies on PPO in an ecological context, though few of these have used a transgenic approach. Functional and mechanistic studies on PPO in plant-insect interactions using PPO-modified transgenic plants have recently been reported, providing new insight into the biology of this versatile enzyme.

12.2 The Biochemistry of Plant Polyphenol Oxidase

Polyphenol oxidase (catechol oxidase; E.C. 1.10.3.2) has been purified and characterized from a wide range of plant species and a variety of tissues (Constabel et al. 1996; Mayer 2006), and activity levels using common substrates vary widely (Constabel and Ryan 1993). Key features of PPOs are two conserved copper-binding domains, and N-terminal chloroplast and thylakoid transit peptides (van Gelder et al. 1997; Marusek et al. 2006). The size of the predicted mature PPO proteins is typically 54–62 kD (Constabel et al. 1996; van Gelder et al. 1997). However, some PPOs are processed further, and in some cases only the processed form is fully active (Rathjen and Robinson 1992). Many PPOs are predicted to contain a proteolytic processing site near the C-terminus of the polypeptide (Marusek et al. 2006).

Based on its association with browning reactions in crop plants, PPO has been characterized in a wide variety of food plants including banana, wheat, quince, and avocado, and a number of chemical inhibitors have been identified (reviewed in Mayer 2006; Yoruk and Marshall 2003). Although PPO is found at significant levels in a variety of fruits, vegetables and grains, its biological function in these tissues has rarely been studied. PPO is expressed in many different tissues and organs, including roots, leaves, flowers, and vascular tissue (Constabel et al. 1996). Its presence in chloroplasts led to the proposal that PPO may function in pseudocyclic photophosphorylation or as a modulator of oxygen levels, yet to date little evidence supports these roles (Steffens et al. 1994). A complicating issue regarding PPO function has been the separate localization of its phenolic substrates in plant vacuoles, so that the cell would have to be broken in order for PPO to oxidize phenols; this is most likely to occur following pest or pathogen challenge. Roles of PPO in plant defense are thus commonly discussed, although direct evidence for such roles has become available only recently.

A puzzling feature of PPO has been a variable degree of latency, so that PPOs from some species need to be activated with detergents or proteases for full activity. For example, tomato leaf PPO is extracted in its fully active form (Constabel et al. 1995), but poplar leaf PPO requires activation with protease or detergent (Constabel et al. 2000). Recent experiments suggest that such activating treatments function by removal of a peptide from the active site, via proteolysis or partial unfolding of the polypeptide, respectively (Gandia-Herrero et al. 2005). Treatment with low pH can also activate latent PPOs, presumably via a conformational change of the active site, increasing its accessibility to substrates (Kanade et al. 2006).

PPO latency may be significant for its defensive function; we recently showed that latent poplar PPO is activated by its passage through caterpillar guts (Wang and Constabel 2004).

PPOs are known to have broad substrate specificities. Thus, an enzyme from any given source may be capable of oxidizing a variety of simple *ortho*-diphenolics, such as caffeic acid and its conjugates, catechol derivatives, or dihydroxyphenylalanine (DOPA). However, enzymes from different plant species exhibit distinct preference profiles (reviewed in Constabel et al. 1996). Flavonoids with *ortho*-dihydroxy phenolic rings have been found to be PPO substrates, for example, catechin, (−)-epicatechin, and myricetin (Guyot et al. 1996; Jimenez and Garcia-Carmona 1999). Some reports also describe the oxidation of trihydroxy phenolics, such as gallic acid, by PPO (Shin et al. 1997), but it is not clear how widespread this is. Furthermore, for some PPOs a monophenolase activity has been described (Wuyts et al. 2006). Such enzymes may hydroxylate a monophenol such as tyrosine, which can then undergo further oxidation by the polyphenol oxidase activity to the quinone. Tyrosine hydroxylation by PPO has been described in pokeweed, where this reaction constitutes part of the biosynthetic pathway leading to betalains (Gandia-Herrero et al. 2005).

While the range of substrates accepted by isolated PPOs can be readily defined in vitro, knowledge of the substrates that are utilized *in planta* or during defense reactions is much scarcer. Caffeic acid esters such as chlorogenic acid (caffeoylquinate) are excellent PPO substrates and are very common plant metabolites. In tomato and coffee, chlorogenic acid has been identified as the most likely in vivo PPO substrate (Melo et al. 2006; Li and Steffens 2002). Caffeic acid esters are ubiquitous as lignin precursors (Humphreys and Chapple 2002), but in most species these may not accumulate to sufficient levels to be considered likely PPO substrates. In *Populus tremuloides*, catechol is postulated to be released by the breakdown of the abundant phenolic glycosides (Clausen et al. 1989), and would therefore be available as a substrate in damaged tissues (Haruta et al. 2001). For most PPOs, however, the endogenous substrates are unknown. The importance of identifying the *in planta* PPO substrates and understanding the overall phytochemical context of PPO-containing plants is emphasized by recent reports of PPO-like enzymes with biosynthetic roles as hydroxylases of secondary metabolites (Cho et al. 2003; Nakayama et al. 2000).

PPO has been extensively studied by biologists, plant pathologists, and ecologists interested in mechanisms of defense against pests and pathogens. Based on the browning reactions resulting from the reactive PPO-generated quinones, PPO has often been suggested to function as a defense against pests and pathogens. A most dramatic illustration of the efficacy of PPO in this context comes from work on *Solanum berthaultii* in which extremely high PPO levels (45% of soluble protein) are found in glandular trichomes (Kowalski et al. 1992). Breakage of the trichomes by small-bodied insects such as aphids leads to rapid PPO-mediated oxidation and polymerization of phenolics, ultimately entrapping insects, or occluding their mouthparts with a sticky polymer (Kowalski et al. 1992). In most species, however, leaf PPO is found not in trichomes but in mesophyll cells. Here, the PPO-generated quinones were proposed to alkylate dietary protein during insect feeding, and to degrade essential amino acids in insect guts (Felton et al. 1989, 1992). PPO-mediated protein alkylation has been demonstrated to operate against lepidopteran pests in the presence of oxygen and in artificial diets, and is now being investigated in more detail in midgut fluids (see Section 12.5 below).

12.3 PPO and Induced Herbivore Defense in Tomato and Other Plants

The idea that PPO may act as an anti-nutritive defense against leaf-eating insects was first suggested by G. Felton and S. Duffey, who showed an inverse correlation of *Heliothis zea* growth and PPO levels in tomato plants (Felton et al. 1989). Strong support for an anti-herbivore role of PPO came from the discovery that the herbivore defense-inducing signal molecules systemin and methyl jasmonate (MeJA) induce PPO activity and PPO mRNA levels in tomato leaves (Constabel et al. 1995). Systemin is a short peptide required for systemic wound signaling in tomato which strongly upregulates tomato herbivore defenses (Ryan 2000; Narváez-Vásquez and Orozco-Cárdenas, this volume). Though no longer considered to be the primary systemic signal, it is required for the generation of such a signal (Schilmiller and Howe 2005). PPO and other defenses were also found to be strongly induced by MeJA and oligogalacturonic acid, major plant defense signaling compounds (Constabel et al. 1995). Since PPO induction in tomato by multiple signals occurs in parallel with a suite of other anti-herbivore proteins including several types of protease inhibitors (PIs) and the anti-nutritive enzymes arginase and threonine deaminase (Bergey et al. 1996; Chen et al. 2005), PPO is thought to play a similar role in defense against insects.

In tobacco, PPO and PIs are upregulated by tobacco systemin as well as by MeJA (Constabel and Ryan 1998; Ren and Lu 2006). Likewise, strong herbivore-, wound-, and MeJA-induction of PPO was shown in leaves of several poplar species (Constabel et al. 2000; Haruta et al. 2001). Hybrid poplar (*Populus trichocarpa x P. deltoides*) has a strong systemic inducible defense response, which also includes trypsin inhibitors and chitinases, both with confirmed anti-insect activities (Parsons et al. 1989; Lawrence and Novak 2001, 2006). Recent large scale genomics experiments have underscored the complexity of the herbivore defense response in hybrid poplar, which involves upregulation of many additional putative defense genes (Christopher et al. 2004; Ralph et al. 2006; Major and Constabel 2006). Overall, the co-induction of PPO with other herbivore defense proteins in several plant species has provided support for its anti-herbivore function.

The induction of PPO in tomato by both insects and MeJA has been replicated in both laboratory and field studies (Stout et al. 1998; Thaler et al. 1996, 2002). Furthermore, the early work in tomato stimulated numerous studies on inducible PPO in diverse plant species. The inducibility of PPO by wounding or MeJA treatment has been confirmed in other plants, including both herbaceous crops and trees (Constabel and Ryan 1998). The induction of PPO by herbivory has now been shown

Table 12.1 Studies of PPO induction in plants by herbivores with unknown effects on herbivores

Plant taxa	Insect species (Order)[1]	Induction	Reference
Solanaceae			
Tomato	*Leptinotarsa decemlineata* (C)	~2X	Felton (1992)
Potato	*L. decemlineata* regurgitant	~3–7X	Kruzmane et al. (2002)
Salicaceae			
Poplar (hybrid)	*Malacosoma disstria* (L)	~12X	Constabel et al. (2000)
Betulaceae			
Black alder	*Agelastica alni* (C)	~3X	Tscharntke et al. (2001)
Poaceae			
Buffalograss	*Blissus occiduus* (H)	None	Heng-Moss et al. (2004)
Barley, wheat, oats	*Diuraphis noxia* (Ho)	None	Ni et al. (2001)
Fabaceae			
Common bean	*Melanoplus differentialis* (O)	~2X	Alba-Meraz and Choe (2002)
Soybean	*Helicoverpa zea* (L)	None	Bi and Felton (1995)
Soybean	*Ceratoma trifurcata* (C)	None	Felton et al. (1994)
Soybean	*Spissistilus festinus* (H)	1.6X	Felton et al. (1994)
Malvaceae			
Cotton	*Helicoverpa zea* (L)	Not detectable	Bi et al. (1997)
Theaceae			
Tea	*Helopeltis theivora* (H)	~2–3X	Chakraborty and Chakraborthy (2005)

[1] C = Coleoptera, L = Lepidoptera, H = Heteroptera, Ho = Homoptera, O = Orthoptera.

for a taxonomically diverse group of plants (Table 12.1), and this induction has commonly been interpreted as a direct response against the herbivore. By contrast, studies of the potential impact of induced PPO against the herbivores have largely been done with a more limited range of plants and herbivores, primarily noctuid caterpillars on tomato and other species in the Solanaceae (Table 12.2). The results of 11 of 16 experiments demonstrate PPO induction and are consistent with the hypothesis that induced PPO contributes to defense against herbivores. As with all correlative studies, however, it is not possible to determine the specific impact of induced PPO on herbivores due to the other biochemical changes that occur in damaged plants (e.g., Hermsmeier et al. 2001; Thaler et al. 2001; Chen et al. 2005; Major and Constabel 2006). The direct effects of PPO on insect herbivores can thus best be tested using transgenic plants, where PPO levels are manipulated independently of other traits (see Section 12.4).

Other studies of the association between PPO activity and insect herbivore performance using (1) plant genotypes that vary in resistance to herbivory, (2) ontogenetic variation in PPO activity within the plant, and (3) leaves treated with PPO, yielded mixed results. A potato genotype with high PPO activity had increased resistance to the Colorado potato beetle (*Leptinotarsa decemlineata*; Castanera et al. 1996), while resistance to the coffee leaf miner (*Leucoptera coffeella*; Diptera) was apparently unaffected by higher levels of PPO in coffee

Table 12.2 Induction of PPOs in plants with effects tested on herbivores

Plant taxa	Inducing agent (Order)[1]	Induction	Effect	Herbivore[1] and response[2]	Reference
Solanaceae					
Tomato	*Helicoverpa zea* (L)	~2X	(−)	*Spodoptera exigua* (L) GR	Stout et al. (1998)
Tomato	*Helicoverpa zea* (L)	~2X	(−)	Spider mite (A) numbers	Stout et al. (1998)
Tomato	Jasmonic acid	~3X	(0)	*Manduca sexta* (L) RGR	Thaler et al. (2002)
Tomato	Jasmonic acid	~3X	(−)	*Trichoplusia ni* (L) RGR	Thaler et al. (2002)
Tomato	Jasmonic acid	~3X	(−)	Thrips (T) damage	Thaler et al. (1999, 2002)
Tomato	Jasmonic acid	~3X	(−)	Spider mite (A) numbers	Thaler et al. (2002)
Tomato (wild)	Jasmonic acid	~3X	(0)	*Spodoptera exigua* (L) RGR	Thaler et al. (2002)
Tomato	Jasmonic acid	~5X	(−)	*Spodoptera exigua* (L) RGR	Thaler et al. (1999)
Tomato	*Macrosiphum euphorbiae* (H)	Decreased	(+)	*Spodoptera exigua* (L) RGR	Stout et al. (1998)
Tomato	Wind stress	Decreased	(0)	*Manduca sexta* (L) RGR	Cipollini and Redman (1999)
Tomato	Jasmonic acid	~2X	(−)	*Manduca sexta* (L) RGR	Redman et al. (2001)
Tobacco (wild)	clipped sagebrush	~4X	(−)	Field damage by (L) and (O)	Karban et al. (2000)
Potato (wild)	High constitutive PPO	~4-7X	(−)	*Leptinotarsa decemlineata* (C) RGR	Castanera et al. (1996)
Horsenettle	High N fertilizer	Decreased	(0)	*Manduca sexta* (L) herbivory	Cipollini et al. (2002)
Horsenettle	High N fertilizer	Decreased	(0)	*Epitrix* sp. (C) herbivory	Cipollini et al. (2002)
Betulaceae					
Montain birch	*Epirrita autumnata* (L)	~3X	(−)	*Epirrita autumnata* (L) RGR	Ruuhola and Yang (2006)

[1] C = Coleoptera, L = Lepidoptera, O = Orthoptera, T = Thysanoptera, A = Arachnida.
[2] GR = growth rate (mg/day), RGR = relative growth rate (mg/mg/day).

leaves (Melo et al. 2006). Relative growth rates of *Helicoverpa zea* caterpillars were negatively correlated with tomato leaf and tomato fruit PPO levels (Felton et al. 1989), while *Manduca quinquemaculata* caterpillars showed greater performance on younger tobacco leaves, which contain higher PPO levels (Kessler and Baldwin 2002). *Lymantria dispar* caterpillars were unaffected by an eight-fold increase in PPO levels, using mushroom PPO applied to the leaf surface (Barbehenn et al. 2007).

Correlations of PPO activity with defense may be confounded by the complexity of PPO gene families. For example, the poplar genome contains as many as 10–12 PPO genes (I.T. Major, L. Tran and C.P. Constabel, unpublished data), but to date, only three PPO genes have been studied. *PtdPPO1* is exclusively expressed in damaged leaves, while *PtdPPO2* and *PtdPPO3* are predominantly expressed in stems, petioles, or roots. Both *PtdPPO1* and *PtdPPO2* are inducible, but in different tissues (Wang and Constabel 2004). Tissue-specific PPO expression has been most carefully studied in tomato, where seven PPO genes were characterized by Steffens and co-workers (Hunt et al. 1993; Steffens et al. 1994). In elegant work using promoter-GUS fusions, in situ hybridization, and immunolocalization, a highly complex cell- and tissue-specific pattern of expression was established (Thipyapong 1997; Thipyapong and Steffens 1997). Each PPO showed a distinctive expression profile, but at least one PPO gene showed constitutive expression for any given tissue. Interestingly, only the PPO-F gene was found to be herbivore-inducible. In addition to systemin and MeJA, pathogen and abiotic stress signals such as salicylic acid and ethylene were also shown to regulate PPO-F, consistent with additional roles of PPO in pathogen or other stress resistance. In general, diverse expression patterns of PPO in tomato and poplar in response to both developmental and stress signals indicate that PPO may have additional stress-related functions in different systems or situations.

12.4 Some Defensive Functions of PPO have been Demonstrated in Transgenic Tomato and Poplar Plants

In tomato and poplar, the induction and regulation of PPO in the herbivore defense response, in parallel with many confirmed defense genes, has provided indirect evidence for its role in defense ('guilt by association'). Nevertheless, it is possible that the herbivore-induced PPO contributes to wound healing and defense against opportunistic pathogens, rather than direct defense against insects. This question can best be addressed using transgenic plants. The availability of PPO cDNAs from a diversity of species has facilitated this approach in species susceptible to genetic transformation. Based on the commercial interest in the role of PPO in browning, several studies reported the successful anti-sense suppression of PPO in potato tuber and apple fruit (Bachem et al. 1994; Murata et al. 2001; Coetzer et al. 2001). Conversely, the overexpression of PPO in transgenic sugarcane resulted in darker juice (Vickers et al. 2005). However such PPO-modified plants were not used to address questions of biological functions or plant defense.

The defensive roles of PPO were first directly tested with PPO-overexpressing tomato plants, which showed fewer lesions and increased resistance to the bacterial pathogen *Pseudomonas syringae* pv *tomato* (Li and Steffens 2002), while antisense PPO-suppressed tomato showed greater susceptibility (more bacterial replication and more lesions; Thipyapong et al. 2004a). These plants also provide the strongest support to date for a defensive role of PPO against insect herbivores. Noctuid caterpillars *Heliothis armigera* and *Spodoptera litura* showed negative effects on growth when fed on PPO-overexpressing lines, and conversely, positive effects on growth when fed PPO-suppressed lines (P. Thipyapong, personal communication). In transgenic *Populus*, overexpression of the induced leaf PPO gene in low PPO poplar lines facilitated assays for roles of PPO against tree-feeding caterpillars. These studies have provided mixed results. First-instar caterpillars of *Malacosoma disstria* had decreased growth rates on elevated-PPO poplar (Wang and Constabel 2004), but only when experiments were performed in the fall, presumably a result of decreased caterpillar vigor. Similarly, fourth-instar caterpillars of *Lymantria dispar* had decreased growth rates on elevated-PPO poplar in the winter (Barbehenn et al. 2007). A second species of lymantriid caterpillar, *Orgyia leucostigma*, had decreased growth rates on elevated-PPO poplars in one experiment, but no negative effects were observed in another experiment (both in the winter). Here it appears that the varying and small effects of large increases in PPO activity in poplar (five to 40-fold increases) would make induced PPO alone ineffective as a defense against these tree-feeding caterpillars.

The differences in the apparent effectiveness of overexpressing PPO in tomato vs. poplar may be due to several factors, including the differential susceptibilities of test insects. The cell-specific localization of PPO may also be important; in tomato a significant proportion of overexpressed PPO is found in glandular trichomes (P. Thipyapong, personal communication). This may favor pre-ingestive PPO oxidation of phenolics and avoid the anoxic environment of the gut (see below). Similarly, a rapid oxidation of phenolics in tomato may be favored by the lack of latency for the tomato, but not the poplar enzyme (Constabel et al. 1995, 2000).

PPO-overexpressing transgenic poplars have been useful tools for probing other aspects of PPO, such as its stability after ingestion. Defense proteins are predicted to be relatively stable in the harsh conditions found in insect digestive systems, and the recovery of significant amounts of PPO in frass of forest tent caterpillar feeding on transgenic foliage is consistent with this expectation (Wang and Constabel 2004). Furthermore, PPO was activated by its passage through the insect gut, since unlike PPO extracted from leaves, PPO in frass extracts was fully active. Western blot analysis using PPO-specific antibodies showed that PPO in frass migrated at a lower molecular weight, indicating that proteolytic processing at a discrete site occurred in the gut (Wang and Constabel 2004). As mentioned above, PPOs are frequently observed to have a C-terminal proteolytic processing site (Marusek et al. 2006). The biological significance of PPO activation by gut enzymes is not clear, but it may impact its effectiveness.

12.5 PPO Activity Against Insects: Mechanisms of Action and Limitations

The many studies of PPO induction by herbivores attest to the general belief that PPOs play a key role in defense against herbivores. However, since the work of Felton et al. (1989, 1992), studies have rarely examined the mode of action of ingested PPOs. At least three mechanisms have been proposed by which PPO might affect insect herbivores: (1) PPO-generated quinones could alkylate essential amino acids, decreasing plant nutritional quality, (2) redox cycling of quinones may produce oxidative stress in the gut lumen, and (3) phenolic oxidation products, such as quinones and reactive oxygen species (hydrogen peroxide) generated by quinone redox cycling, could be absorbed and have toxic effects on herbivores. The work that has addressed these mechanisms in insect herbivores is summarized below.

As originally shown by Felton et al. (1989, 1992), PPO can directly reduce protein quality in vitro, when incubated with dietary protein and chlorogenic acid at ambient oxygen at pH 7.0. The alkylation of essential amino acids with quinones under these conditions significantly decreased noctuid caterpillar performance, and up to 50% of radiolabeled chlorogenic acid was found bound to protein in frass of noctuid caterpillars fed on tomato foliage (Felton et al. 1989). High pH environments, such as the lepidopteran midgut, favor protein alkylation, and several essential amino acids (lysine, histidine, cysteine, methionine) are particularly susceptible to quinone alkylation (Felton et al. 1992). Under optimal conditions for PPO activity, PPO clearly has an effect on protein nutritional quality. However, limiting factors in insect digestive systems may be low oxygen levels and the presence of antioxidants such as ascorbate or glutathione (see below). Another potential effect of PPO is elevated oxidative stress in the gut lumen, which we examined recently using PPO-overexpressing poplar foliage (Barbehenn et al. 2007). High PPO levels had little effect on observed levels of oxidized proteins or semiquinone radical production in two tree-feeding caterpillar species. Coating leaf disks with a commercial PPO (fungal tyrosinase) likewise produced no increase in semiquinone radical levels. These data suggest that there is little increase in quinone formation following ingestion of high levels of PPO in poplar, contrary to expectations based on the model of Felton et al. (1989, 1992) and previous results with forest tent caterpillars (Wang and Constabel 2004). By contrast, Thipyapong and coworkers have recently shown decreased growth rates and decreased nutritional indices of some noctuid caterpillars on PPO-overexpressing tomato lines (P. Thipyapong, personal communication), consistent with post-ingestive mechanism(s) of PPO activity. We are unaware of any work that has examined the potential effect of PPO on oxidative stress or toxicity at the tissue level in insects.

The activity of ingested PPO is dependant on the chemical environment of the insect gut, such as oxygen and phenolic substrate levels, reductants, inhibitors, and pH. Surprisingly little work has been done to determine how the physiological conditions present in insect gut fluids influence PPO and other defensive reactions. Phenolic substrates must be present for PPO to be effective, but unfortunately these are typically not analyzed and are assumed to be present at sufficient levels.

Likewise, molecular oxygen is an absolute requirement for PPO, and its activity is halted by purging oxygen from the reaction mixture (Duckworth and Coleman 1970; Fig. 12.1). Significantly, the gut contents of caterpillars and grasshoppers contain low steady-state concentrations of oxygen, and are sometimes anaerobic (Johnson and Barbehenn 2000). Oxygen drops from ambient levels of 150 mm Hg (21.0%) to less than 10 mm Hg (1.4%) over a distance of several mm into the foreguts of some caterpillars. The midgut oxygen levels of 0.1–0.5 mm Hg seen in many species would be expected to decrease PPO activities to less than 1% of maximal PPO rates at ambient oxygen levels (Fig. 12.1). Some caterpillar species have enlarged foreguts with much higher oxygen levels than the midgut. Nevertheless, in our work with one such species, *Lymantria dispar*, we found little evidence for a strong effect of PPO (Barbehenn et al. 2007). The limited oxygen availability in midguts of many insects should be kept in mind when using purified proteins or macerated leaf tissues to model chemical processes within herbivores. Where an impact of PPO on herbivores is found, limiting factors in the gut argue for a preingestive mode of action of PPO.

Although pH optima of PPOs are commonly broad, the reactivity of quinones with amino acids in an acidic medium is greatly reduced. Thus, Felton et al. (1992) concluded that PPOs would likely be ineffective against the Colorado potato beetle due to the low pH (5.5–6.5) of the beetle's midgut. Consistent with this conclusion, no significant decreases in the levels of four essential amino acids were found in the feces of *L. decemlineata* that fed on potato leaves from varieties with higher PPO levels (Castanera et al. 1996). Grasshopper gut pH is also acidic, but we are unaware of work on the effects of PPO in the Orthoptera. The high pH found in lepidopteran midguts (ca. pH 9–10) would be expected to decrease the activities of ingested PPO, but the basic conditions favor protein alkylation (Felton et al. 1989).

Extensive work by food scientists has identified many PPO inhibitors that reduce browning of processed foods. Among these inhibitors, ascorbate is ubiquitous in leaves and present at high levels, and would be co-ingested with PPO. When present in midgut fluid or in vitro reaction mixtures at 0.2–0.5 mM, ascorbate can chemically reduce quinones and semiquinone radicals, thereby limiting the effectiveness of PPO as an oxidative defense (Martinez-Cayuela et al. 1988; Janovitz-Klapp et al. 1990; Felton and Duffey 1992; Barbehenn et al. 2007). Levels of ascorbate

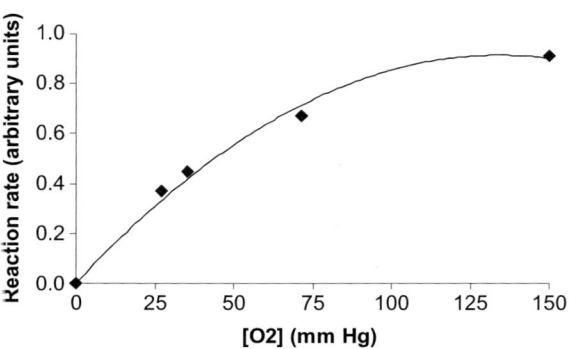

Fig. 12.1 Effect of molecular oxygen concentration on PPO reaction rate [modified from Duckworth and Coleman (1970)]. Reaction mixtures contained PPO (125 μg/ml) and pyrocatechol (0.25 mM) in 0.1 M sodium phosphate buffer (pH 7.0)

in the midgut fluids of tree-feeding caterpillars feeding on young trembling aspen, red oak, and hybrid poplar foliage are sufficiently high to prevent the net production of quinones by PPO (Barbehenn et al. 2003, 2007). Furthermore, an ascorbate recycling system is considered to be central to the antioxidant defenses of tree-feeding caterpillars (Barbehenn et al. 2001). However, it is not known whether the ascorbate ingested by tomato-feeding caterpillars, such as noctuids, is maintained in their gut fluids at sufficiently high levels to inhibit PPO.

Thiols such as glutathione and cysteine also decrease the net production of quinones by PPO and may inhibit the enzyme directly (Negishi and Ozawa 2000). In leaves, glutathione concentrations are roughly 5% as high as ascorbate levels, but can be equal to ascorbate levels in caterpillar midgut fluids (Barbehenn et al. 2001, 2003). Midgut glutathione at 50–100 μM in tree-feeding caterpillars is thus theoretically high enough to inhibit PPO activity (Barbehenn et al. 2003; Nagai and Suzuki 2003; Aydemir and Akkanh 2006), although apparently higher levels of cysteine are necessary to inhibit PPO from some plants (e.g., Janovitz-Klapp et al. 1990). Other potential PPO inhibitors reported include simple phenolics and quercetin (Walker and McCallion 1980; Le Bourvellec et al. 2004; Nerya et al. 2004).

12.6 PPO and Pathogen Defense

A role for PPO in defense against pathogens has been postulated from the earliest days of PPO research. This hypothesis has been supported by many correlative studies, such as the upregulation of PPO in pathogen-challenged plants (reviewed in Constabel et al. 1996; Mayer 2006). Pathogen-induced PPO activity continues to be reported for a variety of plant taxa, including monocots and dicots (e.g., Chen et al. 2000; Deborah et al. 2001). Similarly, studies describing correlations of high PPO levels in cultivars or lines with high pathogen resistance continue to provide support for a pathogen defense role of PPO (Raj et al. 2006). Several groups have also attempted to correlate the protective effects of rhizosphere bacteria with an induction of defense enzymes including PPO, with mixed success (Chen et al. 2000; Ramamoorthy et al. 2002).

Direct evidence for a role of PPO in inhibiting pathogen ingress or growth comes from transgenic tomato plants with enhanced or suppressed PPO levels. When challenged by the bacterial pathogen *Pseudomonas syringae* pv tomato, PPO-overexpressing plants showed reduced bacterial growth, whereas PPO antisense-suppressed lines supported greater bacterial numbers (Li and Steffens 2002; Thipyapong et al. 2004a). These studies are the only direct demonstrations to date of PPO's importance in pathogen defense. Whether such a function extends to other types of pathogens such as fungi remains to be tested. In poplar, infection with *Melampsora medusae* (a foliar rust pathogen) does not induce PPO, but represses its expression together with many other herbivore defense genes (Miranda et al. 2007).

The demonstration that PPO can inhibit bacterial plant pathogens suggests that, in addition to effects on some herbivores, PPO is important for inhibiting microbes introduced into damaged leaves via the mouthparts of feeding insects. These might be opportunistic bacteria or pathogens that are vectored by specific insects, although little is known about the microflora of caterpillar mouthparts and digestive systems. Thus, the distinction between pest and pathogen defenses may be artificial, and these may be seen as synergistic and complementary responses. This contrasts with the view that pest and pathogen defense responses are mutually exclusive. This hypothesis was based on early work on tomato showing that salicylic acid, a potent inducer of pathogen defense and systemic acquired resistance, inhibits the jasmonate-regulated herbivore defense response (Doares et al. 1995; Thaler et al. 2002). Current models of defense signaling outline a complex and overlapping set of responses, regulated by jasmonate, ethylene, and salicylic acid-based signals, leading to several possible outcomes (Devoto and Turner 2005). Functionally, crosstalk is supported by the observation of herbivore-induced resistance of *Arabidopsis* against microbial pathogens, including *P. syringae* (De Vos et al. 2006). In tomato, the PPO-F gene is induced during both the wound response and infection with the pathogen *P. syringae* (Thipyapong 1997, 2004a).

To date, no clear mechanism for the potential anti-pathogen effects of PPO has been demonstrated. Li and Steffens (2002) suggest several possibilities, including (1) general toxicity of PPO-generated quinones to pathogens and plant cells, accelerating cell death, (2) alkylation and reduced bioavailability of cellular proteins to the pathogen, (3) cross-linking of quinones with protein or other phenolics, forming a physical barrier to pathogens in the cell wall, and (4) quinone redox cycling leading to H_2O_2 and other reactive oxygen species (Jiang and Miles 1993). While reactive oxygen species are known to be important factors in plant pathogen interactions and defense signaling, and PPO is implicated in the formation of melanin-like polymers in potato blackspot lesions (Stevens et al. 1998), none of these hypotheses of how PPO might affect pathogens has been tested rigorously so far.

12.7 Conclusions and Future Directions

Although the transgenic approach has led to the direct demonstration of the efficacy of inducible PPO as a defense against some lepidopteran herbivores in some cases, the mechanisms of action against these attackers are still unclear. More detailed analyses of midgut chemistry in a greater variety of insects feeding on high-PPO foliage will begin to address this question. In particular, the extent of quinone binding to protein and amino acids in the gut contents of herbivores needs to be established. The interpretation of feeding studies would be enhanced by testing not only PPO levels and insect performance, but consumption rate and plant nutritional and chemical quality, so that deterrence or compensatory feeding can be detected. In addition, our knowledge of PPO effects would benefit from greater attention to endogenous PPO substrates, rather than simply extrapolating from substrate specificities observed in vitro. In few cases have the probable substrates been identified, yet the biochemical reactivities of PPO-produced quinones are dependent on the specific structure (Jiang and Miles 1993).

Much recent work on PPO has emphasized potential roles in defense. Nevertheless, alternate roles for this enzyme are likely and have been demonstrated. For example, antisense PPO tomato plants have enhanced pathogen resistance and drought tolerance (Thipyapong et al. 2004b), and PPO-like enzymes can act as hydroxylases in secondary metabolism (Steiner et al. 1999; Nakayama et al. 2000; Cho et al. 2003). Given the tremendous variation in PPO expression patterns, activity levels, and potential substrates in different species, similar variation in the adaptive roles played by PPO in defense and other processes may be anticipated.

Acknowledgments The authors would like to thank Piyada Thipyapong for providing unpublished data. Grant support for work in the authors' laboratories is from the Natural Sciences and Engineering Research Council (NSERC) of Canada (CPC), the University of Victoria's Centre for Forest Biology (CPC), and the National Research Initiative of the USDA Cooperative State Research, Education and Extension Service, grant number 2004-35302-14840 to RVB and CPC.

References

Alba-Meraz A, Choe HT (2002) Systemic effects on oxidative enzymes in *Phaseolus vulgaris* leaves that have been wounded by the grasshopper *Melanoplus differentialis* (Thomas) or have had a foliar application of jasmonic acid. Int J Plant Sci 163:317–328

Aydemir T, Akkanh G (2006) Partial purification and characterization of polyphenol oxidase from celery root (*Apium graveolens* L.) and the investigation of the effects on the enzyme activity of some inhibitors. Int J Food Sci Tech 41:1090–1098

Bachem CWB, Speckmann GJ, Vanderlinde PCG, Verheggen FTM, Hunt MD, Steffens JC, Zabeau M (1994) Antisense expression of polyphenol oxidase genes inhibits enzymatic browning in potato tubers. Biotechnology 12:1101–1105

Barbehenn RV, Bumgarner SL, Roosen E, Martin MM (2001) Antioxidant defenses in caterpillars: role of the ascorbate recycling system in the midgut lumen. J Insect Physiol 47:349–357

Barbehenn RV, Walker AC, Uddin F (2003) Antioxidants in the midgut fluids of a tannin-tolerant and a tannin-sensitive caterpillar: effects of seasonal changes in tree leaves. J Chem Ecol 29:1099–1116

Barbehenn RV, Jones CP, Yip L, Tran L, Constabel CP (2007) Does the induction of polyphenol oxidase defend trees against caterpillars? Assessing defenses one at a time with transgenic poplar. Oecologia 154:129–400

Bergey DR, Howe GA, Ryan CA (1996) Polypeptide signaling for plant defensive genes exhibits analogies to defense signaling in animals. Proc Natl Acad Sci USA 93:12053–12058

Bi JL, Felton GW (1995) Foliar oxidative stress and insect herbivory: primary compounds, secondary metabolites, and reactive oxygen species as components of induced resistance. J Chem Ecol 21:1511–1530

Bi JL, Murphy JB, Felton GW (1997) Antinutritive and oxidative components as mechanisms of induced resistance in cotton to *Helicoverpa zea*. J Chem Ecol 23:97–117

Castanera P, Steffens JC, Tingey WM (1996) Biological performance of Colorado potato beetle larvae on potato genotypes with differing levels of polyphenol oxidase. J Chem Ecol 22:91–101

Chakraborty U, Chakraborty N (2005) Impact of environmental factors on infestation of tea leaves by *Helopeltis theivora*, and associated changes in flavonoid flavor components and enzyme activities. Phytoparasitica 33:88–96

Chen C, Belanger RR, Benhamou N, Paulitz TC (2000) Defense enzymes induced in cucumber roots by treatment with plant growth-promoting rhizobacteria (PGPR) and *Pythium aphanidermatum*. Physiol Mol Plant Path 56:13–23

Chen H, Wilkerson CG, Kuchar JA, Phinney BS, Howe GA (2005) Jasmonate-inducible plant enzymes degrade essential amino acids in the herbivore midgut. Proc Natl Acad Sci USA 102:19237–19242

Cho MH, Moinuddin SGA, Helms GL, Hishiyama S, Eichinger D, Davin LB, Lewis NG (2003) (+)-Larreatricin hydroxylase, an enantio-specific polyphenol oxidase from the creosote bush (*Larrea tridentata*). Proc Natl Acad Sci USA 100:10641–10646

Christopher ME, Miranda M, Major IT, Constabel CP (2004) Gene expression profiling of systemically wound-induced defenses in hybrid poplar. Planta 219:936–947

Cipollini DF, Redman AM (1999) Age-dependent effects of jasmonic acid treatment and wind exposure on foliar oxidase activity and insect resistance in tomato. J Chem Ecol 25:271–281

Cipollini ML, Paulk E, Cipollini DF (2002) Effect of nitrogen and water treatment on leaf chemistry in horsenettle (*Solanum carolinense*), and relationship to herbivory by flea beetles (*Epitrix* spp.) and tobacco hornworm (*Manduca sexta*). J Chem Ecol 28:2377–2398

Clausen TP, Reichardt PB, Bryant JP, Werner RA, Post K, Frisby K (1989) Chemical model for short-term induction in quaking aspen *(Populus tremuloides)* foliage against herbivores. J Chem Ecol 15:2335–2346

Coetzer C, Corsini D, Love S, Pavek J, Tumer N (2001) Control of enzymatic browning in potato (*Solanum tuberosum* L.) by sense and antisense RNA from tomato polyphenol oxidase. J Agric Food Chem 49:652–657

Constabel CP, Ryan CA (1998) A survey of wound- and methyl jasmonate-induced leaf polyphenol oxidase in crop plants. Phytochemistry 47:507–511

Constabel CP, Bergey DR, Ryan CA (1996) Polyphenol oxidase as a component of the inducible defense response in tomato against herbivores. In: Romeo JT, Saunders JA, Barbosa P (eds) Phytochemical diversity and redundancy in ecological interactions. Plenum Press, New York, pp 231–252

Constabel CP, Bergey DR, Ryan CA (1995) Systemin activates synthesis of wound-inducible tomato leaf polyphenol oxidase via the octadecanoid defense signaling pathway. Proc Natl Acad Sci USA 92:407–411

Constabel CP, Yip L, Patton JJ, Christopher ME (2000) Polyphenol oxidase from hybrid poplar. Cloning and expression in response to wounding and herbivory. Plant Physiol 124:285–295

De Vos M, Van Zaanen W, Koornneef A, Korzelius JP, Dicke M, Van Loon LC, Pieterse CMJ (2006) Herbivore-induced resistance against microbial pathogens in *Arabidopsis*. Plant Physiol 142:352–363

Deborah SD, Palaniswami A. Vidhyasekaran P, Velazhahan R (2001) Time-course study of the induction of defense enzymes, phenolics and lignin in rice in response to infection by pathogen and non-pathogen. J Plant Dis Prot 108:204–216

Devoto A, Turner JG (2005) Jasmonate-regulated *Arabidopsis* stress signalling network. Physiol Plant 123:161–172

Doares SH, Narváez-Vásquez J, Conconi A, Ryan CA (1995) Salicylic acid inhibits synthesis of proteinase inhibitors in tomato leaves induced by systemin and jasmonic acid. Plant Physiol 108:1741–1746

Duckworth HW, Coleman JE (1970) Physicochemical and kinetic properties of mushroom tyrosinase. J Biol Chem 245:1613–1625

Felton GW, Donato K, Delvecchio RJ, Duffey SS (1989) Activation of plant foliar oxidases by insect feeding reduces nutritive quality of foliage for noctuid herbivores. J Chem Ecol 15:2667–2694

Felton GW, Duffey SS (1992) Avoidance of antinutritive plant defense: role of midgut pH in Colorado potato beetle. J Chem Ecol 18:571–583

Felton GW, Donato KK, Broadway RM, Duffey SS (1992) Impact of oxidized plant phenolics on the nutritional quality of dietary protein to a noctuid herbivore, *Spodoptera exigua*. J Insect Physiol 38:277–285

Felton GW, Summers CB, Mueller AJ (1994) Oxidative responses in soybean foliage to herbivory by bean leaf beetle and three-cornered alfalfa hopper. J Chem Ecol 20:639–650

Gandia-Herrero F, Jimenez-Atienzar M, Cabanes J, Garcia-Carmona F, Escribano J (2005) Evidence for a common regulation in the activation of a polyphenol oxidase by trypsin and sodium dodecyl sulfate. Biol Chem 386:601–607

Guyot S, Vercauteren J, Cheynier V (1996) Structural determination of colourless and yellow dimers resulting from (+)-catechin coupling catalysed by grape polyphenoloxidase. Phytochemistry 42:1279–1288

Haruta M, Pedersen JA, Constabel CP (2001) Polyphenol oxidase and herbivore defense in trembling aspen (*Populus tremuloides*): cDNA cloning, expression, and potential substrates. Physiol Plant 112:552–558

Heng-Moss T, Sarath G, Baxendale F, Novak D, Bose S, Ni XH, Quisenberry S (2004) Characterization of oxidative enzyme changes in buffalograsses challenged by *Blissus occiduus*. J Econ Entomol 97:1086–1095

Hermsmeier D, Schittko U, Baldwin IT (2001) Molecular interactions between the specialist herbivore *Manduca sexta* (Lepidoptera, Sphingidae) and its natural host *Nicotiana attenuata*. I. Large-scale changes in the accumulation of growth- and defense-related plant mRNAs. Plant Physiol 125:683–700

Humphreys JM, Chapple C (2002) Rewriting the lignin roadmap. Curr Op Plant Biol 5:224–229

Hunt MD, Eannetta NT, Yu HF, Newman SM, Steffens JC (1993) cDNA cloning and expression of potato polyphenol oxidase. Plant Mol Biol 21:59–68

Janovitz-Klapp AH, Richard FC, Goupy PM, Nicolas JJ (1990) Inhibition studies on apple polyphenol oxidase. J Agric Food Chem 38:926–931

Jiang Y, Miles PW (1993) Generation of H_2O_2 during enzymatic oxidation of catechin. Phytochemistry 33:29–34

Jimenez M, Garcia-Carmona F (1999) Myricetin, an antioxidant flavonol, is a substrate of polyphenol oxidase. J Sci Food Agric 79:1993–2000

Johnson KS, Barbehenn RV (2000) Oxygen levels in the gut lumens of herbivorous insects. J Insect Physiol 46:897–903

Kanade SR, Paul B, Rao AGA, Gowda LR (2006) The conformational state of polyphenol oxidase from field bean (*Dolichlos lablah*) upon SDS and acid-pH activation. Biochem J 395:551–562

Karban R, Baldwin IT, Baxter KJ, Laue G, Felton GW (2000) Communication between plants: induced resistance in wild tobacco plants following clipping of neighboring sagebrush. Oecologia 125:66–71

Kessler A, Baldwin IT (2002) *Manduca quinquemaculata*'s optimization of intra-plant oviposition to predation, food quality, and thermal constraints. Ecology 83:2346–2354

Kowalski SP, Eannetta NT, Hirzel AT, Steffens JC (1992) Purification and characterization of polyphenol oxidase from glandular trichomes of *Solanum berthaultii*. Plant Physiol 100: 677–684

Kruzmane D, Jankevica L, Ievinsh G (2002) Effect of regurgitant from *Leptinotarsa decemlineata* on wound responses in *Solanum tuberosum* and *Phaseolus vulgaris*. Physiol Plant 115:577–584

Lawrence SD, Novak N (2001) A rapid method for the production and characterization of recombinant insecticidal proteins in plants. Molec Breeding 8:139–146

Lawrence SD, Novak NG (2006) Expression of poplar chitinase in tomato leads to inhibition of development in colorado potato beetle. Biotech Lett 28:593–599

Le Bourvellec C, Le Quere J-M, Sanoner P, Drilleau J-F, Guyot S (2004) Inhibition of apple polyphenol oxidase by procyanidins and polyphenol oxidation products. J Agric Food Chem 52:122–130

Li L, Steffens JC (2002) Overexpression of polyphenol oxidase in transgenic tomato plants results in enhanced bacterial disease resistance. Planta 215:239–247

Major IT, Constabel CP (2006) Molecular analysis of poplar defense against herbivory. Comparison of wound- and insect elicitor-induced gene expression. New Phytol 172:617–635

Martinez-Cayuela M, Faus MJ, Gil A (1988) Effects of some reductants on the activity of cherimoya polyphenol oxidase. Phytochemistry 27:1589–1592

Marusek CM, Trobaugh NM, Flurkey WH, Inlow JK (2006) Comparative analysis of polyphenol oxidase from plant and fungal species. J Inorg Biochem 100:108–123

Mayer AM (2006) Polyphenol oxidases in plants and fungi: going places? A review. Phytochemistry 67:2318–2331

Melo GA, Shimizu MM, Mazzafera P (2006) Polyphenoloxidase activity in coffee leaves and its role in resistance against coffee leaf miner and coffee leaf rust. Phytochemistry 67:277–285

Miranda M, Ralph SG, Mellway R, White R, Heath MC, Bohlmann J, Constabel CP (2007) The transcriptional response of hybrid poplar (*Populus trichocarpa* x *P. deltoides*) to infection by *Melampsora medusae* leaf rust involves induction of flavonoid pathway genes leading to the accumulation of proanthocyanidins. Molec Plant Microbe Interact 20:816–831

Murata M, Nishimura M, Murai N, Haruta M, Homma S, Itoh Y (2001) A transgenic apple callus showing reduced polyphenol oxidase activity and lower browning potential. Biosc Biotech Biochem 65:383–388

Nagai T, Suzuki N (2003) Polyphenol oxidase from bean sprouts (*Glycine max* L.). J Food Sci 68:16–20

Nakayama T, Yonekura-Sakakibara K, Sato T, Kikuchi S, Fukui Y, Fukuchi-Mizutani M, Ueda T, Nakao M, Tanaka Y, Kusumi T, Nishino T (2000) Aureusidin synthase: a polyphenol oxidase homolog responsible for flower coloration. Science 290:1163–1166

Negishi O, Ozawa T (2000). Inhibition of enzymatic browning and protection of sulfhydryl enzymes by thiol compounds. Phytochemistry 54:481–487

Nerya O, Musa R, Khatib S, Tamir S, Vaya J (2004) Chalcones as potent tyrosinase inhibitors: the effect of hydroxyl positions and numbers. Phytochemistry 65:1389–1395

Ni X, Quisenberry SS, Heng-Moss T, Markwell J, Sarath G, Klucas R, Baxendale F (2001) Oxidative responses of resistant and susceptible cereal leaves to symptomatic and nonsymptomatic cereal aphid (Hemiptera: Aphididae) feeding. J Econ Entomol 94:743–751

Parsons TJ, Bradshaw HD, Gordon MP (1989). Systemic accumulation of specific mRNAs in response to wounding in poplar trees. Proc Natl Acad Sci USA 86:7895–7899

Raj SN, Sarosh BR, Shetty HS (2006) Induction and accumulation of polyphenol oxidase activities as implicated in development of resistance against pearl millet downy mildew disease. Funct Plant Biol 33:563–571

Ralph S, Oddy C, Cooper D et al. (2006) Genomics of hybrid poplar (*Populus trichocarpa* x *deltoides*) interacting with forest tent caterpillars (*Malacosoma disstria*): normalized and full-length cDNA libraries, expressed sequence tags, and cDNA microarray for the study of insect-induced defences in poplar. Mol Ecol 15:1275–1297

Ramamoorthy V, Raguchander T, Samiyappan R (2002) Induction of defense-related proteins in tomato roots treated with *Pseudomonas fluorescens* Pf1 and *Fusarium oxysporum* f. sp lycopersici. Plant Soil 239:55–68

Rathjen AH, Robinson SP (1992) Aberrant processing of polyphenol oxidase in a variegated grapevine mutant. Plant Physiol 99:1619–1625

Redman AM, Cipollini DF, Schultz JC (2001) Fitness costs of jasmonic acid-induced defense in tomato, *Lycopersicon esculentum*. Oecologia 126:380–385

Ren F, Lu YT (2006) Overexpression of tobacco hydroxyproline-rich glycopeptide systemin precursor A gene in transgenic tobacco enhances resistance against *Helicoverpa armigera* larvae. Plant Sci 171:286–292

Ruuhola T, Yang S (2006) Wound-induced oxidative responses in mountain birch leaves. Ann Bot 97:29–37

Ryan CA (2000) The systemin signaling pathway: differential activation of plant defensive genes. Biochim Biophys Acta 1477:112–121

Schilmiller AL, Howe GA (2005) Systemic signaling in the wound response. Curr Opin Plant Biol 8:369–377

Shin R, Froderman T, Flurkey WH (1997) Isolation and characterization of a mung bean leaf polyphenol oxidase. Phytochemistry 45:15–21

Steffens JC, Harel E, Hunt MD (1994) Polyphenol oxidase. In: Ellis BE, Kuroki GW, Stafford HA (eds) Genetic engineering of plant secondary metabolism. Plenum Press, New York, pp 276–304

Steiner U, Schliemann W, Bohm H, Strack D (1999) Tyrosinase involved in betalain biosynthesis of higher plants. Planta 208:114–124

Stevens LH, Davelaar E, Kolb RM, Pennings EJM, Smit NPM (1998) Tyrosine and cysteine are substrates for blackspot synthesis in potato. Phytochemistry 49:703–707

Stout MJ, Workman KV, Bostock RM, Duffey SS (1998) Specificity of induced resistance in the tomato, *Lycopersicon esculentum*. Oecologia 113:74–81

Thaler JS, Stout MJ, Karban R, Duffey SS (1996) Exogenous jasmonates simulate insect wounding in tomato plants (*Lycopersicon esculentum*) in the laboratory and field. J Chem Ecol 22:1767–1781

Thaler JS, Fidantsef AL, Duffey SS, Bostock RM (1999) Trade-offs in plant defense against pathogens and herbivores: a field demonstration of chemical elicitors of induced resistance. J Chem Ecol 25:1597–1609

Thaler JS, Stout MJ, Karban R, Duffey SS (2001) Jasmonate-mediated induced plant resistance affects a community of herbivores. Ecol Entomol 26:312–324

Thaler JS, Karban R, Ullman DE, Boege K, Bostock RM (2002) Cross-talk between jasmonate and salicylate plant defense pathways: effects on several plant parasites. Oecologia 131:227–235

Thipyapong P, Steffens JC (1997) Tomato polyphenol oxidase – Differential response of the polyphenol oxidase F promoter to injuries and wound signals. Plant Physiol 115:409–418

Thipyapong P, Joel DM, Steffens JC (1997) Differential expression and turnover of the tomato polyphenol oxidase gene family during vegetative and reproductive development. Plant Physiol 113:707–718

Thipyapong P, Hunt MD, Steffens JC (2004a) Antisense downregulation of polyphenol oxidase results in enhanced disease susceptibility. Planta 220:105–117

Thipyapong P, Melkonian J, Wolfe DW, Steffens JC (2004b) Suppression of polyphenol oxidases increases stress tolerance in tomato. Plant Sci 167:693–703

Tscharntke T, Thiessen S, Dolch R, Boland W (2001) Herbivory, induced resistance, and interplant signal transfer in *Alnus glutinosa*. Biochem Syst Ecol 29:1025–1047

van Gelder CWG, Flurkey WH, Wichers HJ (1997) Sequence and structural features of plant and fungal tyrosinases. Phytochemistry 45:1309–1323

Vickers JE, Grof CPL, Bonnett GD, Jackson PA, Knight DP, Roberts SE, Robinson SP (2005) Overexpression of polyphenol oxidase in transgenic sugarcane results in darker juice and raw sugar. Crop Sci 45:354–362

Walker JRL, McCallion RF (1980) Selective inhibition of ortho-diphenol and para-diphenol oxidases. Phytochemistry 19:373–377

Wang JH, Constabel CP (2004) Polyphenol oxidase overexpression in transgenic *Populus* enhances resistance to herbivory by forest tent caterpillar (*Malacosoma disstria*). Planta 220:87–96

Wuyts N, De Waele D, Swennen R (2006) Extraction and partial characterization of polyphenol oxidase from banana (*Musa acuminata* Grande naine) roots. Plant Physiol Biochem 44:308–314

Yoruk R, Marshall MR (2003) Physicochemical properties and function of plant polyphenol oxidase: a review. J Food Biochem 27:361–422

Chapter 13
Action of Plant Defensive Enzymes in the Insect Midgut

Hui Chen, Eliana Gonzales-Vigil and Gregg A. Howe

Insect feeding activates the expression of host plant defensive proteins that exert direct effects on the attacker. In addition to the well-studied proteinase inhibitors, the plant's defensive protein arsenal includes enzymes that disrupt various aspects of insect digestive physiology. In this chapter, we summarize recent studies on isoforms of arginase and threonine deaminase (TD) that degrade the essential amino acids arginine and threonine, respectively, in the alkaline environment of the lepidopteran gut. We also discuss a vegetative storage protein (VSP2) whose phosphatase activity is responsible for potent insecticidal effects on dipteran and coleopteran insects. A common feature of VSP2, TD, and arginase is their wound-inducible expression via the jasmonate signaling pathway. Jasmonate-regulated defensive enzymes may have evolved from pre-existing housekeeping enzymes that catabolize essential nutrients during normal plant development. The application of proteomics-based approaches to identify plant proteins in the digestive tract of phytophagous insects is facilitating the discovery of novel plant proteins with insecticidal activity.

13.1 Introduction

The optimal growth of insect herbivores depends on their ability to acquire amino acids from dietary protein. Like other animals, leaf-eating insects require the essential amino acids arginine, threonine, isoleucine, leucine, lysine, histidine, methionine, phenylalanine, tryptophan, and valine (Chang 2004). The low protein content of plant tissue poses a major challenge to insects that require these nutrients, as protein is both the major macronutrient and the most commonly limiting nutrient for insect growth (Mattson 1980; Bernays and Chapman 1994). A large body of evidence indicates that many plant defensive compounds impair herbivore performance by restricting the availability of amino acids or other plant-derived nutrients. Much of this literature concerns plant secondary metabolites such as tannins and phenolic

---

G.A. Howe
DOE Plant Research Laboratory, Michigan State University, East Lansing, MI 48824, USA
e-mail: howeg@msu.edu

resins that interfere with the digestibility of dietary protein (Feeny 1976; Rhoades and Cates 1976; Berenbaum 1995).

Higher plants also produce a variety of proteins that disrupt amino acid acquisition and other aspects of insect digestive physiology (Felton 1996). Proteinase inhibitors (PIs) initially described by Ryan and colleagues are perhaps the best example of this type of post-ingestive defense (Ryan 1990; Jongsma and Beekwilder this volume). Upon consumption of plant tissue by the herbivore, PIs bind to and inhibit digestive proteases in the gut lumen. The negative effect of PIs on herbivore growth results from a compensatory response by the insect to overproduce digestive proteinases, which, in turn, depletes pools of essential amino acids (Broadway and Duffey 1986). Because PIs are not catalytic, their ability to thwart insect growth is dependent on accumulation to relatively high concentrations in the digestive system. Plant enzymes have the potential to exert defensive effects at much lower concentrations. This idea has received relatively little attention until recently (Duffey and Stout 1996; Felton 1996; Chen et al. 2005; Felton 2005). Research on plant enzymes that serve a post-ingestive role in defense has focused mainly on polyphenol oxidase and other oxidative enzymes that covalently modify dietary protein and thus impair the digestibility of plant food (Constabel et al. 1995; Duffey and Stout 1996; Felton 1996; Wang and Constabel 2004; Constabel and Barbehenn this volume). Other defensive enzymes, including cysteine proteases, target structural components of the insect digestive apparatus (Pechan et al. 2002; Konno et al. 2004; Mohan et al. 2006). The collective evidence thus indicates that plant enzymes play a pivotal role in anti-insect defense, and broadens the view that secondary metabolites are the major determinant of host plant selection by insects (Fraenkel 1959; Berenbaum 1995).

Plant proteins that serve a post-ingestive role in defense are typically synthesized in response to wounding and herbivore attack. This central tenet of induced resistance was established 35 years ago by Green and Ryan's pioneering work on serine PIs in potato and tomato (Green and Ryan 1972). Induced expression of PIs and other defense-related proteins is controlled in large part by the jasmonate signaling pathway (Walling 2000; Kessler and Baldwin 2002; Gfeller and Farmer 2004; Howe 2004; Schilmiller and Howe 2005). Recent research on jasmonate-signaled processes indicates that enzymes previously thought to be involved in primary metabolism serve an important role in post-ingestive defense. In this chapter, we summarize recent findings on two jasmonate-regulated enzymes, arginase and threonine deaminase (TD), which have the capacity to degrade essential amino acids in the lepidopteran gut. Recent research concerning the role of jasmonate-regulated vegetative storage proteins (VSPs) in plant defense against insect herbivores is also discussed.

13.2 Arginase

L-arginine (Arg) is one of the most functionally diverse amino acids in living cells. In addition to being a building block for proteins, Arg is a precursor for the biosynthesis of polyamines, agmatine, and proline, as well as the cell-signaling molecules glutamate, γ-aminobutyric acid, and nitric oxide (NO; Fig. 13.1). Two of the most

13 Defensive Enzymes in the Insect Midgut

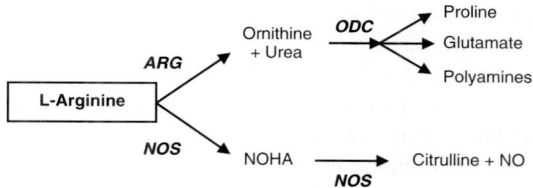

Fig. 13.1 Arginine metabolism in higher plants. Arginine is a substrate for both arginase (ARG) and nitric oxide synthase (NOS). NOS catalyzes the synthesis of nitric oxide (NO) by a two-step reaction involving an N-hydroxy-arginine (NOHA) intermediate. Metabolism of arginine via the arginase-ornithine decarboxylase (ODC) pathway gives rise to proline, glutamate, and simple polyamines. Arginine is also metabolized by arginine decarboxylase to agmatine (not shown), which also is used as a precursor for polyamine synthesis

intensively studied pathways of Arg metabolism are those involving arginase and NO synthase (NOS). Arginase (L-arginine amidino hydrolase, EC 3.5.3.1) is a binuclear manganese metalloenzyme that hydrolyzes Arg to urea and ornithine, the latter of which is a precursor for polyamine biosynthesis. In animal systems, increased arginase expression stimulates the production of polyamines that promote cell proliferation and wound healing (Satriano 2004). In contrast to the growth-promoting effects of polyamines are the cytostatic effects of NO produced by activated macrophages. The switch between the arginase and NOS branches of Arg metabolism is controlled by various signaling and metabolic pathways that regulate Arg availability (Mori and Gotoh 2004; Satriano 2004). Because arginase and NOS compete for a common substrate (Fig. 13.1), increased arginase expression can effectively inhibit the NOS pathway. Many animal pathogens, for example, induce arginase expression as a means of evading NO-mediated host defenses (Vincendeau et al. 2003).

Very little is known about the physiological function of plant arginase. A prerequisite for addressing this question is the unambiguous identification of plant genes that encode arginase. Because the predicted sequences of plant arginase are more similar to agmatinase than to authentic arginases from vertebrates, fungi, and bacteria, some workers have suggested that plant genes annotated as arginase may encode agmatinase or a related amidinohydrolase (Perozich et al. 1998; Sekowska et al. 2000). Recent work on arginases from cultivated tomato (*Solanum lycopersicum*) helped to resolve this issue. The tomato genome contains two arginase genes designated *ARG1* and *ARG2* (Chen et al. 2004). Both genes are predicted to encode 338-amino-acid proteins, which are 89% identical to each other. Heterologous expression studies showed that both enzymes have very similar biochemical properties, including high specificity for Arg. The apparent K_m of Arg is ~30 mM for both isozymes, which is consistent with K_m values reported for arginases purified from various plant tissues. These studies demonstrate that tomato ARG1 and 2, despite their sequence similarity to agmatinase and phylogenetic distinction from non-plant arginases, are genuine arginases (Chen et al. 2004).

Additional insight into the function of plant arginase was obtained by analysis of the enzyme's tissue-specific expression pattern. Tomato *ARG1* and *ARG2* genes, for example, are expressed to their highest levels in reproductive tissues (Chen et al. 2004). This observation agrees with previous studies showing that tomato ovaries and immature fruit contain relatively high levels of arginase activity (Alabadi et al. 1996). Research on soybean indicates that arginase may be involved in mobilizing Arg during early seedling germination (Goldraij and Polacco 1999, 2000). The first clue that arginase may have a role in plant defense came from the observation that expression of tomato *ARG2* (but not *ARG1*) is massively induced in foliar tissue in response to mechanical wounding and methyl jasmonate treatment (Chen et al. 2004). Induced expression of *ARG2* was accompanied by increases in arginase activity, and was abolished in the *jasmonic acid insensitive1* (*jai1*) mutant that is defective in the jasmonate signaling pathway (Li et al. 2004). The absence of detectable ARG2 expression in *jai1* plants suggests that this isoform is not essential for normal plant growth and development.

An unusual feature of plant arginase is its ability to metabolize Arg at high pH. The pH optimum of tomato ARG2 is ~9.5, with little or no activity detected at pH 7.0 (Fig. 13.2). The dependency of ARG2 activity on high pH led to the idea that the enzyme, following its wound-induced accumulation in tomato leaves, acts in the alkaline midgut of lepidopteran insects that feed on tomato foliage; by depleting the pool of Arg available for uptake into the intestine, arginase may reduce the nutritional quality of the plant food. This hypothesis is consistent with the proposal by Broadway and Duffey (1988) that reduced availability of basic amino acids (i.e., Arg and lysine) increases the toxicity of PIs. Several lines of evidence support a role for arginase in post-ingestive defense. First, *jai1* mutant plants that fail to express ARG2 are compromised in defense against herbivore attack (Li et al. 2004; Chen et al. 2005). Second, the midgut content of tobacco hornworm (*Manduca sexta*) larvae reared on foliage from induced tomato plants contained high levels of ARG2 activity (Chen et al. 2005). Finally, increased arginase activity in midgut extracts was associated with reduced levels of free Arg in the same extract (Chen

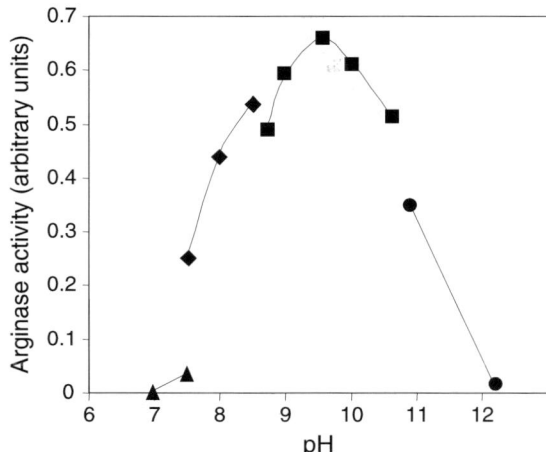

Fig. 13.2 pH optimum of tomato arginase2. A hexahistidine-tagged derivative of tomato arginase2 was expressed in *E. coli* and purified by nickel-affinity chromatography (Chen et al. 2004). The recombinant enzyme was assayed for arginase activity in buffer systems (K-PO$_4$, triangles; Tris, diamonds; glycine, squares; K-PO$_4$, circles) adjusted to the indicated pH

et al. 2005). These results indicate that JA-regulated ARG2 in tomato leaves is active in the *M. sexta* digestive tract, where it catabolizes Arg.

A transgenic approach was used to determine whether increased expression of foliar arginase is sufficient to reduce caterpillar performance (Chen et al. 2005). Transgenic tomato plants that overexpress the *ARG2* cDNA under the control of the cauliflower mosaic virus (CaMV) 35S promoter were generated by *Agrobacterium*-mediated transformation. The constitutive level of arginase activity in unwounded leaves of selected *35S::ARG2* lines far exceeded that in herbivore-damaged wild-type leaves. High arginase activity in these plants did not result in obvious morphological or reproductive phenotypes, nor did it significantly alter the level of free Arg in *35S::ARG2* leaves (Chen and Howe unpublished data). Feeding trials conducted with two independent *35S::ARG2* lines showed that the average weight of larvae grown on transgenic plants was significantly less than that of larvae reared on wild-type plants. These bioassays also showed that larvae consumed more foliage from wild-type plants than from *35S::ARG2* plants. Arginase activity in midgut extracts from *35S::ARG2*-reared larvae was significantly greater than that in wild-type-reared larvae, and this activity was associated with reduced levels of midgut Arg. Thus, ingestion of foliar arginase by *M. sexta* larvae results in the depletion of midgut Arg and reduced larval growth (Chen et al. 2005).

To test the idea that reduced growth of *M. sexta* caterpillars on *35S::ARG2* plants is a direct consequence of Arg deficiency in the diet, feeding trials were performed with plants that were sprayed with a solution of 10 mM Arg (Fig. 13.3). This treatment did not affect larval performance on wild-type plants, suggesting that free Arg is not a limiting factor for insect growth on this host. However, significant improvement of larval growth was observed on Arg-supplemented *35S::ARG2* plants

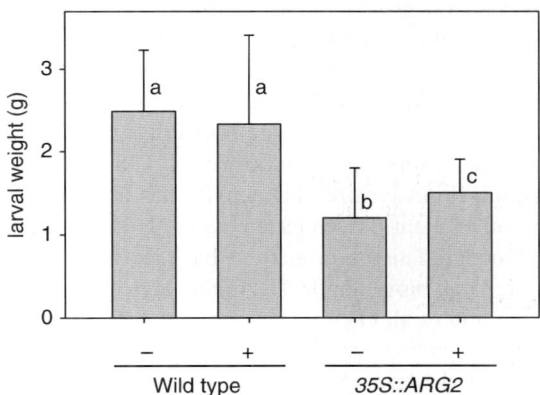

Fig. 13.3 Exogenous arginine partially rescues the reduced growth of *M. sexta* larvae reared on transgenic *35S::ARG2* plants. Plants of the indicated genotype were sprayed with 10 mM arginine (+) or a mock control (−) prior to challenge with newly hatched larvae. Larval weights were determined 10 days after the start of the feeding trial. Data show the mean and SD of at least 23 larvae per host genotype. Lower case letters denote significant differences at $p<0.05$ (Student's *t*-test)

compared to the same genotype treated with a mock control. The ability of exogenous Arg to partially rescue the reduced growth of *M. sexta* larvae on *35S::ARG2* plants supports the hypothesis that Arg is a limiting nutrient for insect growth on this line. These results are also consistent with the notion that reduced Arg availability in the diet increases the toxicity of PIs (Broadway and Duffey 1998). Because larvae reared on Arg-supplemented transgenic plants weighed significantly less than insects grown on wild-type plants (Fig. 13.3), it is possible that the defensive role of arginase involves mechanisms in addition to Arg depletion.

13.3 Threonine Deaminase

Threonine deaminase (EC 4.2.1.16), also known as threonine dehydratase or threonine ammonia-lyase, is a pyridoxal phosphate-dependent enzyme that catalyzes the dehydratation of L-threonine (Thr) to α-ketobutyrate and ammonia. Although TD can also use L-serine as a substrate to produce pyruvate and ammonia, Thr is generally regarded as the enzyme's preferred substrate. Many microorganisms possess distinct biosynthetic and biodegradative forms of TD. The biosynthetic enzyme catalyzes the committed step in the synthesis of Ile and is feedback inhibited by Ile. Biosynthetic TDs from bacteria have been studied extensively as a model for allosteric inhibition (Umbarger 1956). Biodegradative TDs are involved in anaerobic breakdown of Thr to propionate and are insensitive to Ile. Higher plants possess a biosynthetic-type TD whose structure, function, and regulation by Ile is very similar to that of bacterial TDs (Mourad and King 1995; Halgand et al. 2002). This conclusion has been confirmed by molecular cloning and analysis of TD-encoding genes from several plants, including potato (Hildmann et al. 1992), tomato (Samach et al. 1991), chickpea (John et al. 1995), *Arabidopsis* (Mourad and King 1995; Wessel et al. 2000), and the native tobacco *Nicotiana attenuata* (Hermsmeier et al. 2001). The physiological importance of biosynthetic TD in plant growth and development was demonstrated by studies of TD-deficient mutants of *N. plumbaginifolia* and *N. attenuata* (Sidorov et al. 1981; Colau et al. 1987; Kang et al. 2006). Although Ile-insensitive TD activity has been described in senescing tomato leaves and other plant tissues (Szamosi et al. 1993, and citations therein), genes encoding a biodegradative-type enzyme have not been identified in plants.

The first indication that biosynthetic TD might serve a role in plant defense came from work by Hildmann et al. (1992) showing that a *TD* gene in potato is highly expressed in response to mechanical wounding or treatment with the phytohormones abscisic acid (ABA) and MeJA. Subsequent studies showed that TD expression in leaves of tomato and *N. attenuata* is also induced by the jasmonate signaling pathway in response to wounding and herbivory (Samach et al. 1995; Hermsmeier et al. 2001; Strassner et al. 2002; Li et al. 2004). In these solanaceous species, TD is constitutively expressed to high levels in reproductive organs (Hildmann et al. 1992; Kang and Baldwin 2006). In fact, TD is the most abundant protein in tomato flowers (Samach et al. 1991). The overall pattern of TD expression in vegetative and reproductive tissues is very similar to that of PIs and other JA-inducible proteins that have a role in anti-insect defense. The phenomenon of stress-induced TD expression has not been reported for plants outside the Solanaceae.

Recent work from our laboratory indicates that tomato possesses two *TD* genes (designated *SlTD1* and *SlTD2*) that have distinct roles in plant growth and development (Chen et al. 2007). *TD1* is constitutively expressed in all tissues, whereas *TD2* expression is dependent on an intact jasmonate signaling pathway (Li et al. 2004; Chen et al. 2007). Plant TD sequences cluster into two major phylogenetic groups (Fig. 13.4). The sequence of SlTD1 is more similar to that of TDs from rice, poplar, and *Arabidopsis* than it is to SlTD2. The fact that *Arabidopsis*, poplar, and rice each possess a single *TD* gene supports the hypothesis that SlTD1 performs a housekeeping role in Ile biosynthesis. Jasmonate-inducible isoforms from tomato (SlTD2) and potato (StTD2) comprise a distinct subgroup of proteins that are similar to a TD sequence from chickpea. The lack of detectable TD2 expression in *jai1* mutant plants, together with the fact that this mutant does not exhibit symptoms (e.g., stunted growth) of Ile deficiency, indicates that TD1 can produce Ile pools that are utilized for normal growth and development in the absence of TD2 (Li et al. 2004; Chen et al. 2007).

Direct evidence for the idea that TD has a role in anti-insect defense came from the observation that TD2 accumulates in the food bolus and frass of *M. sexta* larvae

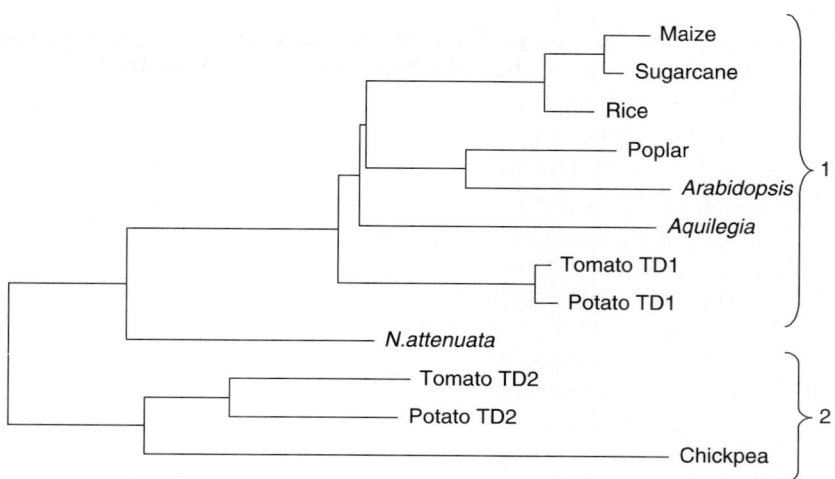

Fig. 13.4 Phylogeny of plant threonine deaminase. Unrooted Neighbor-Joining Tree was constructed with the deduced amino acid sequence of TDs from the indicated plants. Accession numbers: Maize (CO446428), Sugarcane (CA208490), Rice (NP_001051069), Poplar (estExt_fgenesh4_pg.C_280257), *Arabidopsis* (NP_187616), *Aquilegia* (DT735861), Tomato TD1 (ABK20067), Potato TD1 (BI436101), *N. attenuata* (AAX22214), Chickpea (Q39469), Tomato TD2 (P25306), and Potato TD2 (X67846). The sequences cluster into two subgroups designated "1" and "2". Because of its ambiguous position in the tree, *N. attenuata* TD was not assigned to either group

raised on tomato foliage (Chen et al. 2005). High TD2 activity in the midgut was correlated with reduced levels of free Thr. The level of midgut ammonia, which is produced from TD-catalyzed deamination of Thr, was also correlated with TD2 activity. A recent study by Kang et al. (2006) provided direct evidence for a role of TD in anti-insect defense. These workers showed that TD-deficient *N. attenuata* plants are compromised in resistance to *M. sexta* larvae. Supplementation of *N. attenuata* leaves with Thr resulted in increased caterpillar performance, suggesting that Thr availability in the leaf diet is limiting for larval growth. Unlike tomato and potato, *N. attenuata* has a single *TD* gene (Kang et al. 2006). Silencing of this gene thus led to a deficiency in Ile and stunted plant growth. Reduced Ile levels in the silenced lines caused decreased production of jasmonoyl-Ile (JA-Ile), which is an important signal for induced plant defense responses to pathogens (Staswick et al. 1998) and insects (Kang et al. 2006). Thus, TD's defensive role in *N. attenuata* is attributed both to its involvement in JA-Ile synthesis and its function in post-ingestive defense (Kang et al. 2006). The dual role of *N. attenuata* TD may explain the protein's somewhat ambiguous position in the TD phylogenetic tree (Fig. 13.4), and suggests that the evolution of this isoform as an antinutritional enzyme is constrained by its essential function in Ile biosynthesis. Tomato TD2 is presumably not subjected to such constraint because of the presence of TD1. Thus, TD2 may be highly specialized as a midgut-active defensive enzyme.

Biosynthetic TDs contain an N-terminal catalytic domain and a C-terminal regulatory domain (Gallagher et al. 1998). Binding of Ile to the regulatory domain results in an allosteric transition that strongly inhibits catalytic activity. The high level of TD2 activity in the midgut of tomato-reared *M. sexta* suggested that the regulatory properties of the enzyme were altered in a way that confers insensitivity to Ile (Chen et al. 2005). In vitro enzyme assays showed that TD2 isolated from tomato foliage is strongly inhibited by Ile, whereas TD2 activity in midgut and frass extracts is insensitive to Ile. Using LC-MS/MS analysis of midgut extracts, we detected numerous peptides corresponding to the catalytic domain. However, we failed to detect peptides from the regulatory domain. It was thus proposed that the regulatory domain of TD2 is proteolytically cleaved from the catalytic domain following ingestion of leaf material by the insect (Chen et al. 2005). The resulting truncated enzyme can efficiently degrade Thr, which is a dietary requirement for phytophagous insects, in the Ile-rich insect gut.

To prove that proteolytic processing is involved in the production of an Ile-insensitive variant of TD2, it was necessary to establish a product-precursor relationship between the ~55-kDa form of the protein that accumulates in tomato leaves (Samach et al. 1995) and the putatively processed form of the enzyme. Immunoblot studies showed that anti-TD2 antibodies cross-react with a ~55-kDa protein in insect-damaged tomato foliage (Chen et al. 2007). In contrast, a lower molecular-weight (~40-kDa) protein, designated pTD2, was the predominant form of the protein in the food bolus and frass. These results demonstrate that pTD2 is produced by proteolytic processing of TD2 following ingestion of tomato foliage by *M. sexta*. The near absence of unprocessed TD2 in midgut extracts suggests that the processing reaction occurs rapidly in response to maceration and ingestion of leaf tissue. It was also found that pTD2 is the predominant form of the enzyme in frass from tomato-reared *Trichoplusia ni* (cabbage looper), a generalist caterpillar. Consistent with this finding, *T. ni* frass contained TD activity that is insensitive to inhibition by Ile (Chen et al. 2007). Thus, ingestion of foliage by specialist and generalist insect herbivores results in proteolytic removal of the regulatory domain of TD2, resulting in an enzyme that efficiently degrades Thr without being inhibited through feedback by Ile. A proposed model depicting that fate of tomato TD2 in the lepidopteran gut lumen is shown in Fig. 13.5. The structure of TD2 suggests that the processing reaction involves either a plant- or insect-derived endoprotease that cleaves the neck region between the regulatory and catalytic domains. Additional work is needed to test this hypothesis.

Several properties of pTD2 presumably facilitate its post-ingestive role in defense (Chen et al. 2007). First, pTD2 is insensitive to feedback inhibition by Ile. Second, pTD2 is highly resistant to insect digestive proteases as determined by the high level to which the active enzyme is excreted in frass. Third, purified pTD2 is active in an alkaline pH range that matches the lepidopteran midgut; little or no activity was observed at pH values below 6.0. Fourth, pTD2 is active at temperatures exceeding 60°C. This thermostability indicates that the enzyme would be active at elevated body temperatures, which for *M. sexta* caterpillars in natural conditions can exceed 35°C (Casey 1976). Finally, the expression pattern of TD2 is consistent with a function in antiherbivore defense. *TD2* is coordinately induced with other defensive genes in response to wounding and JA treatment (Hildmann et al. 1992; Samach et al. 1995; Li et al. 2004). The high level of constitutive TD2 expression in tomato reproductive tissues (Samach et al. 1991, 1995) is similar to many other JA-regulated defensive proteins, including PIs, arginase, LAP-A, and AtVSP2 (Hildmann et al. 1992; Utsugi et al. 1998; Chao et al. 1999; Chen et al. 2004). These observations support the idea that TD accumulation in floral tissues reflects a mechanism that protects reproductive structures from herbivory.

Functional divergence of SlTD1 and SlTD2 indicates that the latter defensive isozyme may have evolved from a 'housekeeping' TD. It is reasonable to assume, for example, that TD2 arose from a gene duplication event, and that selective pressure imposed by herbivory led to the evolution of TD2. An important feature acquired by TD2 and other post-ingestive defense proteins during evolution was regulation by the jasmonate signaling pathway. It is unclear whether TD2 evolved new biochemical or structural properties that enhance its ability to degrade Thr in the insect gut. Future research aimed at comparing the stability, activity, and structure of TD1 and TD2 may provide additional insight into this question.

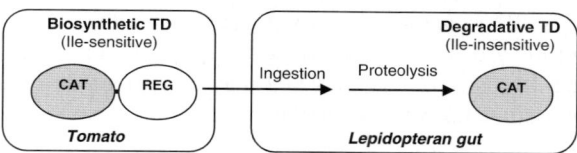

Fig. 13.5 Working model for proteolytic activation of tomato TD2 in the response to herbivore attack. CAT: catalytic domain; REG: regulatory domain

13.4 Vegetative Storage Proteins

Vegetative storage proteins (VSPs) comprise a diverse group of proteins that accumulate to high levels in various vegetative organs of the plant. The dynamics of VSP accumulation and degradation in plant tissues is controlled by both developmental and environmental cues (Berger et al. 1995). The first identified and most extensively characterized VSPs are soybean VSPα and VSPβ (Staswick 1994). A commonly held theory is that accumulated VSPs function as a source of amino acids and other nutrients for developing seeds. However, results from transgenic experiments with VSP-deficient soybean plants indicate that VSPs contribute little if any to plant productivity (Staswick et al. 2001). Thus, alternative roles for VSPs in nutrient storage or other aspects of plant physiology must be considered.

A possible clue to understanding VSP function can be found by studying the enzymatic activity of these proteins. Soybean and *Arabidopsis* VSPs exhibit phosphatase activity (Leelapon et al. 2004; Liu et al. 2005), whereas patatin from potato tubers is a lipid acyl hydrolase (Andrews et al. 1988). The predominant VSPs in bark from mulberry are acalin-like lectins (Van Damme et al. 2002). VSPs from sweet potato and alfalfa are reported to function as trypsin inhibitors and chitinases, respectively (Yeh et al. 1997; Meuriot et al. 2004). These observations raise the possibility that VSPs are enzymes that subsequently acquired a protein storage function (Staswick 1994). Alternatively, the function of VSPs may be intimately linked to their enzymatic activity. In this context, it is noteworthy that many of the above-mentioned activities are typically associated with plant defensive processes. Additional support for a role of VSPs in plant defense comes from the well-established fact that VSP expression in vegetative tissues is strongly promoted by mechanical wounding, herbivory, and jasmonate treatment (Mason and Mullet 1990; Staswick 1990; Franceschi and Grimes 1991; Staswick et al. 1991; Major and Constabel 2006). The absence of VSP expression in vegetative and reproductive tissues of the *Arabidopsis coi1* mutant further demonstrates that expression of this protein is tightly controlled by the jasmonate signaling pathway (Benedetti et al. 1995).

A significant breakthrough in our understanding of VSP function came from the recent analysis of *Arabidopsis* VSP2 (Liu et al. 2005). Purified VSP2, incorporated into artificial diets, exhibited potent anti-insect activity against dipteran and coleopteran insects that have an acidic gut lumen. The growth-inhibiting effect of VSP2 on these insects was at least 10-fold greater than other anti-insect proteins including cystatin and lectin. Significantly, site-directed mutagenesis showed that the phosphatase activity of VSP2 is essential for the protein's insecticidal effects. These results provide compelling evidence that VSP2 is an anti-insect protein and that the protein's phosphatase activity is critical for this function. The mode of action of VSP2 remains to be determined. Given the importance of dietary phosphorous for the growth of leaf-eating caterpillars (Perkins et al. 2004), it is conceivable that the defensive effects of VSP2 may be attributed to the enzyme's ability to perturb phosphate homeostasis in the gut lumen.

13.5 Perspectives

Post-ingestive interactions between plant food and components of the insect digestive system have played a major role in shaping patterns of host plant utilization by insects. The vast majority of research on this topic has focused on plant secondary metabolites that exert toxic or antinutritive effects on insect attackers (Fraenkel 1959). The pioneering work of Ryan and colleagues on defensive PIs introduced the concept that wound-inducible plant proteins act directly in the insect gut as a defense. The recent discovery of jasmonate-regulated isoforms of TD, arginase, and VSP as anti-insect proteins extends this concept to include plant enzymes that metabolize essential nutrients in the insect digestive system. Additional work is needed to clarify the mode of action of these enzymes, and to elucidate the structural features that allow them to function in the extreme environment of the midgut. These and other examples support the idea that selective pressure imposed by insect herbivores led to the evolution of defensive enzymes from pre-existing isoforms that function in primary metabolism. The tremendous diversity of plant enzymes further suggests that many additional anti-insect proteins remain to be discovered. The application of proteomics-based approaches to identify plant proteins that accumulate in the insect food bolus (Chen et al. 2005) and frass (Chen et al. 2007) is expected to facilitate this goal. Future research in this area promises to contribute to the development of sound pest management approaches, including the generation of crop varieties with enhanced pest resistance.

Acknowledgments This work was supported by grants from the USDA National Research Initiative Competitive Grants Program award (2007-35604-17791) and the US Department of Energy (DE-FG02-91ER20021).

References

Alabadi D, Aguero MS, Perez-Amador MA, Carbonell J (1996) Arginase, arginine decarboxylase, ornithine decarboxylase, and polyamines in tomato ovaries. Changes in unpollinated ovaries and parthenocarpic fruits induced by auxin or gibberellin. Plant Physiol 112:1237–1244

Andrews DL, Beames B, Summers MD, Park WD (1988) Characterization of the lipid acyl hydrolase activity of the major potato (*Solanum tuberosum*) tuber protein, patatin, by cloning and abundant expression in a baculovirus vector. Biochem J 252:199–206

Benedetti CE, Xie D, Turner JG (1995) *COI1*-dependent expression of an *Arabidopsis* vegetative storage protein in flowers and siliques and in response to coronatine or methyl jasmonate. Plant Physiol 109:567–572

Berenbaum MR (1995) Turnabout is fair play – Secondary roles for primary compounds. J Chem Ecol 21:925–940

Berger S, Bell E, Sadka A, Mullet JE (1995) *Arabidopsis thaliana AtVsp* is homologous to soybean *VspA* and *VspB*, genes encoding vegetative storage protein acid-phosphatases, and is regulated similarly by methyl jasmonate, wounding, sugars, light and phosphate. Plant Mol Biol 27:933–942

Bernays EA, Chapman RF (1994) Host-plant selection by phytophagous insects. Chapman & Hall, New York

Broadway RM, Duffey SS (1986) Plant proteinase inhibitors – Mechanism of action and effect on the growth and digestive physiology of larval *Heliothis zea* and *Spodoptera exiqua*. J Insect Physiol 32:827–833

Broadway RM, Duffey SS (1988) The effect of plant protein quality on insect digestive physiology and the toxicity of plant proteinase inhibitors. J Insect Physiol 34:1111–1117

Casey TM (1976) Activity patterns, body temperature and thermal ecology in two desert caterpillars (Lepidoptera: Sphingidae). Ecology 57:485–497

Chang CL (2004) Effect of amino acids on larvae and adults of *Ceratitis capitata* (Diptera: Tephritidae). Ann Entomol Soc Am 97:529–535

Chao WS, Gu YQ, Pautot VV, Bray EA, Walling LL (1999) Leucine aminopeptidase RNAs, proteins, and activities increase in response to water deficit, salinity, and the wound signals systemin, methyl jasmonate, and abscisic acid. Plant Physiol 120:979–992

Chen H, Gonzales-Vigil E, Wilkerson CG, Howe GA (2007) Stability of plant defense proteins in the gut of insect herbivores. Plant Physiol 143:1954–1967

Chen H, McCaig BC, Melotto M, He SY, Howe GA (2004) Regulation of plant arginase by wounding, jasmonate, and the phytotoxin coronatine. J Biol Chem 279:45998–46007

Chen H, Wilkerson CG, Kuchar JA, Phinney BS, Howe GA (2005). Jasmonate-inducible plant enzymes degrade essential amino acids in the herbivore midgut. Proc Natl Acad Sci USA 102:19237–19242

Colau D, Negrutiu I, Vanmontagu M, Hernalsteens JP (1987) Complementation of a threonine dehydratase-deficient *Nicotiana plumbaginifolia* mutant after *Agrobacterium-tumefaciens*-mediated transfer of the *Saccharomyces cerevisiae ILV1* gene. Mol Cell Biol 7:2552–2557

Constabel CP, Bergey DR, Ryan CA (1995) Systemin activates synthesis of wound-inducible tomato leaf polyphenol oxidase via the octadecanoid defense signaling pathway. Proc Natl Acad Sci USA 92:407–411

Duffey SS, Stout MJ (1996) Antinutritive and toxic components of plant defense against insects. Arch Insect Biochem Physiol 32:3–37

Feeny P (1976) Plant apparency in chemical defense. Recent Adv Phytochem 10:1–40

Felton GW (1996) Nutritive quality of plant protein: sources of variation and insect herbivore responses. Arch Insect Biochem Physiol 32:107–130

Felton GW (2005) Indigestion is a plant's best defense. Proc Natl Acad Sci USA 102:18771–18772

Fraenkel GS (1959) The raison d'etre of secondary plant substances; these odd chemicals arose as a means of protecting plants from insects and now guide insects to food. Science 129:1466–1470

Franceschi VR, Grimes HD (1991) Induction of soybean vegetative storage proteins and anthocyanins by low-level atmospheric methyl jasmonate. Proc Natl Acad Sci USA 88:6745–6749

Gallagher DT, Gilliland GL, Xiao GY, Zondlo J, Fisher KE, Chinchilla D, Eisenstein E (1998) Structure and control of pyridoxal phosphate dependent allosteric threonine deaminase. Structure 6:465–475

Gfeller A, Farmer EE (2004) Keeping the leaves green above us. Science 306:1515–1516

Goldraij A, Polacco JC (1999) Arginase is inoperative in developing soybean embryos. Plant Physiol 119:297–304

Goldraij A, Polacco JC (2000) Arginine degradation by arginase in mitochondria of soybean seedling cotyledons. Planta 210:652–658

Green TR, Ryan CA (1972) Wound-induced proteinase inhibitor in plant leaves – possible defense mechanism against insects. Science 175:776–777

Halgand F, Wessel PM, Laprevote O, Dumas R (2002) Biochemical and mass spectrometric evidence for quaternary structure modifications of plant threonine deaminase induced by isoleucine. Biochemistry 41:13767–13773

Hermsmeier D, Schittko U, Baldwin IT (2001) Molecular interactions between the specialist herbivore *Manduca sexta* (Lepidoptera, Sphingidae) and its natural host *Nicotiana attenuata*. I. Large-scale changes in the accumulation of growth- and defense-related plant mRNAs. Plant Physiol 125:683–700

Hildmann T, Ebneth M, Pena-Cortes H, Sanchez-Serrano JJ, Willmitzer L, Prat S (1992) General roles of abscisic and jasmonic acids in gene activation as a result of mechanical wounding. Plant Cell 4:1157–1170

Howe GA (2004) Jasmonates as signals in the wound response. J Plant Growth Regul 23:223–237

John SJ, Srivastava V, Guhamukherjee S (1995) Cloning and sequencing of chickpea cDNA coding for threonine deaminase. Plant Physiol 107:1023–1024

Kang JH, Baldwin IT (2006). Isolation and characterization of the threonine deaminase promoter in *Nicotiana attenuata*. Plant Sci 171:435–440

Kang JH, Wang L, Giri A, Baldwin IT (2006) Silencing threonine deaminase and JAR4 in *Nicotiana attenuata* impairs jasmonic acid-isoleucine-mediated defenses against *Manduca sexta*. Plant Cell 18:3303–3320

Kessler A, Baldwin IT (2002) Plant responses to insect herbivory: the emerging molecular analysis. Annu Rev Plant Biol 53:299–328

Konno K, Hirayama C, Nakamura M, Tateishi K, Tamura Y, Hattori M, Kohno K (2004) Papain protects papaya trees from herbivorous insects: role of cysteine proteases in latex. Plant J 37:370–378

Leelapon O, Sarath G, Staswick PE (2004) A single amino acid substitution in soybean VSP alpha increases its acid phosphatase activity nearly 20-fold. Planta 219:1071–1079

Li L, Zhao Y, McCaig BC, Wingerd BA, Wang J, Whalon ME, Pichersky E, Howe GA (2004) The tomato homolog of *CORONATINE-INSENSITIVE1* is required for the maternal control of seed maturation, jasmonate-signaled defense responses, and glandular trichome development. Plant Cell 16:126–143

Liu YL, Ahn JE, Datta S, Salzman RA, Moon J, Huyghues-Despointes B, Pittendrigh B, Murdock LL, Koiwa H, Zhu-Salzman K (2005) *Arabidopsis* vegetative storage protein is an anti-insect acid phosphatase. Plant Physiol 139:1545–1556

Major IT, Constabel CP (2006) Molecular analysis of poplar defense against herbivory: comparison of wound- and insect elicitor-induced gene expression. New Phytol 172:617–635

Mason HS, Mullet JE (1990) Expression of two soybean vegetative storage protein genes during development and in response to water deficit, wounding, and jasmonic acid. Plant Cell 2:569–579

Mattson WJ (1980) Herbivory in relation to plant nitrogen-content. Annu Rev Ecol Syst 11:119–161

Meuriot F, Noquet C, Avice JC, Volenec JJ, Cunningham SM, Sors TG, Caillot S, Ourry A (2004) Methyl jasmonate alters N partitioning, N reserves accumulation and induces gene expression of a 32-kDa vegetative storage protein that possesses chitinase activity in *Medicago sativa* taproots. Physiol Plant 120:113–123

Mohan S, Ma PWK, Pechan T, Bassford ER, Williams WP, Luthe DS (2006) Degradation of the *S. frugiperda* peritrophic matrix by an inducible maize cysteine protease. J Insect Physiol 52:21–28

Mori M, Gotoh T (2004) Arginine metabolic enzymes, nitric oxide and infection. J Nutr 134:2820S–2825S

Mourad G, King J (1995) L-O-Methylthreonine resistant mutant of *Arabidopsis* defective in isoleucine feedback regulation. Plant Physiol 107:43–52

Pechan T, Cohen A, Williams WP, Luthe DS (2002) Insect feeding mobilizes a unique plant defense protease that disrupts the peritrophic matrix of caterpillars. Proc Natl Acad Sci USA 99:13319–13323

Perkins MC, Woods HA, Harrison JF, Elser JJ (2004) Dietary phosphorus affects the growth of larval *Manduca sexta*. Arch Insect Biochem Physiol 55:153–168

Perozich J, Hempel J, Morris SM Jr (1998) Roles of conserved residues in the arginase family. Biochim Biophys Acta 1382:23–37

Rhoades DF, Cates RG (1976) Toward a general theory of plant antiherbivore chemistry. Recent Adv Phytochem 10:168–213

Ryan CA (1990) Protease inhibitors in plants – Genes for improving defenses against insects and pathogens. Annu Rev Phytopath 28:425–449

Samach A, Broday L, Hareven D, Lifschitz E (1995) Expression of an amino acid biosynthesis gene in tomato flowers: developmental upregulation and MeJa response are parenchyma-specific and mutually compatible. Plant J 8:391–406

Samach A, Hareven D, Gutfinger T, Ken-Dror S, Lifschitz E (1991) Biosynthetic threonine deaminase gene of tomato: isolation, structure, and upregulation in floral organs. Proc Natl Acad Sci USA 88:2678–2682

Satriano J (2004) Arginine pathways and the inflammatory response: interregulation of nitric oxide and polyamines: review article. Amino Acids 26:321–329

Schilmiller AL, Howe GA (2005) Systemic signaling in the wound response. Curr Opin Plant Biol 8:369–377

Sekowska A, Danchin A, Risler JL (2000) Phylogeny of related functions: the case of polyamine biosynthetic enzymes. Microbiology 146:1815–1828

Sidorov V, Menczel L, Maliga P (1981) Isoleucine-requiring *Nicotiana* plant deficient in threonine deaminase. Nature 294:87–88

Staswick PE (1990) Novel regulation of vegetative storage protein genes. Plant Cell 2:1–6

Staswick PE (1994) Storage proteins of vegetative plant tissue. Annu Rev Plant Physiol Plant Mol Biol 45:303–322

Staswick PE, Huang JF, Rhee Y (1991) Nitrogen and methyl jasmonate induction of soybean vegetative storage protein genes. Plant Physiol 96:130–136

Staswick PE, Yuen GY, Lehman CC (1998) Jasmonate signaling mutants of *Arabidopsis* are susceptible to the soil fungus *Pythium irregulare*. Plant J 15:747–754

Staswick PE, Zhang ZY, Clemente TE, Specht JE (2001) Efficient down-regulation of the major vegetative storage protein genes in transgenic soybean does not compromise plant productivity. Plant Physiol 127:1819–1826

Strassner J, Schaller F, Frick UB, Howe GA, Weiler EW, Amrhein N, Macheroux P, Schaller A (2002) Characterization and cDNA-microarray expression analysis of 12-oxophytodienoate reductases reveals differential roles for octadecanoid biosynthesis in the local versus the systemic wound response. Plant J 32:585–601

Szamosi I, Shaner DL, Singh BK (1993) Identification and characterization of a biodegradative form of threonine dehydratase in senescing tomato (*Lycopersicon esculentum*) leaf. Plant Physiol 101:999–1004

Umbarger HE (1956) Evidence for a negative-feedback mechanism in the biosynthesis of isoleucine. Science 123:848

Utsugi S, Sakamoto W, Murata M, Motoyoshi F (1998) *Arabidopsis thaliana* vegetative storage protein (VSP) genes: gene organization and tissue-specific expression. Plant Mol Biol 38:565–576

Van Damme EJM, Hause B, Hu JL, Barre A, Rouge P, Proost P, Peumans WJ (2002) Two distinct jacalin-related lectins with a different specificity and subcellular location are major vegetative storage proteins in the bark of the black mulberry tree. Plant Physiol 130:757–769

Vincendeau P, Gobert AP, Daulouede S, Moynet D, Mossalayi MD (2003) Arginases in parasitic diseases. Trends Parasitol 19:9–12

Walling LL (2000) The myriad plant responses to herbivores. J Plant Growth Regul 19:195–216

Wang JH, Constabel CP (2004) Polyphenol oxidase overexpression in transgenic *Populus* enhances resistance to herbivory by forest tent caterpillar (*Malacosoma disstria*). Planta 220:87–96

Wessel PM, Graciet E, Douce R, Dumas R (2000) Evidence for two distinct effector-binding sites in threonine deaminase by site-directed mutagenesis, kinetic, and binding experiments. Biochemistry 39:15136–15143

Yeh KW, Chen JC, Lin MI, Chen YM, Lin CY (1997) Functional activity of sporamin from sweet potato (*Ipomoea batatas* Lam): a tuber storage protein with trypsin inhibitory activity. Plant Mol Biol 33:565–570

Chapter 14
Plant Lectins as Part of the Plant Defense System Against Insects

Els J.M. Van Damme

Many plants contain carbohydrate-binding proteins which are commonly designated as lectins. It is believed that some of these lectins play a role in plant defense against insects. This chapter gives an overview of the state of the art in the field of plant lectin research and highlights the most important results with regard to the insecticidal activity of plant lectins belonging to different families. A distinction is made between the 'classical lectins' that are constitutively expressed and the 'inducible lectins'. Furthermore, the possible mode of action and target sites for the lectins in the insect body are discussed.

14.1 Introduction

During evolution plants have developed a wide range of sophisticated defense mechanisms to counteract attack by pathogens and herbivory. These include both mechanical and chemical defenses, and allow the plant to survive in the same environment as its potential attackers. As part of their chemical defense mechanism plants synthesize a whole battery of so-called defense proteins (Carlini and Grossi-de-Sà 2002). One particular class of defense proteins are the plant lectins, a heterogeneous group of proteins that can specifically interact with particular carbohydrates (Peumans and Van Damme 1995; Van Damme et al. 1998a, 2007b). This chapter will focus on the role of carbohydrate-binding proteins (lectins) in plant defense against insects. A detailed discussion on the insecticidal activity of different families of carbohydrate-binding proteins will be given. In addition, some recent results on inducible lectins involved in plant defense against insects will be discussed.

E.J.M. Van Damme
Laboratory of Biochemistry and Glycobiology, Department of Molecular Biotechnology, Ghent University, 9000 Gent, Belgium
e-mail: ElsJM.VanDamme@UGent.be

14.2 Distribution of Plant Lectins

Many plants including important food plants such as wheat, rice, potato, tomato, soybean and bean contain carbohydrate-binding proteins commonly referred to as 'lectins', 'agglutinins' or 'hemagglutinins'. According to a recently proposed definition, this group of proteins comprises 'all plant proteins possessing at least one non-catalytic domain that binds reversibly to specific mono- or oligosaccharides'. Hitherto, about 500 different plant lectins have been isolated and (partially) characterized. On a first glance all these lectins form a heterogeneous group of proteins because of the obvious differences in structure, specificity and biological activities. However, the majority of all plant lectins can be classified on the basis of structural analyses and sequence data in seven families of structurally and evolutionary related proteins. These lectin families are: (1) the amaranthins, (2) the Cucurbitaceae phloem lectins, (3) the chitin binding lectins composed of hevein domains, (4) the GNA-related lectins, (5) the jacalin-related lectins, (6) the legume lectins and (7) the lectins with ricin-B domains (Van Damme et al. 1998a, 2007b). Recently, evidence was reported for the occurrence in plants of homologs of the *Agaricus bisporus* agglutinin (Peumans et al. 2007) and homologs of class V chitinases with sugar-binding activity (Van Damme et al. 2007a).

Although lectins have for a long time been considered typical seed proteins, the occurrence of plant lectins in vegetative plant tissues is now very well documented. There are, however, striking differences for what concerns the location and concentration of the different lectins. Seed lectins are primarily found in the cotyledons (e.g., legume species) or the endosperm, but can also be confined to the primary axis (e.g., wheat). Seed lectins generally account for 0.1%–5% of the total seed protein. Non-seed lectins have been detected in almost all vegetative tissues ranging from leaves, stems, bark, bulbs, rhizomes, roots to flowers, fruits, phloem sap and even nectar. Lectin concentrations in vegetative storage tissues can amount to 50% of the total protein, whereas in other tissues the lectin is a very minor protein (e.g., in leaves of leek). On average, lectins in vegetative tissues also account for 0.1%–5% of the total protein.

14.3 Classical and Inducible Plant Lectins

For many years research on plant lectins has been focused on a group of lectins that occur in reasonable to high concentrations in different plant tissues. Because of the high lectin concentrations and the developmental regulation it was difficult to imagine a specific role for these lectins within the plant cell. Several lines of evidence indicated that the majority of these lectins may be involved in plant defense. In this respect, plant lectins basically differ from animal lectins because the latter are involved in specific recognition processes within the organism itself. Taking into consideration the increasing evidence that protein–carbohydrate interactions are very important for the normal development and functioning of animal organisms, and the fact that different types of lectins are the mediators of these protein–carbohydrate interactions, the question arose whether the same holds true for plants. Recently, this search for potential physiologically active plant lectins resulted in the discovery of different lectin families, which are candidates to mediate essential protein–carbohydrate interactions in the plant itself (Van Damme et al. 2004, 2007b). Evidence has accumulated that plants synthesize well-defined carbohydrate binding proteins (lectins) upon exposure to stress situations like drought, high salt, wounding, or treatment with some plant hormones (Zhang et al. 2000; Chen et al. 2002). Localization studies demonstrated that – in contrast to the 'classical' plant lectins, which are typically found in vacuoles – the 'inducible' lectins are exclusively located in the cytoplasm and the nucleus. It is hypothesized that this new class of lectins may play a specific role within the plant cell. This section will give a brief overview of the classical and inducible lectins that have been identified to date (Table 14.1).

Table 14.1 Overview of different plant lectin families, their location, carbohydrate binding properties and insecticidal activity

Lectin family	Vacuolar lectins	Nucleocytoplasmic lectins	Carbohydrate specificity	Insecticidal activity*
Amaranthin family	No examples known	Documented in *Amaranthus* sp. and *Prunus* sp.	GalNAc/T-antigen	+
Cucurbitaceae phloem lectins/ Lectins with Nictaba domain	No examples known	Wide taxonomic distribution	(GlcNAc)n N-glycans	?
GNA-related lectins	Wide taxonomic distribution	Found in diverse taxa	Man High Man N-glycans	+++
Chitin-binding lectins composed of hevein domains	Wide taxonomic distribution	No examples known	(GlcNAc)n N-glycans	++
Jacalin-related lectins	Only documented in a few Moraceae sp.	Ubiquitous	Gal/T-antigen Man N-glycans	+
Legume lectins	Common in Fabaceae and Lamiaceae	No examples known	Man/Glc Gal/GalNAc (GlcNAc)n Fuc, Siaα2-3Gal/GalNAc Complex	++
Lectins with Ricin-B domains	Wide taxonomic distribution	No examples known	Gal/GalNAc Siaα2-6Gal/GalNAc	+

*Insecticidal activity has been reported for many (+++), several (++), a few (+) or no (?) lectins from this plant lectin family

Amaranthin family: The Amaranthin family, which is called after the *Amaranthus caudatus* seed lectin is a rather small family of closely related lectins found in different *Amaranthus* species. All known amaranthins are homodimers built of 33 kDa subunits. Detailed specificity studies have shown that amaranthin preferentially recognizes the T-antigen disaccharide Galβ(1,3)GalNAc (Rinderle et al. 1989).

Cucurbitaceae phloem lectins: The family of Cucurbitaceae phloem lectins is a small group of chitin-binding agglutinins found in the phloem exudates of a number of Cucurbitaceae species (Sabnis and Hart 1978). These lectins, also called phloem proteins PP2, are dimeric proteins built up of subunits of approximately 22 kDa.

GNA-related lectins: In 1987 a lectin with exclusive specificity towards mannose (Man) was isolated and characterized in snowdrop (*Galanthus nivalis*) bulbs (Van Damme et al. 1987). The so-called *G. nivalis* agglutinin (abbreviated GNA) turned out to be the prototype of an extended family of mannose-binding lectins occurring in numerous monocot plant families (e.g., Alliaceae, Liliaceae, Orchidaceae, Araceae, Bromeliaceae, Ruscaceae, Iridaceae; Yagi et al. 1996; Van Damme et al. 1998a, b; Neuteboom et al. 2002). Therefore this group of lectins was originally referred to as the 'monocot mannose-binding lectins' (Van Damme et al. 1998b). Recently, very similar lectins have been identified in plants other than *Liliopsida* (e.g., in the liverwort, *Marchantia polymorpha*; Peumans et al. 2002) as well as in bacteria (Parret et al. 2003, 2005), and animals (Tsutsui et al. 2003) Therefore, this group of lectins is now referred to as 'GNA-related lectins' after the first identified member.

All plant lectins with GNA-domains are composed of lectin subunits of approximately 12 kDa, that show specificity towards mannose. Nevertheless, GNA-related lectins from different plants exhibit subtle but marked differences in specificity towards oligomannosides and N-glycans (Kaku et al. 1990, 1992; Van Damme et al. 2007b).

Chitin-binding lectins composed of hevein domains: The name 'hevein' refers to a chitin-binding polypeptide of 43 amino acids found in the latex of *Hevea brasiliensis* (rubber tree; Waljuno et al. 1975). Many plant proteins owe their chitin-binding activity to the presence of one or more hevein domain(s). However, it was shown that various lectins with hevein domains also have high affinity for N-glycosylated animal glycoproteins (Goldstein and Poretz 1986). This group of lectins is widespread and occurs both in monocotyledonous (e.g., wheat germ agglutinin (WGA)) as well as dicotyledonous (e.g,. potato lectin) plant species.

Jacalin-related lectins: The family of jacalin-related lectins comprises all proteins with one or more domains that are structurally equivalent to 'jacalin', a galactoside-binding lectin from jack fruit (*Artocarpus integrifolia*) seeds (Sastry et al. 1986). In recent years several lectins related to jacalin but with specificity towards mannose were discovered and characterized in detail (Van Damme et al. 1996). Therefore the family of jacalin-related lectins is now subdivided into two subfamilies with a distinct specificity and molecular structure. The galactose (Gal)-specific jacalin-related lectins are built up of four cleaved protomers comprising a small (β; 20 amino acid residues) and a large (α; 133 amino acid residues) subunit and exhibit a clear preference for galactose over mannose (Bourne et al. 2002).

In contrast, mannose-specific jacalin-related lectins are built up of uncleaved protomers of approximately 150 amino acids each, which exhibit exclusive specificity towards mannose.

Legume lectins: Historically, the term 'legume lectin' refers to proteins that are structurally and evolutionary related to a well-defined type of lectin that was originally discovered in seeds of legumes (*Fabaceae*) like jack bean, common bean, pea, peanut and soybean. Until recently, legume lectins were exclusively found in Fabaceae. However, the isolation and cloning of a structurally and evolutionary closely related agglutinin from leaves of *Glechoma hederacea* demonstrated that closely related lectins also occur in the family *Lamiaceae* (Wang et al. 2003a, b).

The family of legume lectins (from both *Fabaceae* and *Lamiaceae* species) is fairly homogeneous as far as the molecular structure of the native lectins is concerned (Van Damme et al. 1998a, 2007b). All lectins are composed of subunits of approximately 30 kDa and many of them are glycoproteins.

Although the overall fold of the lectin and its carbohydrate-binding sites are highly conserved the legume lectin family exhibits a marked heterogeneity in sugar-binding activity and specificity (Young and Oomen 1992; Sharma and Surolia 1997). Legume lectins with specificity directed against galactose (Gal), N-acetylgalactosamine (GalNAc), mannose/glucose (Man/Glc), fucose (Fuc), N-acetylglucosamine (GlcNAc) and sialic acid have been found as well as numerous lectins that bind exclusively more complex glycan structures (Van Damme et al. 1998a). In contrast, the lectins from *Lamiaceae* species exhibit specificity that is primarily directed against the T-antigen (GalNAcα1-Ser/Thr; Wu 2005; Singh et al. 2006).

Lectins with ricin-B domains: This group of plant lectins was previously referred to as the family of 'ribosome-inactivating proteins' (abbreviated RIPs), due to the fact that these proteins have been first identified as plant proteins that inactivate eukaryotic ribosomes through the removal of a conserved adenine residue from the large ribosomal RNA (Van Damme et al. 2001; Stirpe 2004). At present, RIPs are considered enzymes, namely N-glycosylases (EC 3.2.2.22) that are capable of cleaving adenine residues from different polynucleotide substrates. Based on their molecular structure RIPs are subdivided in type-1, type-2 and type-3 RIPs. Type-1 RIPs consist of a single enzymatically active polypeptide chain. Type-2 RIPs are chimeric proteins built up of protomers consisting of an enzymatically active A-chain and a carbohydrate-binding B-chain, which are both derived from a single precursor and are held together by a disulfide bond. Finally, type-3 RIPs also are chimeric proteins built up of an N-terminal N-glycosylase domain linked to a C-terminal domain with a still unknown activity/function. Taking the definition of plant lectins into account only the B-chain of type-2 RIPs can be classified as a lectin domain. Therefore, the lectin domain is referred to as the ricin-B domain, named after 'ricin' the first and most famous RIP isolated from seeds of *Ricinus communis* (Lord et al. 1994).

Most plant ricin-B domains preferentially bind to galactose or GalNAc. However, some lectins exhibit a clear preference for sialylated glycans (e.g., *Sambucus nigra* lectin I). As already mentioned above, type-2 RIPs are not only lectins but also enzymes capable of (catalytically) inactivating ribosomes. They are therefore extremely potent cytotoxins, at least at the condition that they succeed in entering the cell. The most notorious example of a type-2 RIP is the extremely toxic ricin. It should be emphasized however, that most type-2 RIPs are only moderately or even weakly toxic. The huge difference in toxicity is not linked to differences in the catalytic activity of the A-chain but is primarily determined by the lectin's ability to penetrate into the cell, which in turn depends on the sugar-binding specificity of the B-chain (Peumans et al. 2001; Van Damme et al. 2001; Stirpe 2004).

Inducible lectins: Unambiguous evidence for the existence of jasmonate-induced lectins was obtained recently from experiments with jasmonate-treated tobacco plants (*Nicotiana tabacum* cv Samsun NN; Chen et al. 2002). Untreated plants do not contain any detectable lectin activity. After jasmonate treatment however, leaves of young tobacco plants express low concentrations of a lectin (called *N. tabacum* agglutinin or Nictaba) with specificity towards oligomers of N-acetylglucosamine. Nictaba is a homodimer of 19 kDa subunits. Sequence analysis of Nictaba and comparison with other sequences in the databases revealed that this plant lectin is a representative of a new family of inducible cytoplasmic proteins with sequence similarity to the Cucurbitaceae phloem lectins. Immunolocalisation studies confirmed that Nictaba is expressed in the cytoplasm and the nucleus. Furthermore, evidence was presented for the interaction of the lectin with several proteins in the plant nucleus (Lannoo et al. 2006).

Another example of an inducible lectin is the mannose-binding lectin (called Orysata) that is induced in rice plants by salt stress, desiccation, pathogen infection and the phytohormones jasmonic acid and abscisic acid (ABA). This protein was already described in 1990 as SalT (a salt-inducible protein; Claes et al. 1990) but was later identified as a lectin belonging to the family of mannose-specific jacalin-related lectins (Zhang et al. 2000). Since Orysata is synthesized only under specific stress conditions, and in addition, is located in the cytoplasm and the nucleus, we presume that this lectin plays a role in the plant response to well-defined stress factors.

Recently we have also identified other families of nucleocytoplasmic plant lectins, presumably with specificity towards mannose and galactose (unpublished results). Unfortunately the study of these lectins is hampered by the very low expression levels of the lectins (even after induction).

14.4 Insecticidal Activity of Plant Lectins

During the last two decades, numerous reports have been published on the insecticidal activity of plant lectins against many pest insects belonging to the orders Lepidoptera, Coleoptera, Diptera and Homoptera (Sharma et al. 2004). Both in vitro assays with lectin spiked artificial diets, and in vivo assays with transgenic plants expressing a foreign lectin gene have demonstrated the potential of several plant lectins as insecticidal proteins (Carlini and Grossi-de-Sà 2002). This section will give a brief overview of the insecticidal properties that have been reported for lectins belonging to the different lectin families (Table 14.1).

Amaranthins: Taking into consideration the exclusive location of the amaranthins in the seeds and their high affinity for the T-antigen, it can be envisaged that these lectins are involved in the plant's defense against seed predators. However, at present very little information is available with regard to the physiological role of amaranthins and their involvement in plant defense. Some recent studies with phloem-specific expression of the *A. caudatus* agglutinin (ACA) in tobacco as well as in cotton suggest that the ACA gene can be considered an effective aphid-resistant gene (Guo et al. 2004; Wu et al. 2006). Insect bioassays with *Myzus persicae* and *Aphis gossypii* showed up to 75% and 64% inhibition, respectively, of aphid population growth on transgenic lines, with the highest inhibition on those transgenic lines expressing the highest lectin levels.

Chitin-binding lectins composed of hevein domains: By virtue of their interaction with GlcNAc-oligomers (chitin), lectins with hevein domains were always believed to be associated with the plant's defense against insects. Feeding trials with artificial diets confirmed that several of these chitin-binding lectins interfere with the growth and development of insects. For example, WGA, rice lectin, stinging nettle lectin, and the potato and thorn apple lectins had a moderate inhibitory effect on the development of the larvae of the cowpea weevil (*Callosobruchus maculatus*; Murdock et al. 1990; Huesing et al. 1991). WGA inhibits the larval growth of the Southern corn rootworm (*Diabrotica undecimpunctata*) and kills the neonate *Ostrinia nubilalis* (European corn borer) larvae at fairly low concentrations (Czapla and Lang 1990). In addition, WGA shows larvicidal activity on the blowfly *Lucilia cuprina* (Eisemann et al. 1994). Similarly, the *Phytolacca americana* lectin is lethal to larvae of the Southern corn rootworm (Czapla and Lang 1990). Recently Gupta et al. (2005) also reported an effect of WGA against neonates of American bollworm (*Helicoverpa armigera*). The results of feeding trials leave no doubt that some chitin-binding lectins have deleterious effects on chewing insects when tested in artificial diets. At present, it still remains to be demonstrated that the expression of a chitin-binding lectin in a transgenic plant offers an increased protection against chewing insects. Tests with artificial diets further indicated that WGA negatively affects the growth and reproduction of the phloem feeding rice brown planthopper (*Nilaparvata lugens*; Powell et al. 1995a) and the mustard aphid (*Lipaphis erysimi*). In addition, transformation of Indian mustard with the cDNA encoding WGA resulted in transgenic plants causing high mortality and reduced fecundity of aphids (*L. erysimi*; Kanrar et al. 2002).

GNA-related lectins: GNA-related lectins are undoubtedly the most intensively studied family of plant lectins that are actually being investigated for their role in the plant's defense against phytophagous insects and other invertebrates. One argument in favor of the defensive role against sucking insects is the observation that some of these lectins accumulate in the phloem sap. The fact that a typical monocot mannose-binding lectin is the most abundant protein in the phloem exudate of flowering stalks of onion (*Allium cepa*) plants is in agreement with the presumed defensive role against phloem feeding insects (Peumans et al. 1997).

GNA was the first lectin found to affect the growth and development of sap-sucking insects, and its insecticidal properties have been studied in most detail. Numerous experiments with artificial diets clearly demonstrated that the snowdrop lectin has detrimental effects on the development and reproduction of Homopteran pests belonging to the Aphididae (Rahbé et al. 1995; Sauvion et al. 1996), Cicadellidae and Delphacidae (Powell et al. 1993, 1995b, 1998). In the meantime, studies with transgenic tobacco, wheat, tomato, potato and rice plants have confirmed the insecticidal activity of GNA towards Homoptera (aphids) and demonstrated that ectopically expressed GNA has deleterious effects on aphids, leaf hoppers and plant hoppers (Hilder et al. 1995; Down et al. 1996; Rao et al. 1998; Stoger et al. 1999). In addition to Homoptera, GNA exerts also noxious effects both in artificial diets and transgenic plants on the larval growth and development of several Lepidoptera including the tomato moth (*Lacanobia oleracea*), the Mexican rice borer (*Eoreuma loftini*), the sugarcane borer (*Diatraea saccharalis*) and the bollworm (*H. armigera*; Gatehouse et al. 1997; Sétamou et al. 2003; Shukla et al. 2005).

Though GNA is certainly a very promising insecticidal protein, intensive efforts aim at the identification of superior alternatives within the same lectin family. Comparative analyses indicated that some GNA-related lectins found in *Alliaceae* species are substantially more active than GNA itself in terms of e.g. specific agglutination activity (Van Damme et al. 1993). Since one can reasonably assume that the insecticidal activity of GNA and related lectins is somehow related to their sugar-binding activity/specificity it seemed worthwhile to check whether some GNA-related lectins possibly have a more promising insecticidal potency. Over the last few years numerous studies have shown that not only GNA but also many GNA-related lectins are detrimental especially to sap-sucking insects and also to some caterpillars. For instance bioassays with artificial diets revealed that the garlic (*Allium sativum*) leaf lectin ASAL was highly toxic to the hemipteran pests *N. lugens* and *Nilaparvata virescens* (Bandyopadhyay et al. 2001; Majumder et al. 2004). Analysis of transgenic tobacco and mustard plants expressing ASAL revealed a decrease in survival and fecundity of these phloem feeding aphids (Dutta et al. 2005a, b). In addition, Sadeghi et al. (2008) have shown that transgenic tobacco plants expressing ASAL and ASAII significantly ($p<0.05$) reduced the weight gain of 4th instar larvae of *Spodoptera littoralis*. The lectins retarded the development of the larvae and their metamorphosis, and were also detrimental to the pupal stage resulting in weight reduction and lethal abnormalities. Mortality was 100% with ASAL compared to 60% with ASAII.

Similarly, the *Allium porrum* (leek) lectin (APA) was recently shown to be a potent insecticidal protein. Ectopically expressed APA provides transgenic tobacco plants significant levels of protection against the cotton leafworm *S. littoralis* due to its effect on larval survival, growth and development (unpublished results).

Insecticidal activity was recently also shown for some Araceae lectins. The *Arum maculatum* lectin was tested in an artificial diet against the aphids *L. erysimi* and *Aphis craccivora* (Majumder et al. 2005). When incorporated in an artificial diet *Arisaema jacquemontii* lectin affected the development of *Bactrocera cucurbitae* larvae (Kaur et al. 2006). Transgenic tobacco expressing *Pinellia ternata* agglutinin exhibited enhanced resistance to peach potato aphid (*M. persicae*; Yao et al. 2003).

Jacalin-related lectins: Feeding trials with artificial diets have shown that the Moraceae seed lectins have anti-insect properties. For example, the *Maclura pomifera* lectin exerted a significant inhibitory effect on the larvae of the cowpea weevil (*C. maculatus*) (Murdock et al. 1990). Similarly, jacalin and the *M. pomifera* lectin inhibited larval growth of the Southern corn rootworm (*D. undecimpunctata*; Czapla and Lang 1990). Jacalin also affected the survival of the potato leafhopper (*Empoasca fabae*; Habibi et al. 1993). Based on these observations it seems likely that the galactose-binding Moraceae seed lectins have a defensive function against insects. At present, there is little information about the role of mannose-binding jacalin-related lectins. Chang et al. (2003) have shown that the *Helianthus tuberosus* agglutinin confers resistance to peach potato aphid (*M. persicae*). The average population of aphids on transgenic plants expressing the lectin gene decreased by 53%–70% as compared to control populations during an 11-day assay. In addition, the development of aphids was notably retarded, suggesting that this lectin gene might be another candidate for genetic engineering of plants for resistance against homopteran insect pests (Chang et al. 2003).

Legume lectins: Many legume seeds contain high concentrations of lectins. It has been suggested that under normal conditions these lectins act as genuine storage proteins. However, as soon as the plant is eaten, the lectins end up in the gastrointestinal tract of the predator and act as defense proteins. Some legume lectins were shown to be toxic to insects or interfere with insect development when tested in vitro. For example, the galactose-binding peanut (*Arachis hypogeae*) agglutinin as well as the basic seed lectin from winged bean (*Psophocarpus tetragonolobus*) show an inhibitory effect on development of cowpea weevil larvae (Murdock et al. 1990; Gatehouse et al. 1991). Similarly, the *Bauhinia purpurea* lectin is lethal to neonate larvae of *O. nubilalis* and inhibits growth of *D. undecimpunctata* larvae (Czapla and Lang 1990). Recently, *Bauhinia monandra* leaf lectin was also shown to cause 50% mortality of *C. maculatus* and *Zabrotes subfasciatus* when incorporated into an artificial diet at 0.5% and 0.3% (w/w), respectively but did not significantly decrease the survival of *Anagasta kuehniella* at 1% (w/w; Macedo et al. 2006). African yam bean (*Sphenostylis stenocarpa*) lectins affected nymphal survival, reduced growth and delayed total developmental time of *Clavigralla tomentosicollis* resulting in a strong overall toxicity (Okeola and Machuka 2001). Incorporation of the N-acetylglucosamine-specific lectin GS-II from leaves and seeds of *Griffonia simplicifolia* in an artificial diet doubled the developmental time of the cowpea weevil (*C. maculatus*; Zhu et al. 1996). Similarly, an N-acetylglucosamine-binding lectin from *Koelreuteria paniculata* (golden rain) affected larval development of *C. maculatus* as well as *A. kuehniella* (Macedo et al. 2003).

The effects of Man/Glc-specific lectin from *Canavalia ensiformis* (concanavalin A, ConA; jackbean) have been tested on crop pests from two different orders, Lepidoptera and Homoptera. When fed to larvae of tomato moth (*L. oleracea*) in artificial diet, ConA retarded development and decreased survival with up to 90% mortality at a lectin concentration of 2.0% (w/w) of the total protein in artificial diet. In liquid artificial diet, ConA also reduced the size of peach potato aphids (*M. persicae*) by up to 30%, retarded development to maturity, and reduced fecundity, but had little effect on survival. Bioassays with *L. oleracea* larvae on ConA expressing potato plants showed that the lectin retarded larval development, and decreased larval weights by >45%, but had no significant effect on survival. It also decreased consumption of plant tissue by the larvae. In agreement with the diet bioassay results, ConA expressing potatoes also decreased the fecundity of *M. persicae* by up to 45%. Therefore ConA has potential as a protective agent against insect pests in transgenic crops (Gatehouse et al. 1999).

Tests with transgenic plants have shown that tobacco plants expressing the pea lectin gene have increased resistance to *Heliothis virescens* (Boulter et al. 1990). In addition, transgenic oilseed rape expressing pea lectin reduced the growth rate of pollen beetle larvae (*Meligethes aeneus*; Melander et al. 2003). A negative correlation was observed between lectin concentration and larval growth.

Although some older reports claim toxicity of bean (*Phaseolus vulgaris*) lectin PHA (with complex specificity) to insects, Murdock et al. (1990) have clearly shown that PHA is not toxic to cowpea weevil (*C. maculatus*), and that the toxic effects previously attributed to PHA are due to contamination of the lectin preparation with the α-amylase inhibitor. It was demonstrated, indeed, both by in vitro assays and with transgenic plants that two lectin-related proteins from *P. vulgaris*, namely the α-amylase inhibitor and arcelin, are far more potent anti-insect proteins than the bean lectin itself. For example, low levels of dietary *P. vulgaris* α-amylase inhibitor effectively inhibit larval growth of typical seed predating insects like the cowpea weevil (*C. maculatus*) and the Azuki bean weevil (*Callosobruchus chinensis*) (Ishimoto and Kitamura 1989). Moreover transgenic pea seeds expressing the bean α-amylase inhibitor acquired resistance against the cowpea and Azuki bean weevil (Shade et al. 1994). Although these lectin-related proteins are evolutionary related to the genuine lectin from bean, they do not bind carbohydrates (Ishimoto et al. 1996; Osborn et al. 1988; Paes et al. 2000).

Lectins with ricin-B domains: Although the role of lectins with ricin-B domains is not fully understood yet, there is growing belief that these proteins play a role in plant defense. At present, there is good evidence for a role of these proteins in defense against some viruses and fungi (Van Damme et al. 2001; Vandenbussche et al. 2004a, b), but very little research addressed the effect of these lectins on insects. However, in recent years evidence is accumulating to suggest a role for ribosome-inactivating proteins in protection of plants against insects. For example, toxicity of cinnamomin, a type-2 RIP from *Cinnamomum camphora*, was observed towards cotton bollworm (*H. armigera*) and mosquito (*Culex pipines pallens*) as well as domestic silkworm (*Bombyx mori*) larvae (Zhou et al. 2000; Wei et al. 2004). Recently insecticidal activity of a type-2 RIP from iris was also reported. The *Iris* RIP showed activity towards both *Myzus nicotianae* and *Spodoptera exigua* when expressed in tobacco (Noghabi et al. 2006).

14.5 Insect Herbivory Induces the Expression of Some Lectins in Plants

The expression of the previously mentioned jasmonate-inducible lectin in tobacco leaves in response to many biotic and abiotic factors has been investigated. Of all plant hormones tested only jasmonic acid was able to induce expression of the lectin

(15–20 mg/g leaf after 3 days floating on 50 µM jasmonate solution). Wounding, however, did not induce lectin expression. Since jasmonates are important signaling molecules in plant defense against herbivores (Halitschke and Baldwin 2004; Schaller and Stintzi this volume), the effect of insect herbivory was also tested on 16 week-old greenhouse grown tobacco plants. Leaves were infested with larvae of the cotton leaf worm (*S. littoralis*) resulting in higher jasmonate concentrations in the leaf, which in turn induced lectin expression (150–200 µg/g leaf). When larvae were allowed to feed on a single leaf, systemic induction of lectin activity was observed in all leaves of the plant. Preliminary experiments also showed that the tobacco lectin exerts a repellent effect on chewing insects (unpublished data).

Zhu-Salzman et al. (1998) reported previously that the expression of GlcNAc-binding *G. simplicifolia* lectin II (GS-II) in leaves is systemically (but not locally at the site of treatment) up-regulated after wounding as well as by jasmonic acid treatment. Similarly, insect attack up-regulates GS-II expression only in systemic leaves but not in local leaves. In contrast to the tobacco lectin, which can only be detected after induction, the *Griffonia* lectin is constitutively expressed at low but detectable levels in leaf tissue.

Wheat plants were also shown to respond to insects with the expression of lectin genes (Williams et al. 2002). The feeding of first-instar larvae of specific biotypes of Hessian fly (*Mayetiola destructor*) on wheat caused rapid changes in the expression of several transcripts at the feeding site including the mRNA for Hfr-1, a jacalin-related mannose-binding lectin. It was reported that Hfr-1 expression is also induced by salicylic acid. Sequence analysis revealed that the C-terminal domain of the Hfr-1 deduced amino acid sequence shows sequence similarity with jacalin-related lectins (Williams et al. 2002; Subramanyam et al. 2006).

More recently, Puthoff et al. (2005) and Giovanini et al. (2007) identified the Hfr-2 and Hfr-3 genes, two other Hessian fly-responsive wheat genes containing lectin domains similar to amaranthin and hevein, respectively. Infestation of wheat by virulent Hessian fly larvae resulted in the upregulation of Hfr-2 gene expression by about 80-fold compared to uninfested controls. Sequence analysis of Hfr-2 revealed an N-terminal lectin domain similar to amaranthin fused to a region similar to haemolytic lectins and channel-forming toxins. Expression of Hfr-2 is also upregulated by methyl jasmonate treatment, phloem feeding of bird cherry-oat aphids (*Rhopalosiphum padi*) and fall armyworm (*Spodoptera frugiperda*) chewing, whereas wounding, salicylic acid and abscisic acid treatment only had a slight effect (Puthoff et al. 2005). mRNA levels for Hfr-3 (with four predicted chitin-binding hevein domains) were shown to increase about 3000-fold above the uninfested control in the incompatible interaction three days after egg hatch of Hessian fly. Since the mRNA abundance for Hfr-3 was dependent on the number of larvae per plant it was suggested that resistance is localized rather than systemic (Giovanini et al. 2007). The Hfr-3 gene was also responsive to bird cherry-oat aphid (*R. padi*) but not to fall armyworm (*S. frugiperda*) attack. Treatment with jasmonate, salicylic acid or abscisic acid, or wounding did not influence Hfr-3 expression. All these data suggest the involvement of a set of lectins in resistance of wheat to Hessian fly.

14.6 Mode of Action

14.6.1 Classical Lectins

As defined above, plant lectins comprise all plant proteins that bind reversibly to specific mono- or oligosaccharides. A typical lectin is multivalent, and therefore capable of agglutinating or clumping cells. Over the last two decades many plant lectins have been crystallized and their three-dimensional structure studied, often in complex with a complementary carbohydrate. These structural studies yielded a lot of valuable information regarding the interaction of the sugar-binding site of plant lectins with carbohydrates or glycoproteins. It was shown that within a given family lectins exhibit very similar three-dimensional structures even though there are important differences in their carbohydrate-binding specificities. For example, all legume lectins have the same β-sandwich fold notwithstanding the fact that there are marked differences in specificity within this particular lectin family. It is common belief that the specific interaction of the lectin with a carbohydrate is at the molecular basis for its activity in plant defense. Furthermore, it is obvious that the specificity of plant lectins is not directed against simple sugars but rather against more complex carbohydrates such as the N- and O-glycans present on several glycoproteins (Van Damme et al. 2007b). Most of the 'classical' lectins specifically recognize typical animal glycans (e.g., sialic acid and GalNAc containing O- and N-linked oligosaccharides) that are abundantly present on the surface of the epithelial cells exposed along the intestinal tract of higher and lower animals. Since these glycans are accessible for dietary proteins they represent potential binding sites for dietary plant lectins. If binding to these receptors results in an adverse effect the lectin will exert harmful or toxic effects. Feeding trials with insects and higher animals confirmed that some plant lectins provoke toxic effects ranging from a slight discomfort to a deadly intoxication (Peumans and Van Damme 1995), which leaves little doubt that lectins play a role, indeed, in plant defense against insects and/or predating animals. In this respect the insecticidal activity of the GNA-related lectins is in agreement with their preferential binding to high mannose-type glycan chains, which are typical constituents of insect glycoproteins.

At present, our knowledge about the mechanism of action of plant lectins is still limited. Several types of interactions can be envisaged within the insect body: plant lectins, especially chitin-binding lectins, can interact with the chitin matrix of the peritrophic membrane. In addition, any other lectin with a carbohydrate binding site complementary to glycoconjugates in the peritrophic membrane, or exposed along the surface of the midgut epithelial cells, can bind to these carbohydrate structures. Finally, one can also imagine that lectin binding to certain glycosylated enzymes/proteins could interfere with cellular processes important for normal functioning of the organism (Van Damme and Peuman 1995; Czapla 1997).

In the case of the chitin-binding lectins the detrimental effect on insects has been ascribed to an impairment of the peritrophic membrane after extensive binding of the lectin to the chitin present in this specialized structure (Lehane 1997; Harper et al. 1998; Wang et al. 2004). Additional target sites (proteins and/or oligosaccharides) in the insect body (e.g., fat tissue, hemolymph, reproduction organs) are possible, however (Carlini and Grossi-de-Sà 2002; Murdock and Shade 2002). The peritrophic membrane consists of a matrix of chitin fibres and proteins, and is present as an envelope in the lumen of the insect gut, which physically protects the epithelial cells. This membrane also has an important role in compartimentalisation of the digestive enzymes. It is very likely that chitin-binding lectins can bind to the chitin in the peritrophic membrane and as such disrupt its structure. Any interference with the peritrophic membrane will have devastating effects leading to starvation and finally death of the insect. This hypothesis was confirmed by Harper et al. (1998) who showed that the morphology of the peritrophic membrane of European corn borer larvae changed within 24 h after feeding on a WGA diet. In contrast to the control larvae, those fed on WGA showed a multilayered and unorganized perithrophic membrane with pieces of desintegrated microvilli, and embedded food particles and bacteria. After 72 h, the peritrophic membrane of WGA-fed larvae was multilayered and discontinuous allowing cell wall fragments to penetrate into the microvilli of the epithelium. Scanning electron microscopy futher revealed a desintegration of the chitinous meshwork and a reduced proteinaceous matrix.

There is presently no clear correlation between lectin specificity and toxicity, but binding to glycoproteins in the midgut of the insect appears to be a prerequisite for any toxic effect of lectins (Czapla 1997). Experiments with *G. simplicifolia* lectin (GS-II) for example, revealed a clear correlation between receptor binding and toxicity towards the cowpea bruchid *C. maculatus* (Zhu-Salzman et al. 1996). Furthermore, lectins that caused a significant mortality or a decrease in weight gain in *O. nubilalis* also bound strongly to *O. nubilalis* brush border membrane proteins However, not all lectins that bound strongly were insecticidal, and therefore, lectin binding is not an absolute predictor of toxicity (Harper et al. 1995; Murdock and Shade 2002).

The mannose-binding lectins have been shown to interact with glycosylated receptors in the insect midgut. One of the major receptors for GNA in *N. lugens* was identified as a subunit of ferritin suggesting that GNA may interfere with the insects' iron metabolism (Du et al. 2000). ConA was shown to interact with glycosylated receptors at the cell surface (Sauvion et al. 2004), one of which was later identified as an aminopeptidase in *Acyrthosiphon pisum* (Cristofoletti et al. 2006). In the case of the garlic leaf lectin ASAL, Bandyopadhyay et al. (2001) reported binding to the carbohydrate part of the 55 and 45 kDa brush border membrane vesicle receptor proteins in mustard aphid (*L. erysimi*) and red cotton bug (*D. cingulatus*), respectively. It was suggested that binding of ASAL to these receptors may decrease the permeability of the membrane, as shown for the mannose-binding lectins GNA and ConA that reduced the uptake and absorption of nutrients in peach potato aphids (Sauvion et al. 1996).

In addition to lectin binding to certain carbohydrate structures in the insect, a second prerequisite for insecticidal activity is the resistance of the lectin to proteolytic degradation. Zhu-Salzman et al. (1998) reported a good correlation between the insecticidal activity of the GlcNAc-specific *Griffonia* lectin GS-II and its resistance to proteolytic degradation by *C. maculatus* midgut extracts. Experiments with mutant proteins lacking carbohydrate-binding and insecticidal activity revealed that these proteins are rapidly digested by gut digestive proteases (Zhu-Salzman and Salzman 2001). Similarly, it has been shown that e.g. the mannose-binding lectins GNA and ASAL remain stable and survive the insect gut (Powell et al. 1998; Bandyopadhyay et al. 2001).

Interestingly, Fitches and Gatehouse (1998) demonstrated binding of GNA and ConA to the soluble and brush border membrane enzymes in the midgut of *L. oleracea*, which affected the activities of soluble and brush border membrane enzymes. In short term bioassays, both GNA and ConA increased gut protein levels and brush border membrane amino peptidase activity. The lectins also increased trypsin activity in the gut (ConA) and the faeces (GNA). GNA also increased α-glucosidase activity, but neither lectin had an effect on alkaline phosphatase activity (Fitches and Gatehouse 1998). In long-term bioassays a reduced α-glucosidase activity was seen with both lectins, but almost no changes were observed for the other enzyme activities.

Experiments with rats revealed that in some cases, the eventual discomfort caused by ingestion of lectins is so severe that experimental animals refuse to continue eating a lectin-containing diet, indicating a repellent effect of lectins (Larue-Achagiotis et al. 1992). In recent years a number of behavioral studies have also been performed with insects, to look for sensory mediation of lectin activity, and effects of lectins on insect behavior. A decrease in the quantity of honeydew excreted has been reported for some lectins, such as e.g. GNA and *Psophocarpus* lectin (Powell et al. 1995a; Powell 2001) suggesting that these lectins act as a feeding deterrent. However, Sauvion et al. (2004) have shown that feeding deterrency observed after ConA treatment of *A. pisum* was a consequence of intoxication rather than a sensory mediated process. Sadeghi et al. (2006) have shown that plant lectins can also interfere with oviposition. Analyses were done with a set of 14 plant lectins using a binary choice bioassay for inhibitory activity on cowpea weevil (*C. maculatus*) oviposition. Coating of chickpea seeds (*Cicer arietinum* L.) with a 0.05% (w/v) solution of plant lectins caused a significant reduction in egg laying. However, no clear correlation could be established between deterrent activity and sugar-binding specificity/molecular structure of the lectins.

14.6.2 Inducible Lectins

Plants have different constitutive as well as inducible defense mechanisms at their disposal to protect themselves against e.g. insect attack. The inducible systems, which are thought to be more durable on the long term, play an important role in resistance and result in different effects on phytophagous pest insects such as a higher toxicity, retardation of larval development or enhanced susceptibility to natural enemies.

Over the last few years it has become very clear that the jasmonate pathway plays an important role in plant resistance against a wide variety of herbivorous pest insects. Within three to five hours after insect attack of *Arabidopsis* plants the expression levels of hundreds of transcripts change (Gfeller and Farmer 2004). According to Reymond et al. (2004), the expression of 67%–84% of these proteins is controlled by the jasmonate pathway. At present it is not known how the jasmonate inducible tobacco lectin exerts its insecticidal activity. However, it is clear that the defense system based on (inducible) lectins -as exemplified by the tobacco lectin- is different from other known plant defense proteins such as the proteinase inhibitors. These proteinase inhibitors are known to exert their effect through inactivation of dietary enzymes and -in contrast to the inducible tobacco lectin- can be induced by a whole range of stress factors (such as wounding, infection by pathogens, insect attack, treatment with UV light, or salicylic acid; Haq et al. 2004).

14.7 Perspectives

Although there are still a lot of unanswered questions regarding the physiological role of plant carbohydrate-binding proteins, substantial progress has been made during the last decade in our general understanding of the role of at least those plant lectins, which are constitutively expressed in reasonable quantities. Biochemical and molecular studies of numerous lectins eventually demonstrated that only a limited number of carbohydrate-binding motifs evolved in plants (Peumans et al. 2000). Since the specificity of these binding motifs is primarily directed against foreign glycans, it is generally accepted now that many plant lectins are involved in the recognition and binding of glycans from foreign organisms, and accordingly play a role in plant defense (Peumans and Van Damme 1995; Van Damme et al. 1998a). Animal and insect feeding studies with purified lectins and experiments with transgenic plants confirmed that at least some lectins enhance the plant's resistance against herbivorous higher animals or phytophagous invertebrates. To reconcile the presumed defensive role with the high concentration, the concept was developed that many plant lectins are storage proteins which serve additional roles in defense in case the plant is challenged by a predator.

In the last decade, protective proteins of higher plants have received much attention from plant breeders, because of their potential to render crop plants resistant to stresses. Genetically modified plants that express defensive proteins at elevated levels may thus be of great agricultural (and commercial) value. Therefore, these defensive proteins are also being investigated as a new type of agrochemical. In principal, any plant lectin with insecticidal activity can be used to make a whole plant or a specific tissue unpalatable to insects. However, the toxicity of some lectins restricts their use as resistance factors. While most lectins are only moderately toxic, there are a few examples of lectins that are highly toxic or even lethal (e.g., ricin, abrin). One should also keep in mind that many food plants and crops contain lectins (Peumans and Van Damme 1996). The fact that a given lectin occurs naturally in crop plants will certainly improve its acceptability as an insect resistance factor in transgenic crop plants. Based on these considerations, GNA-related lectins in leek (*A. porrum* agglutinin or APA), and garlic (*A. sativum* agglutinin or ASAL) have been suggested to be more appropriate candidate insecticidal proteins than GNA itself.

However, at present there is little concern about the use of GNA in crop protection. GNA is considered non-toxic to mammals because of its very low binding capacity to the jejunum (Pusztai et al. 1990). A rice variety expressing the *G. nivalis* lectin gene (1.25% GNA of the total soluble protein in the seed) was recently selected as a model crop and subjected to a 90-day feeding study in order to assess the safety of this transgenic food crop in rats. In this animal study different clinical, biological, immunological, microbiological and pathological parameters were examined. Although a number of significant differences were seen between groups fed the genetically modified plant and control diets, none of them were considered to be adverse (Poulsen et al. 2007).

Besides repellent effects against herbivores, several recent publications report that plant defense proteins and lectins can have additional positive effects on plant resistance by attracting beneficial insects, such as predators and parasitic wasps (Agrawal 2003; Gatehouse 2002; Kessler and Baldwin 2002; Musser et al. 2002). These researchers confirm that plant lectins and insect-resistant transgenic plants can play an important role in the tritrophic relations among plants, pest insects, and beneficial insects.

Many phloem feeding insects are vectors of viral transmission. Saha et al. (2006) found that transgenic rice plants expressing ASAL (1.01% of total soluble protein) under a phloem-specific promoter exhibited less or no incidence of tungro disease after attack of green leafhopper (*N. virescens*). These results show that phloem expression of the lectin not only helps to control the phloem feeding insects but also the viral diseases they transmit.

The discovery that following ingestion, snowdrop lectin remains stable and active within the insect gut, and is able to cross the gut epithelium (Powell et al. 1998; Fitches et al. 2001), offered the opportunity to use GNA as a carrier molecule to deliver other peptides to the circulatory system. Fitches et al. (2002) made fusion proteins with allostatin to deliver this insect neuropeptide hormone to the hemolymph of *L. oleracea*. A similar strategy was used to deliver a venom protein of the spider *Segestria florentina* to the hemolymph of *L. oleracea* (Fitches et al. 2004). The latter fusion protein was also shown to affect survival of *N. lugens* and *M. persicae* when incorporated into artificial diets (Down et al. 2006). Finally a fusion protein consisting of a toxin from the red scorpion (*Mesobuthus tamulus*) N-terminally fused to GNA was acutely toxic when fed to *L. oleracea* larvae (Pham Trung et al. 2006). These results show that GNA can successfully be exploited as a carrier for toxins that are normally only toxic when injected into the insect hemolymph, across the insect gut.

As already mentioned above several mannose-binding GNA-related lectins proved to be promising candidates for the control of hemipteran pests. Because of their different modes of action, these lectins may be used in combination with other insecticidal proteins like *Bacillus thuringiensis* (Bt) toxins and protease inhibitors (Christou et al. 2006). Bano-Maqbool et al. (2001) reported on transgenic rice plants expressing two Cry Bt proteins in combination with GNA. Whereas the Bt genes target the rice leaf folder (*Cnephalocrocis medinalis*) and the yellow stemborer (*Scirpophaga incertulas*), the GNA gene targets the leafhopper (*N. lugens*). The triple transgenic lines were more resistant as compared to the binary counterparts, indicating that the simultaneous introduction of multiple resistance genes can be advantageous.

References

Agrawal AA (2003) Mechanisms, ecological consequences and agricultural implications of tritrophic interactions. Curr Opin Plant Biol 3:329–335

Bandyopadhyay S, Roy A, Das S (2001) Binding of garlic (*Allium sativum*) leaf lectin to the gut receptors of homopteran pests is correlated to its insecticidal activity. Plant Sci 161:1025–1033

Bano-Maqbool S, Riazuddin S, Loc NT, Gatehouse AMR., Gatehouse JA, Christou P (2001) Expression of multiple insecticidal genes confers broad resistance against a range of different rice pests. Mol Breed 7:85–93

Boulter D, Edwards GA, Gatehouse AMR, Gatehouse JA, Hilder VA (1990) Additive protective effects of different plant-derived insect resistance genes in transgenic tobacco plants. Crop Prot 9:351–354

Bourne Y, Astoul CH, Zamboni V, Peumans WJ, Menu-Bouaouiche L, Van Damme EJM, Barre A, Rougé P (2002) Structural basis for the unusual carbohydrate-binding specificity of jacalin towards galactose and mannose. Biochem J 364:173–180

Carlini CR, Grossi-de-Sà MF (2002) Plant toxic proteins with insecticidal properties. A review on their potentialities as bioinsecticides. Toxicon 40:1515–1539

Chang T, Chen L, Chen S, Cai H, Liu X, Xiao G, Zhu Z (2003) Transformation of tobacco with genes encoding *Helianthus tuberosus* agglutinin (HTA) confers resistance to peach-potato aphid (*Myzus persicae*). Transgenic Res 12:607–614

Chen Y, Peumans WJ, Hause B, Bras J, Kumar M, Proost P, Barre A, Rougé P, Van Damme EJM (2002) Jasmonic acid methyl ester induces the synthesis of a cytoplasmic/nuclear chitooligosaccharide-binding lectin in tobacco leaves. FASEB J 16:905–907

Christou P, Capell T, Kohli A, Gatehouse JA, Gatehouse AMR (2006) Recent developments and future prospects in insect pest control in transgenic crops. Trends Plant Sci 11:302–308

Claes B, Dekeyser R, Villarroel R, Van den Bulcke M, Bauw G, Van Montagu M, Caplan A (1990) Characterization of a rice gene showing organ-specific expression in response to salt stress and drought. Plant Cell 2:19–27

Cristofoletti PT, de Sousa FA, Rahbé Y, Terra WR (2006) Characterization of a membrane-bound aminopeptidase purified from *Acyrthosiphon pisum* midgut cells. A major binding site for toxic mannose lectins. FEBS J 273:5574–5588

Czapla TH (1997) Plant lectins as insect control proteins in transgenic plants. In: Carozzi N, Koziel M (eds) Advances in insect control: the role of transgenic plants. Taylor and Francis, London, pp 123–138

Czapla TH, Lang BA (1990) Effect of plant lectins on the larval development of European corn borer (Lepidoptera: Pyralidae) and southern corn rootworm (Coleoptera: Chrysomelidae). J Econ Entomol 83:2480–2485

Down RE, Fitches EC, Wiles DP, Corti P, Bell HA, Gatehouse JA, Edwards JP (2006) Insecticidal spider venom toxin fused to snowdrop lectin is toxic to the peach-potato aphid, *Myzus persicae* (Hemiptera: Aphididae) and the rice brown planthopper, *Nilaparvata lugens* (Hemiptera: Delphacidae). Pest Manag Sci 62:77–85

Down RE, Gatehouse AMR, Hamilton WD, Gatehouse JA (1996) Snowdrop lectin inhibits development and decreases fecundity of the glasshouse potato aphid (*Aulacorthum solani*) when administered in vitro and via transgenic plants both in laboratory and glasshouse trials. J Insect Physiol 42:1035–1045

Du J, Foissac X, Carss A, Gatehouse AMR, Gatehouse JA (2000) Ferritin acts as the most abundant binding protein for snowdrop lectin in the midgut of rice brown planthoppers (*Nilaparvata lugens*). Insect Biochem Mol Biol 30:297–305

Dutta I, Majumder P, Saha P, Sakar A, Ray K, Das S (2005a) Constitutive and phloem specific expression of *Allium sativum* leaf agglutinin (ASAL) to engineer aphid (*Lipaphis erysimi*) resistance in transgenic Indian mustard (*Brassica juncea*). Plant Sci 169:996–1007

Dutta I, Saha P, Majumder, P, Sakar A, Chakraborti D, Banerjee S, Das S (2005b) The efficacy of a novel insecticidal protein, *Allium sativum* leaf lectin (ASAL), against homopteran insects monitored in transgenic tobacco. Plant Biotechnol J 3:601–611

Eisemann CH, Donaldson RA, Pearson RD, Cadogan LC, Vuocolo T, Tellam RL (1994) Larvicidal activity of lectins on *Lucilia cuprina* – Mechanism of action. Entomol Exp Appl 72:1–10

Fitches E, Audsley N, Gatehouse JA, Edwards JP (2002) Fusion proteins containing neuropeptides as novel insect contol agents: snowdrop lectin delivers fused allatostatin to insect haemolymph following oral ingestion. Insect Biochem Mol Biol 32:1653–1661

Fitches E, Edwards MG, Mee C, Grishin E, Gatehouse AM, Edwards JP, Gatehouse JA (2004) Fusion proteins containing insect-specific toxins as pest control agents: snowdrop lectin delivers fused insecticidal spider venom toxin to insect haemolymph following oral ingestion. J Insect Physiol 50:61–71

Fitches E, Gatehouse JA (1998) A comparison of the short and long term effects of insecticidal lectins on the activities of soluble and brush border enzymes of tomato moth larvae (*Lacanobia oleracea*). J Insect Physiol 44:1213–1224

Fitches E, Woodhouse SD, Edwards JP, Gatehouse, J.A. (2001) In vitro and in vivo binding of snowdrop (*Galanthus nivalis* agglutinin; GNA) and jack bean (*Canavalia ensiformis*; ConA) lectins within tomato moth (*Lacanobia oleracea*) larvae: mechanisms of insecticidal action. J Insect Physiol 47:777–787

Gatehouse AMR, Davison GM, Stewart JN, Gatehouse LN, Kumar A, Geoghegan IE, Birch ANE, Gatehouse JA (1999) Concanavalin A inhibits development of tomato moth (*Lacanobia oleracea*) and peach-potato aphid (*Myzus persicae*) when expressed in transgenic potato plants. Mol Breed 5:153–165

Gatehouse AMR, Davison GM, Newell CA, Merryweather A, Hamilton WDO, Burgess EPJ, Gilbert RJC, Gatehouse JA (1997) Transgenic potato plants with enhanced resistance to the tomato moth, *Lacanobia oleracea*: growth room trials. Mol Breed 3:49–63

Gatehouse AMR, Howe DS, Flemming JE, Hilder VA, Gatehouse JA (1991) Biochemical basis of insect resistance in winged bean (*Psophocarpus tetragonolobus*) seeds. J Sci Food Agric 55:63–74

Gatehouse JA (2002) Plant resistance towards insect herbivores: a dynamic interaction. New Phytol 156:145–169

Gfeller A, Farmer EE (2004). Keeping the leaves green above us. Science 306:1515–1516

Giovanini MP, Saltmann KD, Puthoff DP, Gonzalo M, Ohm HW, Williams CE (2007) A novel wheat gene encoding a putative chitin-binding lectin is associated with resistance against Hessian fly. Mol Plant Pathol 8:69–82

Goldstein IJ, Poretz RD (1986) Isolation, physicochemical characterization, and carbohydrate-binding specificity of lectins. In: Liener IE, Sharon N, Goldstein IJ (eds) The lectins, properties, functions, and applications in biology and medicine. Academic Press, Orlando, USA, pp 33–247

Guo HN, Jia YT, Zhou YG, Zhang ZS, Ouyang Q, Jiang Y, Tian YC (2004) Effects of transgenic tobacco plants expressing ACA gene from *Amaranthus caudatus* on the population development of *Myzus persicae*. Acta Bot Sin 46:1100–1105

Gupta GP, Birah A, Rani S (2005) Effect of plant lectins on growth and development of American bollworm (*Helicoverpa armigera*). Indian J Agric Sci 75:207–212

Habibi J, Backus EA, Czapla TH (1993) Plant lectins affect survival of the potato leafhopper (Homoptera: Cicadellidae). J Econ Entomol 86:945–951

Halitschke R, Baldwin IT (2004) Jasmonates and related compounds in plant-insect interactions. J Plant Growth Regul 23:238–245

Haq SK, Atif SM, Khan RH (2004) Protein proteinase inhibitor genes in combat against insects, pests, and pathogens: natural and engineered phytoprotection. Arch Biochem Biophys 431:145–159

Harper SM, Crenshaw RW, Mullins MA, Privalle LS (1995) Lectin binding to insect brush border membranes. J Econ Entomol 88:1197–1202

Harper MS, Hopkins TL, Czapla TH (1998) Effect of wheat germ agglutinin on formation and structure of the peritrophic membrane in European corn borer (*Ostrinia nubilalis*) larvae. Tissue Cell 30:166–176

Hilder VA, Powell KS, Gatehouse AMR, Gatehouse JA, Gatehouse LN, Shi Y, Hamilton WDO, Merryweather A, Newell C, Timans JC, Peumans WJ, Van Damme EJM, Boulter D (1995) Expression of snowdrop lectin in transgenic tobacco plants results in added protection against aphids. Transgenic Res 4:18–25

Huesing JE, Murdock LL, Shade RE (1991) Effect of wheat germ isolectins on development of cowpea weevil. Phytochemistry 30:785–788

Ishimoto M, Kitamura K (1989) Growth inhibitory effects of an α-amylase inhibitor from kidney bean, *Phaseolus vulgaris* (L.) on three species of bruchids (Coleoptera: Bruchidae). Appl Ent Zool 24:281–286

Ishimoto M, Sato T, Chrispeels MJ, Kitamura K (1996) Bruchid resistance of transgenic azuki bean expressing seed alpha-amylase inhibitor of common bean. Entomol Exp Appl 79:309–315

Kaku H, Van Damme EJM, Peumans WJ, Goldstein IJ (1990) Carbohydrate-binding specificity of the daffodil (*Narcissus pseudonarcissus*) and amaryllis (*Hippeastrum* hybr.) bulb lectins. Arch Biochem Biophys 279:298–304

Kaku H, Van Damme EJM, Peumans WJ, Goldstein IJ (1992) New mannose-specific lectins from garlic (*Allium sativum*) and ramsons (*Allium ursinum*) bulbs. Carbohydr Res 229:347–353

Kanrar S, Venkateswari J, Kirti PB, Chopra VL (2002) Transgenic Indian mustard (*Brassica juncea*) with resistance to the mustard aphid (*Lipaphis erysimi* Kalt.). Plant Cell Rep 20:976–981

Kaur M, Singh K, Rup PJ, Kamboj SS, Saxena AK, Sharma M, Bhagat M, Sood SK, Singh J (2006) A tuber lectin from *Arisaema jacquemontii* Blume with anti-insect and anti-proliferative properties. J Biochem Mol Biol 39:432–440

Kessler A, Baldwin IT (2002). Plant responses to insect herbivory: the emerging molecular analysis. Ann Rev Plant Biol 53:299–328

Lannoo N, Peumans WJ, Van Pamel E, Alvarez R, Xiong TC, Hause G, Mazars C, Van Damme EJM (2006) Localization and in vitro binding studies suggest that the cytoplasmic/nuclear tobacco lectin can interact in situ with high-mannose and complex N-glycans. FEBS Lett 580:6329–6337

Larue-Achagiotis C, Picard M, Louis-Sylvestre J (1992) Feeding behavior in rats on a complete diet containing Concanavalin A. Reprod Nutr Dev 32:343–350

Lehane MJ (1997) Peritrophic membrane, structure and function. Ann Rev Entomol 42:525–550

Lord JM, Roberts LM, Robertus JD (1994) Ricin: structure, mode of action, and some current applications. FASEB J 8:201–208

Macedo MLR, Damico DCS, Freire MD, Toyama MH, Marangoni S, Novello JC (2003) Purification and characterization of an N-acetylglucosamine-binding lectin from *Koelreuteria paniculata* seeds and its effect on the larval development of *Callosobruchus maculatus* (Coleoptera: Bruchidae) and *Anagasta kuehniella* (Lepidoptera: Pyralidae). J Agric Food Chem 51:2980–2986

Macedo ML, Freire MD, da Silva MB, Coelho LC (2006) Insecticidal action of *Bauhinia monandra* leaf lectin (BmoLL) against *Anagasta kuehniella* (Lepidoptera: Pyralidae), *Zabrotes subfasciatus* and *Callosobruchus maculatus* (Coleoptera: Bruchidae). Comp Biochem Physiol A Mol Integr Physiol 146:486–498

Majumder P, Banerjee S, Das S (2004) Identification of receptors responsible for binding of the mannose specific lectin to the gut epithelial membrane of the target insects. Glycoconj J 20:525–530

Majumder P, Mondal HA, Das S (2005) Insecticidal activity of *Arum maculatum* tuber lectin and its binding to the glycosylated insect gut receptors. J Agric Food Chem 53:6725–6729

Melander M, Ahman I, Kamnert I, Stromdahl AC (2003) Pea lectin expressed transgenically in oilseed rape reduces growth rate of pollen beetle larvae. Transgenic Res 12:555–567

Murdock LL, Huesing JE, Nielsen SS, Pratt RC, Shade RE (1990) Biological effects of plant lectins on the cowpea weevil. Phytochemistry 29:85–89

Murdock LL, Shade RE (2002) Lectins and protease inhibitors as plant defenses against insects. J Agric Food Chem 50:6605–6611

Musser RO, Hum-Musser SM, Eichenseer H, Peiffer M, Ervin G, Murphy JB, Felton GW (2002). Caterpillar salvia beats plant defenses. Nature 416:599–600

Neuteboom LW, Kunimitsu WY, Webb D, Christopher DA (2002) Characterization and tissue-regulated expression of genes involved in pineapple (*Ananas comosus* L.) root development. Plant Sci 163:1021–1035

Noghabi SS, Van Damme EJM, Smagghe G (2006) Bioassays for insecticidal activity of iris ribosome-inactivating proteins expressed in tobacco plants. Commun Agric Appl Biol Sci 71:285–289

Okeola OG, Machuka J (2001) Biological effects of African yam bean lectins on *Clavigralla tomentosicollis* (Hemiptera: Coreidae). J Econ Entomol 94:724–729

Osborn TC, Alexander DC, Sun SSM, Cardona C, Bliss FA (1988) Insecticidal activity and lectin homology of arcelin seed protein. Science 240:207–210

Paes NS, Gerhardt IR, Coutinho MV, Yokoyama M, Santana E, Harris N, Chrispeels MJ, de Sa MFG (2000) The effect of arcelin-1 on the structure of the midgut of bruchid larvae and immunolocalization of the arcelin protein. J Insect Physiol 46:393–402

Parret AH, Schoofs G, Proost P, De Mot R (2003) Plant lectin-like bacteriocin from a rhizosphere-colonizing *Pseudomonas* isolate. J Bacteriol 185:897–908

Parret AH, Temmerman K, De Mot R (2005) Novel lectin-like bacteriocins of biocontrol strain *Pseudomonas fluorescens* Pf-5. Appl Environ Microbiol 71:5197–5207

Peumans WJ, Barre A., Bras J, Rougé P, Proost P, Van Damme EJM (2002) The liverwort contains a lectin that is structurally and evolutionary related to the monocot mannose-binding lectins. Plant Physiol 129:1054–1065

Peumans WJ, Barre A, Hao Q, Rougé P, Van Damme EJM (2000) Higher plants developed structurally different motifs to recognize foreign glycans. Trends Glycosci Glycotechnol 12: 83–101

Peumans WJ, Fouquaert E, Jauneau A, Rougé P, Lannoo N, Hamada H, Alvarez R, Devreese B, Van Damme EJM (2007) The liverwort *Marchantia polymorpha* expresses orthologs of the fungal *Agaricus bisporus* agglutinin family. Plant Physiol 144:637–647

Peumans WJ, Hao Q, Van Damme EJM (2001) Ribosome-inactivating proteins from plants: more than RNA N-glycosidases? FASEB J 15:1493–1506

Peumans WJ, Smeets K, Van Nerum K, Van Leuven F, Van Damme EJM. (1997) Lectin and alliinase are the predominant proteins in the nectar from leek (*Allium porrum*) flowers. Planta 201:298–302

Peumans WJ, Van Damme EJM (1995) Lectins as plant defense proteins. Plant Physiol 109: 347–352

Peumans WJ, Van Damme EJM (1996) Prevalence, biological activity and genetic manipulation of lectins in foods. Trends Food Sci Technol 7:132–138

Pham Trung N, Fitches E, Gatehouse JA (2006) A fusion protein containing a lepidopteran-specific toxin from the South Indian red scorpion (*Mesobuthus tamulus*) and snowdrop lectin shows oral toxicity to target insects. BMC Biotechnol 6:34–42

Poulsen M, Kroghsbo S, Schroder M, Wilcks A, Jacobsen H, Miller A, Frenzel T, Danier J, Rychlik M, Shu Q, Emami K, Sudhakar D, Gatehouse A, Engel KH, Knudsen I (2007) A 90-day safety study in Wistar rats fed genetically modified rice expressing snowdrop lectin *Galanthus nivalis* (GNA). Food Chem Toxicol 45:350–363

Powell KS (2001) Antimetabolic effects of plant lectins towards nymphal stages of the planthoppers *Tarophagous proserpina* and *Nilaparvata lugens*. Entomol Exp Appl 99:71–77

Powell KS, Gatehouse AMR, Hilder VA, Gatehouse JA (1993) Antimetabolic effects of plant lectins and plant and fungal enzymes on the nymphal stages of two important rice pests, *Nilaparvata lugens* and *Nephotettex cinciteps*. Entomol Exp Appl 66:119–126

Powell KS, Gatehouse AMR, Hilder VA, Gatehouse JA (1995a) Antifeedant effects of plant lectins and an enzyme on the adult stage of the rice brown planthopper, *Nilaparvata lugens*. Entomol Exp Appl 75:51–59

Powell KS, Gatehouse AMR, Peumans WJ, Van Damme EJM, Boonjawat J, Horsham K, Gatehouse JA (1995b) Different antimetabolic effects of related plant lectins towards nymphal stages of *Nilaparvata lugens*. Entomol Exp Appl 75:61–65

Powell KS, Spance J, Bharathi M, Gatehouse JA, Gatehouse AMR (1998) Immnuohistochemical and development studies to elucidate the mechanism of action of the snowdrop lectin on the rice brown planthopper, *Nilaparvata lugens* (Stal). J Insect Physiol 67:529–539

Pusztai A, Ewen SWB, Grant G, Peumans WJ, Van Damme EJM, Rubio L, Bardocz S (1990) The relationship between survival and binding of plant lectins during small intestinal passage and their effectiveness as growth factors. Digestion 46:308–316

Puthoff DP, Sardesai N, Subramanyam S, Nemacheck JA, Williams CE (2005) Hfr-2, a wheat cytolytic toxin-like gene, is upregulated by virulent Hessian fly larval feeding. Mol Plant Pathol 6:41–423

Rahbé Y, Sauvion N, Febvay G, Peumans WJ, Gatehouse AMR (1995) Toxicity of lectins and processing of ingested proteins in the pea aphid *Acyrthosiphon pisum*. Entomol Exp Appl 76:143–155

Rao KV, Rathore KS, Hodges TK, Fu X, Stoger E, Sudhakar D, Williams S, Christou P, Bharathi M, Bown DP, Powell KS, Spence J, Gatehouse AMR, Gatehouse JA (1998) Expression of snowdrop lectin (GNA) in transgenic rice plants confers resistance to rice brown planthopper. Plant J 15:469–477

Reymond P, Bodenhausen N, Van Poecke RM, Krishnamurthy V, Dicke M, Farmer EE (2004) A conserved transcript pattern in response to a specialist and a generalist herbivore. Plant Cell 16:3132–3147

Rinderle SJ, Goldstein IJ, Matta KL, Ratcliffe RM (1989) Isolation and characterization of amaranthin, a lectin present in the seeds of *Amaranthus caudatus*, that recognizes the T- (or cryptic T)-antigen. J Biol Chem 264:16123–16131

Sabnis DD, Hart JW (1978) The isolation and some properties of a lectin (haemagglutinin) from *Cucurbita* phloem exudate. Planta 142:97–101

Sadeghi A, Smagghe G, Broeders S, Hernalsteens JP, De Greve H, Peumans WJ, Van Damme EJM (2008) Ectopically expressed leaf and bulb lectins from garlic (*Allium sativum* L.) protect transgenic tobacco plants against cotton leafworm (*Spodoptera littoralis*). Transgenic Res 17:9–18

Sadeghi A, Van Damme EJM, Peumans WJ, Smagghe G (2006) Deterrent activity of plant lectins on cowpea weevil *Callosobruchus maculatus* (F.) oviposition. Phytochemistry 67:2078–2084

Saha P, Majumder P, Dutta I, Ray T, Roy SC, Das S (2006) Transgenic rice expressing *Allium sativum* leaf lectin with enhanced resistance against sap-sucking insect pests. Planta 223:1329–1343

Sastry MVK, Banerjee P, Patanjali SR, Swamy MJ, Swarnalatha GV, Surolia A (1986) Analysis of the saccharide binding to *Artocarpus integrifolia* lectin reveals specific recognition of T-antigen (β-D-Gal(1,3)D-GalNAc). J Biol Chem 261:11726–11733

Sauvion N, Charles H, Febvay G, Rahbé Y (2004) Effects of jackbean lectin (ConA) on the feeding behaviour and kinetics of intoxication of the pea aphid, *Acyrthosiphon pisum*. Entomol Exp Appl 110:31–44

Sauvion N, Rahbé Y, Peumans WJ, Van Damme EJM, Gatehouse JA, Gatehouse AMR (1996) Effects of GNA and other mannose binding lectins on development and fecundity of the peach potato aphid *Myzus persicae*. Entomol Exp Appl 79:285–293

Sétamou M, Bernal JS, Mirkov TE, Legaspi JC (2003) Effects of snowdrop lectin on Mexican rice borer (Lepidoptera: Pyralidae) life history parameters. J Econ Entomol 96:950–956

Shade RE, Schroeder HE, Pueyo JJ, Tabe LM, Murdock LL, Higgins TJV, Chrispeels MJ (1994) Transgenic pea seeds expressing the α-amylase inhibitor of the common bean are resistant to bruchid beetles. Biotechnology 12:793–796

Sharma HC, Sharma KK, Crouch JH (2004) Genetic transformation of crops for insect resistance: potential and limitations. Crit Rev Plant Sci 23:47–72

Sharma V, Surolia A (1997) Analyses of carbohydrate recognition by legume lectins: size of the combining site loops and their primary specificity. J Mol Biol 267:433–445

Shukla S, Arora R, Sharma HC (2005) Biological activity of soybean trypsin inhibitor and plant lectins against cotton bollworm/legume pod borer, *Helicoverpa armigera*. Plant Biotechnol 22:1–6

Singh T, Wu JH, Peumans WJ, Rougé P, Van Damme EJM, Alvarez RA, Blixt O, Wu AM (2006) Carbohydrate specificity of an insecticidal lectin isolated from the leaves of *Glechoma hederacea* (ground ivy) towards mammalian glycoconjugates. Biochem J 393:331–341

Stirpe F (2004) Ribosome-inactivating proteins. Toxicon 44:371–383

Stoger E, William S, Christou P, Down RE, Gatehouse JA (1999) Expression of the insecticidal lectin from snowdrop (*Galanthus nivalis* agglutinin; GNA) in transgenic wheat plants: effects on predation by the grain aphid *Sitobion avenae*. Mol Breed 5:65–73

Subramanyam S, Sardesai N, Puthoff DP, Meyer JM, Nemacheck JA, Gonzalo M, Williams CE (2006) Expression of two wheat defense-response genes, Hfr-1 and Wci-1, under biotic and abiotic stress. Plant Sci 170:90–103

Tsutsui S, Tasumi S, Suetake H, Suzuki Y (2003) Lectins homologous to those of monocotyledonous plants in the skin mucus and intestine of pufferfish, *Fugu rubripes*. J Biol Chem 278:20882–20889

Van Damme EJM, Allen AK, Peumans WJ (1987) Isolation and characterization of a lectin with exclusive specificity towards mannose from snowdrop (*Galanthus nivalis*) bulbs. FEBS Lett 215:140–144

Van Damme EJM, Barre A, Rougé P, Peumans WJ (2004). Cytoplasmic/nuclear plant lectins: a new story. Trends Plant Sci 9:484–489

Van Damme EJM, Barre A, Verhaert P, Rougé P, Peumans WJ (1996) Molecular cloning of the mitogenic mannose/maltose-specific rhizome lectin from *Calystegia sepium*. FEBS Lett 397: 352–356

Van Damme EJM, Culerrier R, Barre A, Alvarez R, Rougé, P, Peumans WJ (2007a) A novel family of lectins evolutionarily related to class V chitinases: an example of neofunctionalization in legumes. Plant Physiol 144:662–672

Van Damme EJM, Hao Q, Chen Y, Barre A, Vandenbussche F, Desmyter S, Rougé P, Peumans WJ (2001) Ribosome-inactivating proteins: a family of plant proteins that do more than inactivate ribosomes. Crit Rev Plant Sci 20:395–465

Van Damme EJM, Peumans WJ, Barre A, Rougé P (1998a) Plant lectins: a composite of several distinct families of structurally and evolutionary related proteins with diverse biological roles. Crit Rev Plant Sci 17:575–692

Van Damme EJM, Peumans WJ, Pusztai A, Bardocz S (1998b) Handbook of plant lectins: properties and biomedical applications. John Wiley & Sons, Chichester, UK

Van Damme EJM, Rougé P, Peumans WJ (2007b) Carbohydrate–protein interactions: plant lectins. In: Kamerling JP, Boons GJ, Lee YC, Suzuki A, Taniguchi N, Voragen AGJ (eds) Comprehensive glycoscience – from chemistry to systems biology. Elsevier, Oxford, UK, vol 3, pp 563–599

Van Damme EJM, Smeets K, Engelborghs I, Aelbers H, Balzarini J, Pusztai A, Van Leuven F, Goldstein IJ, Peumans WJ (1993) Cloning and characterization of the lectin cDNA clones from onion, shallot and leek. Plant Mol Biol 23:365–376

Vandenbussche F, Desmyter S, Ciani M, Proost P, Peumans WJ, Van Damme EJM (2004a) Analysis of the *in planta* antiviral activity of elderberry ribosome-inactivating proteins. Eur J Biochem 271:1508–1515

Vandenbussche F, Peumans WJ, Desmyter S, Proost, P, Ciani M, Van Damme EJM (2004b) The type-1 and type-2 ribosome-inactivating proteins from Iris confer transgenic tobacco plants local but not systemic protection against viruses. Planta 220:211–221

Waljuno K, Scholma RA, Beintema J, Mariono A, Hahn AM (1975) Amino acid sequence of hevein. In: Proceedings of the International Rubber Conference, Kuala Lumpur, vol 2. Rubber Research Institute of Malaysia, Kuala Lumpur, pp 518–531

Wang P, Li G, Granados RR (2004). Identification of two new peritrophic membrane proteins from larval *Trichoplusia ni*: structural characteristics and their functions in the protease rich insect gut. Insect Biochem Mol Biol 34:215–227

Wang W, Hause B, Peumans WJ, Smagghe G, Mackie A, Fraser R, Van Damme EJM (2003b) The Tn antigen-specific lectin from ground ivy is an insecticidal protein with an unusual physiology. Plant Physiol 132:1322–1334

Wang W, Peumans WJ, Rougé P, Rossi C, Proost P, Chen J, Van Damme EJM (2003a) Leaves of the *Lamiaceae* species *Glechoma hederacea* (ground ivy) contain a lectin that is structurally and evolutionary related to the legume lectins. Plant J 33:293–304

Wei GQ, Liu RS, Wang Q, Liu WY (2004) Toxicity of two type II ribosome-inactivating proteins (cinnamomin and ricin) to domestic silkworm larvae. Arch Insect Biochem Physiol 57:160–165

Williams CE, Collier CC, Nemacheck JA, Liang C, Cambron SE (2002) A lectin-like wheat gene responds systemically to attempted feeding by avirulent first-instar Hessian fly larvae. J Chem Ecol 28:1411–1428

Wu AM (2005) Lectinochemical studies on the glyco-recognition factors of a Tn (α->Ser/Thr) specific lectin isolated from the seeds of *Salvia sclarea*. J Biomed Sci 12:167–184

Wu J, Luo X, Guo H, Xiao J, Tian Y (2006) Transgenic cotton, expressing *Amaranthus caudatus* agglutinin, confers enhanced resistance to aphids. Plant Breed 125:390–394

Yagi F, Hidaka M, Minami Y, Tadera K (1996) A lectin from leaves of *Neoregelia flandria* recognizes D-glucose, D-mannose and N-acetylglucosamine, differing from the mannose-specific lectins from other monocotyledonous species. Plant Cell Physiol 37:1007–1012

Yao J, Pang Y, Qi H, Wan B, Zhao X, Kong W, Sun X, Tang K (2003) Transgenic tobacco expressing *Pinellia ternata* agglutinin confers enhanced resistance to aphids. Transgenic Res 12:715–722

Young NM, Oomen RP (1992) Analysis of sequence variation among legume lectins. A ring of hypervariable residues forms the perimeter of the carbohydrate-binding site. J Mol Biol 228:924–934

Zhang W, Peumans WJ, Barre A, Houles-Astoul C, Rovira P, Rougé P, Proost P, Truffa-Bachi P, Jalali AAH, Van Damme EJM (2000) Isolation and characterization of a jacalin-related mannose-binding lectin from salt-stressed rice (*Oryza sativa*) plants. Planta 210:970–978

Zhou X, Li XD, Yuan JZ, Tang ZH, Liu WY (2000) Toxicity of cinnamomin – a new type II ribosome-inactivating protein to bollworm and mosquito. Insect Biochem Mol Biol 30:259–264

Zhu K, Huesing JE, Shade RE, Bressan RA, Hasegawa PM, Murdock LL (1996) An insecticidal N-acetylglucosamine-specific lectin gene from *Griffonia simplicifolia* (Leguminosae). Plant Physiol 110:195–202

Zhu-Salzman K, Salzman RA (2001) Functional mechanics of the plant defensive *Griffonia simplicifolia* lectin II: resistance to proteolysis is independent of glycoconjugate binding in the insect gut. J Econ Entomol 94:1280–1284

Zhu-Salzman K, Salzman RA, Koiwa H, Murdock LL, Bressan RA, Hasegawa PM (1998) Ethylene negatively regulates local expression of plant defense lectin genes. Physiol Plant 104:365–372

Zhu-Salzman K, Shade RE, Koiwa H, Salzman RA, Narasimhan M, Bressam RA, Hasegawa PM, Murdock LL (1996) Carbohydrate binding and resistance to proteolysis control insecticidal activity of *Griffonia simplicifolia* lectin II. Proc Natl Acad Sci USA 95:15123–15128

Section III
Defense Signaling

Part A
Activation of Plant Defenses

Chapter 15
Systemins and AtPeps: Defense-Related Peptide Signals

Javier Narváez-Vásquez and Martha L. Orozco-Cárdenas

Plants have evolved different families of functionally related peptide signals (15–23 amino acids) that serve to amplify the defense response against insect/herbivore and pathogen attacks through the activation of the octadecanoid signaling pathway. The peptides are derived from larger precursor proteins and released into the extracellular space to interact with membrane receptors by still unknown processing mechanisms. Besides scarce sequence similarities among defense-related peptide signals, their functional properties still suggest some common ancestral origin for defense peptide signaling systems in plants and animals.

15.1 Introduction

In the Solanaceae, a family of defense-related peptide hormones called systemins are involved in the activation of defense genes in response to wounding by herbivore attacks (Ryan and Pearce 2003). Tomato systemin (Sys) is an 18-amino acid oligopeptide located at the C-terminus of a cytosolic precursor protein of 200 amino acids called prosystemin (ProSys; Fig. 15.1A; Ryan and Pearce 1998). Upon wounding, Sys seems to be proteolytically processed from ProSys and released into the cell wall apoplast, where Sys interacts with a membrane receptor to activate defense signaling. In addition to Sys, tomato plants possess three functionally related hydroxyproline-rich glycopeptides (called TomHypSys I, II, and III) of 15–20 residues that are synthesized from a single polyprotein precursor unrelated to ProSys (Pearce and Ryan 2003; Fig. 15.1C). The ProHypSys precursor proteins are synthesized with a signal peptide through the secretory pathway and localized in the cell wall (Narváez-Vásquez et al. 2005).

Recently, a novel wound-inducible, defense-related peptide signal has been identified in *Arabidopsis* (Huffaker et al. 2006a). The *Arabidopsis* peptide1 (AtPep1; Fig. 15.1D) is part of a small family of peptide signals involved in the innate immune

---

J. Narváez-Vásquez
Department of Botany and Plant Sciences, University of California Riverside, Riverside, CA 92521, USA
e-mail: jnarvaez@ucr.edu

A

```
1   MGTPSYDIKNKGDDMQEEPKVKLHHEKGGDEKEKIIEKETPSQDINNKDTISSYVLRDDTQEIPK
66  MEHEEGGYVKEKIVEKETISQYIIKIEGDDDAQEKLKVEYEEEYEKEKIVEKETPSQDINNKGD
131 DAQEKPKVEHEEGDLKETPSQDIIKMEGEGALEITKVVCEKIIVREDL**AVQSKPPSKRDPPKMQTD**
197 **NNKL**
```

B

```
1   *MRVLFLIYLILSPFGAEA*RTLLENHEGLNVGSGYG**RGANLPPPSPASSPPSKE**VSNSVSPTRTDE
66  KTSENTELVMTTIAQGENINQLFSFPTSADNYYQLASFKKLFISYLLPVSYVWNLIGSSSFDHDL
131 VDIFDSKSDERYW**NRKPLSPPSPKPADGQRP**LHSY
```

C

```
1   *MISFFRAFFLIIIISFLIFVGAQ*ARTLLGNYHDDEMLIELKLESGNYG**RTPYKTPPPPTSSSPTHQ**
67  EIVN**GRHDSVLPPPSPKTD**PIIGQLTTITTTPHHDDTVAAPPVG**GRHDYVASPPPPKPQDEQRQ**II
133 ITSSSSTLPLQASY
```

D

```
1   MEKSDRRSEESHLWIPLQCLDQTLRAILKCLGLFHQDSPTTSSPGTSKQPKEEKEDVTME
61  KEEVVVTSR**ATKVKAKCRGKEKVSSGRPGQHN**
```

Fig. 15.1 Deduced amino acid sequences of the precursors of defense-related peptides in plants. (**A**) Tomato systemin precursor protein, prosystemin (200 residues). The systemin sequence (18 residues) at the C-terminus is bold and underlined. Five imperfect repeats (underlined) suggest that the ProSys gene has resulted from gene duplication and elongation events (see text for discussion). (**B–C**) Tobacco (**B**, 165 residues) and tomato (**C**, 146 residues) hydroxyproline-rich systemin polyprotein precursors. Sequences corresponding to the bioactive peptides (15–20 residues) are bold and underlined. The putative signal peptide sequence in both precursors is shown in *italics*. A short conserved sequence at the N-terminus is also underlined. (**D**) *Arabidopsis* peptide1 (AtPep1) precursor protein (92 residues). The sequence of AtPep1 (23 residues) is bold and underlined

response against pathogens, with orthologs in species from several plant families, including the Solanaceae. Like Tomato Sys, AtPep1-related peptides are synthesized from larger precursor proteins that lack a classical signal sequence (Huffaker et al. 2006).

The activation of defense genes by systemins and AtPep1, including the genes encoding their own precursor proteins, is mediated by the octadecanoid signaling pathway (Ryan and Pearce 2003; Huffaker et al. 2006). In this pathway (Schaller et al. 2005), linolenic acid is released from cell membranes by the action of phospholipases, and transformed by a series of enzymatic reactions in the oxylipins phytodienoic acid (OPDA) and jasmonic acid (JA). These oxylipins are potent inducers of defense gene activation (see other contributions in this volume). Recent genetic evidence and localization studies suggest that JA, or JA-derived oxylipins function as the systemic signals involved in the systemic wound response against herbivores in tomato (Howe 2005; Wasternack et al. 2006), and in the pathogen-induced systemic acquired resistance (SAR) in *Arabidopsis* plants (Truman et al. 2007).

The cumulative evidence thus indicates that plants have evolved a peptide-based signaling system that activates and amplifies the synthesis of widely spread oxilipin signals in plants for defense against pathogen and herbivore attacks. This is analogous to the activation of the innate immune and inflammatory defense responses mediated by polypeptide cytokines in animals through the production of eicosanoid lipid signals (Bergey et al. 1996), and suggest that defense-related peptide signals may have evolved from a similar ancestor gene in eukaryotes. The genetic, biochemical, and functional properties, and the evolution of the aforementioned plant defense peptide signals are further discussed in this chapter.

15.2 Tomato Systemin (Sys)

Sys was the first extracellular signal peptide discovered in plants (Pearce et al. 1991), and it is one of the best studied (Ryan and Pearce 1998). Sys is involved in the activation of more than 20 wound-inducible genes in some species of the Solanaceae family, and shares several characteristics with peptide hormones in animals and yeast (Harris 1989). First, Sys is synthesized as a larger precursor protein, suggesting that it should be released by proteolytic cleavage (Fig. 15.1A; McGurl et al. 1992). Second, Sys is active at very low concentrations. In a plant bioassay, femtomol amounts of Sys supplied to young excised tomato plants through their cut stems cause the leaves to accumulate defense proteinase inhibitor proteins (PIs) within 24 hours (Pearce et al. 1991). In addition, subnanomolar concentrations of Sys cause a rapid increase in the pH of the extracellular medium of cell suspension cultures of *Lycopersicon peruvianum*, known as the alkalinization response (Felix and Boller 1995). Third, Sys action is mediated by its interaction with a membrane receptor at the cell surface, indicating that Sys should be released into the extracellular space to exert its biological activity (Scheer and Ryan 1999, 2002).

15.2.1 The ProSys Gene and Protein

The tomato genome possesses a single copy of the ProSys gene. In addition to Sys located at the C-terminus, there are five imperfect repeats of unknown function in ProSys, indicating that the ProSys gene resulted from several cycles of gene duplication and elongation events (McGurl and Ryan 1992). ProSys homologs have only been found in species of the Solaneae subtribe of the Solanaceae family, including tomato, potato, bell pepper, and nightshade, but it is not found in tobacco or *Arabidopsis* (Constabel et al. 1998). Pairwise analysis of the deduced amino acid sequences of ProSys orthologs indicate that there is a high degree of conservation among them (73%–88% identity). Synthetic Sys from all investigated Solaneae species are capable of inducing defense PIs in the tomato plant bioassay (Constabel et al. 1998).

The ProSys gene is under tissue/cell-type specific, developmental, and environmental regulation (Ryan and Pearce 1998). Low constitutive levels of ProSys mRNA are found within the vascular phloem parenchyma (PP) cells of leaves, petioles, stems, but not in roots (Narváez-Vásquez and Ryan 2004). ProSys is also developmentally regulated in all floral organs, where high levels of ProSys mRNA and protein accumulate. In addition, ProSys gene expression is also upregulated in aerial vegetative organs by mechanical wounding, chewing insects, JA and its methyl ester (MeJA), plant and pathogen-derived oligosaccharides, and other pathogen elicitors (Ryan and Pearce 1998).

ProSys does not comprise a predicted signal sequence, transmembrane domains, or ER/Golgi-dependent post-translational modifications, suggesting that the Sys precursor is synthesized on free ribosomes in the cytoplasm (McGurl et al. 1992). The amino acid composition of ProSys (Fig. 15.1A) is characterized by an abundance of acidic amino acids (27.5%), but also a large number of basic Lys residues (15%). ProSys contain stretches of alternating Lys (K) and Glu (E) residues, resembling the so-called 'KEKE' motifs (Realini et al. 1994). Similar motifs have been shown to be involved in protein–protein interactions through the formation of 'polar zippers' (Zhang et al. 1997).

15.2.2 ProSys Processing

Sys is an internal peptide that should be released from its protein precursor by proteolytic cleavage by a processing peptidase, similar to other peptide hormones in eukaryotes (Seidah and Chretien 1997). The processing sites and the responsible enzyme(s) that cleave(s) Sys from ProSys are currently unknown. Unlike most prohormone precursors in animals and yeast, ProSys is not synthesized through the secretory pathway.

ProSys does not have classical dibasic sites for processing by subtilisin/kexin-like, prohormone convertases (PCs; Seidah and Chretien 1997), except for a dibasic KR sequence present in the middle of Sys (Fig. 15.1A and Table 15.1). A Sys-binding protease in tomato cell membranes related to furin in the PC family may cleave Sys at this site (Schaller and Ryan 1994). The protease may be involved in Sys turnover, since the proteolytic products have significantly reduced biological activities (Pearce et al. 1993). Accordingly, substitution of these basic residues with chemically-modified amino acids increases the half-life and activity of the modified Sys (Schaller 1998).

ProSys is also enriched in acidic amino acid doublets (E/D-E/D, Fig. 15.1). In fact, the Sys sequence is flanked by Asp residues at both the N- and the C-termini. Interestingly, acidic pairs of amino acids are also present in the protein precursors of *Arabidopsis* AtPep peptides (Fig. 15.1D; Huffaker et al. 2006). Plant endoproteases with specificities for P_1 Asp residues have been reported (D'Hont et al. 1993; Francois et al. 2002; Rojo et al. 2004; Coffeen and Wolpert 2004), and several peptidases are known to be wound-inducible (Pautot et al. 1993; Bergey et al. 1996; Schaller and Ryan 1996; Horn et al. 2005). Notably, the processing enzyme of the signal-less precursor protein of the potent inflammatory cytokine interleukin-1β (IL-1β) in animal cells is a caspase-like Cys protease that requires substrates with Asp adjacent

Table 15.1 Sequence comparison and posttranslational modifications of defense-related peptides in plants. Proline (P), hydroxyproline (O), threonine (T), and serine (S) residues are bold

Peptide[1]	Sequence	Pentose units
Sys	AVQ**S**K**PP**K**R**D**PP**KMQ**T**D	0
TomHypSys I	R**TP**YK**TOOOOTSSSP**T**H**Q	8–17
TomHypSys II	GRHDYVA**SOOOO**K**P**QDEQRQ	12–16
TomHypSys III	GRHD**S**VL**OOOSO**K**T**D	10
TobHypSys I	RGANL**POOSO**A**SSOO**S**K**E	9
TobHypSys II	NRK**P**L**SOOSO**K**P**ADGQR**P**	6
AtPep1	A**T**KVKAKQRGKEKV**SS**GR**P**GQHN	0

[1] Sys, tomato systemin; TomHypSys, tomato hydroxyproline-rich glycopeptides I, II and III; TobHypSys, tobacco hydroxyproline-rich glycopeptides I and II, AtPep1, Arabidopsis peptide1

and N-terminal to the scissile peptide bond (P_1 residue; Howard et al. 1991). Thus, it is tempting to speculate that the ProSys precursor is processed by a protease that cleaves at the COOH side of Asp (P_1), releasing N-terminal extended Sys peptides that could be trimmed by aminopeptidases in the cytosol (Walling 2006).

The possibility that ProSys is secreted into the apoplast and subsequently processed by extracellular proteases was investigated (Dombrowski et al. 1999). Recombinant ProSys was incubated with apoplastic wash fluids of tomato leaves and the culture medium of *L. peruvianum* cells. At different times the integrity of ProSys was assessed by SDS-PAGE and protein blotting using anti-Sys antibodies. There was no evidence for ProSys processing enzymes in the apoplast or media (Dombrowski et al. 1999).

Evidence that ProSys is synthesized and processed within PP cells comes from direct immunocytochemical studies, using affinity-purified antibodies raised against full-length ProSys, Sys, and Δ ProSys (ProSys lacking the Sys sequence; Narváez-Vásquez and Ryan 2004). The studies indicated that ProSys is localized in the cytosol and the nucleus of PP cells in leaves, petioles, stems, and floral organs of tomato and potato. The amount of labeling increased significantly in leaves upon wounding or MeJA treatment of tomato plants (Fig. 15.2A). The nucleo-cytoplasmic localization pattern was also observed in all cell types of leaf sections from transgenic tomato plants overexpressing the ProSys gene (Narváez-Vásquez and Ryan 2004).

Indirect evidence for intracellular proteolytic processing of ProSys, comes from a recent study in which a chimeric gene encoding a modified ProSys protein, containing a decapeptide sequence of the mosquito trypsin-modulating oostatic factor (TMOF) in the place of Sys, was expressed in transgenic tobacco plants (Tortiglione et al. 2003). The ProSys-TMOF chimeric gene was transcribed and translated into the modified protein within transgenic tissue, and the TMOF decapeptide was detected by HPLC electrospray mass spectrometry (HPLC/ESI-MS) in extracts from transgenic tobacco tissue and protoplasts, but not in extracts from wild type control plants. Only traces of TMOF peptide were detected in the incubating media of the protoplasts, indicating that most if not all of the foreign peptide was being compartmentalized within the cell protoplasts, and was not secreted to the external medium (Tortiglione et al. 2003). Even though tobacco does not have a ProSys gene,

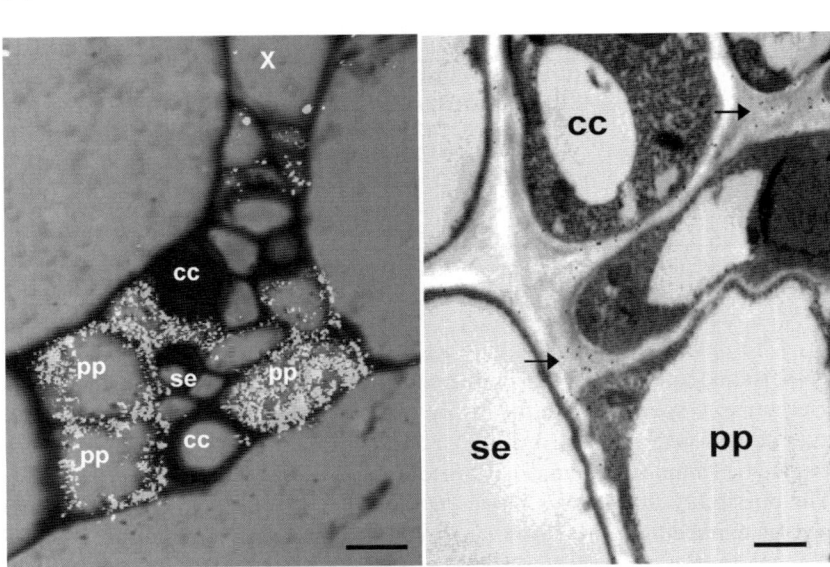

Fig. 15.2 Immuno-cytochemical localization of ProSys and ProHypSys in vascular bundles of tomato leaves. Leaf samples from two-week-old tomato plants exposed for 18 hours to methyl jasmonate vapors were fixed and immuno-gold labeled as previously described (Narváez-Vásquez and Ryan 2004). (**A**) Silver-enhanced, immuno-gold labeled ProSys localized in vascular phloem parenchyma (PP) cells visualized under the confocal microscope. (**B**) Transmission electron micrograph showing immuno-gold labeled ProHypSys in the cell walls of PP cells (Narváez-Vásquez et al. 2005). Arrows point to gold particles. pp, phloem parenchyma; cc, companion cell; se, sieve elements; x, xylem vessel. Bars = A, 5.0 μm; B, 0.5 μm

this result suggests that ProSys is synthesized, processed and/or stored intracellularly in tobacco leaves, most probably in a similar fashion as in tomato.

Genetic screens have been conducted in tomato to find mutants in the Sys signaling pathway (Howe and Ryan 1999). These efforts have led to the identification of several mutants incapable to respond to Sys, wounding, or JA (Howe 2005). However, mutations that affect the processing of ProSys have not been identified yet. The failure of these screens to identify genes required for processing, compartmentalization and/or transport of ProSys or Sys suggests that the ProSys processing pathway involves genes that are either redundant or required for cell/plant viability.

15.2.3 *The Sys Signaling Pathway*

A Sys receptor protein, SR160, has been isolated and its gene cloned (Scheer and Ryan 1999, 2002). SR-160 belongs to a family of Ser/Thr receptor kinases with extracellular Leu-rich repeats, which was found to be highly similar to the *Arabidopsis* brassinolide receptor BRI1 (Scheer and Ryan 2002). Brassinolides are a class of plant steroid hormones with different roles in plant development, but not in plant defense. Isolation of the tomato brassinolide receptor confirmed that it is also the Sys receptor (Montoya et al. 2002). The function of SR160 as the Sys receptor was reconfirmed by expressing *SR160* cDNA in tobacco plants (Scheer et al. 2003). Tobacco plants lack a ProSys gene and Sys perception. In transgenic tobacco expressing *SR-160* mRNA constitutively, the Sys receptor is membrane localized, binds Sys and activates early defense responses (Scheer et al. 2003). This is the first example of a protein receptor in plants having a dual function in defense and development.

Ala substitution and deletion analyses of the Sys polypeptide indicated that the entire 18 amino acids are necessary for full biological activity, and that the N-terminal region might be important for recognition/binding to the Sys receptor, while the C-terminal residues are more essential for activity (Pearce et al. 1993; Meindl et al. 1998). Upon binding to its receptor, Sys activates a complex cascade of signaling events leading to the activation of early (signaling-related) and late (defense-related) wound response genes, and the production of H_2O_2 (Ryan 2000). In tomato leaves, Sys causes a rapid (within minutes) depolarization of the plasma membrane in mesophyll cells, the alkalinization of the extracellular medium or cell wall apoplast, transient increases in cytoplasmic free Ca^{2+} concentrations, the activation of a mitogen-activated protein kinase (MAPK), and the release of linolenic acid (18:3 fatty acid) from cell membranes by the activation of a phospholipase A_2, leading to the biosynthesis of OPDA and JA through the octadecanoid pathway (Ryan 2000).

15.2.4 Systemin and the Systemic Wound Signal

The plant wound defense response against insects or herbivore attacks, discovered by Green and Ryan (1972), is characterized by the accumulation of defense chemicals not only at the site of damage (local response), but also in other undamaged parts of the plant (systemic response). This implies that 'wound' signal(s) are generated at the sites of damage to activate the transcription of defense genes and the synthesis of antinutritional compounds throughout the plant (Green and Ryan 1972).

Sys has the characteristics of a systemic wound signal. When radiolabeled Sys is placed in wounds made on leaves, Sys is loaded into the phloem and transported systemically throughout the plants within a few hours, in a similar fashion as the transport of ^{14}C-sucrose (Pearce et al. 1991; Narváez-Vásquez et al. 1994, 1995). In addition, the overexpression of the *ProSys* precursor gene under the regulation of the CaMV 35S promoter in transgenic tomato plants led to the constitutive accumulation of high levels of several defensive proteins in leaves (McGurl et al. 1994; Bergey et al. 1996). The ProSys overexpressing plants produced a graft-transmissible signal that activates defense genes in a wild type scion (McGurl et al. 1994). Conversely, when the *35S:ProSys* transgene was expressed in the antisense orientation in transgenic tomato plants, the systemic activation of wound-inducible defense genes was strongly reduced and the plants were more susceptible to insect damage (McGurl et al. 1992; Orozco-Cárdenas et al. 1993). Thus, the cumulative results supported the hypothesis that Sys is an important component of a mobile systemic signal.

Recent genetic evidence, however, indicated that JA or a JA-derived oxylipin, functions as the systemic wound signal in tomato, and that Sys main role is to up-regulate JA biosynthesis (Howe 2005). This was demonstrated using reciprocal grafting experiments with tomato mutants deficient in JA biosynthesis and JA perception, which were also defective in systemic wound signaling (Howe 2005; see also Howe and Schaller this volume). Accordingly, several enzymes of the octadecanoid pathway have been localized within the vascular bundle cells of tomato leaves, including LOX, AOS, and AOC (Wasternack et al. 2006). Tomato AOC is a rate-limiting enzyme required for wound-induced JA biosynthesis during Sys signaling (Stenzel et al. 2003). Therefore, Sys synthesis and JA synthesis have now been associated with the phloem. Altogether, the results of the grafting and localization studies indicated a new role for both Sys and JA in the amplification of the systemic signaling process as they are translocated through the plant vasculature (Ryan and Moura 2002; Howe 2005; Wasternack et al. 2006).

15.3 Hydroxyproline-Rich Systemin Glycopeptides (HypSys)

The search for a functional homologue of Sys in tobacco, using a cell alkalinization assay (Felix and Boller 1995), resulted in the identification of two bioactive 18-aa glycopeptides in purified HPLC fractions obtained from tobacco leaf extracts (Pearce et al. 2001). The isolated peptides (called TobHypSys I and II) were capable of inducing the medium alkalinization of tobacco and tomato cell suspension cultures, and also of activating a 48-kDa MAP kinase and defense gene expression when supplied to excised tobacco leaves at nM concentrations (Pearce et al. 2001).

TobHypSys I and II are hydroxyproline-rich glycopeptides, decorated with nine and six pentose units respectively (Table 15.1). The glycosyl moieties are required for full biological activity. TobHypSys I and II are derived from a single polyprotein precursor of 165 amino acids that has a signal peptide (Pearce et al. 2001), representing the first example in plants of multiple signaling peptides derived from the same protein precursor, a common feature found in animals (Tanaka 2003).

Three HypSys glycopeptides were subsequently purified from tomato leaves, using the same medium alkalinization assay with *L. peruvianum* suspension cell cultures (Pearce and Ryan 2003). Tomato HypSys I, II, and III of 20, 18, and 15 residues respectively, are also derived from a single precursor protein of 146 amino acids, which is likely to be secreted by virtue of a predicted signal peptide (Fig. 15.1C). The tomato glycopeptides have between 8 and 16 pentose units, which appear to be arabinose residues bonded to hydroxyproline residues (Table 15.1; Pearce and Ryan 2003). The amino acid sequences of TomHypSys II and III peptides are very similar, with nine identical amino acids (Table 15.1), suggesting they have resulted from gene duplication and elongation events similar to ProSys.

The presence of a signal peptide in both the tobacco and tomato HypSys precursors and the post-translational hydroxylation of proline and glycosylation indicates that the bioactive peptide precursors are synthesized through the secretory pathway. Immunolocalization studies using affinity-purified antibodies that only recognize epitopes in the tomato ProHypSys precursor indicates that the proprotein is localized in the cell walls of leaf vascular parenchyma cells (Fig. 15.2B; Narváez-Vásquez et al. 2005).

Multiple HypSys glycopeptide signals derived from single protein precursors have also been found in other species of the Solanaceae, including petunia, nightshade, and potato (Pearce et al. 2006). In both tobacco and tomato, the *ProHypSys* genes are upregulated in the leaves by wounding, Sys, MeJA, and the HypSys peptides themselves (Pearce et al. 2001; Pearce and Ryan 2003). The Tobacco *ProHypSys* gene is also induced by ABA, chewing and phloem-feeding insects (Rocha-Granados et al. 2005). In tomato, low constitutive levels of HypSys mRNAs are found in leaves, but the basal levels are higher in other organs, including roots (Narváez-Vásquez and Ryan, unpublished results). Overexpression of the native *HypSys* genes in tobacco (Ren and Lu 2006) and tomato (Narváez-Vásquez et al. 2007) caused the constitutive upregulation of both signaling and defense-related genes in the transgenic plants, similar to wounding and Sys in tomato. The *35S:HypSys* transgenic tobacco plants were more resistant to insect attacks (Ren and Lu 2006). In addition, downregulation of *TomHypSys* gene expression by antisense technology also diminished the systemic wound inducibility of defense genes, suggesting that Sys and HypSys work cooperatively to upregulate the systemic wound-defense response in tomato (Narváez-Vásquez et al. 2007).

The processing events responsible for the release of the HypSys peptides in the cell wall in response to wounding are still unknown. The presence of the –N/GR sequence at the N-termini of tobacco and tomato HypSys peptides suggests that these amino acids may be substrates for specific proteases in both species (Table 15.1). Wounding may cause intracellular proteases to mix with the precursor of HypSys peptides in the cell wall matrix as a consequence of cell damage. Wound- or Sys-induced ion fluxes and pH changes in leaf cells may as well induce secretion and/or activation of processing enzymes in the extracellular matrix (Schaller and Oecking 1999; Tanaka 2003). Alternatively, pathogen-derived peptidases may also be involved in processing of bioactive peptides leading to the activation of defense signaling cascades. In any event, the proteolytic release of multiple peptide signals from polyprotein precursors located in the cell walls may be among the plant earliest events in response to insect and pathogen attacks.

15.4 *Arabidopsis* Defense-Related Peptides (AtPeps)

AtPep1 is the first endogenous defense-related peptide purified from leaves of *Arabidopsis thaliana* (Huffaker et al. 2006a). The AtPep1 peptide has 23 amino acids and is derived from the carboxy terminus of a signal-less 92-amino acid precursor protein (ProAtPep1; Fig. 15.1D). AtPep1 regulates the expression of defense genes induced by pathogens through the octadecanoid signaling pathway. The gene encoding ProAtPep1 is part of a small family of at least seven members in *Arabidopsis*, with orthologs in species of many other families, including important crop plants (Table 15.2). The size range of ProAtPep1-related precursor proteins reported to date is between 75 and 154 amino acids (Table 15.2). Based on their chromosomal location the *Arabidopsis AtPep1* gene family has been classified in three subfamilies. Members of subfamilies I (*ProAtPep1, 2*, and *3*) and II (*ProAtPep4, 5,* and *7*) are grouped in two different locations in Chromosome V, and only one gene (*ProAtPep6*) belongs to the subfamily III located on chromosome II (Huffaker et al. 2006b). The *Arabidopsis* paralogs share the 10-amino acid motif SSG(K/R)xGxxN that appear to be important for full biological activity. Notably, Gly17 is the only conserved amino acid in all the other known AtPep1 orthologs (Table 15.2). Ala-substitution studies indicated that Gly17 is essential for AtPep1 biological activity (Huffaker et al. 2006b).

In *Arabidopsis*, the *ProAtPep1* gene is expressed at low constitutive levels in all tissues, with higher basal levels observed in flowers and roots (Huffaker et al. 2006a). In leaves, the *ProAtPep1* gene is induced by wounding, MeJA and ethylene. Other gene family members show similar expression patterns in a wide range of tissues, and appear to be upregulated by MeJA (Huffaker et al. 2006a, b). Analysis of microarray gene expression data in *Arabidopsis* indicated that some of the *ProAtPep1* paralogs are differentially regulated by bacterial and fungal pathogens and pathogen-derived elicitors (Huffaker et al. 2006a). AtPep1 supplied to excised *Arabidopsis* leaves through the cut petioles causes the activation of several genes involved in pathogen defense, such as those encoding for defensin (PDF-1.2), other pathogenesis-related (PR) proteins, *ProAtPep1*, and some of the *ProAtPep1* paralogs (Huffaker et al. 2006b). Feeding AtPep1 to leaves also caused the production of H_2O_2 in the vascular bundles (Huffaker et al. 2006a).

When the ProAtPep1 gene was overexpressed in *Arabidopsis* under the regulation of the 35S promoter, the pathogen-related defensin gene was constitutively expressed and the transgenic plants were more resistant to infection by the soil borne pathogen *Pythium irregulare* (Huffaker et al. 2006a). Infected wild-type control plants grew smaller, especially their roots, than the transgenic plants. This result indicates that the overproduced ProAtPep1 has been processed to release the bioactive peptide to constitutively activate the pathogen-defense response in the transgenic tissue (Huffaker et al. 2006a). No endogenous peptides have been reported before that could activate genes associated with innate immunity in plants.

An AtPep1-binding protein ($\sim$170-kD) was isolated from microsomal membranes of *Arabidopsis* suspension cultured cells by photoaffinity labeling (Yamaguchi et al. 2006). The protein binds an iodinated-AtPep1 analog with a K_d of 0.25 nM, similar to the binding constant of the iodinated Sys analog to its SR-160 receptor (K_d of 0.17 nM; Scheer and Ryan 1999). Mass spectroscopy of tryptic digests identified the binding protein as a receptor-like protein kinase of 1123 amino acids, encoded by the gene At1g73080 in *Arabidopsis* (Yamaguchi et al. 2006). The AtPep1 receptor protein, called AtPepR1, contains 27 leucine-rich repeats, a transmembrane region, and an intracellular Ser/Thr protein kinase

Table 15.2 Comparison of the C-terminal amino acid sequences of AtPep1 paralogs and orthologs containing putative AtPep sequences in plants. The sequence of AtPep1 (23 residues) is aligned with the C-termini deduced from six paralogs and seven orthologs. The amino acids of the conserved SSGR/KxGxxN motif present in all AtPep1 paralogs are bold. Adapted from Huffaker et al. (2006a, b)

Peptide	Precursor gene code	C-terminal amino acid sequence	Precursor length (amino acids)
AtPep1	At5g64900	69-ATKVKAKQRGKEKV**SSGR**P**GQ**H**N***	92
AtPep2	At5g64890	74-DNKAKSKKRDKEKP**SSGR**P**GQ**T**N**SVPNAAIQVYKED*	109
AtPep3	At5g64905	73-EIKARGKNKTKPTP**SSGK**G**GK**H**N***	96
AtPep4	At5g09980	55-GLPGKKNVLKKSRE**SSGK**P**GG**T**N**KKPF*	81
AtPep5	At5g09990	59-SLNVMRKGIRKQPV**SSGK**R**GG**V**N**DYDN*	86
AtPep6	At2g22000	53-ITAVLRRRPRPPY**SSGR**P**GQ**N**N***	75
AtPep7	At5g09975	81-VSGNVAARKGKQQT**SSGK**G**GG**T**N***	104
Canola	CV505388	74-VARLTRRRPRPP-Y**SSG**Q**PG**Q**IN***	95
Potato	CV23975	93-PTERRGRPPSRPKVG**SG**P**PP**Q**NN***	116
Poplar	BI311441	94-DAAVSALARRTPPV**SR**G**GG**G**QTN**TTTS*	121
Medicago	CD401281	87-LSSMGRGGPRRTPLT**QG**P**PP**Q**HN***	111
Soybean	CF333408	93-ASLMATRGSRGSKIT**SDG**S**GP**Q**HN***	115
Rice1		131-ARLRPKPPGNPREG**SG**G**NG**G**HH***	154
Rice2	AK111113	65-DDSKPTRPGAPAEG**SG**G**NG**G**A**IHTAASS*	93

domain. Radiolabeled AtPep1 analog did not bind to microsomal membranes from two insertional mutants of At1g73080, and tobacco suspension cells transformed with the AtPepR1 gene were capable of binding to AtPep1 and showed the typical alkalinization response. These results demonstrate that PepR1 is a functional receptor (Yamaguchi et al. 2006). Four out of six AtPep1 homologs synthesized based on the deduced C-terminal amino acid sequence of ProPep1 paralogs in *Arabidopsis* were biologically active in the alkalinization assay with *Arabidopsis* cells. The bioactive synthetic peptides were also able to compete with the radiolabeled AtPep1 analog for binding to PepR1. This result suggested that PepR1 might actually be the receptor for at least five of the homologous AtPep1 peptides possibly generated during pathogen attacks (Yamaguchi et al. 2006).

15.5 Evolutionary Considerations

The discovery of genetically different defense-related peptide signals in the Solanaceae, acting through the same octadecanoid-signaling pathway prompted Prof. C.A. Ryan and collaborators to 'broaden the definition of systemins to include any signaling peptide from plants that activates plant defensive genes' (Ryan and Pearce 2003). Thus, this definition should be extended to the *Arabidopsis* AtPep peptides, which are also upregulated by wounding and pathogens to activate genes involved in pathogen defense through the octadecanoid pathway. AtPep orthologs are also present in the Solanaceae (Table 15.2; Huffaker et al. 2006a).

There is no sequence homology between the protein precursors of Sys, HypSys, and AtPep1 peptides (Fig. 15.1). The Sys and HypSys peptides have only been found in species of the Solanaceae, although the systemic wound defense response is ubiquitous in plants. Even the tomato and tobacco HypSys precursors only share a limited amino acid identity (compare Fig. 15.1B and C). However, the bioactive peptides show some degree of conservation suggestive of a common origin. For instance, they are enriched in proline/hydroxyproline, serine, and threonine residues, flanked by charged amino acids (Table 15.1). The HypSys glycopeptides contain –OOS- motifs, which are posttranslational modifications derived from –PPS- primary translation motifs also present in Sys (Table 15.1). AtPep1 lacks this motif, but some AtPep1 paralogs and orthologs are proline-rich and highly charged peptides (Table 15.2). Proline residues are known to confer structural conformations in the backbone chains of bioactive peptides that are important for the interactions of peptide ligands with their receptors (Rath et al. 2005).

Like tomato Sys, AtPep1-related peptides are synthesized from larger precursor proteins that lack a classical signal sequence. In addition, the AtPep precursors are highly charged proteins, with stretches of lysine (K) and glutamic acid (E) residues that resemble the KEKE-like motifs (Realini et al. 1994), also found in ProSys (compare Fig. 15.1A and D). Because of this, and the similarities mentioned above, it has been proposed that peptide signals that activate defense responses against pathogen and herbivore attacks may have evolved from a common ancestor gene (Ryan and

Pearce 2003; Huffaker et al. 2006a, b). Mutational events, such as gene insertions, followed by duplication and elongation events appear to be involved in the evolution of peptide signal precursors in plants (Ryan and Pearce 2003). This might explain why the prohormone precursors of defense related peptides have diverged more quickly than the actual bioactive peptides.

Since AtPep1 peptides are widely spread in the plant kingdom, they may have evolved prior to the systemins in the Solanaceae. Since HypSys peptides are found in most species of the Solanaceae, and Sys only in the Solaneae subtribe, HypSys peptides appear to have evolved earlier than Sys. It is possible that the brassinolide receptor evolved initially and then was later recruited to facilitate systemin signaling. It would be interesting to see whether the HypSys peptides interact with homologs of the systemin receptor or have entirely different receptors. Notably, overexpression of the tomato ProSys gene in transgenic potato causes the accumulation of developmentally regulated proteins in the tubers (Narváez-Vásquez and Ryan 2002). The link between signaling pathways involved in plant development and defense is not well understood yet.

15.6 Perspectives

The evidence suggests that defense-related peptide hormones play a primary role in the amplification of signals generated by insects and pathogen attacks in plants through the activation of a lipid-based signaling pathway in the phloem. The main purpose being the rapid systemic accumulation of deterrents and antinutritional factors throughout the plant that will diminish food quality, slow down growth, and/or prevent the colonization of invading organisms. This peptide-based signaling mechanism for plant defense is analogous to the activation of the inflammatory and acute-wound defense response mediated by peptide cytokines such as the interleukins and TNF-α, through the eicosanoid signaling pathway in animals (Bergey et al. 1996).

Although thoroughly studied in animal systems and yeast, the regulation of synthesis, processing, storage and secretion of peptide signals in plants is poorly understood. The processing sites that release the active peptides are not known, nor is the nature of the peptidases involved. The mechanism by which Sys and AtPep1 are released into the cell wall apoplast is still unknown. About half a dozen families of plant peptide hormones identified to date are synthesized from precursors lacking a signal sequence (Matsubayashi and Sakagami 2006), suggesting that more often than in animals and yeast, plant regulatory peptides are processed and transported through non-classical mechanisms to the extracellular space (Nickel 2003). Further studies on the biogenesis of defense-related peptide hormones, and the discovery of new peptide signals and their receptors in plants, will probably provide more evidence for the evolution of defense peptide signaling systems in eukaryotes from a common ancestor.

References

Bergey DR, Howe GA, Ryan CA (1996) Polypeptide signaling for plant defensive genes exhibits analogies to defense signaling in animals. Proc Natl Acad Sci USA 93:12053–12058

Coffeen WC, Wolpert TJ (2004) Purification and characterization of serine proteases that exhibit caspase-like activity and are associated with programmed cell death in *Avena sativa*. Plant Cell 16:857–873

Constabel CP, Yip L, Ryan CA (1998) Prosystemin from potato, black nightshade, and bell pepper: primary structure and biological activity of predicted systemin polypeptides. Plant Mol Biol 36:55–62

D'Hont K, Bosch D, Van Damme J, Goethals M, Vanderkerkhove J, Krebbers E (1993) An aspartic endoproteinase present in seed cleaves *Arabidopsis* 2S albumins in vitro. J Biol Chem 268:20884–20891

Dombrowski JE, Pearce G, Ryan CA (1999) Proteinase inhibitor-inducing activity of the prohormone prosystemin resides exclusively in the C-terminal systemin domain. Proc Natl Acad Sci USA 96:12947–12952

Felix G, Boller T (1995) Systemin induces rapid ion fluxes and ethylene biosynthesis in *Lycopersicon peruvianum* cells. Plant J 7:381–389

Francois I, De Bolle MFC, Dwyer G, Goderis I, Woutors PFJ, Verhaert PD, Proost P, Schaaper WMM, Cammue BPA, Broekaert WF (2002) Transgenic expression in *Arabidopsis* of a polyprotein construct leading to production of two different antimicrobial proteins. Plant Physiol 128:1346–1358

Green TR, Ryan CA (1972) Wound-induced proteinase inhibitor in plant leaves – possible defense mechanism against insects. Science 175:776–777

Harris R (1989) Processing of pro-hormone precursor proteins. Arch Biochem Biophys 275:315–333

Horn M, Patankar AG, Zavala JA, Wu JQ, Doleckova-Maresova L, Vujtechova M, Mares M, Baldwin IT (2005) Differential elicitation of two processing proteases controls the processing pattern of the trypsin proteinase inhibitor precursor in *Nicotiana attenuata*. Plant Physiol 139:375–388

Howard AD, Kostura MJ, Thornberry N, Ding GJF, Limjuco G, Weidner J, Salley JP, Hogquist KA, Chaplin DD, Mumford RA, Schmidt JA, Tocci MJ (1991) IL-1β-Converting enzyme requires aspartic-acid residues for processing of the IL-1β precursor at 2 distinct sites and does not cleave 31-kDa IL-1α. J Immunol 147:2964–2969

Howe GA (2005) Jasmonates as signals in the wound response. J Plant Growth Regul 23:223–237

Howe GA, Ryan CA (1999) Suppressors of systemin signaling identify genes in the tomato wound response pathway. Genetics 153:1411–1421

Huffaker A, Pearce G, Ryan CA (2006a) An endogenous peptide signal in *Arabidopsis* activates components of the innate immune response. Proc Natl Acad Sci USA 103:10098–10103

Huffaker A, Yamaguchi Y, Pearce G, Ryan CA (2006b) AtPep1 peptides. In: Kastin AJ (ed) Handbook of biologically active peptides. Academic Press, San Diego, pp 5–8

Matsubayashi Y, Sakagami Y (2006) Peptide hormones in plants. Ann Rev Plant Biol 57:649–674

McGurl B, Orozco-Cárdenas M, Pearce G, Ryan CA (1994) Overexpression of the prosystemin gene in transgenic tomato plants generates a systemic signal that constitutively induces proteinase-inhibitor synthesis. Proc Natl Acad Sci USA 91:9799–9802

McGurl B, Ryan CA (1992) The organization of the prosystemin gene. Plant Mol Biol 20:405–409

McGurl B, Pearce G, Orozco-Cárdenas M, Ryan CA (1992) Structure, expression, and antisense inhibition of the systemin precursor gene. Science 255:1570–1573

Meindl T, Boller T, Felix G (1998) The plant wound hormone systemin binds with the N-terminal part to its receptor but needs the C-terminal part to activate it. Plant Cell 10:1561–1570

Montoya T, Nomura T, Farrar K, Kaneta T, Yokota T, Bishop GJ (2002) Cloning the tomato *curl3* gene highlights the putative dual role of the leucine-rich repeat receptor kinase tBRI1/SR160 in plant steroid hormone and peptide hormone signaling. Plant Cell 14:3163–3176

Narváez-Vásquez J, Orozco-Cárdenas ML, Ryan CA (1994) A sulfhydryl reagent modulates systemic signaling for wound-induced and systemin-induced proteinase-inhibitor synthesis. Plant Physiol 105:725–730

Narváez-Vásquez J, Orozco-Cárdenas ML, Ryan CA (2007) Systemic wound signaling in tomato leaves is cooperatively regulated by multiple plant peptides. Plant Mol Biol 65:711–718

Narváez-Vásquez J, Pearce G, Orozco-Cárdenas ML, Franceschi VR, Ryan CA (1995) Autoradiographic and biochemical-evidence for the systemic translocation of systemin in tomato plants. Planta 195:593–600

Narváez-Vásquez J, Pearce G, Ryan CA (2005) The plant cell wall matrix harbors a precursor of defense signaling peptides. Proc Natl Acad Sci USA 102:12974–12977

Narváez-Vásquez J, Ryan CA (2002) The systemin precursor gene regulates both defensive and developmental genes in *Solanum tuberosum*. Proc Natl Acad Sci USA 99:15818–15821

Narváez-Vásquez J, Ryan CA (2004) The cellular localization of prosystemin: a functional role for phloem parenchyma in systemic wound signaling. Planta 218:360–369

Nickel W (2003) The mystery of nonclassical protein secretion – A current view on cargo proteins and potential export routes. Eur J Biochem 270:2109–2119

Orozco-Cárdenas ML, McGurl B, Ryan CA (1993) Expression of an antisense prosystemin gene in tomato plants reduces resistance toward *Manduca sexta* larvae. Proc Natl Acad Sci USA 90:8273–8276

Pautot V, Holzer FM, Reish B, Walling LL (1993) Leucine aminopeptidase: an inducible component of the defense response in *Lycopersicon esculentum* (tomato). Proc Natl Acad Sci USA 90:9906–9910

Pearce G, Johnson S, Ryan CA (1993) Structure-activity of deleted and substituted systemin, an 18-amino acid polypeptide inducer of plant defensive genes. J Biol Chem 268:212–216

Pearce G, Moura DS, Stratmann J, Ryan CA (2001) Production of multiple plant hormones from a single polyprotein precursor. Nature 411:817–820

Pearce G, Narváez-Vásquez J, Ryan CA (2006) Systemins. In: Kastin AJ (ed) Handbook of biologically active peptides. Academic Press, San Diego, pp 49–53

Pearce G, Ryan CA (2003) Systemic signaling in tomato plants for defense against herbivores – isolation and characterization of three novel defense-signaling glycopeptide hormones coded in a single precursor gene. J Biol Chem 278:30044–30050

Pearce G, Strydom D, Johnson S, Ryan CA (1991) A polypeptide from tomato leaves activates the expression of proteinase inhibitor genes. Science 253:895–898

Rath A, Davison AR, Deber CM (2005) The structure of 'unstructured' regions in peptides and proteins: role of the polyproline II helix in protein folding and recognition. Biopolymers 80: 179–185

Realini C, Rogers SW, Rechsteiner M (1994) KEKE motifs – proposed roles in protein–protein association and presentation of peptides by MHC Class-I receptors. FEBS Lett 348:109–113

Ren F, Lu Y-T (2006) Overexpression of tobacco hydroxyproline-rich glycopeptide systemin precursor A in transgenic tobacco enhances resistance against *Helicoverpa armigera* larvae. Plant Sci 171:286–292

Rocha-Granados MC, Sánchez-Hernández C, Sánchez-Hernández C, Martínez-Gallardo NA, Ochoa-Alejo N, Délano-Frier JP (2005) The expression of the hydroxyproline-rich glycopeptide systemin precursor A in response to (a)biotic stress and elicitors is indicative of its role in the regulation of the wound response in tobacco (*Nicotiana tabacum* L.). Planta 222:794–810

Rojo E, Martin R, Carter C, Zouhar J, Pan SQ, Plotnikova J, Jin HL, Paneque M, Sanchez-Serrano JJ, Baker B, Ausubel FM, Raikhel NV (2004) VPE gamma exhibits a caspase-like activity that contributes to defense against pathogens. Curr Biol 14:1897–1906

Ryan CA (2000) The systemin signaling pathway: differential activation of plant defensive genes. Biochim Biophys Acta 1477:112–121

Ryan CA, Moura DS (2002) Systemic wound signalling in plants: a new perception. Proc Natl Acad Sci USA 99:6519–6520

Ryan CA, Pearce G (1998) Systemin: a polypeptide signal for plant defensive genes. Ann Rev Cell Dev Biol 14:1–17

Ryan CA, Pearce G (2003) Systemins: a functionally defined family of peptide signal that regulate defensive genes in Solanaceae species. Proc Natl Acad Sci USA 100:14577–14580

Schaller A (1998) Action of proteolysis-resistant systemin analogues in wound signalling. Phytochemistry 47:605–612

Schaller A, Oecking C (1999) Modulation of plasma membrane H^+-ATPase activity differentially activates wound and pathogen defense responses in tomato plants. Plant Cell 11:263–272

Schaller A, Ryan CA (1994) Identification of a 50-KDa systemin-binding protein in tomato plasma-membranes having Kex2p-like properties. Proc Natl Acad Sci USA 91:11802–11806

Schaller A, Ryan CA (1996) Molecular cloning of a tomato leaf cDNA encoding an aspartic protease, a systemic wound response protein. Plant Mol Biol 31:1073–1077

Schaller F, Schaller A, Stintzi A (2005) Biosynthesis and metabolism of jasmonates. J Plant Growth Regul 23:179–199

Scheer JM, Pearce G, Ryan CA (2003) Generation of systemin signaling in tobacco by transformation with the tomato systemin receptor kinase gene. Proc Natl Acad Sci USA 100:10114–10117

Scheer JM, Ryan CA (1999) A 160-kD systemin receptor on the surface of *Lycopersicon peruvianum* suspension-cultured cells. Plant Cell 11:1525–1535

Scheer JM, Ryan CA (2002) The systemin receptor SR160 from *Lycopersicon peruvianum* is a member of the LRR receptor kinase family. Proc Natl Acad Sci USA 99:9585–9590

Seidah NG, Chretien M (1997) Eukaryotic protein processing: endoproteolysis of precursor proteins. Curr Opin Biotechnol 8:602–607

Stenzel I, Hause B, Maucher H, Pitzschke A, Mierch O, Ziegler J, Ryan CA, Wasternack C (2003) Allene oxide cyclase dependence of the wound response and vascular bundle-specific generation of jasmonates in tomato – amplification in wound signaling. Plant J 33:577–589

Tanaka S (2003) Comparative aspects of intracellular proteolytic processing of peptide hormone precursors: studies of procpiomelanocortin processing. Zool Sci 20:1183–1198

Tortiglione C, Fogliano V, Ferracane R, Fanti P, Pennacchio F, Monti LM, Rao R (2003) An insect peptide engineered into the tomato prosystemin gene is released in transgenic tobacco plants and exerts biological activity. Plant Mol Biol 53:891–902

Truman W, Bennett MH, Kubigsteltig I, Turnbull C, Grant M (2007) *Arabidopsis* systemic immunity uses conserved signaling pathways and is mediated by jasmonates. Proc Natl Acad Sci USA 104:1075–1080

Walling LL (2006) Recycling or regulation? The role of amino-terminal modifying enzymes. Curr Opin Plant Biol 9:227–233

Wasternack C, Stenzel I, Hause B, Hause G, Kutter C, Maucher H, Neumerkel J, Feussner I, Miersch O (2006) The wound response in tomato – role of jasmonic acid. J Plant Physiol 163:297–306

Yamaguchi Y, Pearce G, Ryan CA (2006) The cell surface leucine-rich repeat receptor for AtPep1, an endogenous peptide elicitor in *Arabidopsis*, is functional in transgenic tobacco cells. Proc Natl Acad Sci USA 103:10104–10109

Zhang L, Kelley J, Schmeisser G, Kobayashi YM, Jones LR (1997) Complex formation between junction, triadin, calsequestrin, and the ryanodine receptor – proteins of the cardiac junctional sarcoplasmic reticulum membrane. J Biol Chem 272:23389–2339

Chapter 16
MAP Kinases in Plant Responses to Herbivory

Johannes Stratmann

Mitogen-activated protein kinases (MAPKs) are signal transducing enzymes involved in many plant responses to the environment including defense responses to herbivory. Phytophagous insects generate various plant- and insect-derived primary signals such as the plant signaling peptide systemin, mechanical wound signals, and fatty acid-amino acid conjugates. These signals activate a set of specific MAPKs. Reducing the function of these MAPKs impairs the synthesis of secondary stress signals including the key wound signal jasmonic acid, which is required for the upregulation of defense proteins. Consequently, loss-of-MAPK-function results in reduced resistance of plants to herbivorous insects. Insect-responsive MAPKs also function as essential signaling relays in defense responses to pathogens and abiotic stress. It is a major challenge for future research to unravel the molecular mechanisms that perceive and integrate stress signals to generate stimulus-specific MAPK signaling patterns which result in specific cellular defense responses.

16.1 Introduction

A wealth of genomic information in all eukaryotes has revealed the preeminent role of protein kinases for the regulation of cellular processes. PlantsP, an interactive website dedicated to the study of plant phosphorylation, estimated that the *Arabidopsis* genome contains about 1000 putative protein kinases that constitute 5 major classes with a total of 80 distinct families (http://plantsp.genomics.purdue.edu /html). Unlike in animals and yeast, tyrosine-specific phosphorylation is rare in plants. Almost all plant protein kinases are Ser/Thr-specific kinases. They are structurally defined by the presence of 12 conserved subdomains that constitute the catalytic core of the enzyme. The kinase domain is often extended N- and C-terminally by other protein domains (Hanks and Hunter 1995). Many protein kinases become

---

J. Stratmann
Department of Biological Sciences, University of South Carolina, Columbia, SC 29208, USA
e-mail: johstrat@biol.sc.edu

activated by phosphorylation via other kinases and inactivated by dephosphorylation via protein phosphatases, which comprise about 300 members in the *Arabidopsis* genome (PlantP).

Reversible phosphorylation is known to be a part of many signaling pathways that regulate developmental processes and plant responses to the environment. This chapter reviews the role of mitogen-activated protein kinase (MAPK) cascades in plant responses to herbivorous insects and mechanical wounding. To understand the function of MAPKs, it is important to note that many plant stress responses overlap at various levels of signal transduction and cellular responses. Global transcriptome analyses revealed overlaps among gene expression profiles induced by different stress stimuli, but also stimulus-specific gene expression (Cheong et al. 2002; Reymond et al. 2000; Schenk et al. 2000). It appears that a limited number of secondary stress signals, such as ion fluxes (e.g., Ca^{2+} and H^+), jasmonic acid (JA), ethylene, reactive oxygen species (ROS), and salicylic acid (SA), function in multiple signaling pathways that can be induced by structurally and functionally diverse stress signals. MAPKs are known to function as convergence points for multiple stress signals. The interaction and coordination of these signaling components is thought to occur in a signaling network where multiple signaling elements are highly interconnected and mutually affect each other. While this review will focus on signal transduction in plant responses to herbivory, it will take into account that wound-responsive MAPKs also function in other stress signaling pathways.

Only very few studies address the role of MAPKs and other protein kinases in response to damage inflicted by herbivorous insects. Most studies have focused on the role of protein kinases in the response to leaf wounding with mechanical devices, such as a hemostat or forceps for crushing, razor blades for cutting, or carborundum for leaf surface abrasion. It should be noted that mechanical wounding and wounding by chewing insects result in the generation of several primary wound signals, such as systemically propagated rapid electrical and hydraulic signals (Malone 1996; Rhodes et al. 1996; Stratmann and Ryan 1997; Wildon et al. 1992), and peptides such as systemin and hydroxyproline-rich systemins (Pearce et al. 1991, 2001; Pearce and Ryan 2003). In addition, insects deliver fatty acid-amino acid conjugates (Alborn et al. 1997; Halitschke et al. 2003) or other insect-specific elicitors like β-glucosidase (Mattiacci et al. 1995) to wound sites which function as elicitors of direct and indirect defense responses. For some insect species, however, the mechanical stimuli produced by insect mandibles alone may be sufficient to trigger a full-fledged wound response and resistance to insects (Mithofer et al. 2005). Furthermore, unlike wounding by insects, wounding with mechanical devices causes desiccation stress in the wounded leaf with concomitant activation of water stress-related responses (Birkenmeier and Ryan 1998; Reymond et al. 2000). If protein kinases are activated as a consequence of wounding with mechanical devices, it is unclear what the actual activating signal is and whether this is relevant to wound responses elicited by insects.

Chewing insects represent only one feeding guild among phytophagous insects. With regard to the mechanics of plant damage, they do not have much in common with sucking-piercing insects such as aphids, whiteflies, and thrips. This group of insects employs sophisticated strategies to imbibe cell or phloem contents. The actual wound stimulus is more confined and is produced by elaborate specialized stylets that can penetrate single cells. However, sucking-piercing insects seem to activate defense pathways that are also induced by microbial pathogens, nematodes, and chewing insects (Li et al. 2002b, 2006; reviewed in Walling 2000).

Protein phosphatases remove phosphate groups from MAPKs and thereby inactivate them. With a few exceptions (Meskiene et al. 1998; Seo et al. 2007; Xu et al. 1998), not much is known on the role of protein phosphatases in responses to wounding and herbivory. This is also true for the protein substrates of MAPKs and the upstream kinases that activate MAPKs. Therefore, these groups of proteins will not be discussed in detail in this chapter, even though they are of utmost importance for understanding the role and function of MAPKs.

16.2 MAP Kinases in Plant Responses to Herbivory

16.2.1 Mitogen-Activated Protein Kinase (MAPK) Cascades

A MAPK cascade is a three-tiered phosphorelay cascade which is initiated by a MAPK kinase kinase (MAPKKK, also known as MEKK). MAPKKKs are often activated as a consequence of receptor-mediated perception of stress signals. They phosphorylate and thereby activate MAPK kinases (MAPKKs, also known as MEKs or MKKs) on two Ser/Thr residues in the kinase activation loop. MAPKKs are dual-specificity kinases which activate MAPKs (also known as MPKs) through phosphorylation of a Thr and a Tyr residue in a TE/DY phosphorylation motif located in the MAPK activation loop between kinase subdomains VII and VIII (Ichimura et al. 2002). In plants, MAPKKs are the only well-known protein kinases that can phosphorylate Tyr residues. Some MAPKs are transcriptionally regulated, but post-translational activation of preformed MAPKs via phosphorylation often precedes transcriptional activation and is required for kinase activity. This might explain why a microarray analysis found only a relatively small number of kinase transcripts upregulated in response to herbivory by a generalist and a specialist caterpillar in *Arabidopsis* plants (Reymond et al. 2004). MAPKs are Ser/Thr kinases which phosphorylate a range of cytosolic and nuclear substrate proteins that function in metabolism or transcription, at least in metazoan and yeast cells. In plants, not many in vivo substrates have been identified, but a picture is emerging that is consistent with animal models.

Plant genomes contain a large number of MAPKs, e.g., 20 in *Arabidopsis*, 21 in poplar, and 15 in rice (Hamel et al. 2006), but only half the number of MAPKKs. The number of functional MAPKKKs is controversial (Champion et al. 2004) and may exceed 60 kinases from various gene families. The general function and genome representation of MAPK cascade components has been described in excellent reviews (Champion et al. 2004; Hamel et al. 2006; Ichimura et al. 2002) and will not be covered here.

Phylogenetic comparisons of plant MAPKs revealed two major clades, one characterized by a TEY, and the other one by a TDY phosphorylation motif (Champion et al. 2004; Hamel et al. 2006; Ichimura et al. 2002). Among dicots, only members of the TEY clade are known to be wound-responsive. They fall into the three distinct subgroups A1, A2, and B1. None of these MAPKs is exclusively activated by wounding or wound signals. It is a general feature of all characterized MAPKs that they respond to a wide range of stimuli and signals (reviewed in Nakagami et al. 2005). Wound-responsive TDY MAPKs have also been identified in monocots. Possible differences between monocot and dicot MAPK signaling in the wound response were recently reviewed and will not be discussed here (Rakwal and Agrawal 2003).

16.2.2 SIPK and WIPK

The best studied plant MAPKs are the stress-induced MAPK (SIPK) and the wound-induced MAPK (WIPK) from tobacco, and their orthologs in *Arabidopsis*, alfalfa, parsley, and tomato. They belong to the A2 and A1 subgroups, respectively. Most stress signals that activate SIPK also activate WIPK (reviewed in Nakagami et al. 2005; Zhang and Klessig 2000). SIPK activity increases in response to almost any disturbance in the cellular environment, including biotic and abiotic stimuli, and even seemingly innocuous stimuli like touch (Nakagami et al. 2005). SIPK and orthologs are preformed and activated posttranslationally without concomitant changes in transcript levels (Higgins et al. 2007; Ichimura et al. 2000; Zhang and Klessig 1998) In contrast, WIPK and orthologs are transcriptionally and posttranslationally regulated in response to wounding and other stress signals. However, increases in transcript levels do not always result in higher WIPK protein levels. Whether this is due to increased protein turnover is presently unclear (Bögre et al. 1997; Higgins et al. 2007; Hirt 1999; Seo et al. 1999; Zhang and Klessig 2000).

Detailed analyses of MAPKs activated by wounding or wound signals are only available for WIPK and SIPK in tobacco and their orthologs in tomato. Leaf wounding, stem excision, and feeding by *Manduca sexta* (Lepidoptera) larvae as well as the wound signals systemin, hydroxyproline-rich systemins, oligosaccharides, and a novel wound-inducible diterpene ($11E,13E$)-labda-11,13-diene-8α,15-diol, all induced MAPK activity in tobacco and tomato (Higgins et al. 2007; Holley et al. 2003; Seo et al. 1995, 2003, 2007; Stratmann and Ryan 1997). In response to wounding and insect attack, the MAPK activity increased systemically, at least over short distances (Seo et al. 1999; Stratmann and Ryan 1997). The function of systemic increases in MAPK activity is not known.

Seo et al. (1995) presented the first loss-of-function analysis of a plant MAPK. In wild type tobacco plants, wounding leads to increases in JA levels followed by expression of basic pathogenesis-related (PR) genes and proteinase inhibitor II (PI-II). Surprisingly, in a transgenic line in which the WIPK gene was cosuppressed, wounding induced salicylic acid (SA) synthesis and expression of acidic PR genes, while JA synthesis and expression of basic PR genes and PI-II was suppressed. This is characteristic of defense responses to microbial pathogens. Since responses to pathogens and wounding are often mutually exclusive (Felton et al. 1999), the co-suppression results suggested that WIPK is participating in a switch-like mechanism that determines whether the wound response or the response to pathogens will be expressed.

To further test the role of WIPK for JA synthesis in response to wounding, transgenic tobacco plants were generated that overexpressed WIPK (Seo et al. 1999). Surprisingly, unwounded plants exhibited constitutively elevated WIPK activity (1.5 fold), 3–4 fold higher JA levels as compared to control plants, and constitutive expression of the PI-II gene. These results raise several questions. How does overexpression lead to increased kinase activity in the absence of wounding? Were the upstream kinases (MAPKKs and MAPKKKs) also activated? The authors suggested that overexpression may lead to autoactivation of WIPK, which is a well-known phenomenon for recombinant MAPKs. The results were questioned by Kim et al. who showed that expression of a constitutively active mutant of NtMEK2, the common MAPKK for SIPK and WIPK (Yang et al. 2001), leads to increased activity of both MAPKs and ethylene synthesis (Kim et al. 2003). However, they were not able to detect increases in JA levels. These differences may be explained by the different gain-of-function approaches; overexpression of a constitutively active MAPKK which activates at least two MAPKs, versus overexpression of WIPK. In any case, the conflicting data indicate that it is important to complement gain-of-function with loss-of-function studies.

In their most recent study, the Ohashi group employed the RNA interference (RNAi) technique to silence *WIPK* and *SIPK* both individually and simultaneously (Seo et al. 2007). *SIPK* has an almost identical paralog, *Ntf4*, which has recently been shown to be largely functionally redundant with *SIPK*, at least with regard to pathogenesis-related responses (Ren et al. 2006). The RNAi construct that targeted *SIPK* also silenced *Ntf4*. These transgenic plants exhibited normal growth and development, but reduced wound-induced JA levels. Wound-induced ethylene synthesis and accumulation of transcripts for the JA-biosynthetic enzymes ALLENE OXIDE SYNTHASE (AOS) and ALLENE OXIDE CYCLASE (AOC) were at least partially SIPK-dependent. Wounding of *SIPK*-silenced plants also resulted in higher levels of acidic *PR* transcripts, a phenomenon also observed in the *WIPK*-cosuppressed line (Seo et al. 1995). But *WIPK*-RNAi lines did not produce this response. Furthermore, SA levels, which were increased in the *WIPK*-cosuppressed lines, were not increased in either wounded *WIPK*-RNAi or *SIPK*-RNAi lines, but they went up in *SIPK/WIPK*-cosilenced plants. A significant reduction in wound response marker genes was not observed in any of the three RNAi lines.

Taken together, the three papers from Ohashi's group point to the following scenario. In response to wounding, JA and ethylene levels increase in a SIPK- and/or WIPK-dependent manner. Since silencing of the kinases was strong, but the reduction of JA and ethylene levels only partial, additional regulatory mechanisms may be in place. To reconcile conflicting data, the authors suggested that either SIPK or WIPK regulate the synthesis of an unknown wound-induced inhibitor which may act together with JA to suppress SA synthesis. Is the inhibition of JA synthesis in the silenced tobacco plants a consequence of cross-inhibition by SA, a known inhibitor of JA synthesis and signaling (Doares et al. 1995)? The authors point out that this is unlikely since silencing of *SIPK* and *WIPK* alone did not result in SA synthesis, yet the JA levels were much lower than in control plants. It remains a challenge to reconcile the different results obtained by using RNAi or cosuppression of *WIPK*. Since WIPK and SIPK are convergence points of many stress signals that function in a signaling network, the ratio of SIPK to WIPK levels, which probably differs among the various MAPK-silenced lines, may be important to generate appropriate responses (see 16.3 Perspectives).

16.2.3 LeMPK1, LeMPK2, and LeMPK3

In tomato, the wound response is mediated by the wound signaling peptide systemin (Narváez-Vásquez and Orozco-Cárdenas this volume). Systemin is both necessary and sufficient for the successful defense of tomato plants against herbivorous insects as shown by antisense inhibition and prosystemin overexpression experiments (Orozco-Cardenas et al. 1993; Chen et al. 2005; McGurl et al. 1994). In response to leaf wounding or the application of systemin through cut stems, MAPK activity increased within minutes in a systemic manner. The stem excision itself rapidly induced MAPK activity in the leaves indicating that a rapidly propagated mechanical signal, probably of hydraulic nature, caused the systemic MAPK activation (Stratmann and Ryan 1997). Stem excision alone induced a transient increase in MAPK activity which lasted only ~15 min. In the presence of systemin, MAPK activity was stronger and more prolonged (>60 min). Activation of MAPKs by systemin was receptor-dependent. A systemin analog that binds to the systemin receptor SR160 without activating it (Meindl et al. 1998; Scheer and Ryan 1999) prevented systemin-induced MAPK activation (Stratmann and Ryan 1997). In addition, the naphthylurea compound suramin, a known inhibitor of cytokine–cytokine receptor interactions in animals, was shown to prevent binding of systemin to SR160 and blocked MAPK activation (Stratmann et al. 2000a). Systemin- and wound-induced MAPK activation was not altered in the JA-deficient mutant *def1*, indicating that MAPKs function either upstream of JA synthesis or in a parallel pathway. Systemin-induced MAPK activity and PI-II synthesis exhibited the same dose-response curve, indicating that MAPKs are functioning in the signaling pathway that leads to expression of wound response genes (Stratmann and Ryan 1997). Later, it was shown that systemin and wounding activate three MAPKs, called LeMPK1, LeMPK2, and LeMPK3 (for <u>L</u>ycopersico<u>n</u> <u>e</u>sculentum <u>MAPK</u>), which represent the tomato orthologs of the tobacco SIPK, Ntf4, and WIPK, respectively (Higgins et al. 2007; Holley et al. 2003).

Wounding generates various signals, and application of systemin through cut stems involves a wound stimulus. To study systemin-specific responses, we used transgenic plants that overexpress prosystemin under the control of the 35S promoter.

These plants exhibit constitutive accumulation of proteinase inhibitors (McGurl et al. 1994) and had been used earlier to identify JA biosynthetic and signaling mutants as suppressors of prosystemin-mediated responses (*spr* mutants; Howe and Ryan 1999). Silencing the *LeMPK1*, *LeMPK2*, and *LeMPK3* genes in these transgenic plants by virus-induced gene silencing (VIGS) resulted in reduced levels of defensive proteinase inhibitors. This demonstrated that the function of these MAPKs is required for prosystemin-induced gene expression. Cosilencing of *LeMPK1/2* in 35S::*Prosystemin* plants also reduced JA synthesis in response to wounding, and complementation of *LeMPK1/2*-silenced plants with methyl-JA (MeJA) restored proteinase inhibitor synthesis. These data support a role of LeMPK1/2 in regulating JA synthesis (Kaitheri Kandoth et al. 2007). They are consistent with experiments showing that MAPK activity does not increase after application of JA or its precursor 12-oxo-phytodienoic acid (Stratmann and Ryan 1997). Since it is well-established that the LeMPK3 ortholog WIPK is required for JA synthesis in tobacco (Seo et al. 1995, 1999, 2007), it is likely that systemin activates at least three MAPKs which function upstream of JA synthesis.

LeMPK1 and 2 are 95% identical at the amino acid level, and they are activated in a concerted manner by a range of stress signals such as oligosaccharide elicitors, wounding, ultraviolet-B radiation, and the fungal toxin fusicoccin suggesting that they are functionally redundant (Higgins et al. 2007; Holley et al. 2003). Surprisingly, we found that they were both required for defense gene activation by systemin. VIGS of either *LeMPK1* or *LeMPK2* reduced proteinase inhibitor levels in 35S::*Prosystemin* plants (Kaitheri Kandoth et al. 2007). There is evidence that SIPK activates WIPK, possibly by regulating WIPK expression (Liu et al. 2003), but this is not consistent with our results in tomato. Since LeMPK1 and 2 appear to share the same functions in leaf tissue, gene dosage effects may be important for modulating signaling flow and thus cellular responses. Many MAPKs in various plant species evolved as a consequence of recent gene duplication events (Hamel et al. 2006). Most of them are expressed, but it is not understood why the presence of paralogs is adaptive. Additional gene knock-out experiments targeting paralogous MAPKs should help to elucidate this phenomenon.

JA synthesis is initiated in the chloroplasts resulting in the synthesis of the JA precursor 12-oxo-phytodienoic acid which is then further metabolized in peroxisomes to JA (Schaller and Stintzi this volume). It is not known how the systemin signal is transduced from the plasma membrane-bound receptor kinase SR160 via the cytosol to the chloroplast. Most plant MAPKs are cytosolic proteins. Therefore, the LeMPKs are likely to further relay the systemin signal to activate JA biosynthetic processes in the chloroplasts. The identification of LeMPK substrates is essential to understand the highly compartmentalized wound/systemin response in tomato cells (Schilmiller and Howe 2005; Wasternack et al. 2006). Known substrates of the LeMPK1/2 ortholog AtMPK6 are the cytosolic ethylene-forming enzymes 1-aminocyclopropane-1-carboxylate synthase 2 and 6 (ACS2 and ACS6; Liu and Zhang 2004). AtACS2/6 were phosphorylated by AtMPK6 resulting in increased ACS activity and ethylene synthesis in response to the bacterial flagellin elicitor. Ethylene is also an essential second messenger in the wound response (O'Donnell et al. 1996) and is generated in response to systemin in suspension-cultured *L. peruvianum* cells (Felix and Boller 1995). In tobacco, wound-induced ethylene synthesis was partially regulated by SIPK (LeMPK1/2 ortholog). It is conceivable that JA synthesis is indirectly regulated by SIPK via ethylene.

Based on remarkable parallels between the animal inflammatory response and the tomato wound response, it was proposed that a MAPK activates a cytosolic phospholipase A_2 ($cPLA_2$) which would initiate JA synthesis via production of linolenic acid from membrane lipids (Stratmann et al. 2000a). In animal cells, certain growth factors activate a MAPK which in turn activates $cPLA_2$ via phosphorylation (Lin et al. 1993). A $cPLA_2$ activity was identified in tomato leaf extracts in response to wounding and systemin (Narváez-Vásquez et al. 1999). However, a specific wound-responsive $cPLA_2$ has not been identified, and it is now evident that the first steps of JA synthesis occur in the chloroplasts (Schilmiller and Howe 2005; Wasternack et al. 2006). This is difficult to reconcile with the activation of a cytosolic PLA_2 by MAPKs. It is also not known from other model systems that MAPKs can translocate from the cytosol into organelles.

Another intermediate signal in the wound response are reactive oxygen species (ROS), which are thought to function downstream of JA and are generated by a NADPH oxidase activity (Orozco-Cárdenas et al. 2001; Sagi et al. 2004). Since LeMPK1-3 function upstream of JA, they might also regulate ROS synthesis. ROS are known as activators of MAPKs (Kovtun et al. 2000) and might cause oxidative damage to signaling components that regulate MAPKs, e.g., protein phosphatases. But the exact mechanism of MAPK activation by ROS is not known. In response to fungal elicitors which trigger an oxidative burst, MAPKs are activated independently of ROS (Cazale et al. 1999; Ligterink et al. 1997; Romeis et al. 1999). Vice versa, there is evidence for a MAPK-independent oxidative burst in host-specific fungal-plant interactions (Romeis et al. 1999). Together this suggests that, at least in some signaling pathways, MAPK activity and ROS generation are induced in parallel and may function independently of each other. However, the relation of wound-induced ROS and MAPK activity remains to be resolved.

Systemin perception by SR160 induces extracellular alkalinization (EA) and a concomitant membrane depolarization (Felix and Boller 1995; Moyen and Johannes 1996; Schaller and Oecking 1999; Stratmann et al. 2000a). The alkalinization response is not unique to systemin, but induced by almost any stress signal. However, it is not known what the consequences of the EA are for signal transduction. Reversing the proton gradient by activators of the plasma membrane proton ATPase (PMA) compromised systemin-induced gene expression suggesting that the PMA regulates the extracellular pH in response to systemin, and that EA is an essential signaling event for the systemin-induced wound response (Schaller and Oecking 1999). It has been proposed that systemin inactivates the PMA in a calcium-dependent manner via phosphorylation through a calcium-dependent protein kinase (Rutschmann et al. 2002; Schaller and Oecking 1999; Schaller 1999). The fungal toxin fusicoccin (FC), which permanently activates the PMA, caused extracellular acidification and inhibited the systemin-induced wound response. Moreover, FC application resulted in expression of PR-genes. This prompted Schaller and Oecking (1999) to propose that PMA activity functions as a switch that determines whether wound response genes or PR-genes will be expressed.

Since most stress stimuli and signals that induce EA also activate A1- and A2-MAPKs, we wondered whether EA and the consequences thereof are required for MAPK activation or vice versa. We performed a comparative analysis of pH effects and MAPK activity in response to systemin and fusicoccin. Fusicoccin activated the wounding/systemin-responsive LeMPK1, 2, and 3, showing that both extracellular alkalinization and acidification lead to the activation of the same MAPKs. However, the MAPK activation kinetics were distinct, with systemin causing transient and FC causing prolonged MAPK activity. Simultaneous application of FC and systemin did not lead to immediate pH changes but resulted in rapid increases in MAPK activity. To test whether changes in the extracellular pH can be uncoupled from MAPK activation, conditioned growth medium of *L. peruvianum* suspension-cultured cells was exchanged with fresh medium. This resulted in a strong medium acidification without concomitant MAPK activation. Together, these experiments demonstrated that changes in the extracellular pH are neither required nor sufficient for MAPK activation. It could also be excluded that MAPKs regulate EA. Therefore, systemin-induced MAPK activation and EA most likely operate in parallel and are not part of one linear signaling pathway that induces wound response gene expression (Higgins et al. 2007).

Systemin is an essential component in the tomato defense response against herbivorous insects and is required for the regulation of JA synthesis (see above). JA functions not only as an intracellular second messenger but also as the long-distance wound signal (Schilmiller and Howe 2005; Stratmann 2003). JA biosynthetic and signaling mutants are strongly impaired in the resistance to herbivorous insects (Howe et al. 1996; Li et al. 2002a, 2004). Consistent with a role of MAPKs in the regulation of JA synthesis, we showed that cosilencing of *LeMPK1* and *LeMPK2* severely compromised resistance to *M. sexta* larvae in 35S::*Prosystemin* plants. We also showed that *Manduca sexta* larvae systemically induce LeMPK1, LeMPK2, and LeMPK3 activity (Kaitheri Kandoth et al. 2007).

The wound/systemin-responsive LeMPKs also play a role in resistance of tomato plants to sucking-piercing insects such as aphids. The mechanics of the damage inflicted by aphid stylets is different from chewing insects with regard to scale and nature. Resistance of tomato plants to aphids and certain whitefly and nematode species is dependent on the *Mi-1* gene (Li et al. 2006). *Mi-1*-mediated resistance requires salicylic acid (SA) and results in upregulation of SA-inducible PR genes. This is reminiscent of tomato defense responses against pathogens which are regulated by MAPK cascades (Del Pozo et al. 2004; Ekengren et al. 2003; Pedley and Martin 2004). An involvement of MAPK(K)s in aphid resistance was tested by VIGS of *LeMPK1/2*, *LeMPK3*, and *LeMKK2*. All three VIGS constructs resulted in increased aphid survival on *Mi-1*-expressing tomato leaves demonstrating that *Mi-1*-mediated resistance to aphids is regulated by a MAPK cascade (Li et al. 2006). These two reports represent the only loss-of-function analyses showing that specific kinases function as essential signaling components in plant defense responses that confer resistance to herbivorous insects.

As pointed out earlier, MAPKs of the A1 and A2 subgroup are highly promiscuous kinases responding to almost any disturbance in the cellular environment. LeMPK1 and 2 function as convergence points for stress stimuli and signals such as feeding insects, wounding, systemin, hydroxyproline-rich systemins, oligosaccharide elicitors, and ultraviolet-B and -C radiation (Holley et al. 2003; Higgins et al. 2007; Pearce et al. 2001; Stratmann et al. 2000b). The response of tomato leaves to ultraviolet-B (UV-B) radiation is interesting with regard to the specificity and interconnectedness of MAPK-regulated stress responses. While UV-B alone strongly induced MAPK activation, it induced neither 12-oxo-phytodienoic acid and JA formation, nor proteinase inhibitor synthesis. However, a weak wound stimulus given prior to UV-B irradiation resulted in a synergistic systemic accumulation of proteinase inhibitors (Stratmann et al. 2000b). Since proteinase inhibitor gene expression is strictly JA-dependent, this experiment indicates that MAPK activation by UV-B is not sufficient to bring about JA synthesis and defense gene expression. Similar experiments with *Nicotiana attenuata* plants in a field setting demonstrated that UV-B-responsive genes show a strong overlap with insect-induced genes (Izaguirre et al. 2003). Taken together, it seems evident that activation of the same MAPKs by different stress signals and stimuli is associated with both signal-specific responses and overlaps among stress responses. No mechanism has been identified so far that would explain this phenomenon (see 16.3 Perspectives).

16.2.4 The Role of MPK4 in the Wound Response

MAPKs of the A1 and A2 subgroups are not the only wound-responsive MAPKs. In *Arabidopsis*, the B1-MAPK, AtMPK4, can be activated by wounding, touch (gentle moving of the plants with hands) and some abiotic stresses (Ichimura et al. 2000). MAPK mutants are notoriously difficult to identify in standard mutant screens. Therefore, it was surprising that an *mpk4* mutant could be identified (Petersen et al. 2000), especially considering that there is a closely related homolog, AtMPK11, in the same subclade, and MPK5 and MPK12 in two additional subclades of the B1 subgroup (Ichimura et al. 2002). The isolation of *mpk4* also demonstrates that neither the other B1-MAPKs nor the multistress-responsive A-group MAPKs are able to compensate for the loss of MPK4. The *mpk4* mutant is more resistant to bacterial and oomycete pathogens as compared to wild type plants. This is accompanied by constitutive expression of PR genes and elevated SA levels, which is indicative of the pathogen-induced systemic acquired resistance (SAR). SAR is a broad-spectrum resistance to a wide range of microbial pathogens. Mutant *mpk4* plants were MeJA-insensitive, even in transgenic *NahG*-expressing plants that do not accumulate SA. This demonstrated that the JA insensitivity was not caused by SA – JA crosstalk. The *mpk4* mutant showed a dwarf phenotype, but responded normally to multiple abiotic stress treatments and plant hormones. The effect of the mutation was specific as determined by a microarray analysis that found only a small number of genes whose expression was affected by the mutation.

The insensitivity of *mpk4* plants to JA suggests that MPK4 functions in JA signaling downstream of JA synthesis, unlike the A group MAPKs that most likely function upstream of JA synthesis. It will be interesting to unravel the relationship of MPK4 and the E3-ligase SCF^{COI1}, which also functions downstream of JA and has recently been shown to target repressors of JA-induced gene expression for degradation by the proteasome (Chini et al. 2007; Thines et al. 2007). While Petersen et al. (2000) did not test whether the *Arabidopsis mpk4* mutant shows a compromised wound response, this was studied in tobacco by Gomi et al. (2005). The putative tobacco ortholog NtMPK4 can be transiently activated by leaf wounding, similarly to A group MAPKs. *NtMPK4*-RNAi plants showed a phenotype that closely resembled the *mpk4* phenotype in *Arabidopsis* with dwarfed morphology, elevated SA levels, and expression of SA-responsive *PR* genes. Wound-induced expression of the wound response marker gene *PI-II* was reduced, while *AOS* expression was not. JA levels were not measured in the *NtMPK4*-RNAi plants, and it was not tested whether *PI-II* expression can be restored in JA-treated *NtMPK4*-RNAi plants. It will be important to continue these studies to establish whether MPK4 is required for defenses against herbivorous insects, and how MPK4 acts together with the other wound-responsive MAPKs to orchestrate the wound response.

16.3 Perspectives

We know surprisingly little about reversible phosphorylation in responses to feeding insects. The studies reviewed here point to a role of MAPKs in herbivory. But this conclusion is mostly based on experiments that apply mechanical wounding or wound signals such as systemin. MAPKs involved in responses to pathogens have gained much greater attention including multiple loss- and gain-of-function studies and detailed biochemical analyses. To my knowledge, there are only two MAPK loss-of-function studies that directly involve insects (Kaitheri Kandoth et al. 2007; Li et al. 2006). In the future, it will be important to study the role of each MAPK in greater detail by exposing genetically manipulated plants to herbivorous insects. Since different insect feeding guilds cause different types of damage to plants (see 16.1 Introduction), it will also be important to further delineate whether certain MAPKs function in defenses against all or only a limited subset of insect species. The loss-of-function studies regarding the tomato MAPKs showed that LeMPK1, 2, and 3 are important for systemin-mediated resistance to chewing *M. sexta* larvae and for Mi-1-mediated resistance to aphids, two feeding guilds which cause very different types of damage and induce distinct cellular responses (Li et al. 2006; Thompson and Goggin 2006). Some insects are known to induce direct and indirect defenses via fatty acid-amino acid conjugates (FACs) present in oral secretions or regurgitates (Alborn et al. 1997; Halitschke et al. 2003). It will be exciting to find out whether FACs employ the same signaling components as mechanical wounding.

Stress-responsive MAPKs of the A and B1 groups seem to be major convergence points for many stress signals. Loss-of-function of these MAPKs has large consequences for defenses and protection against stress. Silencing of SIPK, WIPK and their tomato orthologs renders plants not only susceptible to pathogens, but also to chewing insects, aphids, and abiotic stress (Del Pozo et al. 2004; Ekengren et al. 2003; Kaitheri Kandoth et al. 2007; Li et al. 2006; Samuel and Ellis 2002; Seo et al. 2007; Sharma et al. 2003). On the other hand, gain-of-function studies revealed a more specific function, e.g., in the induction of the hypersensitive response (Ren et al. 2002; Yang et al. 2001; Zhang and Liu 2001). Since pathogen- and wound-induced signaling components often crosstalk, it is likely that overexpression of certain active MAPKs will not increase resistance to insects. However, this was never directly tested so far.

Defense responses to wounding or herbivory and responses to pathogens are clearly distinct but both mediated by the same MAPKs. This raises a fundamental question. How is signaling fidelity and specificity generated via these highly promiscuous kinases? Moreover, what are the molecular mechanisms that lead to signal-specific cellular responses? This is a much debated problem in animal and yeast signal transduction (Pouyssegur and Lenormand 2003; Sabbagh et al. 2001). Explanations are based on the concept of a signaling network. In such a network, specific signaling components (nodes) are highly interconnected, respond to multiple input signals and relay the signal to various downstream nodes. MAPKs of the A-group represent focal nodes that respond to a high number of input signals. Positive and negative feedback-mechanisms among nodes, switch-like signaling behavior and signal amplification increase the complexity of the network. This complexity enables flexible responses and integration of multiple signals which is important for natural settings in which plants may be successively or synchronously challenged by multiple stressors. Several solutions have been proposed for the problem of signaling fidelity based on work with yeast or animal systems. There is some evidence that similar mechanisms may operate in plants; however, they are generally not well corroborated.

MAPKs can be activated in either a prolonged or a transient manner. For specific mammalian MAPKs it has been established that the kinase activation kinetics determine the cellular response. Nerve growth factor induces prolonged activity of the mammalian MAPKs ERK1 and ERK2 and differentiation of PC12 cells into neuron-like cells. In contrast, epidermal growth factor induces transient ERK1/2 activation and PC12 cell proliferation, but not differentiation. The different activation kinetics were generated through differential interaction of growth factor receptors with interacting proteins which in turn resulted in assembly of stable or short-lived signaling complexes (Kao et al. 2001).

Stress-responsive plant MAPKs are also known to exhibit different activation kinetics in response to different stimuli. It had been suggested that wounding-related signals lead to transient, and pathogen-related signals lead to prolonged MAPK activation (Higgins et al. 2007; Ren et al. 2002). The activation kinetics are likely to be a function of the activating signal, the nature of the ligand-receptor interaction, and transcriptional and posttranslational activation of MAPKs and negative regulators such as protein phosphatases. In addition, negative and positive feedback loops affect activation kinetics. It had been shown that wound- and systemin-inducible MAPK cascades enter a refractory state after an initial activation indicating the presence of transient negative regulators of the cascade (Bögre et al. 1997; Holley et al. 2003; Yalamanchili and Stratmann 2002). Such a behavior indicates that sustained activity and inducibility of MAPKs is actively prevented, at least in response to wound-related signals. However, no mechanism has been revealed in plants that would explain how MAPK activation kinetics are interpreted by the signaling network to result in specific cellular responses.

Interaction with other proteins may modulate the activity of MAPKs or determine signaling specificity. Well known examples from animals and yeast are scaffolding proteins which specifically interact with MAPKs (Breitkreutz and Tyers 2002; Morrison and Davis 2003; Pouyssegur and Lenormand 2003; Sabbagh et al. 2001). A specific MAPK cascade can be assembled by tethering a specific MAPK to the scaffolding protein which also interacts with specific MAPKKs and MAPKKKs. This would establish signaling fidelity by excluding competitor MAPKs. Since homologs of the yeast and animal scaffolding proteins are not present in plants, it is not known whether similar mechanisms are used in plants to combine cascade components. However, A MAPKKK from alfalfa had been shown to interact with a MAPK suggesting that the MAPKKK itself may function as a scaffold (Nakagami et al. 2004).

Another way to generate specific responses via a signaling network is by differential activation and inactivation of multiple nodes that respond to multiple input signals. For example A-group MAPKs, ethylene, reactive oxygen species, Ca^{2+}, and proton fluxes participate in almost any stress signaling pathway. Each input signal has the potential to activate all these intermediate signals and signal transducers in an input-specific way via specific interactions with receptors that generate a specific pattern of network activity. A receptor-mediated input-specific signaling fingerprint is reflected in the kinetics and amplitude of the activity of individual nodes. This scenario can explain seemingly inconsistent MAPK gain- and loss-of-function experiments. Both approaches severely perturb the signaling network by overstimulation or disruption of a central node. Therefore, loss- and gain-of-function approaches might sometimes be more informative with regard to the rank of a node within the cellular signaling network than for the function of individual nodes in response to a particular input signal. It will be essential in the future to complement such studies with detailed biochemical analyses which address the dynamics of interactions among signaling nodes. Recent progress in phosphoproteomics (Peck 2006) and techniques used to study protein-protein interactions should reveal the dynamics of posttranslational modifications as well as protein complex assembly and disassembly in a signaling network in response to various input signals including herbivory. This would also involve the characterization of enzymatic parameters which are largely unknown for plant MAPKs, with only a few exceptions (Zhang and Klessig 1997). Such data were used to model the response of animal MAPK cascades which revealed that MAPK cascades function more like an on-off switch rather than in signal amplification (Ferrell 1996; Kolch et al. 2005).

Compartmentalization and tissue-specific expression may further contribute to signaling specificity. To date, this has not been explored for most MAPKs; however, some MAPKs are known to be expressed in many different cell types (Ren et al. 2006). Upon activation, some MAPKs translocate from the cytosol into the nucleus to activate transcription (Ahlfors et al. 2004; Lee et al. 2004; Ligterink et al. 1997). Others stay in the cytosol to activate cytosolic substrates (Liu and Zhang 2004), still others are permanently present in the nucleus (Munnik et al. 1999), or translocate from the nucleus to the cytosol (Samaj et al. 2002). MAPKs are likely to exhibit a broad substrate specificity (Feilner et al. 2005), but a more specific MAPK-substrate interaction may be determined by compartment-specific substrate availability.

Unfortunately, only a few in vivo plant MAPK substrates have been identified so far. Protein arrays were developed to determine putative substrates for AtMPK3 and AtMPK6. Both MAPKs phosphorylated 48 and 39 functionally diverse proteins, respectively, and many of these substrates were phosphorylated by both MAPKs (Feilner et al. 2005). Similarly, Stulemeijer et al. (2007) carried out a PepChip Kinomics® Slide Analysis and found that LeMPK1, 2, and 3 share many tripeptide phosphorylation motifs. In addition, each LeMPK phosphorylated specific substrates. This shows that the potential substrate range for plant MAPKs is very broad. It also implies that mechanisms which establish signaling fidelity through restricting substrate availability are important for fine tuning signaling networks in order to produce adequate output responses. A recent exciting development, based on the chromatin immunoprecipitation technique, is that yeast MAPKs may directly interact with transcribed regions of their target genes or associated histones and components of the transcriptional machinery (Chow and Davis 2006).

An exciting future task will be to make use of the ever increasing amount of data on plant MAPKs and associated proteins to generate mathematical models of stress signaling networks. The predictive power of such models is increasingly being recognized for MAPK signaling in animal cells. Models revealed unexpected behavior of the signaling network that could not have been identified experimentally (reviewed in Kolch et al. 2005).

Note added in proof: After submission of this article, an important paper was published by Wu et al. showing that FACs from *M. sexta* oral secretions activate SIPK and WIPK in *N. attenuata*. Silencing of these MAPKs revealed their critical role for FAC-induced JA-, SA-, and ethylene biosynthesis, and for the expression of a range of defense and signaling genes. (Wu J, Hettenhausen C, Meldau S, Baldwin IT (2007) Herbivory rapidly activates MAPK signaling in attacked and unattacked leaf regions but not between leaves of *Nicotiana attenuata*. Plant Cell 19:1096–1122).

References

Ahlfors R, Macioszek V, Rudd J, Brosche M, Schlichting R, Scheel D, Kangasjarvi J (2004) Stress hormone-independent activation and nuclear translocation of mitogen-activated protein kinases in *Arabidopsis thaliana* during ozone exposure. Plant J 40:512–522

Alborn HT, Turlings TCJ, Jones TH, Stenhagen G, Loughrin JH, Tumlinson JH (1997) An elicitor of plant volatiles from Beet Armyworm oral secretion. Science 276:945–949

Birkenmeier GF, Ryan CA (1998) Wound signaling in tomato plants. Evidence that ABA is not a primary signal for defense gene activation. Plant Physiol 117:687–693

Bögre L, Ligterink W, Meskiene I, Barker PJ, Heberle-Bors E, Huskisson NS, Hirt H (1997) Wounding induces the rapid and transient activation of a specific MAP kinase pathway. Plant Cell 9:75–83

Breitkreutz A, Tyers M (2002) MAPK signaling specificity: it takes two to tango. Trends Cell Biol 12:254–257

Cazale AC, Droillard MJ, Wilson C, Heberle-Bors E, Barbier-Brygoo H, Lauriere C (1999) MAP kinase activation by hypoosmotic stress of tobacco cell suspensions: towards the oxidative burst response? Plant J 19:297–307

Champion A, Picaud A, Henry Y (2004) Reassessing the MAP3K and MAP4K relationships. Trends Plant Sci 9:123–129

Chen H, Wilkerson CG, Kuchar JA, Phinney BS, Howe GA (2005) Jasmonate-inducible plant enzymes degrade essential amino acids in the herbivore midgut. Proc Natl Acad Sci USA 102:19237–19242

Cheong YH, Chang H-S, Gupta R, Wang X, Zhu T, Luan S (2002) Transcriptional profiling reveals novel interactions between wounding, pathogen, abiotic stress, and hormonal responses in *Arabidopsis*. Plant Physiol 129:661–677

Chow CW, Davis RJ (2006) Proteins kinases: chromatin-associated enzymes? Cell 127: 887–890

Chini A, Fonseca S, Fernández G, Adie B, Chico JM, Lorenzo O, García-Casado G, López-Vidriero I, Lozano FM, Ponce MR, Micol JL, Solano R (2007) The JAZ family of repressors is the missing link in jasmonate signalling. Nature 448:666–671

Del Pozo O, Pedley KF, Martin G (2004) MAPKKKa is a positive regulator of cell death associated with both plant immunity and disease. EMBO J 23:3072–3082

Doares SH, Narváez-Vásquez J, Conconi A, Ryan CA (1995) Salicylic acid inhibits synthesis of proteinase inhbitors in tomato leaves induced by systemin and jasmonic acid. Plant Physiol 108:1741–1746

Ekengren SK, Liu Y, Schiff M, Dinesh-Kumar SP, Martin GB (2003) Two MAPK cascades, NPR1, and TGA transcription factors play a role in Pto-mediated disease resistance in tomato. Plant J 36:905–917

Feilner T, Hultschig C, Lee J, Meyer S, Immink RG, et al. (2005) High throughput identification of potential *Arabidopsis* mitogen-activated protein kinases substrates. Mol Cell Proteomics 4:1558–1568

Felix G, Boller T (1995) Systemin induces rapid ion fluxes and ethylene biosynthesis in *Lycopersicon peruvianum* cells. Plant J 7:381–389

Felton GW, Korth KL, Bi JL, Wesley SV, Huhman DV, Mathews MC, Murphy JB, Lamb C, Dixon RA (1999) Inverse relationship between systemic resistance of plants to microorganisms and to insect herbivory. Curr Biol 9:317–320

Ferrell JE Jr. (1996) Tripping the switch fantastic: how a protein kinase cascade can convert graded inputs into switch-like outputs. Trends Biochem Sci 21:460–466

Gomi K, Ogawa D, Katou S, Kamada H, Nakajima N, Saji H, Soyano T, Sasabe M, Machida Y, Mitsuhara I, Ohashi Y, Seo S (2005) A Mitogen-activated protein kinase NtMPK4 activated by SIPKK is required for jasmonic acid signaling and involved in ozone tolerance via stomatal movement in tobacco. Plant Cell Physiol 46:1902–1914

Halitschke R, Gase K, Hui D, Schmidt DD, Baldwin IT (2003) Molecular interactions between the specialist herbivore *Manduca sexta* (Lepidoptera, Sphingidae) and its natural host *Nicotiana attenuata*. VI. Microarray analysis reveals that most herbivore-specific transcriptional changes are mediated by fatty acid-amino acid conjugates. Plant Physiol 131:1894–1902

Hamel L-P, Nicole M-C, Sritubtim S, Morency M-J, Ellis M, et al. (2006) Ancient signals: comparative genomics of plant MAPK and MAPKK gene families. Trends Plant Sci 11:192–198

Hanks SK, Hunter T (1995) The eukaryotic protein kinase superfamily: kinase (catalytic) domain structure and classification. FASEB J 9:576–596

Higgins R, Lockwood T, Holley S, Yalamanchili R, Stratmann JW (2007) Changes in extracellular pH are neither required nor sufficient for activation of mitogen-activated protein kinases (MAPKs) in response to systemin and fusicoccin in tomato. Planta 225:1535–1546

Hirt H (1999) Transcriptional upregulation of signaling pathways: more complex than anticipated? Trends Plant Sci 4:7–8

Holley SR, Yalamanchili RD, Moura SD, Ryan CA, Stratmann JW (2003) Convergence of signaling pathways induced by systemin, oligosaccharide elicitors, and ultraviolet-B radiation at the level of mitogen-activated protein kinases in *Lycopersicon peruvianum* suspension-cultured cells. Plant Physiol 132:1728–1738

Howe GA, Ryan CA (1999) Suppressors of systemin signaling identify genes in the tomato wound response pathway. Genetics 153:1411–1421

Howe GA, Lightner J, Browse J, Ryan CA (1996) An octadecanoid pathway mutant (JL5) of tomato is compromised in signaling for defense against insect attack. Plant Cell 8:2067–2077

Ichimura K, Mizoguchi T, Yoshida R, Yuasa T, Shinozaki K (2000) Various abiotic stresses rapidly activate *Arabidopsis* MAP kinases ATMPK4 and ATMPK6. Plant J 24:655–666

Ichimura K, Shinozaki K, Tena G, Sheen J, Henry Y, Champion A, et al. (2002) Mitogen-activated protein kinase cascades in plants: a new nomenclature. Trends Plant Sci 7:301–308

Izaguirre M, Scopel AL, Baldwin IT, Ballaré CL (2003) Convergent responses to stress. Solar UV-B radiation and *Manduca sexta* herbivory elicit overlapping transcriptional responses in field-grown plants of *Nicotiana longiflora*. Plant Physiol 132:1755–1767

Kaitheri Kandoth P, Ranf S, Pancholi SS, Jayanty S, Walla MD, Miller W, Howe GA, Lincoln DE, Stratmann JW (2007) Tomato MAPKs LeMPK1, LeMPK2, and LeMPK3 function in the systemin-mediated defense response against herbivorous insects. Proc Natl Acad Sci USA 104:12205–12210

Kao S-C, Jaiswal RK, Kolch W, Landreth GE (2001) Identification of the mechanisms regulating the differential activation of the MAPK cascade by epidermal growth factor and nerve growth factor in PC12 Cells. J Biol Chem 276:18169–18177

Kim CY, Liu Y, Thorne ET, Yang H, Fukushige H, Gassmann W, Hildebrand D, Sharp RE, Zhang S (2003) Activation of a stress-responsive mitogen-activated protein kinase cascade induces the biosynthesis of ethylene in plants. Plant Cell 15:2707–2718

Kolch W, Calder M, Gilbert D (2005) When kinases meet mathematics: the systems biology of MAPK signalling. FEBS Lett 579:1891–1895

Kovtun Y, Chiu W-L, Tena G, Sheen J (2000) Functional analysis of oxidative stress-activated mitogen-activated protein kinase cascade in plants. Proc Natl Acad Sci USA 97:2940–2945

Lee J, Rudd JJ, Macioszek VK, Scheel D (2004) Dynamic changes in the localization of MAPK cascade components controlling pathogenesis-related (PR) gene expression during innate immunity in parsley. J Biol Chem 279:22440–22448

Li C, Williams MM, Loh Y-T, Lee GI, Howe GA (2002a) Resistance of cultivated tomato to cell content-feeding herbivores is regulated by the octadecanoid-signaling pathway. Plant Physiol 130:494–503

Li C, Williams MM, Loh YT, Lee GI, Howe GA (2002b) Resistance of cultivated tomato to cell content-feeding herbivores is regulated by the octadecanoid-signaling pathway. Plant Physiol 130:494–503

Li L, Zhao Y, McCaig BC, Wingerd BA, Wang J, Whalon ME, Pichersky E, Howe GA (2004) The tomato homolog of CORONATINE-INSENSITIVE1 is required for the maternal control of seed maturation, jasmonate-signaled defense responses, and glandular trichome development. Plant Cell 16:126–143

Li Q, Xie Q-G, Smith-Becker J, Navarre DA, Kaloshian I (2006) Mi-1-mediated aphid resistance involves salicylic acid and mitogen-activated protein kinase signaling cascades. Mol Plant-Microbe Interact 19:655–664

Ligterink W, Kroj T, zur Nieden U, Hirt H, Scheel D (1997) Receptor-mediated activation of a MAP kinase in pathogen defense of plants. Science 276:2054–2057

Lin L-L, Wartmann M, Lin AY, Knopf JL, Seth A, Davis RJ (1993) cPLA2 is phosphorylated and activated by MAP kinase. Cell 72:269–278

Liu Y, Jin H, Yang KY, Kim CY, Baker B, Zhang S (2003) Interaction between two mitogen-activated protein kinases during tobacco defense signaling. Plant J 34:149–160

Liu Y, Zhang S (2004) Phosphorylation of 1-Aminocyclopropane-1-carboxylic acid synthase by MPK6, a stress-responsive mitogen-activated protein kinase, induces ethylene biosynthesis in *Arabidopsis*. Plant Cell 16:3386–3399

Malone M (1996) Rapid, long-distance signal transmission in higher plants. In: Callow JA (ed) Advances in Botanical Research. Academic Press, San Diego, pp 163–228

Mattiacci L, Dicke M, Posthumus MA (1995) β-Glucosidase: an elicitor of herbivore-induced plant odor that attracts host-searching parasitic wasps. Proc Natl Acad Sci USA 92:2036–2040

McGurl B, Orozco-Cardenas ML, Pearce G, Ryan CA (1994) Overexpression of the prosystemin gene in transgenic tomato plants generates a systemic signal that constitutively induces proteinase inhibitor synthesis. Proc Natl Acad Sci USA 91:9799–9802

Meindl T, Boller T, Felix G (1998) The plant wound hormone systemin binds with the N-terminal part to its receptor but needs the C-terminal part to activate it. Plant Cell 10:1561–1570

Meskiene I, Bogre L, Glaser W, Balog J, Brandstotter M, Zwerger K, Ammerer G, Hirt H (1998) MP2C, a plant protein phosphatase 2C, functions as a negative regulator of mitogen-activated protein kinase pathways in yeast and plants. Proc Natl Acad Sci USA 95:1938–1943

Mithofer A, Wanner G, Boland W (2005) Effects of feeding *Spodoptera littoralis* on lima bean leaves. II. Continuous mechanical wounding resembling insect feeding is sufficient to elicit herbivory-related volatile emission. Plant Physiol 137:1160–1168

Morrison DK, Davis RJ (2003) Regulation of MAP kinase signaling modules by scaffold proteins in mammals. Ann Rev Cell Dev Biol 19:91–118

Moyen C, Johannes E (1996) Systemin transiently depolarizes the tomato mesophyll cell membrane and antagonizes fusicoccin-induced extracellular acidification of mesophyll tissue. Plant Cell Environ 19:464–470

Munnik T, Ligterink W, Meskiene I, Calderini O, Byerly J, Musgrave A, Hirt H (1999) Distinct osmo-sensing protein kinase pathways are involved in signalling moderate and severe hyperosmotic stress. Plant J 20:381–388

Nakagami H, Kiegerl S, Hirt H (2004) OMTK1, a novel MAPKKK, channels oxidative stress signaling through direct MAPK interaction. J Biol Chem 279:26959–26966

Nakagami H, Pitzschke A, Hirt H (2005) Emerging MAP kinase pathways in plant stress signalling. Trends Plant Sci 10:339–346

Narváez-Vásquez J, Florin-Christensen J, Ryan CA (1999) Positional specificity of a phospholipase A_2 activity induced by wounding, systemin, and oligosaccharide elicitors in tomato leaves. Plant Cell 11:2249–2260

O'Donnell PJ, Calvert C, Atzorn R, Wasternack C, Leyser HMO, Bowles DJ (1996) Ethylene as a signal mediating the wound response of tomato plants. Science 274:1914–1917

Orozco-Cardenas ML, McGurl B, Ryan CA (1993) Expression of an antisense prosystemin gene in tomato plants reduces resistance toward *Manduca sexta* larvae. Proc Natl Acad Sci USA 90:8273–8276

Orozco-Cárdenas ML, Narváez-Vásquez J, Ryan CA (2001) Hydrogen peroxide acts as a second messenger for the induction of defense genes in tomato plants in response to wounding, systemin, and methyl jasmonate. Plant Cell 13:179–191

Pearce G, Strydom D, Johnson S, Ryan CA (1991) A polypeptide from tomato leaves induced wound-inducible proteinase inhibitor proteins. Science 253:895–898

Pearce G, Moura D, Stratmann J, Ryan CA (2001) Production of multiple plant hormones from a single polyprotein precursor. Nature 411:817–820

Pearce G, Ryan CA (2003) Systemic signaling in tomato plants for defense against herbivores. Isolation and characterization of three novel defense-signaling glycopeptide hormones coded in a single precursor gene. J Biol Chem 278:30044–30050

Peck SC (2006) Analysis of protein phosphorylation: methods and strategies for studying kinases and substrates. Plant J 45:512–522

Pedley KF, Martin GB (2004) Identification of MAPKs and their possible MAPK kinase activators involved in the Pto-mediated defense response of tomato. J Biol Chem 279:49229–49235

Petersen M, Brodersen P, Naested H, Andreasso E, Lindhart U, Johansen B, Nielsen HB, Lacy M, Austin MJ, Parker JE, Sharma SB, Klessig DF, Martienssen R, Mattsson O, Jensen AB, Mundy

J (2000) *Arabidopsis* MAP kinase 4 negatively regulates systemic acquired resistance. Cell 103:1111–1120

Pouyssegur J, Lenormand P (2003) Fidelity and spatio-temporal control in MAP kinase (ERKs) signalling. Eur J Biochem 270:3291–3299

Rakwal R, Agrawal GK (2003) Wound signaling-coordination of the octadecanoid and MAPK pathways. Plant Physiol Biochem 41:855–861

Ren D, Yang H, Zhang S (2002) Cell death mediated by MAPK is associated with hydrogen peroxide production in *Arabidopsis*. J Biol Chem 277:559–565

Ren D, Yang KY, Li GJ, Lu Y, Zhang S (2006) Activation of Ntf4, a tobacco mitogen-activated protein kinase, during plant defense response and its involvement in hypersensitive response-like cell death. Plant Physiol 141:1482–1493

Reymond P, Weber H, Damond M, Farmer EE (2000) Differential gene expression in response to mechanical wounding and insect feeding in *Arabidopsis*. Plant Cell 12:707–720

Reymond P, Bodenhausen N, Van Poecke RMP, Krishnamurthy V, Dicke M, Farmer EE (2004) A conserved transcript pattern in response to a specialist and a generalist herbivore. Plant Cell 16:3132–3147

Rhodes JD, Thain JF, Wildon CD (1996) The pathway for systemic electrical signal conduction in the wounded tomato plant. Planta 200:50–57

Romeis T, Piedras P, Zhang S, Klessig DF, Hirt H, Jones JDG (1999) Rapid *avr9*- and *cf9*-dependent activation of MAP kinases in tobacco cell cultures and leaves: convergence of resistance gene, elicitor, wound, and salicylate responses. Plant Cell 11:273–287

Rutschmann F, Stalder U, Piotrowski M, Oecking C, Schaller A (2002) LeCPK1, a calcium-dependent protein kinase from tomato. Plasma membrane targeting and biochemical characterization. Plant Physiol 129:156–168

Sabbagh WJ, Flatauer L, Bardwell J, Bardwell L (2001) Specificity of MAP kinase signaling in yeast differentiation involves transient versus sustained MAPK activation. Mol Cell 8:683–691

Sagi M, Davydov O, Orazova S, Yesbergenova Y, Ophir R, Stratmann JW, Fluhr R (2004) Rboh impinges on wound responsiveness and plant development. Plant Cell 16:616–628

Samaj J, Ovecka M, Hlavacka A, Lecourieux F, Meskiene I, Lichtscheidl I, Lenart P, Salaj J, Volkmann D, Bögre L, Baluska F, Hirt H (2002) Involvement of the mitogen-activated protein kinase SIMK in regulation of root hair tip growth. EMBO J 21:3296–3306

Samuel MA, Ellis BE (2002) Double jeopardy: both overexpression and suppression of a redox-activated plant mitogen-activated protein kinase render tobacco plants ozone sensitive. Plant Cell 14:2059–2069

Schaller A (1999) Oligopeptide signaling and the action of systemin. Plant Mol Bio 40:763–769

Schaller A, Oecking C (1999) Modulation of plasma membrane H^+-ATPase activity differentially activates wound and pathogen defense responses in tomato plants. Plant Cell 11:263–272

Scheer J, Ryan CA (1999) A 160-kD systemin receptor on the surface of *Lycopersicon peruvianum* suspension-cultured cells. Plant Cell 11:1525–1535

Schenk PM, Kazan K, Wilson I, Anderson JP, Richmond T, Somerville SC, Manners JM (2000) Coordinated plant defense responses in *Arabidopsis* revealed by microarray analysis. Proc Natl Acad Sci USA 97:11655–11660

Schilmiller AL, Howe GA (2005) Systemic signaling in the wound response. Curr Opin Plant Biol 8:369–377

Seo S, Okamoto M, Seto H, Ishizuka K, Sano H, Ohashi Y (1995) Tobacco MAP kinase: a possible mediator in wound signal transduction pathways. Science 270:1988–1992

Seo S, Sano H, Ohashi Y (1999) Jasmonate-based wound signal transduction requires activation of WIPK, a tobacco mitogen-activated protein kinase. Plant Cell 11:289–298

Seo S, Seto H, Koshino H, Yoshida S, Ohashi Y (2003) A diterpene as an endogenous signal for the activation of defense responses to infection with tobacco mosaic virus and wounding in tobacco. Plant Cell 15:863–873

Seo S, Katou S, Seto H, Gomi K, Ohashi Y (2007) The mitogen-activated protein kinases WIPK and SIPK regulate the levels of jasmonic and salicylic acids in wounded tobacco plants. Plant J 49:899–909

Sharma PC, Ito A, Shimizu T, Terauchi R, Kamoun S, Saitoh H (2003) Virus-induced silencing of WIPK and SIPK genes reduces resistance to a bacterial pathogen, but has no effect on the INF1-induced hypersensitive response (HR) in *Nicotiana benthamiana*. Mol Genet Genomics 269:583–591

Stratmann JW, Ryan CA (1997) Myelin basic protein kinase activity in tomato leaves is induced systemically by wounding and increases in response to systemin and oligosaccharide elicitors. Proc Natl Acad Sci USA 94:11085–11089

Stratmann J, Scheer J, Ryan CA (2000a) Suramin inhibits initiation of defense signaling by systemin, chitosan and pmg-elicitor in suspension cultured *Lycopersicon peruvianum* cells. Proc Natl Acad Sci USA 97:8862–8867

Stratmann JW, Stelmach BA, Weiler EW, Ryan CA (2000b) UVB/UVA radiation activates a 48 kDa myelin basic protein kinase and potentiates wound signaling in tomato leaves. Photochem Photobiol 71:116–123

Stratmann JW (2003) Long distance run in the wound response – jasmonic acid is pulling ahead. Trends Plant Sci 8:247–250

Stulemeijer IJE, Stratmann JW, Joosten MHAJ (2007) The tomato MAP kinases LeMPK1, -2, and -3 are activated during the Cf-4/Avr4-induced HR, and have overlapping but also different phosphorylation specificities. Plant Physiol 144:1481–1494

Thompson GA, Goggin FL (2006) Transcriptomics and functional genomics of plant defence induction by phloem-feeding insects. J Exp Bot 57:755–766

Thines B, Katsir L, Melotto M, Niu Y, Mandaokar A, Liu G, Nomura K, He SY, Howe GA, Browse J (2007) JAZ repressor proteins are targets of the SCF^{coi1} complex during jasmonate signalling. Nature 448:661–665

Walling LL (2000) The myriad plant responses to herbivores. J Plant Growth Regul 19:195–216

Wasternack C, Stenzel I, Hause B, Hause G, Kutter C, Maucher H, Neumerkel J, Feussner I, Miersch O (2006) The wound response in tomato – Role of jasmonic acid. J Plant Physiol 163:297–306

Wildon DC, Thain JF, Minchin PEH, Gubb IR, Reilly AJ, Skipper YD, Doherty HM, O'Donnell PJ, Bowles DJ (1992) Electrical signalling and systemic proteinase inhibitor induction in the wounded plant. Nature 360:62–65

Xu Q, Fu HH, Gupta R, Luan S (1998) Molecular characterization of a tyrosine-specific protein phosphatase encoded by a stress-responsive gene in *Arabidopsis*. Plant Cell 10:849–857

Yalamanchili RD, Stratmann J (2002) Ultraviolet-B activates components of the systemin signaling pathway in *Lycopersicon peruvianum* suspension-cultured cells. J Biol Chem 277:28424–28430

Yang K-Y, Liu Y, Zhang S (2001) Activation of a mitogen-activated protein kinase pathway is involved in disease resistance in tobacco. Proc Natl Acad Sci USA 98:741–746

Zhang S, Klessig DF (1997) Salicylic acid activates a 48-kD MAP kinase in tobacco. Plant Cell 9:809–824

Zhang S, Klessig DF (1998) The tobacco wounding-activated mitogen-activated protein kinase is encoded by SIPK. Proc Natl Acad Sci USA 95:7225–7230

Zhang S, Klessig DF (2000) Pathogen-induced MAP kinases in tobacco. Results Probl Cell Differ 27:65–84

Zhang S, Liu Y (2001) Activation of salicylic acid-induced protein kinase, a mitogen-activated protein kinase, induced multiple defense responses in tobacco. Plant Cell 13:1877–1889

Chapter 17
Jasmonate Biosynthesis and Signaling for Induced Plant Defense against Herbivory

Andreas Schaller and Annick Stintzi

Jasmonates are a growing class of signaling molecules and plant hormones which are derived from polyunsaturated fatty acids via the octadecanoid pathway, and characterized by a pentacyclic ring structure. Until recently, jasmonic acid has been viewed as the end product of the pathway and as the bioactive hormone. It becomes increasingly clear, however, that biological activity is not limited to jasmonic acid, but extends to, and may even differ between its many metabolites and conjugates as well as its biosynthetic precursors. Like other plant hormones, jasmonates exhibit a broad spectrum of physiological activities, ranging from seed germination, over reproductive development, all the way to senescence. Jasmonates also serve important roles as signaling molecules in plant defense, particularly defense against insect herbivores and necrotrophic patghogens. In this chapter, we will briefly discuss each step of the octadecanoid pathway, emphasizing on those that are relevant for the regulation of jasmonic acid biosynthesis, and on insights derived from the recently solved crystal structures of two of the pathway's enzymes. With respect to jasmonate signaling, we will focus on their role as signal molecules in the systemic defense response against insect herbivores, and on jasmonate-dependent activation of defense gene expression.

17.1 Introduction

Since the identification of methyl jasmonate (MeJA) as a secondary metabolite in essential oils of jasmine in 1962 (Demole et al. 1962), many related compounds have been discovered in a wide range of plants, and are collectively referred to as jasmonates. Jasmonates share their biosynthetic origin from oxygenated polyunsaturated fatty acids, and a substituted pentacyclic ring as a common structural element. In the early 1980s, their widespread occurrence throughout the plant kingdom (Meyer et al. 1984), and their growth-inhibitory (Dathe et al. 1981), and

---

A. Stintzi
University of Hohenheim, Institute of Plant Physiology and Biotechnology, D-70599 Stuttgart, Germany
e-mail: stintzi@uni-hohenheim.de

senescence-promoting activities (Ueda and Kato 1980) were established. First indications for a role of jasmonates in the regulation of gene expression were obtained by Parthier and co-workers who observed the accumulation of jasmonate-inducible proteins (JIPs) in senescing barley leaves (Weidhase et al. 1987; Mueller-Uri et al. 1988). Seminal work of Farmer and Ryan subsequently demonstrated that MeJA and jasmonic acid (JA) induce the accumulation of proteinase inhibitors as a direct defense against insect herbivores (Farmer and Ryan 1990; Farmer et al. 1991). Shortly after, jasmonates were shown to mediate the elicitor-induced accumulation of antimicrobial phytoalexins in cell cultures (Gundlach et al. 1992), and the induction of vegetative storage proteins in soybean and *Arabidopsis* (Franceschi and Grimes 1991; Staswick et al. 1992). These findings greatly stimulated the interest in jasmonates as a new class of signaling molecules in plant defense against both insects and pathogens.

In subsequent years, numerous mutants were characterized that are impaired in either jasmonate synthesis or response clearly establishing their function as plant defense regulators. Mutants impaired in jasmonate perception and signaling, including the *coi1*, *jin1*, and *jai3* mutants in *Arabidopsis* (Feys et al. 1994; Berger et al. 1996; Lorenzo et al. 2004; Lorenzo and Solano 2005), *coi1* in *N. attenuata* (Paschold et al. 2007), and the *jai1* mutant in tomato (Li et al. 2004) fail to mount appropriate defense responses. Also compromised in the induction of defense responses are mutants that are deficient in the polyunsaturated fatty acid precursors of jasmonates (the *fad3fad7fad8* and the *spr2/LeFad7* mutants in *Arabidopsis* and tomato; McConn et al. 1997; Li et al. 2003), and mutants affected in the octadecanoid pathway for jasmonate biosynthesis, i.e. *dad1*, *aos*, *acx1/5*, *jar1* in *Arabidopsis* (Park et al. 2002; Staswick et al. 2002; von Malek et al. 2002; Schilmiller et al. 2007), *def1* in tomato (Howe et al. 1996) and *jar4* in *N. attenuata* (Kang et al. 2006).

In addition to their role as defense regulators, the characterization of biosynthesis and perception mutants established jasmonates as phytohormones in plant reproductive development. Many of the *Arabidopsis* mutants are male sterile due to defects in anther and pollen maturation (Feys et al. 1994; McConn and Browse 1996; Sanders et al. 2000; Stintzi and Browse 2000; Ishiguro et al. 2001; Park et al. 2002; von Malek et al. 2002; Schilmiller et al. 2007) while a defect in the maternal control of seed maturation appears to cause sterility in *jai1* in tomato (Li et al. 2004). Other jasmonate-regulated developmental processes include root growth (Staswick et al. 1992), glandular trichome development (Li et al. 2004), tuber formation (Yoshihara et al. 1989; Pelacho and Mingo-Castel 1991), laticifer differentiation (Hao and Wu 2000), seed germination (Corbineau et al. 1988; Finch-Savage et al. 1996), carbon/nitrogen allocation (Creelman and Mullet 1997), senescence (Ueda and Kato 1980; Creelman and Mullet 1997), and tendril coiling (Falkenstein et al. 1991).

A number of excellent review articles have recently been published on the biosynthesis of jasmonates and on their activity, particularly their contribution to the regulation of plant defense responses (Blee 2002; Liechti and Farmer 2002; Halitschke and Baldwin 2004; Howe 2004; Pozo et al. 2004; Schaller et al. 2004; Browse 2005; Lorenzo and Solano 2005; Schilmiller and Howe 2005; Delker et al. 2006; Wasternack 2006; Cheong and Choi 2007; Wasternack 2007), and we refer the reader to these articles for a comprehensive discussion of the available literature. In this chapter, we will emphasize on the most recent findings with respect to jasmonate biosynthesis, and discuss those aspects of jasmonate signaling that are relevant for the systemic induction of herbivore defenses and defense gene activation.

17.2 Jasmonate Biosynthesis

The synthesis of jasmonates and many other oxylipins is initiated by lipoxygenases (LOXs), which catalyze the regio- and stereoselective dioxygenation of polyunsaturated fatty acids (reviewed by Blee 2002; Feussner and Wasternack 2002; Howe and Schilmiller 2002; Schaller et al. 2004; Wasternack 2007). Linoleic acid (18:2) and linolenic acid (18:3) are oxygenated by specific LOXs at C9 or C13 to result in the corresponding (9S)- or (13S)-hydroperoxy-octadecadi(tri)enoic acids, which feed into at least seven alternative pathways resulting in the formation of a large variety of oxylipins (Blee 2002; Feussner and Wasternack 2002). The first committed step in the two parallel pathways for JA biosynthesis (Fig. 17.1), i.e. the octadecanoid pathway from 18:3 and the hexadecanoid pathway from 16:3 (Weber et al. 1997), is performed by allene oxide synthase (AOS), an unusual cytochrome P450 which uses its hydroperoxide substrate as source for reducing equivalents and as oxygen donor, and is thus independent of molecular oxygen and NAD(P)H. AOS catalyzes the dehydration of 13(S)-hydroperoxy-octadecatrienoic acid (13-HPOT) to form an unstable allene oxide, 12,13(S)-epoxy-octadecatrienoic acid (12,13-EOT). In aqueous media, 12,13-EOT rapidly decomposes to α- and γ-ketols, or undergoes cyclization to form 12-oxo-phytodienoic acid (OPDA). As opposed to spontaneous cyclization which results in a racemic mixture of OPDA enantiomers, allene oxide cyclase (AOC) ensures the formation of the optically pure 9S,13S enantiomer. Dinor-OPDA (dnOPDA) is generated in the parallel pathway from 16:3 (Fig. 17.1). The short half-life of 12,13-EOT in aqueous media (26 s at 0°C and pH 6.7; Hamberg and Fahlstadius 1990; Ziegler et al. 1999) and the optical purity of endogenous OPDA (Laudert et al. 1997) suggest tight coupling of the AOS and AOC reactions in vivo. However, physical contact of AOS and AOC in an enzyme complex does not seem to be required for stereochemical control of the cyclization reaction (Zerbe et al. 2007).

Only 9S,13S-OPDA, i.e. one out of four possible OPDA stereoisomers, is a percursor for biologically active JA. AOC is thus crucially important to establish the enantiomeric structure of the cyclopentenone ring. The crystal structure of *Arabidopsis* AOC2 has recently been solved shedding light on how the enzyme exerts stereochemical control on the cyclization reaction (Hofmann et al. 2006). Considering the fact that cyclization occurs spontaneously in aqueous solution, AOC2 does not need to be much of a catalyst in terms of lowering the activation energy barrier. Indeed, binding of the substrate or the transition state does not involve any induced fit mechanism. The hydrophobic protein environment and very few ionic interactions with a glutamate residue (Glu23) and a tightly bound water molecule,

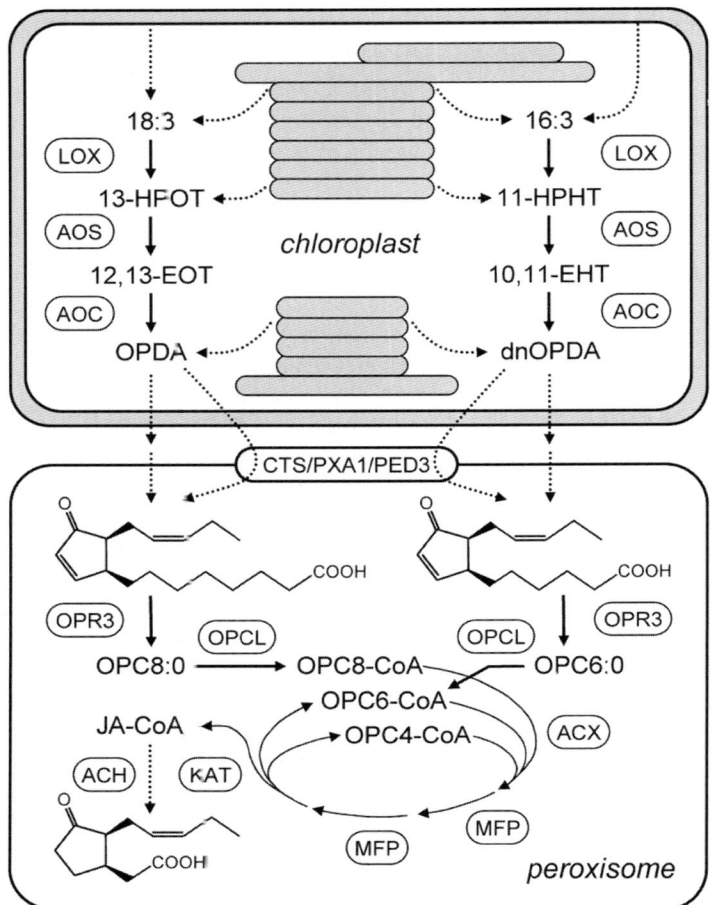

Fig. 17.1 Biosynthesis of jasmonic acid. Polyunsaturated fatty acids (18:3 and 16:3) are precursors for jasmonic acid biosynthesis via the octa- and hexadecanoid pathways, respectively. The first three steps are localized in plastids and lead to the formation of 12-oxophytodienoic acid (OPDA) and dinor 12-oxophytodienoic acid (dnOPDA). The subsequent steps, including the reduction of OPDA (dnOPDA) to OPC8:0 (OPC6:0) followed by three (two) cycles of β-oxidation result in the production of jasmonic acid in peroxisomes. Arrows and broken arrows are used to differentiate between well characterized reactions and those steps that are still hypothetical and for which the corresponding enzymes remain to be identified. Further detail is given in the text

ensure binding and correct positioning of the substrate 12,13-EOT. Steric restrictions imposed by the protein environment enforce the necessary conformational changes of the substrate's hydrocarbon tail resulting in the absolute stereoselectivity of the AOC2-mediated as opposed to the chemical cyclization reaction (Hofmann et al. 2006).

The passive role of AOC2 in the steroselective synthesis of 9S13S-OPDA is reminiscent of dirigent proteins and the way in which they promote stereoselective formation of lignans. Lignan formation involves the regio- and sterochemically controlled oxidative coupling of two phenols, e.g. the formation of (+)-pinoresinol from two molecules of coniferylalcohol (Davin et al. 1997). In absence of dirigent proteins, resonance-stabilized phenoxy radical intermediates couple randomly to form a mixture of racemic lignans. Dirigent proteins, which lack a catalytic center, are believed to bind and orient the free radical intermediates allowing stereoselective coupling to occur (Davin et al. 1997).

Interestingly, both dirigent proteins and AOC2 are distant members of the lipocalin family (Charron et al. 2005; Hofmann et al. 2006; Pleiss and Schaller unpublished observation). Lipocalins are β-barrel proteins comprising a central hydrophobic cavity for binding of small lipophilic molecules like steroids, pheromones, odorants, or retinoids. In fact most lipocalins do not function as enzymes but rather as binding proteins in olfaction, pheromone transport, retinol transport, and invertebrate cryptic coloration (Charron et al. 2005). Likewise, dirigent proteins and AOC2 act as binding proteins of the unstable lipophilic precursors (phenoxy radicals and allene oxides, respectively) to exert stereochemical control in the synthesis of lignans and OPDA.

The formation of 9S,13S-OPDA as the first member of the jasmonate family with signaling activity concludes the plastid-localized part of the octa(hexa)decanoid pathway. Little is known about how OPDA (and/or dnOPDA) is transferred from plastids to peroxisomes, where the final steps of JA biosynthesis occur. Recently, the ABC transporter COMATOSE (CTS, also known as PXA1 (Zolman et al. 2001) or PED3 (Hayashi et al. 2002)) was shown to be involved in this process (Theodoulou et al. 2005). CTS catalyzes the ATP-dependent import of fatty acids into peroxisomes as substrates for β-oxidation. The *cts* mutant has reduced levels of JA, is impaired in wound-induced JA accumulation, and expresses the JA-dependent *VSP1* gene at a lower level suggesting that CTS delivers substrates for JA synthesis into the peroxisomes, most likely (dn)OPDA, or the corresponding CoA esters (Theodoulou et al. 2005). However, other pathways for (dn)OPDA import must exist, as indicated by the residual levels of JA, and the lack of JA-deficiency symptoms (e.g., male sterility) in the *cts* mutant. Additional pathways for fatty acid import into peroxisomes, by diffusion or by an as yet unidentified mechanism, were also suggested by Fulda et al. (2004). Because of the higher pH in peroxisomes as compared to the cytoplasm, weak acids are predicted to be trapped in peroxisomes as the anion, and this may account for some of the (dn)OPDA import (Theodoulou et al. 2005).

Once within the peroxisomes, 9S,13S-OPDA is reduced by 12-oxophytodienoate reductase (OPR3) to yield 3-oxo-2-(2'(Z)-pentenyl)-cyclopentane-1-octanoic acid (OPC-8:0), and dnOPDA is reduced to the corresponding hexanoic acid derivative (OPC-6:0; Schaller et al. 2000; Stintzi and Browse 2000; Fig. 17.1). The signaling properties of jasmonates bearing a cyclopentenone ring (OPDA, dnOPDA) are clearly distinct from JA and its derivatives which are characterized by the reduced cyclopentanone ring (Blechert et al. 1999; Stintzi et al. 2001; Taki et al. 2005). OPR3 may therefore be particularly important for the regulation of the relative levels of these two classes of signaling molecules.

Even though OPR3 belongs to a small family of related flavin-dependent oxidoreductases (at least three in tomato, six genes in *Arabidopsis*, six in pea, eight in maize, ten in rice) it is functionally unique, as evident from JA-deficiency symptoms (male-sterility) of the *OPR3* loss-of-function mutant in *Arabidopsis* (Sanders et al. 2000; Stintzi and Browse 2000) and tomato (Schaller and Stintzi unpublished). The genetic evidence is consistent with biochemical and structural data which suggest that all OPRs catalyze the reduction of α,β-unsaturated carbonyls (conjugated enones) while only OPR3s and the rice ortholog OsOPR7 (Tani et al. 2007) are capable of reducing the $9S,13S$ enantiomer of OPDA (Schaller et al. 2000; Breithaupt et al. 2001, 2006; Strassner et al. 2002). Therefore, these are the only OPRs contributing to JA biosynthesis. The OPR1 isoforms from *Arabidopsis*, tomato, and rice, on the other hand, were shown to reduce numerous conjugated enones including $9R,13R$-OPDA, but they do not accept $9S,13S$-OPDA the precursor of biologically active JA (Schaller and Weiler 1997; Straßner et al. 1999; Schaller et al. 2000; Strassner et al. 2002; Sobajima et al. 2003; Tani et al. 2007).

Recent crystal structure analyses provided insights into the mechanisms of substrate reduction and the remarkable differences in stereospecificity between OPR1 and 3 (Breithaupt et al. 2001, 2006; Fox et al. 2005; Malone et al. 2005). Consistent with the reaction mechanism proposed for the related Old Yellow Enzyme from yeast, the carbonyl moiety of the substrate forms hydrogen bonds with two histidine residues (His187 and His197 in tomato OPR1) which leads to a polarization of the α,β-double bond. Consequently, hydride transfer from the reduced flavin cofactor to the substrate Cβ is facilitated, followed by a protonation of the Cα by a tyrosine residue (Breithaupt et al. 2001). A comparison of the OPR1 and OPR3 structures revealed a more open active site cavity in OPR3, explaining its more relaxed specificity allowing reduction of both the $(9S,13S)$ and $(9R,13R)$ isomers of OPDA. Two residues in OPR1, Tyr78 and Tyr246 seem to act as gatekeepers, narrowing the active site and blocking the entry of 9S,13S-OPDA (Breithaupt et al. 2001, 2006).

The shortening of the hexanoic and octanoic acid side chains in OPC-6:0 and OPC-8:0 to yield JA involves two or three rounds of β-oxidation, respectively. Prior to entry into the β-oxidation cycle, the carboxylic moiety needs to be activated as CoA ester. Co-expression with genes known to be involved in JA biosynthesis suggested At1g20510, a member of the ATP-dependent acyl-activating family of enzymes, as a candidate for the required Acyl-CoA ligase. The recombinant enzyme was found to activate OPDA and OPC-8:0 in vitro, as well as medium- to-long straight-chain fatty acids (Koo et al. 2006). Loss-of-function mutants in *Arabidopsis* accumulated less JA in response to mechanical wounding providing direct evidence for a role in JA biosynthesis. Hyper accumulation of OPC-8:0 in the null mutants supported OPC-8:0 as the physiological substrate, and the enzyme was thus named OPC-8:0 CoA Ligase1 (OPCL1; Koo et al. 2006; Fig 17.1). A closely related enzyme (At4g05160) was suggested by Schneider et al. (2005) as a candidate OPC-6:0-CoA ligase. Consistent with such a role, the enzyme is located in peroxisomes and its expression is induced by MeJA. However, the recombinant enzyme was found to prefer medium-chain fatty acids over OPC-6:0 as substrates in vitro (Schneider et al. 2005), and the corresponding loss-of-function mutant exhibited wild-type levels of JA (Koo et al. 2006). Its contribution to JA biosynthesis is thus questionable.

Beta-oxidation itself involves three core enzymes, acyl-CoA oxidase (ACX), multifunctional protein (MFP; comprising enoyl-CoA hydratase and β-hydroxyacyl-CoA dehydrogenase activities), and 3-ketoacyl-CoA thiolase (Fig. 17.1). Despite early findings implicating β-oxidation in JA biosynthesis (Vick and Zimmerman 1984), direct evidence for the contribution of these enzymes is very recent. ACX1A was shown to catalyze the first step in the β-oxidation of OPC-8:0-CoA, and was found to be responsible for the bulk of wound-induced JA production in tomato. Consistant with its essential role in JA biosynthesis, the *acx1* tomato mutant was impaired in wound-induced defense gene activation and insect resistance (Li et al. 2005). In *Arabidopsis*, ACX1 is responsible for about 80% of JA production after wounding (Cruz Castillo et al. 2004; Schilmiller et al. 2007), and only the *acx1/5* double mutant showed hallmarks of severe JA deficiency, including impaired insect resistance and reduced male reproductive function (Schilmiller et al. 2007). Also the *aim1* mutant in *Arabidopsis*, which is disrupted in one of two MFP genes, is impaired in wound-induced accumulation of JA and expression of JA-dependent genes (Delker et al. 2007). Among the five 3-ketoacyl-CoA thiolase genes in *Arabidopsis*, KAT2 appears to be the one most relevant for JA biosynthesis. In transgenic plants silenced for *KAT2* expression, wound-induced JA accumulation was markedly reduced (65%–80%), and the induction of a JA-dependent marker gene (*JR2*) impaired, both locally at the site of wounding and systemically (Cruz Castillo et al. 2004). As the final step in JA biosynthesis, the JA-CoA ester has to be hydrolyzed to release the free acid. Candidate acyl-thioesterases have been identified in *Arabidopsis*, two of which appear to be peroxisomal (*At*ACH1 and *At*ACH2; Tilton et al. 2000), but a direct involvement in JA biosynthesis remains to be shown.

Traditionally, JA has been viewed as the end product of the pathway and the bioactive hormone. The many different metabolites of JA and the enzymatic steps in their interconversion were thought to contribute to hormone homeostasis to sustain and control the levels of active JA. This view is changing, however, and E.E. Farmer takes the extreme position to consider JA as a mere precursor of bioactive hormones, i.e. JA conjugates (Farmer 2007). Indeed, there is clear evidence showing that some JA metabolites have unique signaling properties, and that some processes are not controlled by JA but rather by JA derivatives (reviewed by Wasternack 2007). For example, the nyctinastic leaf movement in *Albizzia* depends on a specific enantiomer of 12-OH-JA-*O*-glucoside (Nakamura et al. 2006a, b), 12-OH-JA (tuberonic acid) has long been implicated in potato tuber formation (Yoshihara et al. 1989), and in defense signaling, the active signal appears to be the amide-linked isoleucine conjugate JA-Ile, rather than JA itself (Staswick and Tiryaki 2004; Kang et al. 2006).

The relevance of JA-Ile as a signal in its own right was first demonstrated by Wasternack and co-workers who observed that JA and JA-Ile induce different responses when applied exogenously to barley leaves, and that they are both active without being inter-converted (Kramell et al. 1997; Wasternack et al. 1998). Biosynthesis of JA-Ile involves the adenylation of JA, followed by the exchange of

AMP with isoleucine, and is catalyzed by the amino acid conjugate synthetase JAR1 (Staswick et al. 2002; Staswick and Tiryaki 2004). The *Arabidopsis jar1* mutant is insensitive to exogenously applied JA with respect to root growth inhibition, induced resistance to *Phytium irregulare*, induced systemic resistance (ISR), and protection against ozone damage (Staswick and Tiryaki 2004, and references therein). Furthermore, silencing of the *JAR1* ortholog in *N. attenuata* compromised defense gene induction and resistance against *M. sexta* (Kang et al. 2006). Therefore, conjugation to Ile appears to be necessary for at least a subset of the jasmonate-regulated processes. Further support for the specific signaling function of JA-Ile is provided by the recent observation that unlike JA or OPDA, JA-Ile mediates the specific degradation of repressors of jasmonate-dependent gene expression (Thines et al. 2007, see below).

17.3 Jasmonate Signaling in the Systemic Induction of Herbivore Defenses

The systemic induction of defensive proteinase inhibitors in tomato plants in response to local wounding or herbivore attack was discovered by Ryan and coworkers 35 years ago and has since served as a model system to study long-range signaling processes in plants (Green and Ryan 1972). A number of chemical signals were identified in the Ryan lab that are intricately involved in the systemic induction of defense responses. This includes systemin peptides which are derived from larger precursor proteins by proteolytic processing (Pearce et al. 1991; McGurl et al. 1992; Ryan and Pearce 2003; Narváez-Vásquez and Orozco-Cárdenas this volume), oligogalacturonides which are generated by a polygalacturonase systemically induced after wounding (Bergey et al. 1999), and jasmonates which are derived from the octadecanoid pathway (Farmer and Ryan 1990; Farmer et al. 1991). Farmer and Ryan proposed a model according to which primary wound signals like oligogalacturonides and systemin trigger the activation of the octadecanoid pathway resulting in a burst of JA production which ultimately leads to the activation of defense genes (Farmer and Ryan 1992). The wound signaling pathway, its interaction with other hormone and defense signaling pathways, and the role of individual signaling molecules were subject of numerous recent reviews (Pieterse et al. 2006; Ryan and Pearce 2003; Stratmann 2003; Howe 2004; Lorenzo and Solano 2005; Schilmiller and Howe 2005; Wasternack et al. 2006; Cheong and Choi 2007), and systemic wound signaling is discussed in detail by Schaller and Howe (this volume). Here we will focus on the specific role of jasmonates in the activation of defense gene expression.

Since the discovery of the systemic wound response researchers have been fascinated by the question of what the long-distance signal might be (Ryan 1992; Bowles 1998). The search for the systemic wound signal in tomato plants led to the discovery of systemin in the Ryan lab, and its role as the systemically mobile signal was supported by numerous observations: Systemin was shown to be necessary and sufficient for the systemic wound response, it triggers the synthesis and accumulation of JA, it is mobile within the phloem when applied to wound sites, and its precursor protein is specifically expressed in the vasculature. Based on these findings, a model had been proposed, according to which systemin is proteolytically released from its precursor upon wounding, loaded into sieve elements, and delivered along source-to-sink gradients into systemic tissues where it stimulates the synthesis of jasmonic acid as a secondary signal for the induction of defense gene expression (Ryan 2000).

However, this model is not entirely consistent with recent findings from the Howe lab, demonstrating that systemic signaling requires the activity of systemin and the synthesis of JA only in the wounded, not in the systemic leaves. Furthermore, the induction of defense genes in systemic tissues depends on JA perception and signaling, but not on the capacity to synthesize JA (Li et al. 2002, 2003; Schaller and Howe this volume). These findings are in complete agreement with the observation that expression of octadecanoid pathway genes is induced by wounding at the site of tissue damage but not systemically, and that there is a dramatic increase in JA levels in wounded leaves but not in unwounded tissues (Strassner et al. 2002). The data suggest that systemin acts at the site of tissue damage, to strengthen the systemic wound response by boosting the octadecanoid pathway, for the generation of the long-distance signal, maybe JA itself or one of its derivatives (Ryan and Moura 2002; Stratmann 2003; Schilmiller and Howe 2005; Schaller and Howe this volume).

The joint role of (pro)systemin and the octadecanoid pathway in the generation of a phloem-mobile signal for systemic induction of defense genes is consistent with their localization in the vasculature of tomato plants. Several of the octadecanoid pathway enzymes – LOX, AOS, and AOC – are located in the companion cell-sieve element complex (Hause et al. 2000, 2003a), whereas prosystemin accumulates in cells of the phloem parenchyma (Jacinto et al. 1997; Narváez-Vásquez and Ryan 2004; Narváez-Vásquez and Orozco-Cárdenas this volume). The presence of prosystemin and octadecanoid pathway enzymes in different cell types of the vascular bundle suggests a model in which systemin is released from phloem parenchyma cells in response to wounding, is then perceived at the cell surface of neighboring companion cells, where it triggers the activation of the octadecanoid pathway for the production of the systemic signal, and its release into the sieve elements for long-distance transport (Schilmiller and Howe 2005; Wasternack 2006).

The onset of JA accumulation at the site of tissue damage is almost instantaneous and peaks at around 1 hour after wounding (Doares et al. 1995; McConn et al. 1997; Ziegler et al. 2001; Strassner et al. 2002), and must therefore be independent from changes in gene expression. How the initial burst in jasmonate production is controlled, and how the production of jasmonates is limited in unstressed tissues are still open questions. It is obviously not the level of octadecanoid pathway enzymes, which are constitutively expressed and abundant in unstressed tomato leaves (Hause et al. 2003b; Stenzel et al. 2003b; Li et al. 2004). These enzymes do either not have access to their respective substrates, in this case substrate availability would be limiting for jasmonate production, and/or constitutively expressed octadecanoid pathway enzymes may be inactive and require posttranslational modification for activation. Both hypotheses are consistent with the repeated observation that transgenic plants overexpressing octadecanoid pathway enzymes do not have elevated resting levels of JA, but show increased JA production after wounding (Wang et al. 1999; Laudert et al. 2000; Park et al. 2002; Stenzel et al. 2003a).

Primary substrates for the octa(hexa)decanoid pathway are polyunsaturated fatty acids (18:3, 16:3) which are abundant in chloroplast lipids, but not readily available as precursors for JA production in unwounded plants. Additional lipid-bound substrates may include fatty acid hydroperoxides, OPDA, and dnOPDA generated by LOX, AOS, and AOC from esterified rather than free fatty acids (Fig. 17.1; Buseman et al. 2006). The availability of these substrates as precursors for JA biosynthesis is controlled by lipolytic activities, which may include different types of phospholipases and acyl hydrolases (reviewed by Delker et al. 2006). However, in contrast to DAD1 which is required for JA production during male reproductive development (Ishiguro et al. 2001), the lipase(s) involved in the wound-induced burst of JA production have not been identified at the molecular level.

An interesting scenario of how post-translational modification may contribute to the regulation of octadecanoid pathway activity and the initial burst of JA production after wounding has recently been derived from the crystal structure of tomato OPR3. The enzyme was found to crystallize as a homodimer, in which each protomer blocks the active site of the other (Breithaupt et al. 2006). Dimerization of OPR3 concomitant with a loss of activity was also observed in solution, suggesting that OPR3 activity may be controlled in vivo by regulation of the monomer/dimer equilibrium. The crystallized OPR3 dimer was found to be stabilized by a sulfate ion at the dimer interface. Intriguingly, the sulfate is located close to Tyr364 and is positioned perfectly to mimic a phosphorylated tyrosine residue suggesting that dimerization and hence OPR3 activity may be regulated by phosphorylation of Tyr364 in vivo (Breithaupt et al. 2006). This scenario would be consistent with a proposed role for reversible protein phosphorylation in JA biosynthesis and the regulation of the wound response (Rojo et al. 1998; Schaller and Oecking 1999; Stratmann this volume).

Following the production of JA at the site of tissue damage and transmission of the long distance signal, the activation of defense responses in systemic tissues requires COI1 (Li et al. 2002, 2003). COI1 was identified many years ago as an essential component of the JA signaling pathway in a screen for insensitivity to coronatine, a structural analogue of JA-Ile (Feys et al. 1994; Xie et al. 1998). The finding that COI1 is part of the Skp/Cullin/Fbox (SCF)^{COI1} complex, a type of E_3 ubiquitin ligase, suggested that JA signaling is controlled by a negative regulator that is ubiquinated specifically by SCF^{COI1}, targeted for destruction by the 26S proteasome, resulting in the activation of JA responses (Xu et al. 2002). The negative regulators of JA signaling have remained elusive until, very recently, two independent studies led to the identification of JAZ (jasmonate ZIM domain) proteins as targets of SCF^{COI1} (Chini et al. 2007; Thines et al. 2007).

JAI3 was identified in *Arabidopsis* as a member of the JAZ protein family which negatively regulates MYC2, the key transcriptional activator of JA-dependent genes

(Lorenzo et al. 2004; Chini et al. 2007). In the *jai3-1* mutant, a splicing acceptor site mutation results in the formation of a truncated JAI3 (JAZ3) protein, which is no longer subject to COI1-dependent degradation resulting in a JA-insensitive phenotype (Chini et al. 2007). Thines et al. (2007) identified the *JAZ* family in a search for transcripts rapidly upregulated by JA. JAZ1 was shown to be a negative regulator of JA-dependent genes, which is degraded in a COI1-dependent manner in response to JA. They further demonstrated that the physical interaction of COI1 and JAZ1 is promoted by JA-Ile, but not JA, MeJA, or OPDA. This interaction was independent of other protein factors suggesting that the COI1-JAZ1 complex is the site of JA-Ile perception (Thines et al. 2007). These exciting findings explain beautifully how jasmonate signals may activate defense responses in systemic tissues by triggering the COI1-dependent degradation of JAZ proteins, thus relieving MYC transcriptions factors from negative regulation, resulting in the transcriptional activation of defense genes.

17.4 Perspectives

Despite tremendous progress in recent years, many open questions remain with respect to jasmonate biosynthesis and signaling. As far as JA biosynthesis is concerned, surprisingly little is known of how it all starts. The question is not as trivial as it may seem. Obviously, plastid-localized 13-LOXs catalyze the formation of fatty acid hydroperoxides as the first step in JA biosynthesis, but LOX substrates may include free polyunsaturated fatty acids liberated from chloroplast lipids by unidentified lipases (Bachmann et al. 2002), or fatty acids that remain esterified in membrane lipids (Brash et al. 1987). Indeed, octa(hexa)decanoid pathway intermediates are present in complex lipids in *Arabidopsis*. Seventeen complex lipids have been identified already, including mono- and digalactosyldiacylglycerols (MGDGs and DGDGs), as well as phosphatidylglycerol, with OPDA and dnOPDA esterified in the *sn1* and/or *sn2* positions (Stelmach et al. 2001; Hisamatsu et al. 2003, 2005; Buseman et al. 2006). The very rapid (within 15 min) and very large (200–1000 fold) increase of complex lipids containing two oxylipin chains after wounding, as well as the relative abundance and positional specificity of OPDA and dnOPDA in the different galactolipid species which reflects the composition of galactolipids prior to wounding, suggest that they were generated in situ by direct conversion of esterified 18:3 and 16:3 (Buseman et al. 2006). Consistent with this notion, oxylipin-containing galactolipids and AOS are co-localized in the thylakoid membrane fraction of *Arabidopsis* chloroplasts (Böttcher and Weiler 2007). In the light of such a diversity of oxylipin-containing lipids in *Arabidopsis*, the identification of the lipases that are involved in their release, and of the mechanisms in control of their activities are urgent problems to be addressed in the future.

A long-standing question with respect to jasmonate signaling concerns the activities of different jasmonate family members. The traditional view of JA being the bioactive hormone was challenged by Seo et al. who suggested that defense signaling relies on the formation of MeJA by a JA carboxyl methyltransferase (Seo et al. 2001). Furthermore, the JAR1-dependent conjugation to Ile was found to be required for (at least a subset of) JA responses (Staswick and Tiryaki 2004). The identification of the COI1-JAZ1 complex as the perception site for JA-Ile but not for JA, MeJA, or OPDA provided strong support for the importance of JA-Ile as a regulator of jasmonate response genes (Thines et al. 2007). However, this finding does not preclude the possibility that jasmonates other than JA-Ile may control subsets of COI1-dependent genes, as it was shown for OPDA (Stintzi et al. 2001). Conceivably, different members of the JAZ protein family may interact with different transcription factors, and thus, may each regulate part of the response to jasmonates. The degradation of these JAZ proteins and, consequently, transcriptional activation depend on their interaction with COI1, which in case of JAZ1 is stimulated by JA-Ile. Other jasmonates may promote the interaction of COI1 with different JAZ repressors, target them for degradation, and release inhibition of jasmonate-responsive genes.

Despite the progress made in recent years, there are still many open questions and discoveries waiting to be made concerning both, jasmonate biosynthesis and signaling, but also related to the transport of jasmonates and communication between organelles, plant organs, and organisms.

Acknowledgments The authors thank the German Research Foundation (DFG) and the German Academic Exchange Service (DAAD) for support.

References

Bachmann A, Hause B, Maucher H, Garbe E, Vörös K, Weichert H, Wasternack C, Feussner I (2002) Jasmonate-induced lipid peroxidation in barley leaves initiated by distinct 13-LOX forms of chloroplasts. Biol Chem 383:1645–1657

Berger S, Bell E, Mullet JE (1996) Two methyl jasmonate-insensitive mutants show altered expression of AtVSP in response to methyl jasmonate and wounding. Plant Physiol 111:525–531

Bergey DR, Orozco-Cárdenas M, de Moura DS, Ryan CA (1999) A wound- and systemin-inducible polygalacturonase in tomato leaves. Proc Natl Acad Sci USA 96:1756–1760

Blechert S, Bockelmann C, Füßlein M, Von Schrader T, Stelmach B, Niesel U, Weiler EW (1999) Structure-activity analyses reveal the existence of two separate groups of active octadecanoids in elicitation of the tendril-coiling response of *Bryonia dioica* Jacq. Planta 207:470–479

Blee E (2002) Impact of phyto-oxylipins in plant defense. Trends Plant Sci 7:315–321

Böttcher C, Weiler EW (2007) *Cyclo*-Oxylipin-galactolipids in plants: occurrence and dynamics. Planta 226:629–637

Bowles D (1998) Signal transduction in the wound response of tomato plants. Phil Transact Royal Soc 353:1495–1510

Brash AR, Ingram CD, Harris TM (1987) Analysis of a specific oxygenation reaction of soybean lipoxygenase-1 with fatty acids esterified in phospholipids. Biochemistry 26:5465–5471

Breithaupt C, Kurzbauer R, Lilie H, Schaller A, Strassner J, Huber R, Macheroux P, Clausen T (2006) Crystal structure of 12-oxophytodienoate reductase 3 from tomato: self-inhibition by dimerization. Proc Natl Acad Sci USA 103:14337–14342

Breithaupt C, Strassner J, Breitinger U, Huber R, Macheroux P, Schaller A, Clausen T (2001) X-ray structure of 12-oxophytodienoate reductase 1 provides structural insight into substrate binding and specificity within the family of OYE. Structure 9:419–429

Browse J (2005) Jasmonate: an oxylipin signal with many roles in plants. Vitam Horm 72:431–456
Buseman CM, Tamura P, Sparks AA, Baughman EJ, Maatta S, Zhao J, Roth MR, Esch SW, Shah J, Williams TD, et al. (2006) Wounding stimulates the accumulation of glycerolipids containing oxophytodienoic acid and dinor-oxophytodienoic acid in *Arabidopsis* leaves. Plant Physiol 142:28–39
Charron JBF, Ouellet F, Pelletier M, Danyluk J, Chauve C, Sarhan F (2005) Identification, expression, and evolutionary analyses of plant lipocalins. Plant Physiol 139:2017–2028
Cheong J-J, Choi YD (2007) Signaling pathways for the biosynthesis and action of jasmonates. J Plant Biol 50:122–133
Chini A, Fonseca S, Fernández G, Adie B, Chico JM, Lorenzo O, García-Casado G, López-Vidriero I, Lozano FM, Ponce MR, et al. (2007) The JAZ family of repressors is the missing link in jasmonate signalling. Nature 448:666–671
Corbineau F, Rudnicki RM, Côme D (1988) The effects of methyl jasmonate on sunflower (*Helianthus annuus* L.) seed germination and seedling development. Plant Growth Regul 7: 157–169
Creelman RA, Mullet JE (1997) Biosynthesis and action of jasmonates in plants. Annu Rev Plant Physiol Plant Mol Biol 48:355–381
Cruz Castillo M, Martinez C, Buchala A, Metraux JP, Leon J (2004) Gene-specific involvement of beta-oxidation in wound-activated responses in *Arabidopsis*. Plant Physiol 135:85–94
Dathe W, Rönsch H, Preiss A, Schade W, Sembdner G, Schreiber K (1981) Endogenous plant hormones of the broad bean, *Vicia faba* L. (–)-Jasmonic acid, a plant growth inhibitor in pericarp. Planta 153:530–535
Davin LB, Wang HB, Crowell AL, Bedgar DL, Martin DM, Sarkanen S, Lewis NG (1997) Stereoselective bimolecular phenoxy radical coupling by an auxiliary (dirigent) protein without an active center. Science 275:362–366
Delker C, Stenzel I, Hause B, Miersch O, Feussner I, Wasternack C (2006) Jasmonate biosynthesis in *Arabidopsis thaliana* – enzymes, products, regulation. Plant Biol 8:297–306
Delker C, Zolman BK, Miersch O, Wasternack C (2007) Jasmonate biosynthesis in *Arabidopsis thaliana* requires peroxisomal beta-oxidation enzymes – additional proof by properties of *pex6* and *aim1*. Phytochemistry 68:1642–1650
Demole E, Lederer E, Mercier D (1962) Isolement et détermination de la structure du jasmonate de méthyle, constituant odorant charactéristique de l'essence de jasmin. Helv Chim Acta 45: 675–685
Doares SH, Syrovets T, Weiler EW, Ryan CA (1995) Oligogalacturonides and chitosan activate plant defensive genes through the octadecanoid pathway. Proc Natl Acad Sci USA 92: 4095–4098
Falkenstein E, Groth B, Mithöfer A, Weiler EW (1991) Methyljasmonate and linolenic acid are potent inducers of tendril coiling. Planta 185:316–322
Farmer EE (2007) Plant biology: jasmonate perception machines. Nature 448:659–660
Farmer EE, Johnson RR, Ryan CA (1991) Regulation of expression of proteinase inhibitor genes by methyl jasmonate and jasmonic acid. Plant Physiol 98:995–1002
Farmer EE, Ryan CA. (1990) Interplant Communication: Airborne methyl jasmonate induces synthesis of proteinase inhibitors in plant leaves. Proc Natl Acad Sci USA 87:7713–7716
Farmer EE, Ryan CA (1992) Octadecanoid precursors of jasmonic acid activate the synthesis of wound-inducible proteinase inhibitors. Plant Cell 4:129–134
Feussner I, Wasternack C (2002) The lipoxygenase pathway. Annu Rev Plant Biol 53:275–297
Feys BJF, Benedetti CE, Penfold CN, Turner JG (1994) *Arabidopsis* mutants selected for resistance to the phytotoxin coronatine are male sterile, insensitive to methyl jasmonate, and resistant to a bacterial pathogen. Plant Cell 6:751–759
Finch-Savage WE, Blake PS, Clay HA (1996) Desiccation stress in recalcitrant *Quercus robur* L seeds results in lipid peroxidation and increased synthesis of jasmonates and abscisic acid. J Exp Bot 47:661–667

Fox BG, Malone TE, Johnson KA, Madson SE, Aceti M, Bingman CA, Blommel PG, Buchan B, Burns B, Cao J, et al. (2005) X-ray structure of *Arabidopsis* Atlg77680, 12-oxophytodienoate reductase isoform 1. Proteins 61:206–208

Franceschi VR, Grimes HD (1991) Induction of soybean vegetative storage proteins and anthocyanins by low-level atmospheric methyl jasmonate. Proc Natl Acad Sci USA 88:6745–6749

Fulda M, Schnurr J, Abbadi A, Heinz E, Browse J (2004) Peroxisomal Acyl-CoA synthetase activity is essential for seedling development in *Arabidopsis thaliana*. Plant Cell 16:394–405

Green TR, Ryan CA (1972) Wound-induced proteinase inhibitor in plant leaves: a possible defense mechanism against insects. Science 175:776–777

Gundlach H, Müller MJ, Kutchan TM, Zenk MH (1992) Jasmonic acid as a signal transducer in elicitor-induced plant cell cultures. Proc Natl Acad Sci USA 89:2389–2393

Halitschke R, Baldwin IT (2004) Jasmonates and related compounds in plant-insect interactions. J Plant Growth Regul 23:238–245

Hamberg M, Fahlstadius P (1990) Allene oxide cyclase: a new enzyme in plant lipid metabolism. Arch Biochem Biophys 276:518–526

Hao BZ, Wu JL (2000) Laticifer differentiation in *Hevea brasiliensis*: induction by exogenous jasmonic acid and linolenic acid. Ann Bot-London 85:37–43

Hause B, Hause G, Kutter C, Miersch O, Wasternack C (2003a) Enzymes of jasmonate biosynthesis occur in tomato sieve elements. Plant Cell Physiol 44:643–648

Hause B, Stenzel I, Miersch O, Maucher H, Kramell R, Ziegler J, Wasternack C (2000) Tissue-specific oxylipin signature of tomato flowers: allene oxide cyclase is highly expressed in distinct flower organs and vascular bundles. Plant J 24:113–126

Hause B, Stenzel I, Miersch O, Wasternack C (2003b) Occurrence of the allene oxide cyclase in different organs and tissues of *Arabidopsis thaliana*. Phytochemistry 64:971–980

Hayashi M, Nito K, Takei-Hoshi R, Yagi M, Kondo M, Suenaga A, Yamaya T, Nishimura M (2002) Ped3p is a peroxisomal ATP-binding cassette transporter that might supply substrates for fatty acid beta-oxidation. Plant Cell Physiol 43:1–11

Hisamatsu Y, Goto N, Hasegawa K, Shigemori H (2003) Arabidopsides A and B, two new oxylipins from *Arabidopsis thaliana*. Terahedron Lett 44:5553–5556

Hisamatsu Y, Goto N, Sekiguchi M, Hasegawa K, Shigemori H (2005) Oxylipins arabidopsides C and D from *Arabidopsis thaliana*. J Nat Prod 68:600–603

Hofmann E, Zerbe P, Schaller F (2006) The crystal structure of *Arabidopsis thaliana* allene oxide cyclase: insights into the oxylipin cyclization reaction. Plant Cell 18:3201–3217

Howe GA (2004) Jasmonates as signals in the wound response. J Plant Growth Regul 23:223–237

Howe GA, Lightner J, Browse J, Ryan CA (1996) An octadecanoid pathway mutant (JL5) of tomato is compromised in signaling for defense against insect attack. Plant Cell 11:2067–2077

Howe GA, Schilmiller AL (2002) Oxylipin metabolism in response to stress. Curr Opin Plant Biol 5:230–236

Ishiguro S, Kawai-Oda A, Ueda J, Nishida I, Okada K (2001) The *DEFECTIVE IN ANTHER DEHISCENCE1* gene encodes a novel phospholipase A$_1$ catalyzing the initial step of jasmonic acid biosynthesis, which synchronizes pollen maturation, anther dehiscence, and flower opening in *Arabidopsis*. Plant Cell 13:2191–2209

Jacinto T, McGurl B, Franceschi V, DelanoFreier J, Ryan CA (1997) Tomato prosystemin promoter confers wound-inducible, vascular bundle-specific expression of the beta-glucuronidase gene in transgenic tomato plants. Planta 203:406–412

Kang J-H, Wang L, Giri A, Baldwin IT (2006) Silencing threonine deaminase and JAR4 in *Nicotiana attenuata* impairs jasmonic acid-isoleucine-mediated defenses against *Manduca sexta*. Plant Cell 18:3303–3320

Koo AJK, Chung HS, Kobayashi Y, Howe GA (2006) Identification of a peroxisomal acyl-activating enzyme involved in the biosynthesis of jasmonic acid in *Arabidopsis*. J Biol Chem 281:33511–33520

Kramell R, Miersch O, Hause B, Ortel B, Parthier B, Wasternack C (1997) Amino acid conjugates of jasmonic acid induce jasmonate-responsive gene expression in barley (*Hordeum vulgare* L.) leaves. FEBS Lett 414:197–202

Laudert D, Hennig P, Stelmach BA, Müller A, Andert L, Weiler EW (1997) Analysis of 12-oxo-phytodienoic acid enantiomers in biological samples by capillary gas chromatography-mass spectrometry using cyclodextrin stationary phases. Anal Biochem 246:211–217

Laudert D, Schaller F, Weiler EW (2000) Transgenic *Nicotiana tabacum* and *Arabidopsis thaliana* plants overexpressing allene oxide synthase. Planta 211:163–165

Li L, Li C, Lee GI, Howe GA (2002) Distinct roles for jasmonate synthesis and action in the systemic wound response of tomato. Proc Natl Acad Sci USA 99:6416–6421

Li C, Liu G, Xu C, Lee GI, Bauer P, Ling HQ, Ganal MW, Howe GA (2003) The tomato *suppressor of prosystemin-mediated responses2* gene encodes a fatty acid desaturase required for the biosynthesis of jasmonic acid and the production of a systemic wound signal for defense gene expression. Plant Cell 15:1646–1661

Li CY, Schilmiller AL, Liu GH, Lee GI, Jayanty S, Sageman C, Vrebalov J, Giovannoni JJ, Yagi K, Kobayashi Y, et al. (2005) Role of beta-oxidation in jasmonate biosynthesis and systemic wound signaling in tomato. Plant Cell 17:971–986

Li L, Zhao Y, McCaig BC, Wingerd BA, Wang J, Whalon ME, Pichersky E, Howe GA (2004) The tomato homolog of CORONATINE-INSENSITIVE1 is required for the maternal control of seed maturation, jasmonate-signaled defense responses, and glandular trichome development. Plant Cell 16:126–143

Liechti R, Farmer EE (2002) The jasmonate pathway. Science 296:1649–1650

Lorenzo O, Chico JM, Sanchez-Serrano JJ, Solano R (2004) Jasmonate-insensitive1 encodes a MYC transcription factor essential to discriminate between different jasmonate-regulated defense responses in *Arabidopsis*. Plant Cell 16:1938–1950

Lorenzo O, Solano R (2005) Molecular players regulating the jasmonate signalling network. Curr Opin Plant Biol 8:532–540

Malone TE, Madson SE, Wrobel RL, Jeon WB, Rosenberg NS, Johnson KA, Bingman CA, Smith DW, Phillips GN, Markley JL, et al. (2005) X-ray structure of *Arabidopsis* At2g06050, 12-oxophytodienoate reductase isoform 3. Proteins 58:243–245

McConn M, Browse J (1996) The critical requirement for linolenic acid is pollen development, not photosynthesis, in an *Arabidopsis* mutant. Plant Cell 8:403–416

McConn M, Creelman RA, Bell E, Mullet JE, Browse J (1997) Jasmonate is essential for insect defense in *Arabidopsis*. Proc Natl Acad Sci USA 94:5473–5477

McGurl B, Pearce G, Orozco-Cárdenas M, Ryan CA (1992) Structure, expression and antisense inhibition of the systemin precursor gene. Science 255:1570–1573

Meyer A, Miersch O, Büttner C, Dathe W, Sembdner G (1984) Occurrence of the plant growth regulator jasmonic acid in plants. J Plant Growth Regul 3:1–8

Mueller-Uri F, Parthier B, Nover L (1988) Jasmonate-induced alteration of gene expression in barley leaf segments analyzed by in-vivo and in-vitro protein synthesis. Planta 176:241–247

Nakamura Y, Kiyota H, Kumagai T, Ueda M (2006a) Direct observation of the target cell for jasmonate-type leaf-closing factor: genus-specific binding of leaf-movement factors to the plant motor cell. Tetrahedron Lett 47:2893–2897

Nakamura Y, Miyatake R, Matsubara A, Kiyota H, Ueda M (2006b) Enantio-diffferential approach to identify the target cell for glucosyl jasmonate-type leaf-closing factor, by using fluorescence-labeled probe compounds. Tetrahedron 62:8805–8813

Narváez-Vásquez J, Ryan CA (2004) The cellular localization of prosystemin: a functional role for phloem parenchyma in systemic wound signaling. Planta 218:360–369

Park JH, Halitschke R, Kim HB, Baldwin IT, Feldmann KA, Feyereisen R (2002) A knock-out mutation in allene oxide synthase results in male sterility and defective wound signal transduction in *Arabidopsis* due to a block in jasmonic acid biosynthesis. Plant J 31:1–12

Paschold A, Halitschke R, Baldwin IT (2007) Co(i)-ordinating defenses: NaCOI1 mediates herbivore-induced resistance in *Nicotiana attenuata* and reveals the role of herbivore movement in avoiding defenses. Plant J 51:79–91

Pearce G, Strydom D, Johnson S, Ryan CA (1991) A polypeptide from tomato leaves induces wound-inducible proteinase inhibitor proteins. Science 253:895–898

Pelacho AM, Mingo-Castel AM (1991) Jasmonic acid induces tuberization of potato stolons cultured in vitro. Plant Physiol 97:1253–1255

Pieterse CMJ, Schaller A, Mauch-Mani B, Conrath U (2006). Signaling in plant resistance responses: divergence and cross-talk of defense pathways. In: Tuzun S, Bent E (eds) Multigenic and induced systemic resistance in plants. Springer, New York, pp 166–196

Pozo MJ, Van Loon LC, Pieterse CMJ (2004) Jasmonates – signals in plant-microbe interactions. J Plant Growth Regul 23 211–222

Rojo E, Titarenko E, Leon J Berger S, Vancanneyt G, Sanchez-Serrano JJ (1998) Reversible protein phosphorylation regulates jasmonic acid-dependent and -independent wound signal transduction pathways in *Arabidopsis thaliana*. Plant J 13:153–165

Ryan CA (1992) The search for the proteinase inhibitor-inducing factor, PIIF. Plant Mol Biol 19:123–133

Ryan CA (2000) The systemin signaling pathway: differential activation of plant defensive genes. BBA-Protein Struct M 1477:112–121

Ryan CA, Moura DS (2002) Systemic wound signaling in plants: a new perception. Proc Natl Acad Sci USA 99:6519–6520

Ryan CA, Pearce G (2003) Systemins: a functionally defined family of peptide signal that regulate defensive genes in *Solanaceae* species. Proc Natl Acad Sci USA 100:14577–14580

Sanders PM, Lee PY, Biesger C, Boone JD, Beals TP, Weiler EW, Goldberg RB (2000) The *Arabidopsis DELAYED DEHISCENCE1* gene encodes an enzyme in the jasmonic acid synthesis pathway. Plant Cell 12:1042–1061

Schaller F, Biesgen C, Müssig C, Altmann T, Weiler EW (2000) 12-Oxophytodienoate reductase 3 (OPR3) is the isoenzyme involved in jasmonate biosynthesis. Planta 210:979–984

Schaller A, Oecking C (1999) Modulation of plasma membrane H^+-ATPase activity differentially activates wound and pathogen defense responses in tomato plants. Plant Cell 11:263–272

Schaller F, Schaller A, Stintzi A (2004) Biosynthesis and metabolism of jasmonates. J Plant Growth Regul 23:179–199

Schaller F, Weiler EW (1997) Molecular cloning and characterization of 12-oxophytodienoate reductase, an enzyme of the octadecanoid signaling pathway from *Arabidopsis thaliana*. Structural and functional relationship to yeast old yellow enzyme. J Biol Chem 272:28066–28072

Schilmiller AL, Howe GA (2005) Systemic signaling in the wound response. Curr Opin Plant Biol 8:369–377

Schilmiller AL, Koo AJK, Howe GA (2007) Functional diversification of acyl-coenzyme A oxidases in jasmonic acid biosynthesis and action. Plant Physiol 143:812–824

Schneider K, Kienow L, Schmelzer E, Colby T, Bartsch M, Miersch O, Wasternack C, Kombrink E, Stuible HP (2005) A new type of peroxisomal acyl-coenzyme A synthetase from *Arabidopsis thaliana* has the catalytic capacity to activate biosynthetic precursors of jasmonic acid. J Biol Chem 280:13962–13972

Seo HS, Song JT, Cheong JJ, Lee YH, Lee YW, Hwang I, Lee JS, Choi YD (2001) Jasmonic acid carboxyl methyltransferase: a key enzyme for jasmonate-regulated plant responses. Proc Natl Acad Sci USA 98:4788–4793

Sobajima H, Takeda M, Sugimori M, Kobashi N, Kiribuchi K, Cho EM, Akimoto C, Yamaguchi T, Minami E, Shibuya N, et al. (2003) Cloning and characterization of a jasmonic acid-responsive gene encoding 12-oxophytodienoic acid reductase in suspension-cultured rice cells. Planta 216:692–698

Staswick PE, Su W, Howell SH (1992) Methyl jasmonate inhibition of root growth and induction of a leaf protein are decreased in an *Arabidopsis thaliana* mutant. Proc Natl Acad Sci USA 89:6837–6840

Staswick PE, Tiryaki I (2004) The oxylipin signal jasmonic acid is activated by an enzyme that conjugates it to isoleucine in *Arabidopsis*. Plant Cell 16:2117–2127

Staswick PE, Tiryaki I, Rowe ML (2002) Jasmonate response locus *JAR1* and several related *Arabidopsis* genes encode enzymes of the firefly luciferase superfamily that show activity on jasmonic, salicylic, and indole-3-acetic acids in an assay for adenylation. Plant Cell 14:1405–1415

Stelmach BA, Müller A, Hennig P, Gebhardt S, Schubert-Zsilavecz M, Weiler EW (2001) A novel class of oxylipins, *sn*1-*O*-(12-oxophytodienoyl)-*sn*2-*O*-(hexadecatrienoyl)-monogalactosyl diglyceride, from *Arabidopsis thaliana*. J Biol Chem 276:12832–12838

Stenzel I, Hause B, Maucher H, Pitzschke A, Miersch O, Ziegler J, Ryan CA, Wasternack C (2003a) Allene oxide cyclase dependence of the wound response and vascular bundle-specific generation of jasmonates in tomato – amplification in wound signalling. Plant J 33:577–589

Stenzel I, Hause B, Miersch O, Kurz T, Maucher H, Weichert H, Ziegler J, Feussner I, Wasternack C (2003b) Jasmonate biosynthesis and the allene oxide cyclase family of *Arabidopsis thaliana*. Plant Mol Biol 51:895–911

Stintzi A, Browse J (2000) The *Arabidopsis* male-sterile mutant, *opr3*, lacks the 12-oxophytodienoic acid reductase required for jasmonate synthesis. Proc Natl Acad Sci USA 97:10625–10630

Stintzi A, Weber H, Reymond P, Browse J, Farmer EE (2001) Plant defense in the absence of jasmonic acid: the role of cyclopentenones. Proc Natl Acad Sci USA 98:12837–12842

Straßner J, Fürholz A, Macheroux P, Amrhein N, Schaller A (1999) A homolog of old yellow enzyme in tomato. Spectral properties and substrate specificity of the recombinant protein. J Biol Chem 274:35067–35073

Strassner J, Schaller F, Frick UB, Howe GA, Weiler EW, Amrhein NA, Macheroux P, Schaller A (2002) Characterization and cDNA-microarray expression analysis of 12-oxophytodienoate reductases reveals differential roles for octadecanoid biosynthesis in the local versus the systemic wound response. Plant J 32:585–601

Stratmann JW (2003) Long distance run in the wound response – jasmonic acid is pulling ahead. Trends Plant Sci 8:247–250

Taki N, Sasaki-Sekimoto Y, Obayashi T, Kikuta A, Kobayashi K, Ainai T, Yagi K, Sakurai N, Suzuki H, Masuda T, et al. (2005) 12-oxo-phytodienoic acid triggers expression of a distinct set of genes and plays a role in wound-induced gene expression in *Arabidopsis*. Plant Physiol 139:1268–1283

Tani T, Sobajima H, Okada K, Chujo T, Arimura S, Tsutsumi N, Nishimura M, Seto H, Nojiri H, Yamane H (2007) Identfcation of the OsOPR7 gene encoding 12-oxophytodienoate reductase involved in the biosynthesis of jasmonic acid in rice. Planta DOI 10.1007/s00425-007-0635-7

Theodoulou FL, Job K, Slocombe SP, Footitt S, Holdsworth M, Baker A, Larson TR, Graham IA (2005) Jasmonoic acid levels are reduced in COMATOSE ATP-binding cassette transporter mutants. Implications for transport of jasmonate precursors into peroxisomes. Plant Physiol 137:835–840

Thines B, Katsir L, Melotto M, Niu Y, Mandaokar A, Liu G, Nomura K, He SY, Howe GA, Browse J (2007) JAZ repressor proteins are targets of the SCF^{COI1} complex during jasmonate signalling. Nature 448:661–665

Tilton G, Shockey J, Browse J (2000) Two families of acyl-CoA thioesterases in *Arabidopsis*. Biochem Soc T 28:946–947

Ueda J, Kato J (1980) Isolation and identification of a senescence-promoting substance from wormwood (*Artemisia absinthium* L.). Plant Physiol 66:246–249

Vick BA, Zimmerman DC (1984) Biosynthesis of jasmonic acid by several plant species. Plant Physiol 75:458–461

von Malek B, van der Graaff E, Schneitz K, Keller B (2002) The *Arabidopsis* male-sterile mutant *dde2-2* is defective in the *ALLENE OXIDE SYNTHASE* gene encoding one of the key enzymes of the jasmonic acid biosynthesis pathway. Planta 216:187–192

Wang C, Avdiushko S, Hildebrand DF (1999) Overexpression of a cytoplasm-localized allene oxide synthase promotes the wound-induced accumulation of jasmonic acid in transgenic tobacco. Plant Mol Biol 40:783–793

Wasternack C (2006). Oxylipins: biosynthesis, signal transduction and action. In: Hedden P Thomas SG (eds) Plant hormone signalling. Blackwell Publishing, Oxford, UK, pp 185–228

Wasternack C (2007). Jasmonates: an update on biosynthesis, signal transduction and action in plant stress response, growth and development. In Ann Bot, pp 1–17

Wasternack C, Ortel B, Miersch O, Kramell R, Beale M, Greulich F, Feussner I, Hause B, Krumm T, Boland W, et al. (1998) Diversity in octadecanoid-induced gene expression of tomato. J Plant Physiol 152:345–352

Wasternack C, Stenzel I, Hause B, Hause G, Kutter C, Maucher H, Neumerkel J, Feussner I, Miersch O (2006) The wound response in tomato – role of jasmonic acid. J Plant Physiol 163:297–306

Weber H, Vick BA, Farmer EE (1997) Dinor-oxo-phytodienoic acid: a new hexadecanoid signal in the jasmonate family. Proc Natl Acad Sci USA 94:10473–10478

Weidhase RA, Kramell HM, Lehmann J, Liebisch HW, Lerbs W, Parthier B (1987) Methyl jasmonate-induced changes in the polypeptide pattern of senescing barley leaf segments. Plant Sci 51:177–186

Xie D-X, Feys BF, James S, Nieto-Rostro M, Turner JG (1998) *COI1*: an *Arabidopsis* gene required for jasmonate-regulated defense and fertility. Science 280:1091–1094

Xu LH, Liu FQ, Lechner E, Genschik P, Crosby WL, Ma H, Peng W, Huang DF, Xie DX (2002) The SCF^{COI1} ubiquitin-ligase complexes are required for jasmonate response in *Arabidopsis*. Plant Cell 14:1919–1935

Yoshihara T, Omer E-SA, Koshino H, Sakamura S, Kikuta Y, Koda Y (1989) Structure of a tuber-inducing stimulus from potato leaves (*Solanum tuberosum* L.). Agric Biol Chem 53:2835–2837

Zerbe P, Weiler EW, Schaller F (2007) Preparative enzymatic solid phase synthesis of cis(+)-12-oxo-phytodienoic acid – physical interaction of AOS and AOC is not necessary. Phytochemistry 68:229–236

Ziegler J, Keinanen M, Baldwin IT (2001) Herbivore-induced allene oxide synthase transcripts and jasmonic acid in *Nicotiana attenuata*. Phytochemistry 58:729–738

Ziegler J, Wasternack C, Hamberg M (1999) On the specificity of allene oxide cyclase. Lipids 34:1005–1015

Zolman BK, Silva ID, Bartel B (2001) The *Arabidopsis pxa1* mutant is defective in an ATP-binding cassette transporter-like protein required for peroxisomal fatty acid β-oxidation. Plant Physiol 127:1266–1278

Part B
Signals Between Plants and Insects

Chapter 18
Caterpillar Secretions and Induced Plant Responses

Gary W. Felton

> *Saliva and the glands that produce it could be the leading edge of rapid evolutionary adaptation to new food sources.*
>
> –(Tabak and Kuska 2004)

Plants employ both direct inducible defenses (those that directly affect the physiology of the herbivore) and indirect induced defenses (defenses that attract or enhance the effectiveness of natural enemies of herbivores). During the last two decades, there has been a proliferation of studies showing that the oral secretions and saliva of herbivores play important roles in mediating inducible responses. Much of the study has centered on lepidopteran larvae (i.e., moths and butterflies) – no doubt due to their economic importance as agricultural pests, but also due to the relative ease of working with their secretions. In this chapter, I will not attempt a comprehensive review on the subject, but instead will focus this discussion on the secretions of larval Lepidoptera.

18.1 Introduction

Caterpillars feed by chewing and ingesting plant tissue, resulting in tissue damage which is sufficient to trigger plant defense responses. Investigators have attempted to mimic their feeding style by wounding plant tissues with scissors, tracing wheels, cork borers, or most elegantly with the MecWorm (Mithöfer et al. 2005). While each of these treatments causes wound responses in the plant, both locally at the site of wounding and systemically in unwounded tissues, it has become clear that wounding by itself cannot account for the full breadth and the specificity of plant responses to herbivory. Therefore, herbivore-associated factors have to exist which modulate the wound response and/or elicit additional reactions in the host plant (Hartley and Lawton 1987; Turlings et al. 1990; Mattiacci et al. 1995; Korth and Dixon 1997; Baldwin et al. 2001; Gomez et al. 2005). Insect-derived elicitors may be present in the faeces and, most prominently, in oral secretions.

G.W. Felton
Department of Entomology, Pennsylvania State University, University Park, PA 16802, USA
e-mail: gwf10@psu.edu

A. Schaller (ed.), *Induced Plant Resistance to Herbivory*,
© Springer Science+Business Media B.V. 2008

To determine if oral secretions play any additional role in the specificity of the plant wound responses, the secretions are typically applied to artificially wounded plants. The diminutive size of insect herbivores other than caterpillars, such as aphids, whiteflies, thrips, etc. makes their study more intractable. The piercing-sucking feeding style of the aphids, leafhoppers and other insects adds another level of difficulty when trying to mimic their feeding. The majority of the research on their saliva has focused on its roles in detoxifying plant compounds (Miles 1999) and in transmission of plant pathogens (Fereres 2007), and less on its role in induction of plant defenses (Williams et al. 2005). Despite their small size, saliva can be collected from aphids and whiteflies using a feeding membrane (Miles and Harrewijn 1991). The saliva of insect herbivores was the subject of a review by my laboratory (Felton and Eichenseer 1999); however, the review is now over 8 years old and substantial progress has occurred since. Herein, this chapter focuses on recent advances on caterpillar secretions.

18.2 Oral Secretions and Saliva – A Clarification

Larval Lepidoptera possess two pairs of salivary glands: the labial glands and mandibular glands (Borcas 1903; Parthasarathy and Gopinathan 2005). The silk secretions of the labial glands of the silkworm *Bombyx mori* have been known and used for thousands of years for clothing (Yokoyama 1963) and medicine (Lev 2003), yet the fact that the same glands can produce watery saliva is not widely appreciated by many entomologists. I will define saliva *sensu stricto*, to refer only to the secretions of the labial and mandibular salivary glands. The labial glands are generally long and tubular, and converge into a common duct and then are released from the spinneret (Fig. 18.1); the same structure from where silk is released. In the case of the noctuid *Helicoverpa (=Heliothis) zea*, the larval labial glands were mistakenly believed to empty and secrete into the digestive tract at the pharynx (Chauthan and Callahan 1967), although this has been corrected by several authors (Standlea and Yonke 1968; Macgown and Sikorowski 1982; Felton and Eichenseer 1999).

The mandibular glands are tubular but highly variable among species (Parthasarathy and Gopinathan 2005). Frequently they are relatively minute and restricted to the head region (Eichenseer et al. 2002); in other cases they may be nearly as long as the labial glands and extend well into the thorax (Vegliante 2005). The glands open into the lumen of the mandibular adductor apodeme. In some species, a second lobed accessory gland may be found attached to the mandibles, which open through numerous pores on both sides of the mandibles (Vegliante 2005).

The chemical characterization of caterpillar saliva has been barely touched. Proteins are present in labial gland saliva (Liu et al. 2004), but other macromolecules and small molecules have not been reported. Mandibular gland secretions may also contain proteins (Eichenseer et al. 1999), but these secretions have long been noted to be rich in different classes of lipids, such as sterols and triglycerides (Wroniszewska 1966; Mossadegh 1978; Eichenseer et al. 2002; Howard and Baker 2004).

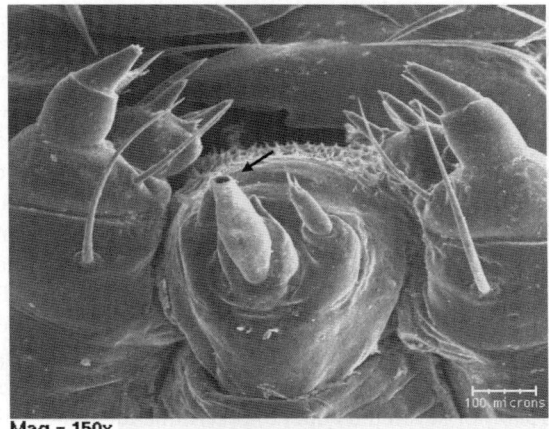

Fig. 18.1 Electron micrograph of frontal view of head of larval *Helicoverpa zea*. Arrow points to spinneret, the external secretory structure for silk and labial gland saliva. Aperture in the spinneret for the release of silk and/or saliva is clearly visible. (Micrograph courtesy of Michelle Peiffer)

The salivary secretions of the labial and mandibular glands are distinct from the regurgitant or 'oral secretions' (=OS) which have been widely used in studies of plant-herbivore interactions (Felton and Eichenseer 1999; Liu et al. 2004). Multiple laboratories [e.g., Baldwin (Roda et al. 2004), Boland (Arimura et al. 2005), De Moraes (De Moraes and Mescher 2004), Dicke (Mattiacci et al. 1995) Korth (Korth and Dixon 1997), Mori (Yoshinaga et al. 2006), Pare (Truitt and Pare 2004), Schmelz (Schmelz et al. 2006), Tumlinson (Tumlinson and Lait 2005), Turlings (Gouinguene et al. 2003) and Wang (Yan et al. 2005)] are working on OS. These secretions arise from the herbivore's digestive system (von Dahl et al. 2006) and may contain insect-derived substances (e.g., digestive enzymes), as well as plant material in various stages of digestion. Because the OS are collected as regurgitant, it is entirely possible that small amounts of labial and mandibular saliva are collected during this process.

The active components of caterpillar OS have most often been shown to be fatty acid-amino acid conjugates (=FACs; Pohnert et al. 1999a, b; Halitschke et al. 2001; Lait et al. 2003; Spiteller and Boland 2003a, b; Yoshinaga et al. 2005; Aboshi et al. 2007) including volicitin, the first FAC identified as an elicitor (Alborn et al. 1997). More recently, plant protein fragments of chloroplastic ATP synthase γ-subunit protein (termed inceptin) found in the OS of the fall armyworm were shown to induce indirect and direct defenses in cowpea (Schmelz et al. 2006, 2007). Also, OS may contain microbes ingested with food or derived from resident microbial populations in the digestive system (Spiteller et al. 2000; Cardoza et al. 2006).

18.3 Experimental Approaches for Studying Oral Secretions and Saliva

The function of saliva in caterpillars can be assessed using surgical, pharmacological, and/or molecular approaches. *Surgical techniques* have been applied to prevent caterpillars from secreting saliva. These techniques involve the extirpation of the salivary glands or cauterization of the spinneret to prevent salivation. Wroniszewska (1966) surgically removed the mandibular glands of *Galleria mellonella* and demonstrated that the loss of the glands causes a considerable decrease in the growth rate of larvae, although the specific cause for slower growth rates was not noted. Similarly Mossadegh (1978) surgically removed mandibular glands to examine the role of their secretions on silk filaments, and also cauterized the spinnerets (using a heated needle) to prevent silk deposition. Cauterization of the spinneret has also been used to examine the role of silk in trail following behavior in social caterpillar species (Fitzgerald 1993, 2003). Cauterization of the spinneret and surgical removal of the labial glands have been used to examine the function of labial saliva (Musser et al. 2002, 2005a, b, 2006; Bede et al. 2006) We have recently used a medical cauterization tool which more consistently accomplishes a uniform cauterization (Peiffer and Felton 2005). Machida (1965) removed the Lyonet's gland (Filippi's gland) or severed its duct to determine the impact of the gland on silk secretion.

Surgical removal of salivary glands, particularly the mandibular glands that are often relatively small, requires a high degree of skill to avoid killing larvae. Furthermore, the interpretation of results using surgically invasive procedures should be cautioned because rupture of the cuticle causes a massive cascade of wound responses (Jarosz 1993; Han et al. 1999). Cauterization of the spinneret avoids this invasive procedure, but is limited to examining labial gland secretions. In some species (e.g., *Pieris rapae*) cauterization may adversely affect larval feeding behavior (Musser et al. 2005). Despite the robust size of *Manduca sexta* larvae, their spinnerets are comparatively minute and difficult to cauterize without damaging the mouthparts (Felton lab, personal observations).

Pharmacological techniques include directly treating plant tissues with OS or salivary secretions. Typically OS is collected by inducing larvae to regurgitate which is then applied to artificial plant wounds. Saliva has been prepared by surgical removal and homogenization of salivary glands with subsequent assay of their biological and/or chemical properties. These techniques have been widely used to study salivary gland components in blood-feeding arthropods (Valenzuela 2002; Ribeiro et al. 2004) and in herbivorous insects (Musser et al. 2002; Shackel et al. 2005). Unfortunately, in many cases it has not been determined which components of OS or salivary gland homogenates are normally secreted during feeding. Without further studies showing that these components are actually deposited during feeding, pharmacological studies are of restricted value.

Fortunately, several approaches are available to identify secreted components. First, collection of saliva can be made directly from the spinneret (Liu et al. 2004) or saliva that is deposited on a feeding substrate can be collected and analyzed. The limitation of this approach is that at present only very minute quantities (ca. <1 nl) of saliva can be collected from individual larvae (personal observations). However, with greatly enhanced sensitivity of chemical and biochemical techniques, this is no longer a major impediment for analyses. Collecting saliva from a substrate after feeding suffers from not knowing where the collected substance was secreted from—e.g., labial glands, mandibular glands, regurgitant, or other exocrine glands. Also, proteins may be degraded because of the presence of co-secreted proteases. In the case of the salivary enzyme glucose oxidase, we used a combination of molecular and surgical techniques to demonstrate that the enzyme was produced primarily in the labial glands with minor contribution from the mandibular glands (Eichenseer et al. 1999). Also, use of an antibody specific for glucose oxidase allowed us to not only quantify the amount of enzyme secreted, but also to visualize where on the leaf surface caterpillars were secreting their saliva (Peiffer and Felton 2005).

To my knowledge, this approach has not been used to determine if specific components of OS are secreted. In one elegant example, Pare and colleagues labeled corn seedlings with $^{14}CO_2$ in situ, and *Spodoptera exigua* larvae were then allowed to feed on the labeled leaves to determine the production of volicitin. The OS was collected from third-instars and contained approximately 120 pmol volicitin per larva. When radio-labeled larvae were placed on unlabeled leaves, the amount of volicitin introduced to the feeding site was estimated at 100 pmol/larva (Truitt and Pare 2004).

Finally *molecular methods* have proven extremely powerful for providing a wealth of information about arthropod saliva (Valenzuela 2002). Preparation and high-throughput sequencing of full length cDNA salivary gland libraries combined with functional genomics will vastly improve our understanding of salivary gland function. More recently, proteomic approaches have been employed to identify proteins that are produced in the glands, many of which had eluded identification by standard biochemical assays (Valenzuela 2002). Analysis of molecular sequences by PSORT analysis (http://psort.hgc.jp/form2.html) can provide strong evidence for proteins that may be secreted which then can be further verified by detection with antibodies. RNA interference (RNAi) techniques are being employed to suppress specific salivary genes and then to examine their function (Aljamali et al. 2003; Karim et al. 2004; Mutti et al. 2006). Although RNAi is an extremely powerful tool for loss-of-function analysis, there are concerns over the possible pleiotropic effects. If direct injection of the interfering RNAs into the hemocoel is required, the induction of undesired wound or immune responses may complicate interpretation of results. In other cases, specific expression of transgenes in the salivary glands has been achieved using a vector such as the piggyBac vector (Royer et al. 2005), and offers great potential for understanding function of gene products.

18.4 Oral Secretions and Induced Defenses

Considerable progress has been made in the last decade in identifying elicitors of induced defenses in the oral secretions of caterpillars. Several groups of elicitors have been found including β-glucosidase (Mattiacci et al. 1995), FACs such as volicitin (Alborn et al. 1997), and most recently chloroplastic peptide fragments termed inceptins (Schmelz et al. 2006, 2007). The inceptins are formed when *S. frugiperda* attack cowpea and partially digest a chloroplast protein, ATP-synthase. How plants recognize these herbivore-specific elicitors, and use defense signal transduction pathways to regulate direct and indirect defenses is not well understood. Caterpillar OS (including FACs) are known to elicit bursts of the signaling compounds jasmonic acid and ethylene (Kahl et al. 2000). The inceptins trigger increases in ethylene, jasmonic acid, and salicylic acid in cowpea (Schmelz et al. 2006). Recently it was shown in *N. attenuata*, that *M. sexta* OS/FACs activate MAP-kinases (an SIPK and WIPK; see Stratmann (this volume) for a more detailed discussion of the role of these two MPKs in defense signaling), which are upstream components regulating salicylic acid, jasmonic acid, and ethylene biosynthesis (Wu et al. 2007). OS and FACs subsequently cause changes in gene expression (Korth and Dixon 1997; Halitschke et al. 2003; Roda et al. 2004), protein expression (Giri et al. 2006), and production of volatile organic compounds, as well as other secondary metabolites (Roda et al. 2004; von Dahl et al. 2006).

Most often OS has been shown to induce indirect defenses (Turlings et al. 1993; Rose and Tumlinson 2005). I refer the reader to Chapter 19 by Tumlinson and Engelberth for a more extensive coverage of FACs. The application of FACs to wound sites frequently, but not always, mimics that observed with insect feeding. Interestingly, in the case of the *S. frugiperda* and cowpea interaction, the FACs are inactive as elicitors (Schmelz et al. 2006). Additional studies in particular systems indicate that OS may not always play a major role in the indirect defenses. In one example, the generalist parasitoid *Campoletis chlorideae*, did not differentiate between maize seedlings infested by the two different noctuid insects (*Helicoverpa armigera* and *Pseudaletia separata*; Yan and Wang 2006). The parasitoid did not show a difference in attraction to mechanically damaged plants treated with caterpillar OS or water, although clear differences in volatile emissions were observed among the various treatments (Yan and Wang 2006). In some instances, tobacco leaves artificially damaged and then treated with caterpillar OS (*H. armigera* and *Helicoverpa assulta*) emitted the same levels of volatiles as leaves treated with water (Yan et al. 2005).

Moreover, although OS may elicit the production of volatiles, their production may not always result in enhanced plant defense. Besides serving as a cue for the attraction of natural enemies, volatiles may also attract other herbivores as was shown in potato where Colorado potato beetles were more attracted to damaged plants (Bolter et al. 1997) or plants treated with beetle or cabbage looper OS than to undamaged plants (Landolt et al. 1999). Similarly fall armyworm larvae were more attracted to odors from herbivore-damaged plants than undamaged plants (Carroll et al. 2006; but see De Moraes et al. 2001). These results suggest that although induced volatiles may attract natural enemies of herbivores, they may inadvertently recruit herbivores to damaged plants in some instances (Carroll et al. 2006). It is thus particularly important to examine the role of induced volatiles in natural ecosystems (Kessler and Baldwin 2001). There are still other examples, where OS do not induce indirect but rather direct defenses (Korth and Dixon 1997; Roda et al. 2004; Major and Constabel 2006; Schmelz et al. 2006). Furthermore, in *N. attenuata*, OS from *M. sexta* was shown to suppress an inducible direct defense, the production of the alkaloid nicotine (Winz and Baldwin 2001). This finding may serve as an example of how the herbivore tries to escape plant defenses.

18.5 Saliva and Induced Defenses

Considering the prominent role that saliva plays in host evasion by blood-feeding arthropods, it would not be surprising if herbivorous insects employed a similar counter defense strategy. Our laboratory began studying the role of saliva in evading induced defenses using the generalist noctuid *H. zea* as a model. We used both cauterization of the spinneret and application of salivary gland homogenates to determine that labial gland saliva from *H. zea* mitigates induced resistance and the induction of a direct defense (i.e., nicotine) in tobacco *Nicotiana tabacum* (Musser et al. 2002, 2005). Further evidence for saliva suppressing induced resistance was obtained by surgical removal of salivary glands (Musser et al. 2006). Fractionation of the salivary gland homogenate showed that glucose oxidase (GOX) was primarily responsible for suppression of resistance (Musser et al. 2002). GOX is an abundant protein in *H. zea* saliva and is expressed at highest levels during larval feeding (Eichenseer et al. 1999). The enzyme mediates the following reaction:

$$\alpha\text{-D-glucose} + O_2 \rightarrow \text{D-glucono-1,5-lactone} + H_2O_2$$

The product of the reaction, H_2O_2, appears to have the most activity in suppression. Zong and Wang (2004) examined the role of saliva from labial gland homogenates of three noctuid species, *Helicoverpa armigera*, *Helicoverpa assulta*, and *Spodoptera litura* in inducing nicotine production in tobacco. They reported that feeding by the two *Helicoverpa* species, or application of their labial gland homogenates to plant wounds suppressed nicotine. GOX concentrations are exceptionally high in *H. armigera* and somewhat lower in *H. assulta*.. In contrast, feeding by *S. litura* or application of its labial gland homogenates in which GOX was not detectable, failed to suppress nicotine induction. Based upon these results and experiments with fungal GOX applied to wounded leaves, the authors concluded that salivary GOX (and H_2O_2) plays an important role in the suppression of wound-induced nicotine (Zong and Wang 2004).

GOX is found in many noctuid caterpillar species (Eichenseer and Felton, unpublished data) and may function in defense suppression in other systems. For instance, when tobacco leaves are fed on by tobacco budworms *Heliothis virescens*, there is no detectable volatile nicotine release; however if their spinnerets are ablated, then a significant burst of volatile nicotine is observed (Delphia et al. 2006). The budworm has comparatively high levels of GOX, but its role in suppressing the release of volatile nicotine was not tested in this case. Ablation of the spinnerets in the beet armyworm *Spodoptera exigua* revealed that their saliva suppresses transcript levels of two genes encoding early enzymes in terpenoid biosynthesis in *Medicago truncatula* (Bede et al. 2006). GOX, which is found in the saliva of *S. exigua*, similarly caused reduction in the transcript levels of these genes. Whether or not the reduction in these transcript levels is sufficient to alter the production of volatiles that may function as indirect defenses is still unknown.

Further studies with *H. zea* suggest that the role of saliva in evading host defenses is more complex than what we previously thought. In the case of tomato, another host of *H. zea*, GOX has a profoundly different effect than in tobacco. In this case, application of fungal GOX (and the resulting production of H_2O_2) causes a massive increase in defenses such as proteinase inhibitors (Orozco-Cardenas et al. 2001). We have confirmed these findings by applying purified GOX to mechanically-damaged leaves and showing that it significantly increases transcript levels of proteinase inhibitor-II (*PI-II*) genes as measured by quantitative real time PCR (qRT-PCR) (Peiffer and Felton, unpublished data). Thus we expected that application of saliva (collected from the spinneret) would cause similar increases when applied to mechanically-damaged leaves; even very small quantities of saliva (<1 nl) strongly induce *PI-II* expression (unpublished data). These findings may provide an example of how a virulence factor of the insect (GOX and H_2O_2, respectively) can be exploited by the plant as an elicitor to induce appropriate counter measures (induction of proteinase inhibitors in tomato).

18.6 Dynamic Nature of Secretions

There may be considerable phenotypic variability in the composition of caterpillar secretions which is dependent upon developmental and environmental influences. The developmental stage of the larva can have a profound effect on the composition of the saliva. The synthesis of some salivary gland proteins such as GOX and lysozyme reaches maximal levels during the feeding interval of each molt (Eichenseer et al. 1999; Liu et al. 2004), whereas others such as chitinase reach peak levels after larvae cease feeding and begin to molt (Zheng et al. 2003). The temporal expression of a salivary gland protein may reveal important cues as to whether the protein is secreted during feeding. As larvae reach the non-feeding prepupal stage, the labial glands undergo apoptosis and synthesis of most proteins ceases (Jochová et al. 1997). Furthermore, some cellular proteins may be sloughed into the saliva during this developmental stage.

Not only the stage of larval development, but also the diet of the larva can strongly impact the composition and secretion of saliva. Labial GOX (and total protein of the gland) in *H. zea* was much higher in larvae reared on tobacco, as compared to tomato or cotton (Peiffer and Felton 2005). The use of a GOX-specific antibody allowed direct quantification of GOX secreted onto the leaf surface, and it was found that larvae secreted variable amounts of GOX with a trend toward less secretion on tomato compared to the other two hosts (Peiffer and Felton 2005). Last instar larvae secreted between 0.4 and 0.7 microgram of GOX per hour (Peiffer and Felton 2005). In contrast to GOX, lysozyme mRNA was much lower in tobacco-fed larvae compared to those fed cotton or tomato (Liu et al. 2004). These findings indicate that the amounts of specific proteins in the saliva are regulated differentially, and are not merely a function of the total protein contained in the glands. Labial glands from tobacco fed larvae are highest in protein, but lysozyme levels were lowest (Peiffer and Felton 2005). Whether the differential rates of secretion for different proteins on varying hosts have any adaptive value should be examined.

The activity of labial GOX in the beet armyworm *S. exigua* and the bertha armyworm *Mamestra configurata* was much greater in artificial diet-fed larvae compared to those ingesting the legume *Medicago truncatula* (Merkx-Jacques and Bede 2004, 2005). In contrast, lysozyme expression was barely detectable in artificial diet fed-*H zea* compared to those reared on host plants (Liu et al. 2004). These studies provide an important precautionary note on generalizing the function of saliva from larvae reared on artificial diets.

The host plant has a profound impact on not only the composition and secretion of saliva, but also on the size of the salivary glands. Eichenseer et al. (2002) reported that the host plant of *H. zea* significantly affected the size of mandibular glands, as well as the amount of carotenoids present in their tissues. Mandibular glands in soybean-fed larvae were 2-fold larger in length and/or diameter compared to larvae reared on tomato or cotton, despite the larvae being of identical developmental stage. The increased size of the glands is presumably due to increased cell size, because in *B. mori*, the cell number of both labial and mandibular glands remain constant in number (330 cells), yet the glands increase 1000x-fold in size during development (Parthasarathy and Gopinathan 2005).

Also the composition of OS may differ depending upon the food ingested. The fact that different hosts respond differentially to *M. sexta* feeding, not only indicates that there may be host-specific differences in defense response, but also suggests that the composition of secretions may vary when feeding on different hosts (Schmidt et al. 2005). While different dietary sources of *M. sexta* or *H. virescens* larvae, including both host and nonhost plants, did not affect the amino acid composition of the FACs in OS (Alborn et al. 2003), quantitative differences or differences in other OS components may occur. For example, *Heliothis subflexa* larvae fail to produce volicitin if they feed exclusively on the fruits of *Physalis angulata* which lack linolenic acid, a necessary component of the FAC (De Moraes and Mescher 2004). Only fall armyworm *Spodoptera frugiperda* larvae that ingested chloroplastic ATP synthase γ-subunit proteins and produced the peptide fragments termed 'inceptins' significantly induced defenses in cowpea (Schmelz et al. 2006, 2007).

18.7 Saliva and Oral Secretions Should be Examined Concurrently

Oral secretions alone may not fully explain the specificity of caterpillar-induced responses. For instance, *S. exigua* feeding on maize inhibited the expression of some volicitin-induced genes, despite the presence of volicitin in their OS (Lawrence and Novak 2004). This and other results suggest that additional components of caterpillar secretions may be involved in these responses. The Korth lab has examined the impact of herbivory, OS, and saliva on genes controlling terpene production in the model system of *Medicago truncatula*. They found that feeding by *S. exigua*, OS, or FACs (N-linolenoyl-glutamate or N-linoleoyl-glutamate) increased transcript levels of a putative terpene synthase (MtTPS; Gomez et al. 2005). In a separate paper, their lab examined the effect of saliva on the terpene biosynthesis pathway, and they found that transcripts coding for early pathway enzymes were lower in tissues that suffered herbivory (Bede et al. 2006). Larvae were ablated as previously described (Musser et al. 2002), and the authors concluded that a salivary factor, possibly GOX, was involved in reducing the transcript levels (Bede et al. 2006). These two studies illustrate, that in addition to wounding, both insect OS and saliva are important in ultimately determining the plant's defensive response.

However, in some caterpillar/host plant systems, saliva and/or GOX may not be equally important. A microarray of 241-genes from *Nicotiana attenuata* differentially expressed in response to attack by *M. sexta* was used to compare the transcriptional response to FACs from *Manduca* regurgitant and fungal GOX (Halitschke et al. 2003). GOX (and glucose) injection of unwounded leaf tissue up-regulated 37 transcripts and down-regulated 41 transcripts within a 2 h period; however only a handful of these genes were similarly regulated 1 h after treatment of wounded leaves with *Manduca* regurgitant. The authors concluded that GOX-derived H_2O_2 plays only a minor role in elicitation of *Manduca*-specific plant responses (Halitschke et al. 2003). Also salivary gland homogenates (both labial and mandibular) from *M. sexta* failed to amplify trypsin inhibitor expression in its host *N. attenuata* (Roda et al. 2004). It is not surprising that fungal GOX did not produce similar *M. sexta*-specific plant responses, considering that this species has virtually no detectable salivary GOX (Eichenseer and Felton, unpublished data).

Delphia et al. (2006) explored the role of *H. virescens* OS and salivary secretions on volatile induction in the commercial tobacco species, *Nicotiana tabacum*. Using the ablation technique previously described to prevent the release of labial saliva, plants damaged by intact caterpillars were found to release 11 volatile compounds, whereas ablated caterpillars induced the same 11 compounds plus an additional eight. Furthermore, plants damaged by ablated caterpillars generally released greater amounts of volatile compounds, most notably volatile nicotine, as compared to plants damaged by intact caterpillars. Apparently, the application of both saliva and OS induced less volatile nicotine as compared to the treatment with OS alone. These results show that both OS and saliva are necessary to produce the volatile profile of *H. virescens* feeding on tobacco, and that saliva may mask the effect of FACs found in the OS (Delphia et al. 2006). These studies suggest that herbivores may minimize their display of elicitors—either through minimizing the production of elicitors or by masking their effects through secretion of salivary components.

18.8 Saliva and Oral Secretions and Host Range

Induced plant responses to insect herbivores are a widespread phenomenon; the degree to which these responses are specific to a given herbivore has been studied in a variety of examples (Hartley and Lawton 1987; Felton et al. 1994; Karban and Baldwin 1997; Stout et al. 1998; Agrawal 2000; Van Zandt and Agrawal 2004). The question whether herbivore diet breadth (or host specialization) is a factor in specificity has been examined in a few instances. The specificity of induction by two specialist (narrow host range) and two generalist (broad host range) caterpillars was compared in wild radish, where no general patterns in specificity between specialists and generalists were observed (Agrawal 2000). The biochemical or molecular bases of the induced responses were not determined in this study.

A few representative examples where oral secretions and the host range of the herbivore may be important in determining the specificity of induced responses will be presented. One example is ground cherry *Physalis angulata*, where the specialist *Heliothis subflexa* larvae suffers lower rates of parasitism by *Cardiochiles nigriceps* than the extreme generalist *H. virescens* (Oppenheim and Gould 2002). The strength of this experimental system is that it is a natural system, and that the two *Heliothis* species are closely related, sibling species representing both a generalist and a specialist herbivore. *H. subflexa* larvae feed on ground cherry fruits lacking linolenic acid, a key component of volicitin, an elicitor of indirect defenses (De Moraes and Mescher 2004). Volicitin is absent in the oral secretions of fruit-feeding caterpillars and thus, the volatile profiles of plants induced by fruit feeding differ from those induced by leaf feeding (De Moraes and Mescher 2004). In contrast, the generalist *H. virescens* is unable to complete development on the fruit due to the lack of linolenic acid. In this system *H. subflexa* gains a selective advantage by feeding on fruit, avoiding the production of an elicitor, and ultimately avoiding parasitism. Oppenheimer and Gould (2002) indicated that differences in parasitoid attraction to herbivore-induced plant volatiles did not contribute to *H. subflexa*'s comparatively low parasitism rate, although their focus was on induction of volatiles by leaf-feeding. Continued study in this system is warranted.

The availability of microarrays for a growing number of plant species provides powerful tools for examining the specificity of induced responses to herbivores in the context of host breadth. Herbivore host range and plant transcriptional responses have been studied in the wild tobacco species *Nicotiana attenuata* where it was found that the transcriptional responses to feeding by the Solanaceous specialist *Manduca sexta* were distinct from feeding by the generalists *Heliothis virescens* and *Spodoptera exigua* (Voelckel and Baldwin 2004). Differences in plant responses were correlated with the profile of FAC elicitors in the regurgitant (Voelckel and Baldwin 2004); however the authors did not consider the possible contribution of salivary factors such as GOX (Voelckel and Baldwin 2004). We found that GOX is at high levels in the noctuid species (*H. virescens* and *S. exigua*), but virtually undetectable in *M. sexta*, as determined by enzyme activity and reactivity to an antibody (Eichenseer and Felton, unpublished data). Interestingly, the number of down-regulated genes correlates well with the presence of GOX: plants fed on by *S. exigua* and *H. virescens* had more down-regulated genes than plants fed on by *M. sexta* (Voelckel and Baldwin 2004).

The power of using the *N. attenuata-Manduca* system is that it is a natural system that has been extensively studied in terms of induced responses to herbivores (Baldwin et al. 2001; Kessler and Baldwin 2002). However, it is limited in that the microarray comprised only a restricted set of 240 *M. sexta*-responsive genes (Voelckel and Baldwin 2004). *Arabidopsis* offers a distinct advantage because of the availability of a large, unbiased microarray containing in excess of 7200 unique genes. Using this microarray, Reymond et al. (2004) found that there were few differences in the transcript profiles between plants fed on by *Pieris rapae* (narrow host range) and *Spodoptera littoralis* (broader host range). The authors found that *Pieris* OS induced several genes not regulated by jasmonic acid and with further experiments concluded that the FACs are not responsible for the induction of *Pieris*-responsive genes (Reymond et al. 2004); standing in direct contrast to the study cited above (Voelckel and Baldwin 2004). The lack of differences between responses to *Pieris* and *Spodoptera* is surprising, considering the likely differences in both OS and saliva between these two species. However both species appear to lack salivary GOX. We were unable to detect GOX in *Pieris* larvae (Eichenseer and Felton, unpublished data) and it was also undetectable in *S. littoralis* (Zong and Wang 2004). Reymond et al. (2004) also reported that mechanical wounding induces several genes that are not activated by *Pieris* feeding, suggesting that OS or saliva may be responsible for suppression as shown with *H. zea* and tobacco (Musser et al. 2002). Nevertheless the transcript profiles show surprisingly few differences between feeding by *S. littoralis* and *P. rapae*.

These cited studies illustrate the difficulty in generalizing across different host plant-herbivore systems. Even *M. sexta*, a specialist on Solanaceous plants, elicits distinct transcript profiles when feeding on different hosts such as *N. attenuata* or black nightshade *Solanum nigrum* (Schmidt et al. 2005). There are several possible explanations for these disparate findings including that composition of saliva or OS may differ when caterpillars change host plants, thus influencing the induction and/or suppression of defenses. Alternatively, there may be differences in how particular plant species (or even cultivars) recognize herbivore secretions.

18.9 Future Directions and Perspectives

There are several critical needs that will shape the future direction of research on herbivore secretions and induced responses. First, considering that insect herbivore species likely number in excess of a million, the secretions of a precious few insect herbivores have been characterized. Improvements in proteomics and metabolomics allowing the analysis of minute volumes will help uncover the biochemical and functional complexity of caterpillar secretions. A continued array of novel compounds will undoubtedly be identified as more species are examined (e.g., grasshoppers; Tumlinson and Engelberth, this volume).

With a few notable exceptions (Mattiacci et al. 1995; Schmelz et al. 2006, 2007), chemical characterization of Lepidopteran OS has mostly focused on FACs, whereas virtually all of the research on saliva in larval Lepidoptera has focused on proteins. As far as I am aware, very little is known regarding small molecules present in their saliva (Howard and Baker 2004), although some have been reported in the salivary glands: ascorbic acid (Kramer et al. 1981), carotenoids (Eichenseer et al. 2002), and even jasmonic acid (Tooker and De Moraes 2006) have been found at high levels in salivary glands; however it is unknown if these molecules are secreted. Jasmonic acid secretion would be of obvious significance in eliciting plant defenses. The most difficult experimental challenge has been collecting a sufficient volume of saliva necessary for biochemical analyses, and for conducting pharmacological-type assays with plants. Nevertheless, the technology is now available to biochemically characterize the small molecules present in saliva using HPLC/MS or GC/MS methods.

Second, very few studies have included a dose-response analysis of OS or saliva when determining effects on induced responses. There have been an inexplicable number of studies that have used an identical quantity of OS for testing, regardless of the relevance of species or stage of insect used. There needs to be greater emphasis on quantifying the amount of specific elicitors released during feeding in order to make further progress.

Third, there is the need for understanding the adaptive value of insect elicitors. If elicitation of induced indirect and direct defenses negatively impacts the fitness of insect herbivores as proposed, then there should be strong selective pressure on herbivores to evade these host defenses. Several scenarios are possible including minimizing the display of elicitors—achieved by producing less of the elicitor, or by masking the effect of the elicitor with other oral or salivary components (Delphia et al. 2006). However, there may be physiological constraints on minimizing the production of elicitors if they perform vital physiological functions. In the case of FACs their function(s) are unknown, but they may act as gut surfactants (Spiteller and Boland 2003b) and/or facilitate nutrient uptake. Elucidation of the genes regulating their biosynthesis will facilitate the application of molecular approaches such as RNA interference to examine their function in vivo. Another strategy for evading host defenses is to elicit a competing pathway—one that regulates negative cross talk with the jasmonate pathway (e.g., salicylate). The latter has been observed with several pathovars of the bacterial pathogen *Psuedomonas syringae* that secrete a jasmonate mimic, coronatine, which activates the jasmonate pathway at the expense of salicylate-dependent systemic resistance (Nomura et al. 2005; Uppalapati et al. 2005; Zhao et al. 2003). Whether herbivores employ such a strategy is unknown, but it has been suggested by some data (Musser et al. 2002).

Fourth, there is a need to understand the mode-of-action and target sites of elicitors. Very little is understood at the molecular level about how plants recognize elicitors such as FACs and then activate transcriptional responses. Evidence for an elicitor-protein interaction was presented from work on maize laves where it was shown that a plasma membrane fractions bind volicitin (Truitt et al. 2004). Treatment of leaves with methyl jasmonate also enhanced the binding of volicitin to the plasma membrane fraction. Further study is needed to determine if this plasma membrane protein(s) functions as a membrane transporter and/or as a receptor to activate secondary messenger cascades.

Finally, an additional glandular structure may be important in secretion during caterpillar feeding. Noctuids are ostensibly the largest family of Lepidoptera; one of their defining larval characters is the ventral eversible gland (=VEG), a secretory structure that is found on the ventral surface of the larval thorax (Fig. 18.2;

Fig. 18.2 Light micrograph of lateral view of head of larval *Helicoverpa zea*. VEG = ventral eversible gland, in distended view. (Micrograph courtesy of Michelle Peiffer)

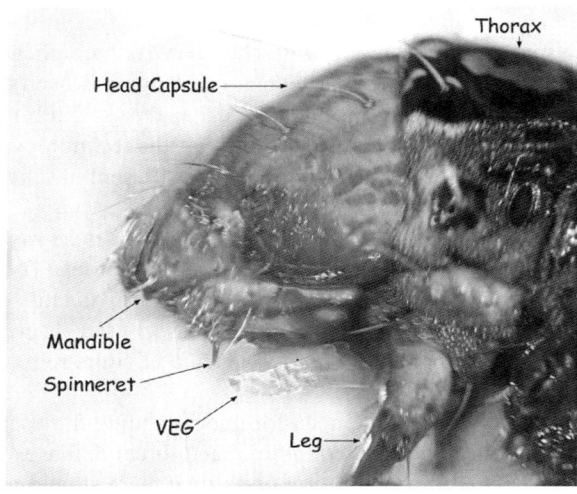

Speidel et al. 1996). The function of this gland is mostly unknown, but it is known to secrete various lipids (Severson et al. 1991) and proteins (unpublished data). Considering its close proximity to the mandibles and buccal cavity when distended, its secretions would presumably be deposited onto the food substrate. Additionally, its secretions could be expected in regurgitant or OS when it is collected. We are in the initial stages of analyzing the VEG components, and testing their effect on induced responses. Our preliminary findings indicate that these secretions will add yet another layer of complexity to the dynamic interactions between caterpillars and their host plants.

Acknowledgments The support of research by the National Science Foundation and the United States Department of Agriculture NRI programs is gratefully acknowledged. Unpublished data from Michelle Peiffer and Herb Eichenseer is appreciated.

References

Aboshi T, Yoshinaga N, Noge K, Nishida R, Mori N (2007) Efficient incorporation of unsaturated fatty acids into volicitin-related compounds in *Spodoptera litura* (Lepidoptera: Noctuidae). Biosci Biotech Biochem 71:607–610

Agrawal AA (2000) Specificity of induced resistance in wild radish: causes and consequences for two specialist and two generalist caterpillars. Oikos 89:493–500

Alborn HT, Brennan MM, Tumlinson JH (2003) Differential activity and degradation of plant volatile elicitors in regurgitant of tobacco hornworm (*Manduca sexta*) larvae. J Chem Ecol 29:1357–1372

Alborn HT, Turlings TCJ, Jones TH, Stenhagen G, Loughrin JH, Tumlinson JH (1997) An elicitor of plant volatiles from beet armyworm oral secretion. Science 276:945–949

Aljamali MN, Bior AD, Sauer JR, Essenberg RC (2003) RNA interference in ticks: a study using histamine binding protein dsRNA in the female tick *Amblyomma americanum*. Insect Mol Biol 12:299–305

Arimura G, Kost C, Boland W (2005) Herbivore-induced, indirect plant defences. Biochim Biophys Acta 1734:91–111

Baldwin IT, Halitschke R, Kessler A, Schittko U (2001) Merging molecular and ecological approaches in plant-insect interactions. Curr Opin Plant Biol 4:351–358

Bede J, Musser R, Felton G, Korth K (2006) Caterpillar herbivory and salivary enzymes decrease transcript levels of *Medicago truncatula* genes encoding early enzymes in terpenoid biosynthesis. Plant Mol Biol 60:519–531

Bolter CJ, Dicke M, vanLoon JJA, Visser JH, Posthumus MA (1997) Attraction of Colorado potato beetle to herbivore-damaged plants during herbivory and after its termination. J Chem Ecol 23:1003–1023

Bordas L (1903) Mandibular glands of lepidoptera larvae. C R Hebd Seances Acad Sci 136:1273–1275

Cardoza YJ, Klepzig KD, Raffa KF (2006) Bacteria in oral secretions of an endophytic insect inhibit antagonistic fungi. Ecol Entomol 31:636–645

Carroll MJ, Schmelz EA, Meagher RL, Teal PEA (2006) Attraction of *Spodoptera frugiperda* larvae to volatiles from herbivore-damaged maize seedlings. J Chem Ecol 32:1911–1924

Chauthan AR, Callahan PS (1967) Developmental morphology of alimentary canal of *Heliothis zea* (Lepidoptera – Noctuidae). Ann Entomol Soc Amer 60:1136–1967

De Moraes CM, Mescher MC (2004) Biochemical crypsis in the avoidance of natural enemies by an insect herbivore. Proc Natl Acad Sci USA 101:8993–8997

De Moraes CM, Mescher MC, Tumlinson JH (2001) Caterpillar-induced nocturnal plant volatiles repel conspecific females. Nature 410:577–580

Delphia C, Mescher M, Felton GW, De Moraes CM (2006) The role of insect-derived cues in eliciting indirect plant defenses in tobacco, *Nicotiana tabacum*. Plant Signal Behav 1:243–250

Eichenseer H, Mathews MC, Bi JL, Murphy JB, Felton GW (1999) Salivary glucose oxidase: multifunctional roles for *Helicoverpa zea*? Arch Insect Biochem Physiol 42:99–109

Eichenseer H, Murphy JB, Felton GW (2002) Sequestration of host plant carotenoids in the larval tissues of *Helicoverpa zea*. J Insect Physiol 48:311–318

Felton GW, Eichenseer H (1999) Herbivore saliva and its effects on plant defense against herbivores and pathogens. In: Agrawal AA, Tuzun S, Bent E (eds) Induced plant defenses against pathogens and herbivores. APS Press, St. Paul, pp 19–36

Felton GW, Summers CB, Mueller AJ (1994) Oxidative responses in soybean foliage to herbivory by bean leaf beetle and 3-cornered alfalfa hopper. J Chem Ecol 20:639–650

Fereres A (2007) The role of aphid salivation in the transmission of plant viruses. Phytoparasitica 35:3–7

Fitzgerald TD (1993) Trail and arena marking by caterpillars of *Archips cerasivoranus* (Lepidoptera, Tortricidae). J Chem Ecol 19:1479–1489

Fitzgerald TD (2003) Role of trail pheromone in foraging and processionary behavior of pine processionary caterpillars *Thaumetopoea pityocampa*. J Chem Ecol 29:513–532

Giri AP, Wunsche H, Mitra S, Zavala JA, Muck A, Svatos A, Baldwin IT (2006) Molecular interactions between the specialist herbivore *Manduca sexta* (Lepidoptera, Sphingidae) and its natural host *Nicotiana attenuata*. VII. Changes in the plant's proteome. Plant Physiol 142:1621–1641

Gomez SK, Cox MM, Bede JC, Inoue K, Alborn HT, Tumlinson JH, Korth KL (2005) Lepidopteran herbivory and oral factors induce transcripts encoding novel terpene synthases in *Medicago truncatula*. Arch Insect Biochem Physiol 58:114–127

Gouinguene S, Alborn H, Turlings TCJ (2003) Induction of volatile emissions in maize by different larval instars of *Spodoptera littoralis*. J Chem Ecol 29:145–162

Halitschke R, Gase K, Hui DQ, Schmidt DD, Baldwin IT (2003) Molecular interactions between the specialist herbivore *Manduca sexta* (Lepidoptera, Sphingidae) and its natural host *Nicotiana attenuata*. VI. Microarray analysis reveals that most herbivore-specific transcriptional changes are mediated by fatty acid-amino acid conjugates. Plant Physiol 131:1894–1902

Halitschke R, Schittko U, Pohnert G, Boland W, Baldwin IT (2001) Molecular interactions between the specialist herbivore *Manduca sexta* (Lepidoptera, Sphingidae) and its natural host *Nicotiana attenuata*. III. Fatty acid-amino acid conjugates in herbivore oral secretions are necessary and sufficient for herbivore-specific plant responses. Plant Physiol 125:711–717

Han YS, Chun JS, Schwartz A, Nelson S, Paskewitz SM (1999) Induction of mosquito hemolymph proteins in response to immune challenge and wounding. Develop Comp Immunol 23:553–562

Hartley SE, Lawton JH (1987) Effects of different types of damage on the chemistry of birch foliage, and the responses of birch feeding insects. Oecologia 74:432–437

Howard RW, Baker JE (2004) Stage-specific surface chemicals of *Plodia interpunctella*: 2-acyl-1,3-cyclohexanediones from larval mandibular glands serve as cuticular lipids. Comp Biochem Physiol B: Biochem Mol Biol 138:193–206

Jarosz J (1993) Induction kinetics of immune antibacterial proteins in pupae of *Galleria mellonella* and *Pieris brassicae*. Comp Biochem Physiol B Biochem Mol Biol 106:415–421

Jochová J, Quaglin D, Zakeri Z, Woo K, Sikorska M, Weaver M, Lockshin RA (1997) Protein synthesis, DNA degradation, and morphological changes during programmed cell death in labial glands of *Manduca sexta*. Dev Genet 21:249–257

Kahl J, Siemens DH, Aerts RJ, Gabler R, Kuhnemann F, Preston CA, Baldwin IT (2000) Herbivore-induced ethylene suppresses a direct defense but not a putative indirect defense against an adapted herbivore. Planta 210:336–342

Karban R, Baldwin IT (1997) Induced responses to herbivory. The University of Chicago Press

Karim S, Ramakrishnan VG, Tucker JS, Essenberg RC, Sauer JR (2004) *Amblyomma americanum* salivary glands: double-stranded RNA-mediated gene silencing of synaptobrevin homologue and inhibition of PGE2 stimulated protein secretion. Insect Biochem Mol Biol 34:407–413

Kessler A, Baldwin IT (2001) Defensive function of herbivore-induced plant volatile emissions in nature. Science 291:2141–2144

Kessler A, Baldwin IT (2002) Plant responses to insect herbivory: the emerging molecular analysis. Ann Rev Plant Biol 53:299–328

Korth KL, Dixon RA (1997) Evidence for chewing insect-specific molecular events distinct from a general wound response in leaves. Plant Physiol 115:1299–1305

Kramer KJ, Spiers RD, Lookhart G, Seib PA, Liang YT (1981) Sequestration of ascorbic acid by the larval labial gland and hemolymph of the tobacco hornworm, *Manduca sexta* L (Lepidoptera, Sphingidae). Insect Biochem 11:93–96

Lait CG, Alborn HT, Teal PEA, Tumlinson JH (2003) Rapid biosynthesis of N-linolenoyl-L-glutamine, an elicitor of plant volatiles, by membrane-associated enzyme(s) in *Manduca sexta*. Proc Natl Acad Sci USA 100:7027–7032

Landolt PJ, Tumlinson JH, Alborn DH (1999) Attraction of Colorado potato beetle (Coleoptera: Chrysomelidae) to damaged and chemically induced potato plants. Environ Entomol 28: 973–978

Lawrence SD, Novak NG (2004) Maize genes induced by herbivory and volicitin. J Chem Ecol 30:2543–2557

Lev E (2003) Traditional healing with animals (zootherapy): medieval to present-day Levantine practice. J Ethnopharmacol 85 107–118

Liu F, Cui L, Cox-Foster D, Felton GW (2004) Characterization of a salivary lysozyme in larval *Helicoverpa zea*. J Chem Ecol 30:2439–2457

Macgown MW, Sikorowski PP (1982) Anatomy of the digestive system of *Heliothis zea* larvae. Mississippi Agricultural and Forestry Experiment Station Bulletin 11:1–15

Machida Y (1965) Studies on the silk glands in the silkworm, *Bombyx mori* L. Morphological and functional studies of Filippi's glands in the silkworm. Science Bulletin of Faculty of Journal of Chemical Ecology Agriculture, Kyushu University 22:95–108

Major IT, Constabel CP (2006) Molecular analysis of poplar defense against herbivory: comparison of wound- and insect elicitor-induced gene expression. New Phytol 172:617–635

Mattiacci L, Dicke M, Posthumus MA (1995) Beta-Glucosidase: an elicitor of herbivore-induced plant odor that attracts host-searching parasitic wasps. Proc Natl Acad Sci USA 92:2036–2040

Merkx-Jacques M, Bede JC (2004) Caterpillar salivary enzymes: 'eliciting' a response. Phytoprotection 85:33–37

Merkx-Jacques M, Bede JC (2005) Influence of diet on the larval beet armyworm *Spodoptera exigua*, glucose oxidase. J Insect Sci 5:48
Miles PW (1999) Aphid saliva. Biological Reviews 74:41–85
Miles PW, Harrewijn P (1991) Discharge by aphids of soluble secretions into dietary sources. Entomol Exp Appl 59:123–134
Mithöfer A, Wanner G, Boland W (2005) Effects of feeding *Spodoptera littoralis* on lima bean leaves. II. Continuous mechanical wounding resembling insect feeding is sufficient to elicit herbivory-related volatile emission. Plant Physiol 137:1160–1168
Mossadegh MS (1978) Mechanism of secretion of the contents of the mandibular glands of *Plodia interpunctella* larvae. Physiol Entomol 3:335–340
Musser RO, Cipollini DF, Hum-Musser SM, Williams SA, Brown JK, Felton GW (2005a) Evidence that the caterpillar salivary enzyme glucose oxidase provides herbivore offense in solanaceous plants. Arch Insect Biochem Physiol 58:128–137
Musser RO, Farmer E, Peiffer M, Felton GW (2006) Ablation of caterpillar labial salivary glands: technique for determining the role of saliva in insect-plant interactions. J Chem Ecol 32:981–992
Musser RO, Hum-Musser SM, Eichenseer H, Peiffer M, Ervin G, Murphy JB, Felton GW (2002) Herbivory: caterpillar saliva beats plant defences – A new weapon emerges in the evolutionary arms race between plants and herbivores. Nature 416:599–600
Musser RO, Kwon HS, Williams SA, White CJ, Romano MA, Holt SM, Bradbury S, Brown JK, Felton GW (2005a) Evidence that caterpillar labial saliva suppresses infectivity of potential bacterial pathogens. Arch Insect Biochem Physiol 58:138–144
Mutti NS, Park Y, Reese JC, Reeck GR (2006) RNAi knockdown of a salivary transcript leading to lethality in the pea aphid, *Acyrthosiphon pisum*. J Insect Sci 6:38
Nomura K, Melotto M, He SY (2005) Suppression of host defense in compatible plant-*Pseudomonas syringae* interactions. Curr Opin Plant Biol 8:361–368
Oppenheim SJ, Gould F (2002) Is attraction fatal? The effects of herbivore-induced plant volatiles on herbivore parasitism. Ecology 83:3416–3425
Orozco-Cardenas ML, Narvaez-Vasquez J, Ryan CA (2001) Hydrogen peroxide acts as a second messenger for the induction of defense genes in tomato plants in response to wounding, systemin, and methyl jasmonate. Plant Cell 13:179–191
Parthasarathy R, Gopinathan KP (2005) Comparative analysis of the development of the mandibular salivary glands and the labial silk glands in the mulberry silkworm, *Bombyx mori*. Gene Exp Patterns 5:323–339
Peiffer M, Felton GW (2005) The host plant as a factor in the synthesis and secretion of salivary glucose oxidase in larval *Helicoverpa zea*. Arch Insect Biochem Physiol 58:106–113
Pohnert G, Jung V, Haukioja E, Lempa K, Boland W (1999a) New fatty acid amides from regurgitant of lepidopteran (Noctuidae, Geometridae) caterpillars. Tetrahedron 55:11275–11280
Pohnert G, Koch T, Boland W (1999b) Synthesis of volicitin: a novel three-component Wittig approach to chiral 17-hydroxylinolenic acid. Chem Commun 12:1087–1088
Reymond P, Bodenhausen N, Van-Poecke RMP, Krishnamurthy V, Dicke M, Farmer EE (2004) A conserved transcript pattern in response to a specialist and a generalist herbivore. Plant Cell 16:3132–3147
Ribeiro JMC, Andersen J, Silva-Neto MAC, Pham VM, Garfield MK, Valenzuela JG (2004) Exploring the sialome of the blood-sucking bug *Rhodnius prolixus*. Insect Biochem Mol Biol 34:61–79
Roda A, Halitschke R, Steppuhn A, Baldwin IT (2004) Individual variability in herbivore-specific elicitors from the plant's perspective. Mol Ecol 13:2421–2433
Rose USR, Tumlinson JH (2005) Systemic induction of volatile release in cotton: how specific is the signal to herbivory? Planta 222:327–335
Royer C, Jalabert A, Da Rocha M, Grenier AM, Mauchamp B, Couble P, Chavancy G (2005) Biosynthesis and cocoon-export of a recombinant globular protein in transgenic silkworms. Transgenic Res 14:463–472

Schmelz EA, Carroll MJ, LeClere S, Phipps SM, Meredith J, Chourey PS, Alborn HT, Teal PEA (2006) Fragments of ATP synthase mediate plant perception of insect attack. Proc Natl Acad Sci USA 103:8894–8899

Schmelz EA, LeClere S, Carroll MJ, Alborn HT, Teal PEA (2007) Cowpea chloroplastic ATP synthase is the source of multiple plant defense elicitors during insect herbivory. Plant Physiol 144:793–805

Schmidt DD, Voelckel C, Hartl M, Schmidt S, Baldwin IT (2005) Specificity in ecological interactions. attack from the same Lepidopteran herbivore results in species-specific transcriptional responses in two Solanaceous host plants. Plant Physiol 138:1763–1773

Severson RF, Rogers CE, Marti OG, Gueldner RC, Arrendale RF (1991) Ventral eversible gland volatiles from the larvae of the fall armyworm, *Spodoptera frugiperda* (J.E. Smith) (Lepidoptera: Noctuidae). Agric Biol Chem 55:2527–2530

Shackel KA, de la Paz Celorio-Mancera M, Ahmadi H, Greve LC, Teuber LR, Backus EA, Labavitch JM (2005) Micro-injection of *Lygus* salivary gland proteins to simulate feeding damage in alfalfa and cotton flowers. Arch Insect Biochem Physiol 58:69–83

Speidel W, Fanger H, Naumann CM (1996) The phylogeny of the Noctuidae (Lepidoptera). Syst Entomol 21:219–251

Spiteller D, Boland W (2003a) N-(15,16-Dihydroxylinoleoyl)-glutamine and N-(15,16-epoxylinoleoyl)-glutamine isolated from oral secretions of lepidopteran larvae. Tetrahedron 59:135–139

Spiteller D, Boland W (2003b) N-(17-acyloxy-acyl)-glutamines: novel surfactants from oral secretions of lepidopteran larvae. J Org Chem 68:8743–8749

Spiteller D, Dettner K, Boland W (2000) Gut bacteria may be involved in interactions between plants, herbivores and their predators: microbial biosynthesis of N-acylglutamine surfactants as elicitors of plant volatiles. Biol Chem 381:755–762

Standlea PP, Yonke TR (1968) Clarification of description of digestive system of *Heliothis zea*. Ann Entomol Society Amer 51:1478–1481

Stout MJ, Workman KV, Bostock RM, Duffey SS (1998) Specificity of induced resistance in the tomato, *Lycopersicon esculentum*. Oecologia 113:74–81

Tabak LA, Kuska R (2004) Mouth to mouth. Nat His 113:33–37

Tooker JF, De Moraes CM (2006) Jasmonate in lepidopteran larvae. J Chem Ecol 32:2321–2326

Truitt CL, Pare PW (2004) In situ translocation of volicitin by beet armyworm larvae to maize and systemic immobility of the herbivore elicitor *in planta*. Planta 218:999–1007

Truitt CL, Wei HX, Pare PW (2004) A plasma membrane protein from *Zea mays* binds with the herbivore elicitor volicitin. Plant Cell 16:523–532

Tumlinson JH, Lait CG (2005) Biosynthesis of fatty acid amide elicitors of plant volatiles by insect herbivores. Arch Insect Biochem Physiol 58:54–68

Turlings TCJ, McCall PJ, Alborn HT, Tumlinson JH (1993) An elicitor in caterpillar oral secretions that induces corn seedlings to emit chemical signals attractive to parasitic wasps. J Chem Ecol 19:411–425

Turlings TCJ, Tumlinson JH, Lewis WJ (1990) Exploitation of herbivore-induced plant odors by host-seeking parasitic wasps. Science 250:1251–1253

Uppalapati SR, Ayoubi P, Weng H, Palmer DA, Mitchell RE, Jones W, Bender CL (2005) The phytotoxin coronatine and methyl jasmonate impact multiple phytohormone pathways in tomato. Plant J 42:201–217

Valenzuela JG (2002) High-throughput approaches to study salivary proteins and genes from vectors of disease. Insect Biochem Mol Biol 32:1199–1209

Valenzuela JG, Francischetti IM, Pham VM, Garfield MK, Mather TN, Ribeiro JM (2002) Exploring the sialome of the tick *Ixodes scapularis*. J Exp Biol 205:2843–2864

Van Zandt PA, Agrawal AA (2004) Specificity of induced plant responses to specialist herbivores of the common milkweed *Asclepias syriaca*. Oikos 104:401–409

Vegliante F (2005) Larval head anatomy of *Heterogynis penella* (Zygaenoidea, Heterogynidae), and a general discussion of caterpillar head structure (Insecta, Lepidoptera). Acta Zool 86:167–194

Voelckel C, Baldwin IT (2004) Generalist and specialist lepidopteran larvae elicit different transcriptional responses in *Nicotiana attenuata*, which correlate with larval FAC profiles. Ecol Lett 7:770–775

von Dahl CC, Havecker M, Schlogl R, Baldwin IT (2006) Caterpillar-elicited methanol emission: a new signal in plant-herbivore interactions? Plant J 46:948–960

Williams L, Rodriguez-Saona C, Pare PW, Crafts-Brandner SJ (2005) The piercing-sucking herbivores *Lygus hesperus* and *Nezara viridula* induce volatile emissions in plants. Arch Insect Biochem Physiol 58:84–96

Winz RA, Baldwin IT (2001) Molecular interactions between the specialist herbivore *Manduca sexta* (Lepidoptera, Sphingidae) and its natural host *Nicotiana attenuata*. IV. Insect-induced ethylene reduces jasmonate-induced nicotine accumulation by regulating putrescine N-methyltransferase transcripts. Plant Physiol 125:2189–2202

Wroniszewska A (1966) Mandibular glands of the wax moth larva, *Galleria mellonella* (L.). J Insect Physiol 12:509–514

Wu J, Hettenhausen C, Meldau S, Baldwin IT (2007) Herbivory rapidly activates MAPK signaling in attacked and unattacked leaf regions but not between leaves of *Nicotiana attenuata*. Plant Cell 19:1096–1122

Yan ZG, Wang CZ (2006) Similar attractiveness of maize volatiles induced by *Helicoverpa armigera* and *Pseudaletia separata* to the generalist parasitoid *Campoletis chlorideae*. Entomol Exp Appl 118:87–96

Yan ZG, Yan YH, Wang CZ (2005) Attractiveness of tobacco volatiles induced by *Helicoverpa armigera* and *Helicoverpa assulta* to *Campoletis chlorideae*. Chinese Sci Bull 50:1334–1341

Yokoyama T (1963) Sericulture. Annual Review Entomol 8:287–302

Yoshinaga N, Kato K, Kageyama C, Fujisaki K, Nishida R, Mori N (2006) Ultraweak photon emission from herbivory-injured maize plants. Naturwissenschaften 93:38–41

Yoshinaga N, Morigaki N, Matsuda F, Nishida R, Mori N (2005) In vitro biosynthesis of volicitin in *Spodoptera litura*. Insect Biochem Mol Biol 35:175–184

Zheng Y-P, Retnakaran A, Krell P-J, Arif B-M, Primavera M, Feng Q-L (2003) Temporal, spatial and induced expression of chitinase in the spruce budworm, *Choristoneura fumiferana*. J Insect Physiol 49:241–249

Zhao Y, Thilmony R, Bender C L, Schaller A, He SY, Howe GA (2003) Virulence systems of *Pseudomonas syringae* pv. tomato promote bacterial speck disease in tomato by targeting the jasmonate signaling pathway. Plant J 36:485–99

Zong N, Wang CZ (2004) Induction of nicotine in tobacco by herbivory and its relation to glucose oxidase activity in the labial gland of three noctuid caterpillars. Chinese Sci Bull 49:1596–1601

Chapter 19
Fatty Acid-Derived Signals that Induce or Regulate Plant Defenses Against Herbivory

James H. Tumlinson and Juergen Engelberth

Jasmonic acid and other derivatives of linolenic acid produced in plants by the octadecanoid pathway, as well as the six-carbon fatty acid derivatives called green leaf compounds, play major roles in regulating plant defenses against herbivores. So do also conjugates of linolenic acid with glutamine and glutamate, found in the regurgitant of several lepidopteran larvae, as well as in crickets and *Drosophila* larvae. In all these cases, the linolenic acid precursor of the regulating compound is produced by the plant on which the herbivore feeds. More recently elicitors of plant volatiles containing a 16-carbon fatty acid moiety with either a saturated chain or a double bond at carbon six in the chain have been isolated and identified from the spit of grasshoppers. The origin of the grasshopper elicitors is not yet known. Also not known is how broadly active the herbivore-produced elicitors are across a range of plant species. Further, the mechanisms by which the green leaf compounds and the elicitors induce defensive reactions in plants are not understood, although it appears that they affect the octadecanoid pathway in different ways.

19.1 Introduction

When a caterpillar chews on a leaf, a series of reactions are triggered that enable the plant to defend against herbivory and limit damage. These biochemical defensive mechanisms of plants are highly sophisticated. They may act directly against the herbivore by deterring feeding or inhibiting digestion. Alternatively, indirect defenses may be induced that result in the release of complex blends of volatile organic compounds (VOCs) for the recruitment of natural enemies, parasites and predators, of the herbivores. The specificity of plant responses to herbivory is amazing. Mechanical damage alone, for example, induces plant defensive reactions which in most cases are distinctly different from herbivory-induced defenses which involve the deposition of insect-derived substances on the damaged plant

---

J.H. Tumlinson
Department of Entomology, Pennsylvania State University, University Park, PA 16802, USA
e-mail: jht2@psu.edu

tissues (Arimura et al. 2004; Degenhardt and Lincoln 2006; De Moraes et al. 2001; Dugravot et al. 2005; Litvak and Monson 1998; Paré and Tumlinson 1997a; Röse and Tumlinson 2005; Rodrigues-Saona et al. 2002; Schmelz et al. 2003; Turlings et al. 1998, 2000; Wei et al. 2006; Williams et al. 2005). The herbivore-induced blends of VOCs released by plants during the light period differ dramatically from the similarly induced nocturnal blends (Loughrin et al. 1994). While the diurnal blends attract natural enemies, the nocturnal blends may deter oviposition by insect herbivores (De Moraes et al. 2001). Attack by different species of herbivores on the same species/variety of plant induces release of distinctly different blends of VOCs, and natural enemies can distinguish between plants attacked by a host and those attacked by non-hosts on the basis of the blend composition (De Moraes et al. 1998). Further, plants respond systemically to herbivore attack, with undamaged leaves releasing VOC blends that may differ in composition from the blends released by herbivore-damaged leaves (Röse et al. 1996).

Attack and counterattack are orchestrated by numerous chemical signals that regulate the battle between plants and herbivores. Insect herbivores produce chemicals in their saliva and/or regurgitant that trigger plant defensive reactions. The herbivores may also produce substances that counteract or alter plant defensive reactions (Felton this volume). Plants under attack by herbivores synthesize (Paré and Tumlinson 1997) and release VOC blends which may be perceived by and attract natural enemies of the herbivores. These blends, or components thereof, may also be perceived by neighboring plants and act as alarm signals that induce the receiving plants to prime their defenses against impending attack (Engelberth et al. 2004; Kessler et al. 2006; Ton et al. 2007). Internal signals also play a role in initiating and regulating plant defenses. In particular, the octadecanoid pathway is thought to play a major role in responding to herbivory and initiating and regulating biochemical defensive mechanisms.

In each of these cases, fatty acid derivatives or metabolites function as key signals, or components of the signals, that mediate the complex biochemical reactions and interactions. The importance of fatty acids in these processes may simply be a result of their ubiquity in nature and their roles as sources of energy and precursors for biosynthesis of many important compounds in plants and animals. Their ability to be modified by conjugation with other molecules like amino acids, or by oxidation or reduction to change their solubility, reactivity, and other characteristics make them excellent substrates. As we illustrate in the following examples, they are components of both internal and external signals that regulate plant defensive reactions against attack by insect herbivores.

19.2 Insect Herbivore-Produced Elicitors of Plant Defenses

19.2.1 Fatty Acid–Amino Acid Conjugates

The first non-protein elicitor of plant volatiles discovered in insect herbivores was isolated and identified from the regurgitant of beet armyworm, *Spodoptera exigua*, caterpillars. The isolation and identification of this compound, *N*-(17-hydroxylinolenoyl)-L-glutamine, named volicitin (Fig. 19.1), was based on its ability

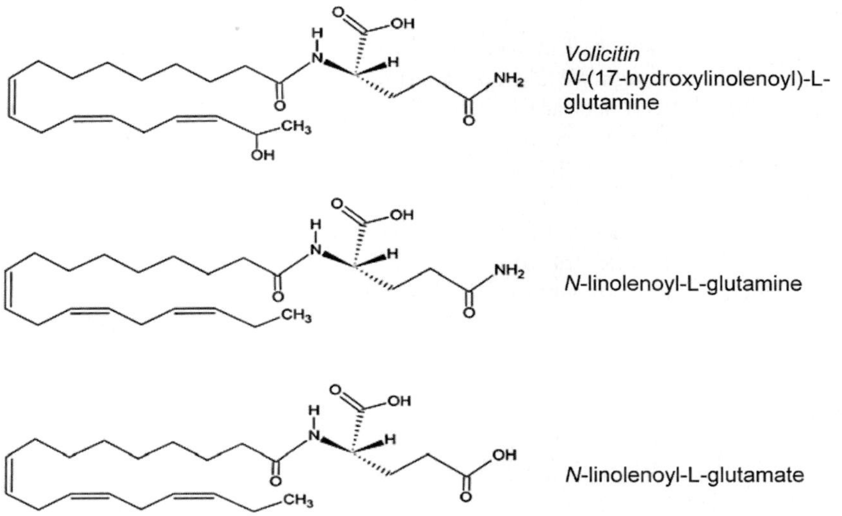

Fig. 19.1 Structures of linolenic acid–amino acid conjugates that have been found to be active as elicitors of VOC release in corn seedlings. These compounds and their linoleic acid analogs have been found in the regurgitant of the larvae of numerous lepidopteran species, and more recently in crickets and *Drosophila* larvae. The configuration at carbon 17 is reported to be (*S*) in volicitin isolated from insects, but both (17*R*)- and (17*S*)-volicitin, as well as the racemic compound, are equally active when applied to corn seedlings

to induce corn seedlings to release volatiles (Alborn et al. 1997, 2000; Turlings et al. 2000). Subsequently, other fatty acid–amino acid conjugates (FACs) have been discovered in the oral secretion or regurgitant of several caterpillar species (Paré et al. 1998; Pohnert et al. 1999; Alborn et al. 2000, 2003; Halitschke et al. 2001; Mori et al. 2001, 2003). Common features of the insect-derived FACs discovered thus far are the presence of either L-glutamine or L-glutamic acid linked via an amide bond to linolenic acid, 17-hydroxylinolenic acid, or the corresponding linoleic acid derivatives. The most potent elicitors in this group of FACs, based on bioassays conducted with corn seedlings and wild tobacco, are the linolenic acid analogs (Fig. 19.1).

The three compounds shown in Fig. 19.1 were all demonstrated to be elicitors of plant volatiles, albeit with varying activity (Alborn et al. 1997, 2000, 2003; Halitschke et al. 2001). On corn seedlings, we found the conjugates lacking the hydroxyl on carbon 17 of linolenic acid to be less than half as active as volicitin (Alborn et al. 2000). Consistently, Sawada et al. (2006) reported that *N*-linolenoyl-L-glutamine has only about 30% of the activity of volicitin, supporting the importance of the C_{17}-hydroxyl for bioactivity. There is little information available as to the specific activity of these compounds or analogs on other plants, but in some species they have been reported to have little if any activity (Schmelz et al. 2006; Spiteller et al. 2001). The most potent elicitor on corn seedlings, volicitin, contains two asymmetric carbons. Interestingly, synthetic volicitin made with D-glutamine did not induce volatile emission when applied to plants (Alborn et al. 2000). In contrast, the configuration of carbon 17 in the linolenic acid chain, to which a hydroxyl group is attached in volicitin, does not seem critical for activity. In the original work, although we did not determine the configuration at carbon 17 of natural volicitin from beet armyworm caterpillars, we found that synthetic volicitin that is racemic at this position appears equal in potency to the natural elicitor as an inducer of corn seedling volatile release (Tumlinson et al. 2000). While the absolute configuration of volicitin from several species of lepidopteran larvae has recently been determined to be (S) (Spiteller et al. 2001; Sawada et al. 2006), this does not seem to be relevant for bioactivity. Sawada et al. (2006) found no significant difference in the amount of volatiles released by corn seedlings induced with $(17S)$- or $(17R)$-volicitin. It is thus clear that in corn seedlings the configuration of the asymmetric carbon in glutamine is more critical than that of the hydroxylated C_{17} of linolenic acid in inducing biosynthesis and release of volatile organic compounds.

In most cases described thus far, plants damaged by insect herbivores or treated with caterpillar regurgitant release greater quantities and/or different proportions of VOCs than plants receiving only mechanical damage (Arimura et al. 2004; Degenhardt and Lincoln 2006; De Moraes et al. 2001; Dugravot et al. 2005; Litvak and Monson 1998; Paré and Tumlinson 1997; Röse and Tumlinson 2005; Rodrigues-Saona et al. 2002; Schmelz et al. 2003; Turlings et al. 1998, 2000; Wei et al. 2006; Williams et al. 2005). For some species however, like lima bean, extensive mechanical damage has been reported to result in release of substantial quantities of VOCs (Mithöfer et al. 2005), and this plant seems unresponsive to insect-derived elicitors (Spiteller et al. 2001). Nevertheless, the fact that specialist parasitoids can distinguish volatiles induced by host larvae from those induced by non-host larvae, even when both species of larvae are feeding on the same species and variety of plant in the same location (De Moraes et al. 1998), shows that insect herbivore-produced substances must play a critical role in induction of plant defenses and herbivore-specific volatile blends.

While FACs appear to be found in most Lepidoptera studied thus far, it is not yet known how widespread these elicitors are among other orders of insects. Very recently Yoshinaga et al. (2007) discovered N-linolenoyl- and N-linoleoyl-glutamates, as well as hydroxylated FACs and glutamine conjugates in the regurgitant of two closely related cricket species (Orthoptera: Gryllidae) and in *Drosophila* larvae. Although Halitschke et al. (2001) reported that the FACs found in the oral secretions of *Manduca sexta* were 'necessary and sufficient for herbivore specific plant responses', FAC elicitors are clearly not the only compounds produced by chewing caterpillars with the potential to induce defensive reactions by plants or otherwise affect their biochemistry. Schmelz et al. (2006) demonstrated induced volatile release in cowpea seedlings fed on by fall armyworm caterpillars. While volicitin and N-linolenoyl-L-glutamine were found in the regurgitant of these caterpillars, they failed to stimulate any significant response. This finding led to the discovery of inceptin, a new peptide elicitor of cowpea volatiles from fall armyworm regurgitant (Schmelz et al. 2006). When a caterpillar feeds on a plant, FACs, enzymes, and other caterpillar-derived, as well as plant-derived compounds that may not yet have been identified come into play, and have a concerted effect on the plant's response. These factors may vary considerably among the different Lepidoptera species and this variability undoubtedly accounts for at least some of the diversity in plant responses, including observed differences in blends of emitted volatiles. Recent evidence indicates that glucose oxidase (GOX) occurs at higher concentration in the saliva of *Helicoverpa zea* than in *Heliothis virescens*, and this may affect the FAC-induced biosynthesis of volatile organic compounds in plants (De Moraes and Felton personal communication, this volume). Oxidation of glucose in plants by *H. zea* salivary GOX leads to an increase in gluconic acid and also H_2O_2 (Eichenseer et al. 1999), a compound that triggers increases in free SA and ethylene (Chamnongpol et al. 1998). Ethylene has been shown to alter induced emission of volatiles (Schmelz et al. 2003). Therefore, high levels of GOX activity in *H. zea* saliva could partially explain how its feeding stimulates plants to produce blends of volatiles that differ from those induced by *H. virescens* feeding (Mori et al. 2001). Thus, although the FAC elicitors are key factors, ultimately all the mediators of plant-insect interactions must be elucidated and their concerted effect on plant biochemistry and gene expression determined before we can fully understand these interactions.

Considering the potency of some FACs as elicitors of volatile emission in plants, thus attracting the caterpillar's natural enemies, the unanswered question arises as to why caterpillars make FACs at all. It has been postulated that FACs might simply act as surfactants or emulsifiers in the caterpillar gut (Spiteller et al. 2000; Halitschke et al. 2001). However, it is unclear why caterpillars would specifically synthesize *N*-linolenoyl-L-glutamine when other amino acids (Yoshinaga et al. 2003) or fatty acids could potentially form suitable FAC surfactants without eliciting a defensive response in plants. Manifestly, the glutamine/glutamate and linolenic acid-based FACs must play a critical role in caterpillar metabolism or physiology, or else they would likely have been eliminated through evolution. It is interesting to note that dietary linolenic acid and glutamine are both essential for the survival of most lepidopterous larvae (Stanley-Samuelson 1994; Kutlesa and Caveney 2001). The larvae can produce eicosanoid fatty acids from exogenous 18-carbon polyunsaturated fatty acids (PUFAs) such as linolenic acid (Stanley-Samuelson et al. 1987). These 20-carbon PUFAs are crucial structural components of cellular and sub-cellular membranes and eicosanoid metabolites are biologically active signaling and regulatory molecules (Stanley-Samuelson 1994, and references therein).

Recently, De Moraes and Mescher (2004) showed interesting differences in the FAC composition of oral secretions of *Heliothis subflexa* caterpillars fed on the fruit of their host plant, *Physalis angulata*, and conspecific larvae fed on leaves of the same plant species. The fruit lacks detectable linolenic acid, and larvae fed on fruit diet do not produce detectable amounts of volicitin or *N*-linolenoyl-L-glutamine, while those fed on leaves of the same plant, which do contain linolenic acid, produce both of these FAC elicitors. Thus, the composition of the FAC mixture is strongly influenced by the caterpillar diet, which is consistent with previous research showing that the fatty acid moiety of FAC elicitors is derived from the diet upon which caterpillars feed (Paré et al. 1998). The origin of the fatty acid and amino acid moieties of FAC elicitors was determined by feeding *S. exigua* caterpillars on corn seedlings uniformly labeled with ^{13}C by growing in a $^{13}CO_2$ atmosphere (Paré et al. 1998). Volicitin in oral secretions from caterpillars that had fed on the labeled seedlings was found to be synthesized by the insect by adding a hydroxyl group and glutamine to linolenic acid obtained directly from the plant on which it fed (Paré et al. 1998). While the fatty acid precursor of volicitin is plant-derived, the bioactive product is formed specifically only in the caterpillar. This strongly suggests that these molecules play an important, yet still unknown role either in the metabolism or some other process critical to the life of herbivorous insects.

It has been suggested that one of several species of bacteria present in the gut of caterpillars may be responsible for synthesizing the FAC's. It was in fact shown that such microorganisms are capable of conjugating 12-phenyldodecanoic acid with glutamine in vitro, and they may therefore play a role in the synthesis of fatty acid amides in general (Spiteller et al. 2000). However, the rate of biosynthesis was extremely slow, yielding only a small amount of *N*-12-phenyldodecanoylglutamine within the first hours, and requiring days for quantities to increase significantly (Spiteller et al. 2000). It was also reported that *N*-linolenoyl-L-glutamine was synthesized by a few of the bacteria isolated from *S. exigua* regurgitant, although no data were provided as to the rate of biosynthesis or quantities produced. However, the slow rate (>24 hour) of bacterial volicitin biosynthesis (Spiteller et al. 2000) cannot account for the rapid biosynthesis required for elicitor accumulation. The rate of biosynthesis would need to be high enough to overcome the simultaneous, and relatively rapid (<8 hour), enzymatic FAC hydrolysis inherent to the midgut lumen and regurgitant of caterpillars (Mori et al. 2001). All attempts in our laboratory to synthesize *N*-linolenoyl-L-glutamine using bacterial cultures derived from *M. sexta* caterpillar regurgitant, gut isolates, and tissue homogenates yielded no evidence to support a role for bacteria in FAC biosynthesis (Lait et al. 2003).

In the Lepidoptera species we have studied thus far, it is clear that FACs are produced by membrane embedded enzymes found in the crop and anterior midgut tissues of the larvae (Lait et al. 2003; Tumlinson and Lait 2005). Since the crop is essentially an extension of the oral cavity, it seems likely that the elicitors produced in the crop are recirculated in some way during the feeding process to contact the leaf tissues that are damaged by the feeding caterpillar. In fact, in a recent study, in which *S. exigua* larvae were allowed to feed on ^{14}C-labeled corn seedlings and then transferred to unlabeled seedlings, approximately 100 pmol of volicitin per caterpillar was extracted from damaged leaves after nine hours of feeding (Truitt and Paré 2004). This study established that FACs are 'applied' to damaged leaf tissues by the feeding caterpillar, but that they apparently do not serve as mobile messengers within the plant to trigger systemic emission of volatile organic compounds.

While there is much work yet to be done to fully understand the role of FACs in plant-insect interactions, it is clear that they play an important, yet undisclosed, role in caterpillar metabolism. Further, they are active in inducing volatile biosynthesis in some, but not all plants. It remains to be seen how widely active they are across the spectrum of plant families and species, and whether the relative activity of the different FACs varies with different species of plants.

19.2.2 The Caeliferins from Grasshoppers

A new class of insect herbivore-produced elicitors of plant volatiles has been isolated and identified from the oral secretions of the grasshopper species *Schistocerca americana* (Alborn et al. 2007). Thus far, they have only been found in the Orthoptera suborder Caelifera, and thus have been named caeliferins. While the basic structure of the caeliferins is also a fatty acid, the structural differences from the FACs are very significant (Fig. 19.2). The caeliferin fatty acid chain may vary

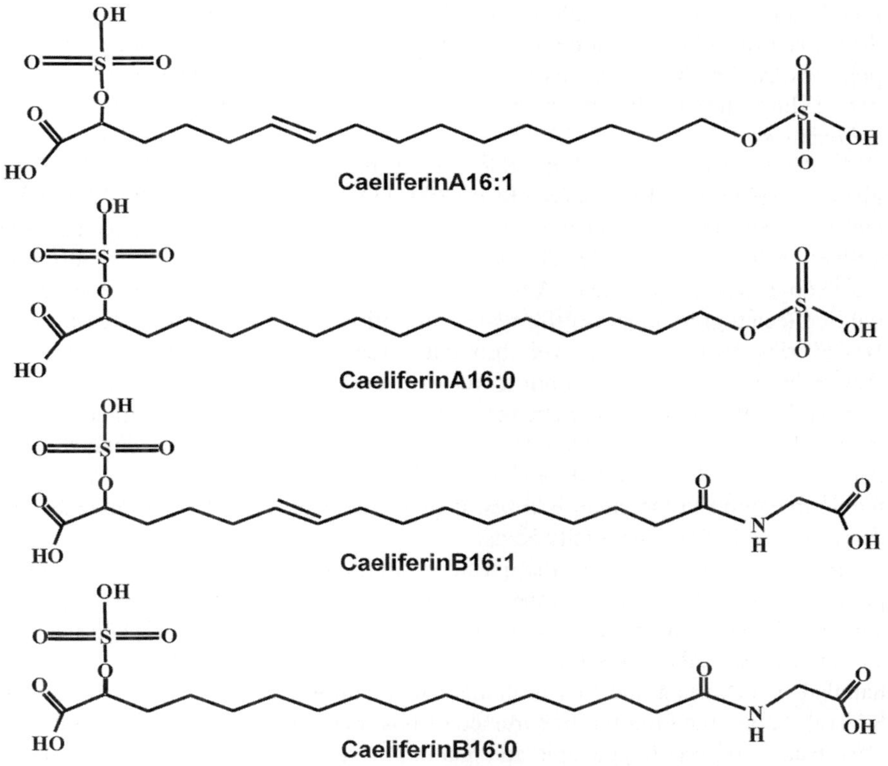

Fig. 19.2 Structures of 16-carbon caeliferins isolated and identified from regurgitant of *Schistocerca americana*. The most abundant component in *S. americana* regurgitant is caeliferin A16:0. The most active elicitor of VOC from corn seedlings is caeliferin A16:1

in length from 15 to 19 carbons and thus far, only saturated and monounsaturated caeliferins have been discovered. Caeliferin A has sulfated hydroxyls on the α- and ω-carbons of the fatty acid chain, while the caeliferin B molecules are α-hydroxylated diacids with the ω-carboxyl conjugated to glycine, similar to the FACs. Further, the oral secretions of the grasshoppers contain a series of compounds of each type with chain lengths varying from 15 to 19 carbons, but in *S. americana* the 16-carbon compounds are most abundant in each class (Fig. 19.2).

The isolation and purification of the active caeliferins was monitored with a corn seedling volatile bioassay (Alborn et al. 2007). Caeliferin A16:1 is most active in inducing VOC emission, while caeliferin A16:0 and caeliferin B16:0 are considerably less active. Other compounds in this family, including caeliferin A17:0 and 18:0, also appeared to have a low level of activity on corn seedlings. The biological activity of caeliferins on other plant species has not been tested yet. Given the polyphagous nature of grasshoppers, and the occurrence of the caeliferins in all grasshopper species investigated so far, it will be interesting to see if these compounds have a broad spectrum of activity across a range of plant species. It is also not known whether volatiles emitted by plants on which the grasshoppers feed attract natural enemies of the grasshoppers, and if so, what the impact of induced volatile-attracted natural enemies would be on grasshopper populations. An alternative possibility is that plant volatiles induced by grasshopper feeding attract other members of the species, and thus aid in grasshopper aggregation.

In the Lepidoptera, the fatty acid moiety of the FACs is obtained from the plant or diet on which the larvae feed (Paré and Tumlinson 1997a, b; De Moraes and Mescher 2004), and enzymes in the crop and midgut join the fatty acids to glutamine (Lait et al. 2003). The origin of the fatty acid moiety in caeliferins in grasshoppers is not yet clear. Alborn et al. (2007) reported that regurgitant from field-collected grasshoppers differs in composition from that of lab-reared insects. However, when wild insects were brought to the lab, the composition changed to that of lab-reared animals within a week, regardless of the diet with which they were supplied. Whether these changes are due to unknown changes in diet or other factors, like crowding, is not known. Since the known caeliferins have chain lengths from 15 to19 carbons, including chains with an odd number of carbons, and since the double bond in the chain is *trans*, it seems unlikely that these compounds are derived directly from plant fatty acids.

As with the FAC elicitors, the questions arises as to what function these compounds might serve in the grasshoppers. They may play a role in digestion since sulfated and sulfonated alcohols and fatty acids are known to be good surfactants and emulsifiers. Also, grasshoppers produce copious amounts of frothy spit when handled or disturbed and this is thought to function as a defensive mechanism. Several reports mention the importance of host plant compounds in the defensive effectiveness of grasshopper spit, and that unknown water soluble compounds have deterrent activity against natural enemies (Ortego et al. 1997; Sword 2001; Calcagno et al. 2004). The emulsifying properties of the caeliferins may be important in allowing the aqueous spit to carry lipophilic toxic compounds.

Preliminary evidence indicates that the spit of several, if not all species of grasshoppers in the suborder Caelifera contains substantial quantities of one or more caeliferins. Therefore, these compounds must have some importance in the physiology and/or behavior of this group of insects, but their function(s) remains to be determined.

19.3 The Green Leaf Volatiles

The blend of induced volatiles strongly depends on the herbivore and the plant species. However, the major components of these bouquets are mono- and sesquiterpenes as well as green leaf volatiles (GLVs). While most volatiles serve as signals in tritrophic interactions, some also seem to exhibit other, more direct functions in the plant defense response. In this context GLVs have emerged as a class of compounds, which now appears to play important roles in diverse plant defense strategies. When GLVs were first characterized at the beginning of the last century (Curtius and Franzen 1911), they were considered as a shunt from a pathway leading to traumatin, the first wound hormone described for plants (English and Bonner 1937). This view has changed and the biosynthetic pathway for GLV production is now well understood (Hatanaka 1993; Matsui 2006). GLVs are fatty acid-derived products formed from linolenic and linoleic acids, which serve as substrates for a pathway-specific 13-lipoxygenase. The resulting 13-hydroperoxy C_{18} fatty acid is then cleaved by the enzyme hydroperoxide lyase (HPL) producing (Z)-3-hexenal (from 18:3 fatty acids) or hexanal (from 18:2 fatty acids) as well as 12-oxo-(Z)-9-decenoic acid. Further processing of (Z)-3-hexenal by alcohol dehydrogenase, acetylation and isomerization leads to the production of the remaining C_6-components, like (Z)-3-hexenol, Z-3-hexenyl acetate, and the respective (E)-2-enantiomers (Fig. 19.3). Recently, D'Auria et al. (2007) characterized a BADH acyltransferase in *Arabidopsis*, which is responsible for the acetylation of (Z)-3-hexenol. GLVs are almost immediately released locally after wounding

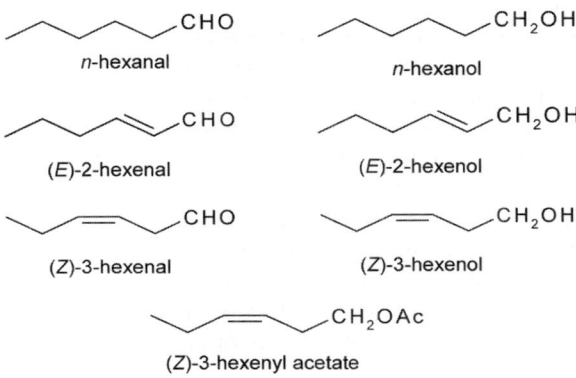

Fig. 19.3 Structures of common green leaf volatiles (GLVs) emitted by plants

(Hatanaka 1993), but can also be produced and released systemically in response to herbivore damage (Turlings and Tumlinson 1992; Paré and Tumlinson 1997; Röse et al. 1996). Besides their association with vegetative tissues, GLVs are also major constituents of volatiles emitted by flowers and fruits.

19.3.1 GLVs and Plant Defense

The diverse functions of GLVs in plant defense include direct effects on attacking pathogens and herbivores, as well as roles as volatile signals between and within plants. Several studies have shown that GLVs can act as anti-bacterial and anti-fungal agents, either directly or through the induction of phytoalexins (Matsui 2006, and references therein). For example, Prost et al. (2005) showed that (Z)-3- and (E)-2-hexenal are potent growth inhibitors of certain plant pathogens. Also, in a study by Shiojiri et al. (2006) it was clearly demonstrated that GLVs affect the plant pathogen *Botrytis cinerea* by reducing its growth rate on *Arabidopsis*. Additionally, GLVs have been shown to affect aphid performance (Hildebrand et al. 1993). It was found that C_6 aldehydes have a direct effect on aphid fecundity and leaf chemistry. C_6 alcohols, on the other hand, did not affect the aphids directly, but caused changes in the chemical composition of the leaves. This result was confirmed by Vancanneyt et al. (2001), who used genetically modified potato plants depleted in HPL to study the effect of GLVs on aphid performance.

In addition to direct effects on microorganisms and aphids, GLVs play a significant role in plant–plant signaling. Communication between plants through the emission of certain volatiles was first described by Rhoades (1983) and Baldwin and Schultz (1983). They found independently that plants exposed to volatiles released from mechanically damaged plants are less attractive to insect herbivores. Later, Arimura et al. (2000) demonstrated convincingly for the first time that volatiles released from herbivore-infested plants can signal the induction of defense-related genes in uninfested neighboring plants. Among the volatiles that induced defensive genes were certain terpenoids and GLVs. GLVs, applied as pure chemicals, had already been shown to induce defense-related genes in *Arabidopsis* (Bate and Rothstein 1998). The potential of GLVs to induce the production of volatiles, even in systemic leaves, was first reported by Farag and Paré (2002) for tomato. Finally in 2005, Ruther and Kleier showed that GLVs applied together with ethylene, which is also released upon herbivory, induced the release of a volatile blend comparable to that released in response to actual insect damage. While these examples clearly demonstrate the potential of GLVs to act as signals in the plant defense response, not much is known about the signaling pathways involved. It appears however, that the octadecanoid signaling pathway plays an important role in this process. A closer look at the publicly available gene expression data reveals that most genes affected by GLVs are also responsive to JA application. More direct evidence for the involvement of jasmonates in the GLV response was provided by Kishimoto et al. (2006), who showed that JA signaling mutants are less responsive to GLVs.

19.3.2 Priming

While GLVs obviously induce gene expression and volatile release in plants, these responses are either incomplete or less pronounced than after actual herbivore attack or application of JA. The question therefore arises whether or not this weak induction of defensive genes, or the low levels of volatiles released, are sufficient to effectively reduce insect herbivore damage. To answer that question, Engelberth et al. (2004) performed a comprehensive study of the effect of GLV treatment on subsequent insect herbivory. Corn seedlings exposed to GLV responded with the rapid but transient accumulation of JA which returned to resting levels after only two to three hours. This accumulation of JA was accompanied by a very moderate release of volatiles, and thus confirmed previous studies. In a second set of experiments, the physiological effects of GLV exposure on subsequent elicitation by insect-derived elicitors was tested. Plants were exposed to GLVs overnight and then treated with elicitors, at a time when JA levels had reached normal resting levels again. These plants responded to elicitor treatment more rapidly and strongly, with respect to JA accumulation and volatile release, as compared to unexposed but otherwise similarly treated control plants. Interestingly, pre-treatment with GLVs specifically affected elicitor-induced JA accumulation, but had no effect on other aspects of the wound response. This was the first demonstration of a priming effect of GLVs against insect herbivory and shed new light on the potential function of these compounds in a plant community. Priming, meaning a process in which the response to a certain challenge is accelerated or enhanced by prior stimulation, is well established for plant-pathogen interactions (Prime-A-Plant Group et al. 2006; Conrath et al. 2002). In the context of plant-insect interactions GLVs seem to serve as priming agents by preparing plants for impending herbivory.

Since its initial discovery, the priming effect of GLVs has been confirmed in a more natural environment by Kessler et al. (2006). Using a microarray enriched in tobacco genes related to insect herbivory they found transcriptional responses in wild tobacco plants growing in the vicinity of clipped sagebrush but no increase in direct defenses like nicotine and proteinase inhibitor production was observed. However, when *M. sexta* caterpillars started feeding on these primed plants, an accelerated production of trypsin proteinase inhibitor occurred. This primed response of wild tobacco plants exposed to clipped sagebrush also resulted in lower herbivore damage as well as a higher mortality rate of young *Manduca* caterpillars. Among the volatiles responsible for this priming effect were (*E*)-2-hexenal, but also methacrolein and methyl jasmonate.

In a more recent study by Ton et al. (2007) the effect of priming by herbivore-induced volatiles on direct and indirect defenses in corn was shown on a molecular, chemical, and behavioral level. By use of a differential hybridization screen they were able to identify 10 defense-related genes, which were inducible by caterpillar feeding, mechanical wounding, application of caterpillar regurgitant, and JA. Exposure to volatiles from herbivore-infested plants did not activate these genes directly, but primed a subset of them for stronger and/or earlier induction upon subsequent defense elicitation, resulting in reduced caterpillar damage, caterpillar development, and increased attraction of the natural enemies of the caterpillar, the parasitic wasp *Cotesia marginiventris*.

While most studies addressed the signaling role of volatiles, including GLVs, in inter-plant communication, it seems obvious that the same volatiles can also serve as signals in intra-plant communication (Karban et al. 2006). Karban et al. investigated the role of volatiles in induced resistance between individual sagebrush (*Artemisia tridentata*) plants, but also among branches within an individual plant. It was found that airflow was essential for the induction of induced resistance. Sagebrush, like many other desert plants, is highly sectorial, which does not allow for a free exchange of material, including signaling compounds, through vascular connections and thus, the coordinated induction of resistance to herbivores. Instead, volatiles are used to overcome this obstacle and take the role of the systemic signal preparing other parts of the plant to better fend off insect herbivores. The role GLVs in this process is unclear, however, since the activity of individual volatiles has not been tested so far. Priming of indirect defenses through volatiles has also been shown for extrafloral nectar production in lima beans (Heil and Kost 2006). Extrafloral nectar secretion attracts carnivorous arthropods thereby causing increased predator pressure on attacking herbivores. In a field study, lima beans were exposed to artificial blends of synthetic volatiles. It was found that those plants primed by volatile exposure produced almost three times as much extrafloral nectar in response to mechanical damage when compared to unprimed plants. The authors also report that the signaling function of herbivore-induced volatiles is not limited to inter-plant signaling, but extends to a role as systemic signal within the plant (Heil and Bueno 2007), as originally described by Karban et al. (2006).

Because complex volatile blends were used in the studies by Ton et al. (2007), Karban et al. (2006), and Heil and Kost (2006), which contained GLVs among other typical volatiles, it has to be clarified whether the whole blend or individual compounds are responsible for the observed priming effect. It can not be excluded that other components of the volatile blend like certain terpenes can also function as signals in inter-plant communication.

As mentioned before, GLVs appear to activate JA-related signaling pathways and the GLV-induced production of JA was shown in corn seedlings. Yet, not much is known about the GLV signaling pathway in other plant species. However, the activation of typical JA-dependent defense genes supports a role for JA-related signaling pathways in the priming process initiated by GLVs. For corn, Engelberth et al. (2007) demonstrated that genes for the biosynthesis of JA are significantly induced further suggesting that the octadecanoid signaling pathway plays an important role in GLV-induced processes.

19.4 The Octadecanoid Signaling Pathway

When insect herbivores attack their host plant most of the countermeasures are signaled through the octadecanoid pathway with jasmonic acid (JA) and JA-conjugates as the major internal plant signals. The octadecanoid signaling cascade is initiated by tissue damage resulting in the formation of 12-oxo-phytodienoic acid (OPDA) from polyunsaturated fatty acids in plastids. The final steps of the pathway occur in peroxisomes, including the reduction of the olefinic bond in the pentacyclic ring of OPDA by 12-oxo-phytodienoate-10,11-reductase (OPR), followed by three cycles of beta-oxidation resulting in JA formation (Schaller et al. 2004; Schaller and Stintzi this volume).

While much of the early work on octadecanoid signaling focused on JA, particularly on its role in the activation of defense genes, it is now clear that pathway intermediates like OPDA, and JA-conjugates exhibit their own distinct activity in the plant. For example, OPDA has been shown to induce tendril coiling in *Bryonia dioica* more effectively than JA (Stelmach et al. 1998). Also, the OPDA induced changes in gene expression are distinctly different from those obtained with JA (Stintzi et al. 2001; Taki et al. 2005). Direct proof for the distinct activities of OPDA and JA came from the characterization of the *Arabidopsis opr3* mutant, which is defective in the reduction of OPDA and no longer able to form JA. The mutant is male-sterile, indicating that JA is strictly required for male reproductive development. In contrast, octadecanoid-dependent defense responses are not impaired in *opr3*, showing that OPDA can substitute for JA in the induction of defense gene expression (Stintzi and Browse 2000; Stintzi et al. 2001; Schaller and Stintzi this volume). Like JA, OPDA is massively up-regulated at the site of mechanical damage and it appears that its function is not only to serve as a precursor for JA, but also to play an independent and important part in the immediate wound response.

It also has become clear in recent years, that mechanical wounding may not always be a good mimic of herbivory. While wounding is a natural component of insect herbivore damage, significant differences were shown to occur when mechanically damaged plants were compared to those actually damaged by herbivores (Reymond et al. 2000). These differences can be attributed to elicitors like volicitin (and other fatty acid amides), inceptins, and the caeliferins, which are essential components of the respective insect regurgitant. These compounds were first described when analyses of VOC revealed significant differences between plants that were only mechanically damaged and those plants, where spit from caterpillars was applied to the wound site. Qualitative and quantitative differences were observed and could be attributed to factors in the regurgitant. By using purified or synthesized elicitors, Schmelz et al. (2003) showed that those elicitors significantly affected JA levels and VOC release. Engelberth et al. (2007) showed that in corn leaves not only local JA and OPDA levels were increased after elicitor application, when compared to mechanical wounding, but also that JA distal from the elicitor application site was up-regulated without measurable delay. Interestingly, while local OPDA is highly induced, no increases of OPDA were detected in the distal parts of the corn leaf. This strongly suggests that OPDA and JA biosynthesis are differentially regulated and first results in our lab indicate that free OPDA is not involved in JA biosynthesis in the wound response, which would further support the role of OPDA as an independent signaling compound. The activity of GLVs towards the induction of JA adds to the list of stimuli that interact with the octadecanoid signaling pathway (Engelberth et al. 2004). Not only did exposure to GLV induce JA accumulation in corn seedlings, it also prepared them against impending insect herbivory expressed as increased JA production and VOC release when induced with insect elicitors. On the other hand, no effect of priming by GLVs on wound-induced JA accumulation was observed, and JA levels were not increased further by a second application of GLVs, indicating a certain specificity of the GLV signal (Engelberth et al. 2004, 2007). Additionally, the fact that GLV-induced elevated JA levels were not accompanied by a concomitant increase of OPDA bears close resemblance to the observed JA accumulation in distal parts of the leaf after insect elicitor application, and suggests a linking between these two events. It also suggested that either a highly regulated turn-over, or a conjugated precursor exists within the pathway leading to the biosynthesis of JA, which does not allow for any free intermediate of the pathway to be detected. So far all experiments performed to detect conjugated OPDA or JA, e.g. in the form of lipids as described by Stelmach et al. (2001) for OPDA or amino acid conjugates of JA, failed to produce any evidence for this kind of precursor.

Besides inducing increases of *AOS* and *AOC* transcript levels, GLV also appear to act as a major regulator for the induction of *OPR* genes in corn. Also, OPR activity is significantly increased after 15 hours exposure to GLV, at a time when the priming effect is most prominent (Engelberth et al. 2004, 2007). This increase in activity might account for a higher turn-over activity when challenged with insect-derived elicitors resulting in elevated JA levels when compared to equally challenged non-GLV exposed plants. Based on sequence identity with AtOPR3 and LeOPR3, only *ZmOPR8* seems likely to be involved in JA biosynthesis. However, the actual functions and the physiological consequences of distinct *OPR* induction in corn by GLVs and other stimuli like insect herbivore damage or mechanical wounding have yet to be established and a thorough biochemical characterization of individual OPR proteins in corn is needed to allow for the correct pathway-specific allocation of the respective enzyme.

The activation of the octadecanoid pathway appears to be the major signaling event in plants with respect to insect herbivory. Wounding, insect-derived elicitors, and GLVs affect this pathway in a very sophisticated way, each inducing a distinctly different response. These differences in the regulation of this pathway seem to be of critical importance not only for our understanding of how plants cope with their environment, but also with regard to the development of future, environmentally sound pest management strategies.

19.5 Future Directions

There are several questions that need to be addressed to increase our understanding of the mechanisms by which plants defend themselves against herbivore attack. The knowledge gained from investigation of the basic biochemical and ecological plant/insect interactions may enable us to exploit these mechanisms to increase the capabilities of crop plants to defend against insect pests, and thus create more sustainable pest management methods.

Whether the various types of insect-derived elicitors are active across a broad range of plant species is a question that has not yet been answered. Are certain types of elicitors active on certain types of plants only, and can this be correlated with the types of plants that the herbivores producing these elicitors feed on? As yet, a clear pattern of activity is not apparent. Certainly, there are species of plants, like cowpeas, that do not appear to react to FAC elicitors. Also, the relative activity of the FAC elicitors has been determined on corn seedlings, but on no other plant species. Is volicitin more active than N-linolenoyl-L-glutamine and -glutamate on all species of plants that react to these compounds, or does the relative activity of the different FACs vary across the broad array of plant species. Similarly, is the relative activity of the caeliferins the same across all species of plants that respond to the grasshopper-produced elicitors? The answers to these questions may shed some light on the question of the mechanism by which the elicitors induce plant defenses. Is this mechanism the same for all herbivore-produced elicitors or does it differ with different types of elicitors?

Similar questions can be asked about priming. Is this a general phenomenon or is it restricted to a few species? How effective is it, and does it play a significant role in protecting plants from herbivore attack or reducing plant damage in natural ecosystems? Are there other priming molecules in addition to the GLVs? Is the relative activity of the different GLV molecules the same for all plants, or does it change with different plant species? What is the mechanism by which priming occurs? Can we develop a method to use priming to protect crop plants against impending attack by insect pests?

Finally, there is the question of why the insect herbivores make the elicitors and what role they play in insect herbivore physiology, metabolism, or defense. Since the FACs have been discovered in a broad range of Lepidopteran species, and more recently in Orthoptera and Diptera, it seems possible that they play a critical role in these insects' digestive processes and metabolism. Several experiments are planned or in progress to decipher this role, including labeling studies to determine the metabolic processes with which the FAC elicitors are involved. In addition, the enzymes in the gut of Lepidopteran caterpillars that synthesize and metabolize these FACs are being purified and the genes cloned. Gene silencing experiments may provide information as to the importance of these molecules in digestion and other metabolic processes.

References

Alborn HT, Brennan MM, Tumlinson JH (2003) Differential activity and degradation of plant volatile elicitors present in the regurgitant of tobacco hornworm (*Manduca sexta*) larvae. J Chem Ecol 29:1357–1372

Alborn HT, Hansen TV, Jones TH, Bennett DC, Tumlinson JH, Schmelz EA, Teal PEA (2007) Disulfooxy fatty acids from the American bird grasshopper *Schistocerca americana*, elicitors of plant volatiles. Proc Natl Acad Sci USA 104:12976–12981

Alborn HT, Jones TH, Stenhagen GS, Tumlinson JH (2000) Identification and synthesis of volicitin and related components from beet armyworm oral secretion. J Chem Ecol 26:203–220

Alborn HT, Turlings TCJ, Jones TH, Stenhagen G, Loughrin JH, Tumlinson JH (1997) An elicitor of plant volatiles identified from beet armyworm oral secretion. Science 276:945–949

Arimura G, Ozawa R, Kugimiya S, Takabayaashi J, Bohlman J (2004) Herbivore-induced defense response in a model legume. Two-spotted spider mites induce emission of (E)-beta-ocimene and transcript accumulation of (E)-beta ocimene synthase in *Lotus japonicus*. Plant Physiol 135:1976–1983

Arimura G, Ozawa R, Shimoda T, Nishioka T, Boland W, Takabayashi J (2000) Herbivory-induced volatiles elicit defense genes in lima bean leaves. Nature 406:512–515

Baldwin IT, Schultz JC (1983) Rapid changes in tree leaf chemistry induced by damage – evidence for communication between plants. Science 221:277–279

Bate NJ, Rothstein SJ (1998) C_6-volatiles derived from the lipoxygenase pathway induce a subset of defense-related genes. Plant J 16:561–569

Calcagno MP, Avila JL, Rudman I, Otero LD, Alonso-Amelot ME (2004) Food-dependent regurgitate effectiveness in the defence of grasshoppers against ants: the case of bracken-fed *Abracris flavolineata* (Orthoptera: Acrididae). Physiol Entomol 29:123–128

Chamnongpol S, Willekens H, Camp WV (1998) Defense activation and enhanced pathogen tolerance induced by H_2O_2 in transgenic tobacco. Proc Natl Acad Sci USA 95:5818–5823

Conrath U, Pieterse CM, Mauch-Mani B (2002) Priming in plant-pathogen interactions. Trends Plant Sci 7(5):210–216

Curtius T, Franzen H (1911) Aldehyde aus gruenen Pflanzenteilen. Chem Zentr II:1142–1143

D'Auria JC, Pichersky E, Schaub A, Hansel A, Gershenzon J (2007) Characterization of a BAHD acyltransferase responsible for producing the green leaf volatile (Z)-3-hexen-1-yl acetate in *Arabidopsis thaliana*. Plant J 49:194–207

Degenhardt DC, Lincoln DE (2006) Volatile emissions from an odorous plant in response to herbivory and methyl jasmorate exposure. J Chem Ecol 32:725–743

De Moraes CM, Lewis WJ, Paré PW, Alborn HT, Tumlinson JH (1998) Herbivore-infested plants selectively attract parasitoids. Nature 393:570–573

De Moraes CM, Mescher MC (2004) Biochemical crypsis in the avoidance of natural enemies by an insect herbivore. Proc Natl Acad Sci USA 101:8993–8997

De Moraes CM, Mescher MC, Tumlinson JH (2001) Caterpillar-induced nocturnal plant volatiles repel conspecific females. Nature 410:577–580

Dugravot S, Mondy N, Mandon N, Thibout E (2005) Increased sulfur precursors and volatiles production by the leek *Allium porrum* in response to specialist insect attack. J Chem Ecol 31:1299–1314

Eichenseer H, Mathews MC, Bi JL, Murphy JB, Felton GW (1999) Salivary glucose oxidase: multifunctional roles for *Helicoverpa zea*? Arch Insect Biochem Physiol 42:99–109

Engelberth J, Alborn HT, Schmelz EA, Tumlinson JH (2004) Airborne signals prime plants against insect herbivore attack. Proc Natl Acad Sci USA 101:1781–1785

Engelberth J, Seidl-Adams I, Schulz JC, Tumlinson JH (2007) Insect elicitors and exposure to green leafy volatiles differentially up-regulate major octadecanoids and transcripts of 12-oxo phytodienoic acid reductases in *Zea mays*. Mol Plant Microbe Interact 20:707–716

English J, Bonner J (1937) The wound hormones of plants. I. Traumatin, the active principle of the bean test. J Biol Chem 121:791–799

Farag MA, Paré PW (2002) C_5-green leaf volatiles trigger local and systemic VOC emissions in tomato. Phytochemistry 61 545–554

Halitschke R, Schittko U, Pohnert G, Boland W, Baldwin IT (2001) Molecular interactions between the specialist herbivore *Manduca sexta* (Lepidoptera, Sphingidae) and its natural host *Nicotiana attenuata*. III. Fatty acid–amino acid conjugates in herbivore oral secretions are necessary and sufficient for herbivore-specific plant responses. Plant Physiol 125:711–717

Hatanaka A (1993) The biogeneration of green odour by green leaves. Phytochemistry 34:1201–1218

Heil M, Bueno CS (2007) Within-plant signaling by volatiles leads to induction and priming of an indirect plant defense in nature. Proc Natl Acad Sci USA 104:5467–5472

Heil M, Kost C (2006) Priming of indirect defenses. Ecol Lett 9:813–817
Hildebrand DF, Brown GC, Jackson DM, Hamilton-Kemp TR (1993) Effects of some leaf-emitted volatile compounds on aphid population increase. J Chem Ecol 19:1875–1887
Karban R, Shiojiri K, Huntzinger M, McCall AC (2006) Damage-induced resistance in sagebrush: volatiles are key to intra- and interplant communication. Ecology 87:922–930
Kessler A, Halitschke R, Diezel C, Baldwin IT (2006) Priming of plant defense responses in nature by airborne signaling between *Artemisia tridentata* and *Nicotiana attenuata*. Oecologia 148:280–292
Kishimoto K, Matsui K, Ozawa R, Takabayashi J (2006) ETR1-, JAR1- and PAD2-dependent signaling pathways are involved in C6-aldehyde-induced defense responses of *Arabidopsis*. Plant Sci 171:415–423
Kutlesa NJ, Caveney S (2001) Insecticidal activity of glufosinate through glutamine depletion in a caterpillar. Pest Manag Sci 57:25–32
Lait CG, Alborn HT, Teal PEA, Tumlinson JH (2003) Rapid biosynthesis of N-linolenoyl-L-glutamine, an elicitor of plant volatiles, by membrane associated enzyme(s) in *Manduca sexta*. Proc Natl Acad Sci USA 100:7027–7032
Litvak ME, Monson RK (1998) Patterns of induced and constitutive monoterpene production in conifer needles in relation to insect herbivory. Oecologia 114:531–540
Loughrin JH, Manukian A, Heath RR, Turlings TCJ, Tumlinson JH (1994) Diurnal cycle of emission of induced volatile terpenoids by herbivore-injured cotton plants. Proc Natl Acad Sci USA 91:11836–11840
Matsui K (2006) Green leaf volatiles: hydroperoxide lyase pathway of oxylipin metabolism. Curr Opin Plant Biol 9:274–280
Mithöfer A, Wanner G, Boland W (2005) Effects of feeding *Spodoptera littoralis* on lima bean leaves. II. Continuous mechanical wounding resembling insect feeding is sufficient to elicit herbivory-related volatile emission. Plant Physiol 137:1160–1168
Mori N, Yoshinaga N, Sawada Y, Fukui M, Shimoda M, Fujisaki K, Nishida R, Kuwahara Y (2003) Identification odvolicitin-related compounds from the regurgitant of lepidopteran caterpillars. Biosci Biotech Biochem 67:1168–1171
Mori N, Alborn HT, Teal PEA, Tumlinson JH (2001) Enzymatic decomposition of elicitors of plant volatiles in *Heliothis virescens* and *Helicoverpa zea*. J Insect Physiol 47:749–757
Ortego F, Evans PH, Bowers WS (1997) Enteric and plant-derived deterrents in regurgitate of American bird grasshopper, *Schistocerca americana*. J Chem Ecol 23:1941–1950
Paré PW, Alborn HT, Tumlinson JH (1998) Concerted biosynthesis of an insect elicitor of plant volatiles. Proc Natl Acad Sci USA 95:13971–13975
Paré PW, Tumlinson JH (1997a) Induced synthesis of plant volatiles. Nature 385:30–31
Paré PW, Tumlinson JH (1997b) De novo biosynthesis of volatiles induced by insect herbivory in cotton plants. Plant Physiol 114:1161–1167
Pohnert G, Jung V, Haukioja E, Lempa K, Boland W (1999) New fatty acid amides from regurgitant of lepidopteran (Noctuidae, Geometridae) caterpillars. Tetrahedron 55: 11275–11280
Prime-A-Plant Group, Conrath U, Beckers GJ, Flors V, Garcia-Agustin P, Jakab G, Mauch F, Newman MA, Pieterse CM, Poinssot B, Pozo MJ, Pugin A, Schaffrath U, Ton J, Wendehenne D, Zimmerli L, Mauch-Mani B (2006) Priming: getting ready for the battle. Mol Plant Microbe Interact 19:1062–1071
Prost I, Dhondt S, Rothe G, Vicente J, Rodriguez MJ, Kift N, Carbonne F, Griffiths G, Esquerre-Tugaye MT, Rosahl S, Castresana C, Hamberg M, Fournier J (2005) Evaluation of the antimicrobial activities of plant oxylipins supports their involvement in defense against pathogens. Plant Physiol 139:1902–1913
Reymond P, Weber H, Damond M, Farmer EE (2000) Differential gene expression in response to mechanical wounding and insect feeding in *Arabidopsis*. Plant Cell 12:707–719
Rhoades DF (1983) Responses of alder and willow to attack by tent caterpillars and webworms – Evidence for pheromonal sensitivity of willows. ACS Symp Ser 208:55–68

Rodrigues-Saona C, Crafts-Brandner SJ, Williams L, Paré PW (2002) *Lygus hesperus* feeding and salivary gland extracts induce volatile emissions in plants. J Chem Ecol 28: 1733–1747

Röse USR, Manukian A, Heath RR, Tumlinson JH (1996) Volatile semiochemicals released from undamaged cotton leaves: a systemic response of living plants to caterpillar damage. Plant Physiol 111:487–495

Röse USR, Tumlinson JH (2005) Systemic induction of volatile release in cotton: how specific is the signal to herbivory. Planta 222:327–335

Ruther J, Kleier S (2005) Plant–plant signaling: ethylene synergizes volatile emission in *Zea mays* induced by exposure to (Z)-3-hexen-1-ol. J Chem Ecol 31:2217–2222

Sawada Y, Yoshinaga N, Fujisaka K, Nishida R, Kuwahara Y, Mori N (2006) Absolute configuration of volicitin from the regurgitant of lepidopteran caterpillars and biological activity of volicitin-related compounds. Biosci Biotechnol Biochem 70:2185–2190

Schaller F, Schaller A, Stintzi A (2004) Biosynthesis and metabolism of jasmonates. J Plant Growth Regul 23(3):179–199

Schmelz EA, Alborn HT Tumlinson JH (2003) Synergistic interactions between volicitin, jasmonic acid and ethylene mediate insect-induced volatile emission in *Zea mays*. Physiol Plant 117:403–412

Schmelz EA, Carroll MJ, LeClere S, Phipps SM, Meredith J, Chourey PJ, Alborn HT, Teal PEA (2006) Fragments of ATP synthase mediate plant perception of insect attack. Proc Natl Acad Sci USA 103:8894–8899

Shiojiri K, Kishimoto K, Ozawa R, Kugimiya S, Urashimo S, Arimura G, Horiuchi J, Nishioka T, Matsui K, Takabayashi J (2006) Changing green leaf volatile biosynthesis in plants: an approach for improving plant resistance against both herbivores and pathogens. Proc Natl Acad Sci USA 103:16672–16676

Spiteller D, Dettner K, Boland W (2000) Gut bacteria may be involved in interactions between plants, herbivores and their predators: microbial biosynthesis of N-acylglutamine surfactants as elicitors of plant volatiles. Biol Chem 381:755–762

Spiteller D, Pohnert G, Boland W (2001) Absolute configuration of volicitin, an elicitor of plant volatile biosynthesis from lepidopteran larvae. Tetrahedron Lett 42:1483–1485

Stanley-Samuelson DW (1994) Prostaglandins and related eicosanoids in insects. Adv Insect Physiol 24:115–212

Stanley-Samuelson DW, Jurenka RA, Loher W, Blomquist GJ (1987) Metabolism of polyunsaturated fatty acids by larvae of the waxmoth, *Galleria mellonella*. Arch Insect Biochem Physiol 6:141–149

Stelmach BA, Muller A, Hennig P, Gebhardt S, Schubert-Zsilavecz M, Weiler EW (2001) A novel class of oxylipins, sn1-O-(12-oxophytodienoyl)-sn2-O-(hexadecatrienoyl)-monogalactosyl diglyceride, from *Arabidopsis thaliana*. J Biol Chem 276:28628–28628

Stelmach BA, Muller A, Hennig P, Laudert D, Andert L, Weiler EW (1998) Quantitation of the octadecanoid 12-oxo-phytodienoic acid, a signaling compound in plant mechanotransduction. Phytochemistry 47:539–546

Stintzi A, Weber H, Reymond P, Browse J, Farmer EE (2001) Plant defense in the absence of jasmonic acid: the role of cyclopentenones. Proc Natl Acad Sci USA 98: 12837–12842

Stintzi A, Browse J (2000) The *Arabidopsis* male-sterile mutant, *opr3*, lacks the 12-oxophytodienoic acid reductase required for jasmonate synthesis. Proc Natl Acad Sci USA 97:10625–10630

Sword GA (2001) Tasty on the outside, but toxic in the middle: grasshopper regurgitation and host plant-mediated toxicity to a vertebrate predator. Oecologia 128:416–421

Taki N, Sasaki-Sekimoto Y, Obayashi T, Kikuta A, Kobayashi K, Ainai T, Yagi K, Sakurai N, Suzuki H, Masuda T Takamiya K, Shibata D, Kobayashi Y, Ohta H (2005) 12-Oxophytodienoic acid triggers expression of a distinct set of genes and plays a role in wound-induced gene expression in *Arabidopsis*. Plant Physiol 139:1268–1283

Ton J, D'Alessandro MD, Jourdie V, Jakab G, Karlen D, Held M, Mauch-Mani B, Turlings TCJ (2007) Priming by airborne signals boosts direct and indirect resistance in maize. Plant J 49:16–26

Truitt CL, Paré PW (2004) In situ translocation of volicitin by beet armyworm larvae to maize and systemic immobility of the herbivore elicitor *in planta*. Planta 218:999–1007

Tumlinson JH, Alborn HT, Loughrin JH, Turlings TCJ, Jones TH (2000) Plant volatile elicitor from insects. US Patent # 6,054,483; April 25, 2000 (M-3499); US Patent # 6,207,712 B1; March 27, 2001

Tumlinson JH, Lait CG (2005) Biosynthesis of fatty acid amide elicitors of plant volatiles by insect herbivores. Arch Insect Biochem Physiol 58:54–68

Turlings TCJ, Alborn HT, Loughrin JH, Tumlinson JH (2000) Volicitin, an elicitor of maize volatiles in oral secretion of *Spodoptera exigua*: isolation and bioactivity. J Chem Ecol 26:189–202

Turlings TCJ, Bernasconi M, Bertossa R, Bigler F, Caloz G, Dorn S (1998) The induction of volatile emissions in maize by three herbivore species with different feeding habits: possible consequences for their natural enemies. Biol Control 11:122–129

Turlings TC, Tumlinson JH (1992) Systemic release of chemical signals by herbivore-injured corn. Proc Natl Acad Sci USA 89:8399–8402

Vancanneyt G, Sanz C, Farmaki T, Paneque M, Ortego F, Castanera P, Sanchez-Serrano JJ (2001) Hydroperoxide lyase depletion in transgenic potato plants leads to an increase in aphid performance. Proc Natl Acad Sci USA 98:8139–8144

Wei JN, Zhu JW, Kang L (2006) Volatiles released from bean plants in response to agromyzid flies. Planta 224:279–287

Williams L, Rodriguez-Saona C, Paré PW, Crafts-Brandner SJ (2005) Arch Biochem Phys 58:84–96

Yoshinaga N, Aboshi T, Ishikawa C, Fukui M, Shimoda M, Nishida R, Lait CG, Tumlinson JH, Mori N (2007) Fatty acid amides, previously identified in caterpillars, found in the cricket *Teleogryllus taewanemma* and fruit fly *Drosophila melanogaster* larvae. J Chem Ecol 33:1376–1381

Yoshinaga N, Sawada Y, Nishida R, Kuwahara Y, Mori N (2003) Specific incorporation of L-glutamine into volicitin in the regurgitant of *Spodoptera litura*. Biosci Biotechnol Biochem 67:2655–2657

Chapter 20
Aromatic Volatiles and Their Involvement in Plant Defense

Anthony V. Qualley and Natalia Dudareva

More than one percent of secondary metabolites is represented by volatile compounds which are involved in plant reproduction and defense. About 20% of this volatile subset are aromatic constituents consisting of phenylpropanoids, benzenoids, phenylpropenes, and nitrogen-containing aromatics. Despite the wide diversity of aromatic compounds only a small number are released, often at low levels, in response to and anticipation of attacks by herbivorous insect species. Scattered amounts of information show that aromatic volatiles have roles in repulsion and intoxication of attacking organisms as well as in the attraction of beneficial insect species that control pest insect populations through predation and parasitization. In this chapter, we overview the current body of knowledge regarding biosynthesis of aromatic volatiles emitted from plants as well as their roles in plant-insect and plant–plant interactions and highlight the major unsolved questions awaiting further exploration.

20.1 Introduction

Volatile compounds represent approximately one percent of plant secondary metabolites whose primary functions include plant defense against herbivores and pathogens. These volatiles, when emitted from plant tissues as part of a defense system, can directly repel (De Moraes et al. 2001; Kessler and Baldwin 2001) or intoxicate (Vancanneyt et al. 2001) attacking organisms, and even attract natural enemies of herbivores, indirectly protecting the plant via tritrophic interactions (Mercke et al. 2004; Arimura et al. 2004; Degen et al. 2004). By releasing volatiles, an herbivore-damaged plant can simultaneously reduce the number of attackers (Kessler and Baldwin 2001) while alerting neighboring plants of impending danger (Shulaev et al. 1997). In neighboring plants, volatile warning signs induce the expression of defense genes and emission of volatiles (Arimura et al. 2000; Birkett

---

N. Dudareva
Department of Horticulture and Landscape Architecture, Purdue University, West Lafayette, IN 47907, USA
e-mail: dudareva@purdue.edu

et al. 2000; Ruther and Kleier 2005; Farag et al. 2005) and also, in some cases, prime nearby plants to respond faster to future herbivore attack (Engelberth et al. 2004; Kessler et al. 2006). Plant volatiles are mainly represented by terpenoids, fatty acid derivatives, and amino acid derivatives, including those from L-phenylalanine (Phe), the phenylpropanoids, and benzenoids (Dudareva et al. 2006). Although the number of volatile phenylpropanoids and benzenoids identified from plants is comparable to the volatile terpenoids, progress in research regarding their involvement in plant-plant and plant-insect interactions has been far from equal. In the last ten years some progress has been made elucidating the biosynthesis of volatile aromatic compounds, however, still very little is known about their contribution to plant defense. Our focus is on the biosynthesis of volatile aromatics and their involvement in plant defense.

20.2 Biosynthesis of Volatile Aromatic Compounds

Aromatic metabolites are recognized as an important class of widespread secondary metabolites possessing antiherbivory properties that actively participate in plant direct defense. Their structural diversity ranges from hydroxycinnamic acids to flavonoids and stilbenes (Dixon et al. 2002). These compounds are mainly derived from Phe with more complex metabolites (flavonoids, isoflavonoids, and stilbenes) formed by condensation of a phenylpropane unit with a unit derived from acetate via malonyl coenzyme A. While most of these aromatic compounds are usually non-volatile, the volatile subset is represented by benzenoid (C_6–C_1), phenylpropanoid (C_6–C_3) and phenylpropanoid-related compounds (C_6–C_2) and by products synthesized directly from the shikimic acid pathway (e.g., indole or methyl anthranilate; Fig. 20.1). Although some of these compounds are already fairly volatile (aldehydes, alcohols, alkanes/alkenes, ethers, and esters), further modifications such as hydroxylation, acylation and methylation often enhance their volatilities as well as those of non-volatile compounds (Dudareva et al. 2004).

20.2.1 Aromatic Volatiles Derived from Phenylalanine

The first committed step in the biosynthesis of the majority of phenylpropanoid and benzenoid compounds is catalyzed by a well-known and widely distributed enzyme, L-phenylalanine ammonia-lyase (PAL). PAL catalyzes the deamination of Phe to produce *trans*-cinnamic acid. Formation of the benzenoids (C_6–C_1) from cinnamic acid requires shortening of the side chain by a C_2 unit, for which several routes have been proposed (Fig. 20.1). This process could occur via a CoA-dependent – β-oxidative pathway, a CoA-independent –non-β-oxidative pathway, or a combination of both routes. The β-oxidative branchway is analogous to that underlying β-oxidation of fatty acids and proceeds through the formation of four CoA-ester intermediates. Shortening of the *trans*-cinnamic acid side-chain by

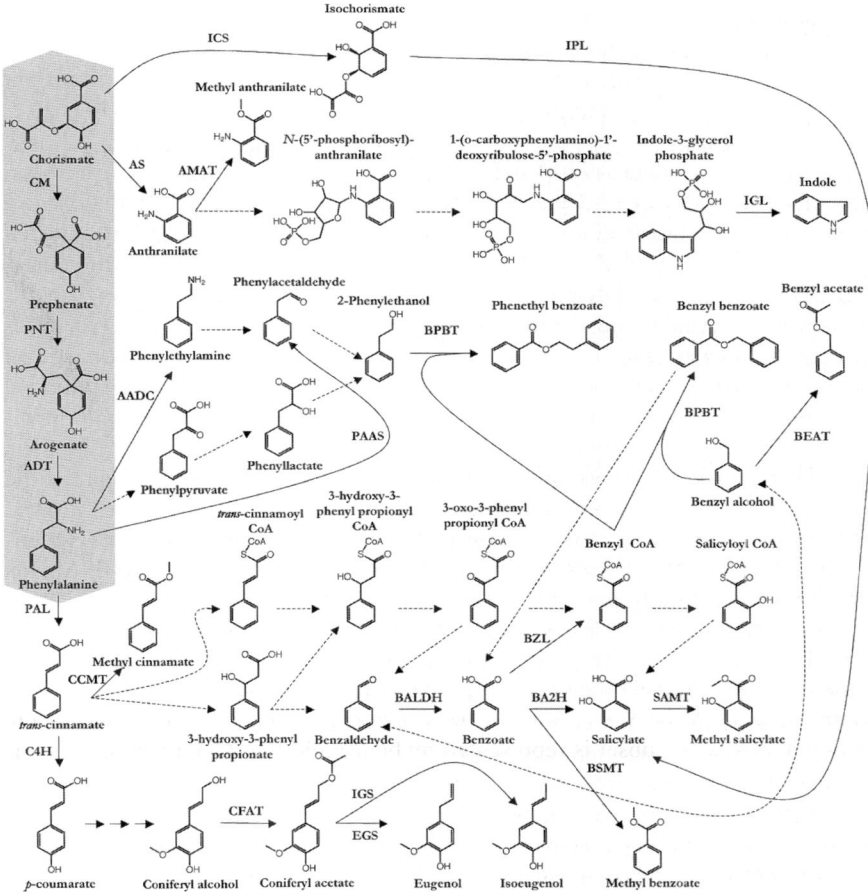

Fig. 20.1 Proposed biosynthetic pathways leading to biosynthesis of volatile aromatic compounds in plants. Solid arrows indicate established biochemical reactions, whereas broken arrows indicate possible steps not yet identified. AADC, amino-acid decarboxylase; ADT, arogenate dehydratase; AMAT, anthraniloyl-coenzyme A (CoA):methanol acyltransferase; AS, anthranilate synthase; BA2H, benzoic acid 2-hydroxylase; BALDH, benzaldehyde dehydrogenase; BEAT, acetyl-CoA:benzyl alcohol acetyltransferase; BPBT, benzoyl-CoA:benzyl alcohol/phenylethanol benzoyltransferase; BSMT, S-adenosyl-l-Met:benzoic acid/salicylic acid carboxyl methyltransferase; BZL, benzoate:CoA ligase; C4H, cinnamate 4-hydroxylase; CCMT, S-adenosyl-L-Met:coumarate/cinnamate carboxyl methyltransferase; CFAT, acetyl-CoA:coniferyl alcohol acetyltransferase; CM, chorismate mutase; EGS, eugenol synthase; ICS, isochorismate synthase; IGL, indole:glycerol-3-phosphate lyase; IGS, isoeugenol synthase; IPL, isochorismate pyruvate lyase; PAL, L-phenylalanine ammonia lyase; PAAS, phenylacetaldehyde synthase; PNT, prephenate aminotransferase; SAMT, S-adenosyl-L-Met:salicylate carboxyl methyltransferase

β-oxidation has been shown in cucumber (*Cucumis sativus*) and *Nicotiana attenuata* in feeding experiments using stable isotope-labeled ($^2H_6$, ^{18}O) 3-hydroxy-3-phenylpropionic acid, which resulted in labeling of benzoic and salicylic acids but not benzaldehyde, a key intermediate in the non-β-oxidative pathway (Fig. 20.1; Jarvis et al. 2000).

The CoA-independent – non-β-oxidative pathway involves hydration of free *trans*-cinnamic acid to 3-hydroxy-3-phenylpropionic acid and side-chain degradation via a reverse aldol cleavage, leading to the formation of benzaldehyde which is oxidized to benzoic acid by an $NADP^+$- dependent enzyme, benzaldehyde dehydrogenase (BALDH). A non-β-oxidative mechanism for the formation of *p*-hydroxybenzoic acid was demonstrated using in vitro studies carried out with cell-suspension cultures of carrot (Schnitzler et al. 1992), cell-free extracts of *Lithospermum erythrorhizon* (Yazaki et al. 1991), and potato (French et al. 1976). This pathway is characterized by the presence of *p*-hydroxybenzaldehyde as an important metabolic intermediate formed prior to oxidation to *p*-hydroxybenzoic acid, and requires an aldehyde dehydrogenase for the conversion of an aldehyde to its corresponding carboxylic acid. A non-oxidative conversion of *trans*-cinnamic acid to benzaldehyde followed by oxidation to benzoic acid was also shown in *Hypericum androsaemum* (Hypericaceae) cell culture using pathway-specific radiolabeled precursors (Abd El-Mawla et al. 2001). Interestingly, this pathway was not found in *Centaurium erythraea* (Gentianaceae) cell culture wherein 3-hydroxybenzoic acid appears to originate directly from the shikimate pathway (Abd El-Mawla et al. 2001). The contribution of both β-oxidative and non-β-oxidative pathways towards the formation of benzenoid compounds was recently shown in petunia (*Petunia hybrida*) flowers using in vivo stable isotope labeling and computer-assisted metabolic flux analysis (Boatright et al. 2004; Orlova et al. 2006). In addition, a recent discovery in the biosynthesis of salicylic acid suggests that it may be formed directly from isochorismate during the *Arabidopsis thaliana* pathogen response, raising the possibility of benzenoid biosynthesis from shikimate/chorismate pathway intermediates (Wildermuth et al. 2001). Overall, these results show that different routes to benzenoid compounds exist in different plant taxa and even in conspecifics depending on physiological conditions. The mechanism of side chain shortening remains an important unsolved question and surprisingly little is known to date about the enzymes and genes responsible for biosynthesis of phenylpropanoid/benzenoid compounds. However, significant progress has been made in the discovery of enzymes and genes involved in the final steps of benzenoid volatile formation. Two enzyme superfamilies were found to greatly contribute to the final biosynthetic steps of volatile benzenoids. They include the BAHD superfamily of acyltransferases named according to the first four biochemically characterized enzymes of this family, BEAT (acetyl-CoA:benzyl alcohol acetyltransferase), AHCT (anthocyanin *O*-hydroxycinnamoyltransferase), HCBT (anthranilate *N*-hydroxycinnamoyl/ benzoyltransferase) and DAT (deacetylvindoline 4-*O*-acetyltransferase; D'Auria, 2006) and the SABATH family of methyltransferases, also named based on the first identified genes belonging to this family, SAMT (SAM:salicylic acid carboxyl methyltransferase), BAMT (SAM:benzoic acid carboxyl methyltransferase), and theobromine synthase (D'Auria et al. 2003).

Acyltransferases belonging to BAHD superfamily catalyze the most common modification of secondary metabolites using alcohols and acyl CoAs as substrates to form acyl esters, many of which are volatile. Biosynthetic enzymes forming benzyl acetate (Dudareva et al. 1998), benzyl benzoate (D'Auria et al. 2002; Boatright et al. 2004), phenylethyl benzoate (Boatright et al. 2004) and cinnamyl acetate (Beekwilder et al. 2004) have been isolated and characterized from different plant species including *Clarkia breweri*, petunia and banana (*Musa* sp.).

Members of SABATH family of methyltransferases form the volatile methyl esters of salicylate, benzoate, cinnamate and *p*-coumarate. These enzymes use *S*-adenosyl-L-methionine to methylate the carboxyl groups of small molecules (Effmert et al. 2005). Eleven such carboxyl methyltransferases have been isolated and characterized at the molecular, biochemical and structural levels within the past decade. Two types of enzymes were distinguished based on their substrate preferences: the SAMT-type enzymes isolated from *C. breweri*, *Stephanotis floribunda*, *Antirrhinum majus*, *Hoya carnosa*, and petunia, which have higher catalytic efficiency and preference for salicylic acid, and the BAMT-type enzymes from *A. majus*, *Arabidopsis thaliana*, *Arabidopsis lyrata*, and *Nicotiana suaveolens* which prefer benzoic acid (reviewed in Effmert et al. 2005). Recently, a carboxyl methyltransferase that can methylate *trans*-cinnamate and *p*-coumarate with high efficiency was isolated from basil (*Ocimum basilicum*; Kapteyn et al. 2007). Surprisingly, this newly isolated carboxyl methyltransferase does not cluster with SAMT- and BAMT-type enzymes, having instead *Arabidopsis* indole-3-acetic acid carboxyl methyltransferase as its closest relative of known function (Kapteyn et al. 2007).

To date, none of the genes encoding the array of enzymes within the benzenoid network that lead to the formation of volatile compound precursors have been isolated. However, the activities of two enzymes, benzoyl-coenzyme A ligase (BZL) and benzoic acid 2-hydroxylase (BA2H), responsible for the formation of benzoyl-CoA and salicylic acid, have been shown in *C. breweri* and tobacco, respectively (Beuerle and Pichersky 2002; Leon et al. 1995).

Formation of phenylpropanoid-related (C_6–C_2) compounds such as phenylacetaldehyde and 2-phenylethanol from Phe does not occur via *trans*-cinnamic acid and competes with PAL for Phe utilization (Boatright et al. 2004; Kaminaga et al. 2006; Tieman et al. 2006). Moreover, a quantitative explanation of the labeling kinetics of phenylacetaldehyde and 2-phenylethanol from deuterium-labeled Phe suggest that phenylacetaldehyde in petunia flowers is not the only precursor of 2-phenylethanol and that the major flux to the latter goes through a different route (Orlova et al. 2006), possibly through phenylpyruvate and phenyllactic acid as recently reported in rose flowers (Watanabe et al. 2002). In contrast, neither phenylpyruvate nor phenyllactate were detected in tomato (*Solanum lycopersicum*) fruits implying that in this system, 2-phenylethanol is mainly formed from phenylacetaldehyde via the action of recently identified phenylacetaldehyde reductases (Tieman et al. 2006, 2007). Biosynthesis of phenylacetaldehyde from Phe requires the removal of both the carboxyl and amino groups, which in petunia occurs by the action of phenylacetaldehyde synthase (PAAS; Kaminaga et al. 2006). In tomato, Phe is first converted to phenylethylamine by aromatic amino acid decarboxylase

(AADC) and requires the action of a hypothesized amine oxidase, dehydrogenase, or transaminase for phenylacetaldehyde formation (Tieman et al. 2006).

The volatile phenylpropenes (C_6–C_3), such as eugenol, isoeugenol, methyleugenol, isomethyleugenol, chavicol, and methylchavicol share the initial biosynthetic steps with the lignin biochemical pathway up to the phenylpropenol (monolignol) stage and then require two enzymatic reactions to eliminate the oxygen functionality at C-9 position (Koeduka et al. 2006; Dexter et al. 2007). Phenylpropene-forming enzymes have been isolated and characterized from basil, petunia and *C. breweri* (Koeduka et al. 2006, 2007). In these species, coniferyl alcohol is first converted to coniferyl acetate by coniferyl alcohol acetyltransferase (CFAT; Dexter et al. 2007; D.R. Gang, personal communication) prior to its reduction by eugenol synthase or isoeugenol synthase (EGS or IGS) to eugenol and isoeugenol, respectively. However, CFAT and its homologs have been identified only in basil and petunia thus far. Although eugenol and isoeugenol differ exclusively by the position of the double bond in the propene side chain, in petunia their formation is mediated by two different and highly diverged NADPH-dependent reductases (Koeduka et al. 2007). In contrast, *C. breweri* contains three distinct NADPH-dependent reductases, two of which are responsible for eugenol formation and the third possessing isoeugenol synthase activity. While one eugenol synthase (EGS) isoform is closely related to isoeugenol synthase (IGS), the other is highly diverged (Koeduka et al. 2007). Similar to the role of coniferyl acetate in eugenol and isoeugenol formation in petunia, coumaryl acetate serves as the biosynthetic precursor of chavicol in basil (Vassao et al. 2006). Often eugenol, isoeugenol, and chavicol undergo further methylation and *O*-methyltransferases responsible for the downstream production of methyl eugenol, isomethyl eugenol and methyl chavicol have been identified and characterized from *C. breweri* and basil plants (Wang et al. 1997; Gang et al. 2002).

20.2.2 Aromatic Volatiles Derived from Chorismate

One volatile benzenoid compound formed independently from Phe is methyl anthranilate, whose precursor anthranilate is synthesized from chorismate (Radwanski and Last 1995). While anthranilate synthase has been isolated and characterized from several plant species (reviewed in Radwanski and Last 1995), an enzyme responsible for anthranilate carboxyl methylation has not yet been identified. It is possible that a carboxyl methyltransferase belonging to SABATH family of methyltransferases is responsible for this process in some species, however a member of the BAHD superfamily of acyltransferases, anthraniloyl-coenzyme A:methanol acyltransferase (AMAT), was found to be responsible for the formation of methyl anthranilate in Concord grape (*Vitis labrusca*; Wang and De Luca 2005). Anthraniloyl-CoA substrate used by AMAT could be synthesized by a BZL analog as was demonstrated for partially purified BZL from *C. breweri*, which possessed anthraniloyl-CoA ligase activity (Beuerle and Pichersky 2002).

Aromatic volatile plant defense compounds formed independently of Phe also include the nitrogen containing compound indole, a heterocyclic organic metabolite common to higher plants. Indole formation in plants occurs from chorismate via the well-characterized tryptophan biosynthesis pathway, preceding tryptophan formation (Radwanski and Last 1995). The final step of indole biosynthesis is catalyzed by indole-3-glycerol phosphate lyase (IGL), which cleaves phosphorylated glycerol from indole-3-glycerol phosphate with concurrent formation of the volatile indole (Frey et al. 2004). Maize IGL was likely derived from tryptophan synthase whose alpha subunit is responsible for formation of indole that is channeled to a beta subunit, converting it to tryptophan. During evolution, divergence of IGL from tryptophan synthase has allowed it to function efficiently independent of the beta subunit allowing for the release of volatile indole, an event that does not occur in tryptophan biosynthesis. Moreover, the *IGL* gene is transcriptionally activated in maize in response to herbivory and exogenous application of methyl jasmonate or volicitin from pest regurgitant. Its induced expression pattern parallels the level of volatile indole emission (Frey et al. 2000).

20.3 Aromatic Volatiles and Plant Defense

In the past two decades it has been well documented that plants emit blends of volatile compounds from their tissues in response to herbivore damage. Odor blends emitted by attacked plants are diverse; they are composed of more than 200 different compounds often present as minor constituents (Dicke and van Loon 2000). In many cases these minor constituents are aromatic compounds. Volatiles emitted can directly affect herbivore physiology and behavior due to their potentially toxic, repellent, or deterrent properties (Bernasconi et al. 1998; De Moraes et al. 2001; Kessler and Baldwin 2001; Vancanneyt et al. 2001; Aharoni et al. 2003) and attract enemies of attacking herbivores such as parasitic wasps, flies or predatory mites, to protect the signaling plant from further damage (Dicke et al. 1990; Turlings et al. 1990; Vet and Dicke 1992; Paré and Tumlinson 1997; Drukker et al. 2000; Kessler and Baldwin 2001). Moreover, some volatile compounds can mediate both direct and indirect defenses, as is shown for methyl salicylate (see below).

20.3.1 Aromatic Volatiles in Plant Direct Defense

The involvement of volatile compounds in plant direct defense is widely accepted, though, to date, surprisingly little is known about the role of aromatic volatiles in the intoxication, repulsion, or deterrence of herbivores. Only a small portion of 329 known volatile phenylpropanoids (Knudsen and Gershenzon 2006) were found in the headspaces of the herbivore-damaged plants and those were released from a limited range of plant taxa and often at very low levels (Tables 20.1 and 20.2). Moreover, biological activity toward insects has been shown for even fewer aromatic

Table 20.1 Herbivore-induced volatile aromatics. Entries are grouped first by compound, then plant species, pest species, and publication date

Compound	Plant Species	Herbivore	Reference
Indol	Arachis hypogaea	Spodoptera exigua	Cardoza et al. (2002)
	Gossypium herbaceum	Spodoptera littoralis	Gouinguene et al. (2005)
	Gossypium hirsutum	Spodoptera exigua	Loughrin et al. (1994, 1995a); Röse et al. (1996); Paré and Tumlinson (1997, 1998) and Rodriguez-Saona et al. (2003)
	Phaseolus lunatus	Spodoptera littoralis	Mithöfer et al. (2005)
	Vigna unguiculata	Spodoptera littoralis	Gouinguene et al. (2005)
	Zea mays	Rhopalosiphum maidis	Bernasconi et al. (1998)
	Zea mays	Spodoptera exigua	Frey et al. (2000); Schmelz et al. (2003) and D'Alessandro et al. (2006)
	Zea mays	Spodoptera littoralis	D'Alessandro and Turlings (2005) and Gouinguene et al. (2005)
Methylsalicylate	Capsicum annuum	Tetranychus urticae	Van Den Boom et al. (2004)
	Datura stramonium	Tetranychus urticae	Van Den Boom et al. (2004)
	Glycine max	Tetranychus urticae	Van Den Boom et al. (2004)
	Humulus lupulus	Tetranychus urticae	Van Den Boom et al. (2004)
	Medicago truncatula	Spodoptera littoralis	Leitner et al. (2005)
	Robinia pseudoacacia	Tetranychus urticae	Van Den Boom et al. (2004)
	Vicia faba	Aphis fabae	Hardie et al. (1994)
	Vigna unguiculata	Spodoptera littoralis	Gouinguene et al. (2005)
	Vigna unguiculata	Tetranychus urticae	Van Den Boom et al. (2004)
	Vitis vinifera	Tetranychus urticae	Van Den Boom et al. (2004)
Methylanthranilate	Gossypium herbaceum	Spodoptera littoralis	Gouinguene et al. (2005)
	Zea mays	Rhopalosiphum maidis	Bernasconi et al. (1998)

Table 20.1 (continued)

Compound	Plant Species	Herbivore	Reference
	Zea mays	Spodoptera exigua	D'Alessandro et al. (2006)
	Zea mays	Spodoptera littoralis	Degen et al. (2004); D'Alessandro and Turlings (2005) and Gouinguene et al. (2005)
2-phenylethyl acetate	Zea mays	Rhopalosiphum maidis	Bernasconi et al. (1998)
	Zea mays	Spodoptera exigua	D'Alessandro et al. (2006)
	Zea mays	Spodoptera littoralis	Degen et al. (2004); D'Alessandro and Turlings (2005) and Gouinguene et al. (2005)
Benzyl acetate	Zea mays	Rhopalosiphum maidis	Bernasconi et al. (1998)
	Zea mays	Spodoptera exigua	D'Alessandro et al. (2006)
	Zea mays	Spodoptera littoralis	Degen et al. (2004) and Gouinguene et al. (2005)
Methylbenzoate	Gossypium herbaceum	Spodoptera littoralis	Gouinguene et al. (2005)
	Malus sp.	Tetranychus urticae	Takabayashi et al. (1991)
	Vigna unguiculata	Spodoptera littoralis	Gouinguene et al. (2005)
	Zea mays	Spodoptera littoralis	Gouinguene et al. (2005)
2-phenylethanol	Malus sp.	Popilla japonica	Loughrin et al. (1995b)
	Robinia pseudoacacia	Tetranychus urticae	Van Den Boom et al. (2004)
3-hexenyl benzoate	Malus sp.	Panonychus ulmi	Takabayashi et al. (1991)
	Malus sp.	Tetranychus urticae	Takabayashi et al. (1991)
Cresol	Medicago truncatula	Tetranychus urticae	Leitner et al. (2005)
	Phaseolus lunatus	Tetranychus urticae	Hopke et al. (1994)
Trimethylbenzene	Medicago truncatula	Spodoptera littoralis	Leitner et al. (2005)
	Medicago truncatula	Tetranychus urticae	Leitner et al. (2005)

Table 20.1 (continued)

Compound	Plant Species	Herbivore	Reference
3,5 dimethoxytoluene	*Medicago truncatula*	*Spodoptera littoralis*	Leitner et al. (2005)
3,5 dimethylanisole	*Medicago truncatula*	*Tetranychus urticae*	Leitner et al. (2005)
Benzene acetonitrile	*Robinia pseudoacacia*	*Tetranychus urticae*	Van Den Boom et al. (2004)
Benzyl alcohol	*Phaseolus lunatus*	*Tetranychus urticae*	Hopke et al. (1994)
Ethyl benzoate	*Malus* sp.	*Tetranychus urticae*	Takabayashi et al. (1991)
Phenylacetonitrile	*Malus* sp.	*Popilla japonica*	Loughrin et al. (1995)

volatiles. The repellent properties of methyl salicylate toward several aphid species have been observed both in olfactometer and field experiments. In behavioral studies, methyl salicylate repelled the black bean aphid (*Aphis fabae*) and also inhibited its attraction to host-specific cues from broad bean (*Vicia faba*; Hardie et al. 1994). In the field, exogenous application of methyl salicylate delayed the immigration and setting of bird cherry oat aphids (*Rhopalosiphum padi*) on barley plants (Pettersson et al. 1994; Ninkovic et al. 2003).

Another aromatic compound possessing repellent activity is eugenol, a major constituent of cloves and some basil species. Eugenol is highly repellent to the four beetle species *Sitophilus granaries, Sitophilus zeamais, Tribolium castaneum,* and *Prostephanus truncates* and inhibits the development of *S. granaries* and *S. zeamais* eggs, larvae, and pupae inside grain kernels (Obeng-Ofori and Reichmuth 1997). In addition, insecticidal activities have been shown for a broad range of other aromatic compounds such as benzyl benzoate, benzyl salicylate, isoeugenol, methyl eugenol, methyl cinnamate, cinnamaldehyde, safrole, and isosafrole (Ngoh et al. 1998; Bin Jantan et al. 2005). While the emission of these compounds in response to herbivory has yet to be reported in plants, many other aromatic compounds (summarized in Table 20.1) are emitted from herbivore-attacked plants, but their biological activities towards pest and beneficial insects remain to be determined.

20.3.2 Aromatic Volatiles in Plant Indirect Defense

In addition to their direct defenses, herbivore-attacked plants defend themselves indirectly by releasing volatiles that attract predatory or parasitic anthropods, natural enemies of the attacking herbivores. The tritrophic plant-herbivore-predator/ parasitoid interactions, since first suggested in 1980 (Price et al. 1980), are widely

Table 20.2 Volatile aromatics involved in tritrophic interactions. Entries are grouped first by compound, then plant species, pest species, predator/parasitoid species, and publication date

Compound	Plant	Herbivore	Predator/Parasitoid	Reference
Methylsalicylate	Arabidopsis thaliana	Pieris rapae	Cotesia rubecula	Van Poecke et al. (2001)
	Cucumus sativus	Tetranychus urticae	Phytoseiulus persimilis	Takabayashi et al. (1994)
	Gerbera jamesonii	Tetranychus urticae	Phytoseiulus persimilis	Gols et al. (1999)
	Glycine max	Aphis glycines	Coccinella septempunctata	Zhu and Park (2005)
	Lotus japonica	Tetranychus urticae	Phytoseiulus persimilis	Ozawa et al. (2004)
	Nicotiana attenuata	Manduca quinquemaculata	Geocoris pallens	Kessler and Baldwin (2001)
	Nicotiana attenuata	Dicyphus minimus	Geocoris pallens	Kessler and Baldwin (2001, 2004)
	Nicotiana attenuata	Epitrix hertipennis	Geocoris pallens	Kessler and Baldwin (2001)
	Nicotiana attenuata	Manduca sexta	Geocoris pallens	Kessler and Baldwin (2001)
	Phaseolus lunatus	Tetranychus urticae	Phytoseiulus persimilis	Dicke et al. (1990, 1999); Takabayashi and Dicke (1996); De Boer et al. (2004) and De Boer and Dicke (2004a)
	Phaseolus lunatus	Tetranychus urticae	Amblyseius potentillae	Dicke et al. (1990)
	Solanum lycopersicum	Tetranychus urticae	Phytoseiulus persimilis	Kant et al. (2004)
Indole	Gerbera jamesonii	Tetranychus urticae	Phytoseiulus persimilis	Gols et al. (1999)
	Phaseolus lunatus	Tetranychus urticae	Phytoseiulus persimilis	Dicke et al. (1999)
	Zea mays	Heliothis virescens	Cardiochiles nigriceps	De Moraes et al. (1998)
	Zea mays	Pseudaletia separata	Cotesia kariyai	Takabayashi et al. (1995)
	Zea mays	Spodoptera exigua	Cotesia marginiventris	Turlings et al. (1990, 1991, 1993); Hoballah-Fritsche et al. (2002); D'Alessandro et al. (2006) and Ton et al. (2007)
	Zea mays	Spodoptera exigua	Microplitis croceipes	Turlings et al. (1993)
	Zea mays	Spodoptera littoralis	Cotesia marginiventris	Ton et al. (2007)

Table 20.2 (continued)

Compound	Plant	Herbivore	Predator/Parasitoid	Reference
2-phenylethanol	Glycine max	Aphis glycines	Chrysoperla carnea	Zhu and Park (2005)
	Glycine max	Aphis glycines	Chrysoperla carnea	Zhu and Park (2005)
	Solanum tuberosum	Leptinotarsa decemlineata	Perillus bioculatus	Weissbecker et al. (1999)
2-phenylethyl acetate	Zea mays	Pseudaletia separata	Cotesia kariyai	Takabayashi et al. (1995)
	Zea mays	Spodoptera littoralis	Cotesia marginiventris	Hoballah-Fritsche et al. (2002)
	Zea mays	Spodoptera littoralis	Cotesia marginiventris	Ton et al. (2007)
Benzyl acetate	Zea mays	Pseudaletia separata	Cotesia kariyai	Takabayashi et al. (1995)
	Zea mays	Spodoptera littoralis	Cotesia marginiventris	Hoballah-Fritsche et al. (2002)
Benzyl cyanide	Phaseolus lunatus	Tetranychus urticae	Phytoseiulus persimilis	Dicke et al. (1999)

spread in the plant kingdom. To date, this phenomenon has been reported in more than 23 plant species with a diverse combination of plants, herbivores and natural enemies (Dicke 1999), and aromatic volatiles have been reported in more than half of these examples (Table 20.2). One of the best-studied tritrophic interactions is between lima bean plants (*Phaseolus lunatus*), herbivorous spider mites (*Tetranychus urticae*), and carnivorous mites (*Phytoseiulus persimilis*). Infestation of lima bean plants by spider mites triggers the release of volatiles that attract predatory mites that prey on the pests (Takabayashi and Dicke 1996). One component of the volatile blend released by *T. urticae*-infested lima bean leaves is methyl salicylate, an aromatic volatile commonly detected in the headspace of herbivore-infested plants (Table 20.2; Dicke and van Poecke 2002). Methyl salicylate was not detected from untreated or mechanically damaged lima bean plants, in contrast to some other volatile compounds (Dicke et al. 1990). In addition, a dense spider mite infestation (40/leaf) leads to a larger methyl salicylate emission than infestation by smaller pest populations (10/leaf; De Boer et al. 2004). Thus, the level of methyl salicylate released from infested lima beans appears to be positively correlated with the severity of *T. urticae* attack, allowing predator mites to select the plants hosting the largest numbers of their prey.

Methyl salicylate attracts predatory mites in a dose-dependent manner, but at biologically unrealistic amounts (200 μg) repels the carnivores (De Boer and Dicke 2004a). Predators reared on spider mites infesting cucumber, which emits only a trace amount of methyl salicylate (Takabayashi et al. 1994), chose prey-infested lima bean plants over cucumber when tested in an olfactometer. The addition of exogenous methyl salicylate to the cucumber blend made it attractive to the carnivorous mites (De Boer and Dicke 2004b). Moreover, when this single compound was offered as an alternative odor source in a dual-choice assays, the preference of the predators for a methyl salicylate-containing volatile blend from *T. urticae*-infested lima bean plants was lost suggesting an important role of this volatile in the foraging behavior of natural enemies of herbivorous arthropods (De Boer et al. 2004).

Interestingly, lima bean plants release unique volatile blends upon infestation by different pests, thus allowing natural enemies to avoid unproductive foraging if only non-prey species are present. When lima bean plants were infested by the prey herbivore *T. urticae* or nonprey caterpillar *Spodoptera exigua*, carnivorous mites *P. persimilis* were able to locate their prey by detecting differences in the volatile blends emitted in response to herbivory. Addition of methyl salicylate to the volatile blend released by plants infested with nonprey species eliminated the predatory mites' ability to make this distinction, suggesting a specific role for methyl salicylate in this interaction (De Boer et al. 2004). The opposite situation was found in another legume *Medicago truncatula*, in which spider mite (*T. urticae*) feeding failed to induce methyl salicylate emission whereas *Spodoptera littoralis* herbivory did, highlighting the flexible roles of this ester in plant defense (Leitner et al. 2005).

In addition to *P. lunatus* and *M. truncatula*, at least 12 other plant species show documented release of methyl salicylate upon herbivore damage, although at different levels (Tables 20.1 and 20.2). Slight induction of methyl salicylate emission following herbivory has been found in cucumber (*Cucumus sativus*) and gerbera (*Gerbera jamesonii*), where the levels of methyl salicylate were low compared to the other emitted volatiles (Takabayashi et al. 1994; Gols et al. 1999). High levels of methyl salicylate were detected relative to 38 total compounds released from *Arabidopsis* leaves infested with cabbage white butterfly larvae (*Pieris rapae*) but not from artificially damaged or undamaged plants (Van Poecke et al. 2001). *Arabidopsis* is the only cruciferous plant to date shown to emit methyl salicylate in response to herbivore attack, with subsequent attraction of the larval parasitoid *Cotesia rubecula*. In dual-choice bioassays, *C. rubecula* prefers volatile blends containing methyl salicylate from infested *Arabidopsis* plants versus mixtures from artificially damaged or undamaged plants (Van Poecke et al. 2001). Interestingly, caterpillar induced methyl salicylate emission was correlated with an increased transcript abundance of *AtPAL1* suggesting that flux through the pathway towards this compound was induced by herbivore infestation (Van Poecke et al. 2001). In *N. attenuata* the level of methyl salicylate was significantly elevated in volatile blends emitted from plants attacked by three prominent pest species present during the field experiments: the caterpillars of *Manduca quinquemaculata* (Lepidoptera, Sphingidae), the leaf bug *Dicyphus minimus* (Heteroptera, Miridae), and the flea beetle *Epitrix hirtipennis* (Coleoptera, Chrysomelidae; Kessler and Baldwin 2001, 2004), thus attracting the generalist predator *Geocoris pallens* (Kessler and Baldwin 2004).

Infestation of soybean (*Glycine max*), tomato, and *Lotus japonicus* with the herbivorous mite *T. urticae* resulted in release of volatile blends which contained methyl salicylate and were more attractive to the predatory mite *P. persimilis* than volatiles from uninfested or artificially damaged plants (Ozawa et al. 2000; De Boer and Dicke 2004a, b; Kant et al. 2004; Van Den Boom et al. 2004). In soybean plants, methyl salicylate is also released after attack by soybean aphid (*Aphis glycines*), the single aphid species known to develop large colonies on soybean in North America. This is the only compound from the headspace of aphid-infested plants shown to elicit a significant gas chromatography-electroantennography (GC-EAG) response in lady beetle *Coccinella septempunctata*, a species known to attack soybean aphids in the field. In field tests, traps baited with methyl salicylate were highly attractive to *C. septempunctata*, but not to the other common lady beetle *Harmonia axyridis*, suggesting that the former may use this volatile ester as an olfactory cue for prey location (Zhu and Park 2005).

Infestation of cowpea (*Vigna unguiculata*) and hops (*Humulus lupulus*) with *T. urticae* also led to either a large increase or a novel release of methyl salicylate while in other species including black locust (*Robinia pseudo-acacia*), sweet pepper (*Capsicum annuum*), thorn apple (*Datura stramonium*), and grapevine (*Vitis vinifera*) its emission was induced to a lesser extent (Van Den Boom et al. 2004). When attractiveness of methyl salicylate was tested in field experiments, it was seen that the grape- and hop-yards containing sticky cards baited with this volatile ester captured significantly greater numbers of predatory insects including *Chrysopa nigricornis*, *Hemerobius* sp , *Deraeocoris brevis*, *Stethorus punctum picipes*, and

Orius tristicolor than the yards with unbaited sticky cards (James and Price 2004). Moreover, this increase of predatory insects coincided with a dramatic reduction in the population of spider mites, the major arthropod pest of hops (James and Price 2004).

Taken together, these results show that methyl salicylate can be a general indicator of herbivore damage in the context of the other volatiles also emitted following infestation. Surprisingly, in many of the tritrophic interactions described above, the herbivore-predator participants are the same while the attacked plants are different and they emit different volatile blends with the exception of methyl salicylate. Analysis of chemoreceptors from parasitoid wasps and predatory mites demonstrated that insects are able to detect this compound within the volatile blend (De Bruyne et al. 1991; Van Poecke and Dicke 2002). However, methyl salicylate is not emitted by all plant taxa and moreover, its emission decreases in maize leaves after infestation by *Pseudaletia separata* larvae (Takabayashi et al. 1995), suggesting niche roles for methyl salicylate in select tritrophic interactions.

Indole is another aromatic compound found in the volatile blends of some herbivore-damaged plants. Although the role of indole in tritrophic interactions lacks extensive study, induced indole emission subsequent to herbivory is well established and has been reported from many plant species including maize (Frey et al. 2000), cotton (*Gossypium birsutum*; Turlings et al. 1995; Paré and Tumlinson 1997), gerbera (Gols et al. 1999), lima bean (Mithöfer et al. 2005), and peanut (Cardoza et al. 2002). In maize, indole becomes a major constituent of the volatile spectrum released a few hours after feeding by the beet armyworm (*S. exigua*; Turlings et al. 1990). Its biosynthesis occurs de novo in response to insect damage (Paré and Tumlinson 1998) and positively correlates with the level of infestation (Schmelz et al. 2003). The level of indole emission is also affected by the developmental stage of the infesting herbivores, as was shown during infestation by *P. separata* larvae (Takabayashi et al. 1995), suggesting that it is more advantageous for the plant to attract parasitoids while the pest larvae are still small. Interestingly, a broad genetic variability was found with respect to herbivore-induced indole emission among 31 maize inbred lines ranging from trace to >70% of total volatile emissions (Degen et al. 2004).

The most prominent and well-studied tritrophic systems involving indole consist of maize plants, armyworms, and parasitic wasps of the genus *Cotesia*. The *Cotesia* sub-species *kariyai* (Takabayashi et al. 1995) and *marginiventris* (Turlings et al. 1991) are known to visit maize plants emitting indole in response to damage by *P. separata* larvae and *S. exigua* larvae, respectively. However, the emissions of indole and other herbivore-induced volatile compounds are not always pest-species specific. In maize, regurgitates from five different caterpillar species as well as grasshoppers induce the same set and relative ratio of volatile compounds including indole, but at different total amounts, indicating that these highly detectable cues may not always reliably signal the presence of a suitable host to foraging predators and parasitoids (Turlings et al. 1993). Surprisingly, attraction of *C. marginiventris* was not affected by the presence or absence of indole in studies where caterpillar-infested maize plants and those treated with glyphosate, an inhibitor of the 5-enolpyruvylshikimate-3-phosphate, were studied together. Furthermore, indole seemed to repel rather than attract the parasitoid *Microplitis rufiventris* (D'Alessandro et al. 2006) suggesting that two parasitoids with a comparable biology may employ different strategies in their use of plant-provided cues to locate hosts and indole might play different roles in inducible defenses across plant and insect taxa.

Coincidence of indole emissions and *C. marginiventris* parasitoid activity has also been documented in cotton plants following damage by *S. littoralis* (Paré and Tumlinson 1997) or *S. exigua* (Röse et al. 1996; Loughrin et al. 1994, 1995a). Similar to the situation in maize, indole emissions following herbivory have been shown to differ greatly between cotton cultivars (Loughrin et al. 1995a). However, in contrast to maize, release of indole from cotton plants occurs only at the site of caterpillar damage and not systemically (Röse et al. 1996; Turlings and Tumlinson 1992). Interestingly, simultaneous infestation of cotton plants by *S. exigua* and the whitefly *Bemisia tabaci* led to a reduction of indole emission possibly weakening the attraction strengths between parasitoids and their hosts (Rodriguez-Saona et al. 2003).

When two types of attacking herbivores infested lima bean plants, only feeding by *S. exigua* induced indole emission but not *T. urticae* (De Boer et al. 2004). Even a high-density infestation by spider mites led to only low indole emission whereas feeding by moderate numbers of *S. exigua* larvae (two/leaf) resulted in drastic increases of indole release compared to control and spider mite test groups (De Boer et al. 2004). Neither artificial mechanical damage nor damage done by the snail *Cepaea hortensis* caused a release of indole (Mithöfer et al. 2005). Interestingly, only the jasmonic acid-related signaling pathway is involved in the production of caterpillar-induced volatiles in lima bean plants, while both the salicylic acid- and jasmonic acid-related signaling pathways are involved in the production of *T. urticae*-induced volatiles (Ozawa et al. 2000).

The aromatic volatiles benzylacetate, 2-phenethyl acetate, benzyl cyanide, and methyl anthranilate were also released following herbivore attack (Table 20.2). These compounds were emitted from maize after armyworm infestation with the exception of benzylcyanide and 2-phenylethanol, which were detected in lima beans and potato plants following spider mite and potato beetle damage, respectively (Takabayashi et al. 1995; Bernasconi et al. 1998; Hoballah-Fritsche et al. 2002; D'Alessandro and Turlings 2005; Dicke et al. 1999; Weissbecker et al. 1999). Moreover, it has been shown that 2-phenylethanol, emitted exclusively following potato beetle feeding, elicits very high GC-EAG responses from *Perillus bioculatus*, a predator of potato beetle (Weissbecker et al. 1999).

Release after herbivory of a majority of aromatic compounds at relatively low levels raises the question about their roles in the interactions between plants and arthropods. GC-EAG recordings show that the relative quantities of compounds within a volatile blend do not always correlate with their elicited EAG responses, indicating that minor compounds may provoke some of the strongest responses (Gouinguene et al. 2005; Zhu and Park 2005) and the key insect attractants remain to be determined.

20.4 Aromatic Volatiles and Plant–Plant Interactions

Volatiles released from herbivore-infested plants also mediate plant–plant interactions and may induce expression of defense genes and emission of volatiles in healthy leaves of the same plant or on neighboring, unattacked plants, thus increasing their attractiveness to carnivores and decreasing their susceptibility to the damaging herbivores (Dicke et al. 1990; Arimura et al. 2002, 2004; Ruther and Kleier 2005). Exposure of maize to the volatile blend emitted from *S. littoralis*-infested conspecifics reduces caterpillar feeding and development while significantly increasing attractiveness of the volatile-exposed plants to parasitic *C. marginiventris* wasps, thus boosting both their direct and indirect defenses (Ton et al. 2007). This effect could partly be due to the enhanced production of the aromatic compounds indole and 2-phenethyl acetate in plants following their exposure to caterpillar-induced volatiles. Also, neighboring lima bean plants became more attractive to predatory mites and less susceptible to spider mites when exposed to volatiles emitted from conspecific leaves infested with *T. urticae* but not volatiles emitted by artificially wounded leaves (Arimura et al. 2000). *T. urticae*-infestation induced transcript accumulation of six defense-related genes in lima bean leaves, while exposure of undamaged plants to volatiles from infested plants upregulated transcript levels of only five out of six genes investigated including *PAL*. Treatment of undamaged plants with gaseous methyl salicylate reproduced the expression patterns of six defense genes observed in the *T. urticae*-infested leaves, indicating the potential importance of this compound in plant–plant interactions (Arimura et al. 2000).

In addition to direct elicitation of defenses, exposure to volatile compounds from attacked plants may lead to priming of plant defensive responses in their neighbors. Priming by volatiles prepares the plant to respond more rapidly and intensely against subsequent attack by herbivorous insects (Engelberth et al. 2004; Kessler et al. 2006; Tumlinson and Engelberth this volume). Priming by volatile compounds provides a different way of responding to the threat of insect herbivory via the incomplete activation of defense-related processes, thus reducing biochemical investment towards defenses for the receiver plants until the onset of actual attack (Engelberth et al. 2004; Kessler et al. 2006). While priming by aromatic volatile compounds could be one of the mechanisms involved in plant–plant signaling in nature, the roles of the aromatic volatiles in and the underlying molecular mechanisms and ecological relevance of these particular interactions still remain to be determined.

20.5 Future Directions

To date, it is widely accepted that airborne volatile compounds play important roles in plant defense, however, the contribution of aromatics to these processes still requires further investigation. The most interesting unsolved question is why aromatic volatiles lack the broad representation seen with terpenoids in herbivore-induced volatile blends. The ability of plants to synthesize a wide spectrum of volatile aromatic compounds, often at high levels, has been shown in floral organs

(Verdonk et al. 2003; Boatright et al. 2004). However, in the limited number of plant species studied, aromatic volatile compounds are comparatively underrepresented in plant defense. This discrepancy will be clarified through investigation of plant-insect interactions occurring in a greater number of plants. Emission of lower levels of aromatic compounds may also indicate that their actions are synergistic and may incorporate a level of specificity to the airborne signal, possibilities requiring further exploration at the sensory levels and in field experiments.

The importance of aromatic volatiles should not be overlooked based on their subtle presences since insects have an astonishing ability to detect low levels of volatile compounds that may provoke some of the strongest responses. Previous attempts to determine the effects of individual aromatic volatiles on insect behavior were typically accomplished by the addition of exogenous compounds to existing odor mixtures. Recent breakthroughs in gene discovery and metabolic engineering now make it possible to manipulate the levels of specific volatiles *in planta*, either omitting or increasing their emission. Although the contributions of some volatile terpenoids to plant defense have been explored via this method (Aharoni et al. 2003; Kappers et al. 2005; Schnee et al. 2006), it has yet to be applied to volatile aromatics and will likely corroborate and clarify their previously conceived roles in plant-insect interactions. Of particular interest is how flux through the shikimate pathway, which provides precursors to aromatic volatiles, is regulated during plant defense with the goal of determining whether higher emission of aromatics would be beneficial for plant fitness and defense. Recent discovery of the MYB transcription factor AtMYB15, which activates the shikimate pathway in response to wounding (Chen et al. 2006), will enable us to solve this question via metabolic engineering.

Another important unsolved question concerns the factors determining specificity of plant response to a particular threat. An understanding of signal transduction pathways that lead to the herbivore-induced plant defenses will shed some light on possible mechanisms for the exclusive release of certain compounds in response to given stimuli. In addition, the use of -omics-based approaches applied to both plants and pests will be useful in elucidating general mechanisms involved in plant-insect and plant-plant interaction. Understanding of the mechanisms underlying plant-insect interaction and in particular the role of volatiles in biological control of pests will allow us in the future to supplement and optimize plant volatile signaling and generate crops with enhanced attractiveness to natural enemies during herbivory, providing a reduced need for chemical pesticide application in agriculture.

Acknowledgments This work was supported by grants from the National Science Foundation (Grant No. MCB-0615700), the USDA Cooperative State Research, Education, and Extension Service (Grant No. 2005-35318-16207) and the Fred Gloeckner Foundation.

References

Abd El-Mawla AMA, Schmidt W, Beerhues L (2001) Cinnamic acid is a precursor of benzoic acids in cell cultures of *Hypericum cndrosaemum* L. but not in cell cultures of *Centaurium erythraea* RAFN. Planta 212:288–293

Aharoni A, Giri AP, Deuerlein S, Griepink F, De Kogel WJ, Verstappen FWA, Verhoeven HA, Jongsmaa MA, Schwab W, Bouwmeester HJ (2003) Terpenoid metabolism in wild-type and transgenic Arabidopsis plants. Plant Cell 15:2866–2884

Arimura G, Ozawa R, Kugimiya S, Takabayashi J, Bohlmann J (2004) Herbivore-induced defense response in a model legume: two-spotted spider mites, *Tetranychus urticae*, induce emission of (E)-β-ocimene and transcript accumulation of (E)-β-ocimene synthase in *Lotus japonicus*. Plant Physiol 135:1976–1983

Arimura G, Ozawa R, Nishioka T, Boland W, Koch T, Kuhnemann F, Takabayashi J (2002) Herbivore-induced volatiles induce the emission of ethylene in neighboring lima bean plants. Plant J 29:87–98

Arimura G, Ozawa R, Shomoda T, Nishioka T, Boland W, Takabayashi J (2000) Herbivory-induced volatiles elicit defense genes in lima bean leaves. Nature 406:512–515

Beekwilder J, Alvarez-Huerta M, Neef E, Verstappen FWA, Bouwmeester HJ, Aharoni A (2004) Substrate usage by recombinant alcohol acyltransferases from various fruit species. Plant Physiol 135:1865–1878

Bernasconi M, Turlings TCJ, Ambrosetti L, Bassetti P, Dorn S (1998) Herbivore-induced emissions of maize volatiles repel the corn leaf aphid, *Rhopalosiphum maidis*. Entomol Exp Appl 87: 133–142

Beuerle T, Pichersky E (2002) Purification and characterization of benzoate: coenzyme A ligase from *Clarkia breweri*. Arch Biochem Biophys 400:258–264

Bin Jantan I, Yalvema MF, Ahmed NW, Jamal JA (2005) Insecticidal activities of the leaf oils of eight *Cinnamomum* species against *Aedes aegypti* and *Aedes albopictus*. Pharm Biolog 43: 526–532

Birkett MA et al (2000) New roles for *cis*-jasmone as an insect semiochemical and in plant defense. Proc Natl Acad Sci USA 97:9329–9334

Boatright J, Negre F, Chen X, Kish CM, Wood B, Peel G, Orlova I, Gang D, Rhodes D, Dudareva N (2004) Understanding in vivo benzenoid metabolism in petunia petal tissue. Plant Physiol 135:1993–2011

Cardoza YJ, Alborn HT, Tumlinson JH (2002) In vivo volatile emissions from peanut plants induced by simultaneous fungal infection and insect damage. J Chem Ecol 28:161–174

Chen Y, Xiangbo Z, Wei W, Chen Z, Gu H, Qu LJ (2006) Overexpression of the wounding-responsive gene *AtMYB15* activates the shikimate pathway in *Arabidopsis*. J Int Plant Biol 48:1084–1095

D'Alessandro M, Held M, Triponez Y, Turlings TCJ (2006) The role of indole and other shikimic acid derived maize volatiles in the attraction of two parasitic wasps. J Chem Ecol 32:2733–2748

D'Alessandro M, Turlings TCJ (2005) In situ modification of herbivore-induced plant odors: a novel approach to study the attractiveness of volatile organic compounds to parasitic wasps. Chem Senses 30:739–753

D'Auria JC (2006) Acyltransferases in plants: a good time to be BAHD. Curr Opin Plant Biol 9:331–340

D'Auria JC, Chen F, Pichersky E (2002) Characterization of an acyltransferase capable of synthesizing benzylbenzoate and other volatile esters in flowers and damaged leaves of *Clarkia breweri*. Plant Physiol 130:466–476

D'Auria JC, Chen F, Pichersky E (2003) The SABATH family of methyltransferases in *Arabidopsis thaliana* and other plant species. Rec Adv Phytochem 37:253–283

De Boer JG, Dicke MA (2004a) The role of methyl salicylate in prey searching behavior of the predatory mite *Phytoseiulus persimilis*. J Chem Ecol 30:255–271

De Boer JG, Dicke MA (2004b) Experience with methyl salicylate affects behavioural responses of a predatory mite to blends of herbivore-induced plant volatiles. Entomol Exp Appl 110: 181–189

De Boer JG, Posthumus MA, Dicke MA (2004) Identification of volatiles that are used in discrimination between plants infested with prey or nonprey herbivores by a predatory mite. J Chem Ecol 30:2215–2230

De Bruyne M, Dicke MA, Tjallingii WF (1991) Receptor cell responses in the anterior tarsi of *Phytoseiulus persimilis* to volatile kairomone components. Exp Appl Acarol 13:53–58

Degen T, Dillmann C, Marion-Poll F, Turlings TCJ (2004) High genetic variability of herbivore-induced volatile emission within a broad range of maize inbred lines. Plant Physiol 135: 1928–1938

De Moraes CM, Mescher MC, Tumlinson JH (2001) Caterpillar-induced nocturnal plant volatiles repel conspecific females. Nature 210:577–580

Dexter R, Qualley A, Kish CM, Ma CJ, Koeduka T, Nagegowda DA, Dudareva N, Pichersky E, Clark D (2007) Characterization of a petunia acetyltransferase involved in the biosynthesis of the floral volatile isoeugenol. Plant J 49:265–275

Dicke MA (1999) Are herbivore-induced plant volatiles reliable indicators of herbivore identity to foraging carnivorous arthropods? Entomol Exp Appl 91:131–142

Dicke MA, Abelis MW, Takabayashi J, Bruin J, Posthumus MA (1990) Plant strategies of manipulating predator-prey interactions through allelochemicals: prospects for application in pest control. J Chem Ecol 16:3091–3117

Dicke MA, Gols R, Ludeking D, Posthumus MA (1999) Jasmonic acid and herbivory differentially induce carnivore attracting plant volatiles in lima bean plants. J Chem Ecol 25:1907–1922

Dicke MA, Van Beek TA, Posthumus MA, Van Bokhoven H, De Groot AE (1990) Isolation and identification of volatile kairomone that affects acarine predatorprey interactions Involvement of host plant in its production. J Chem Ecol 16:381–396

Dicke MA, Van Loon JJA (2000) Multitrophic effects of herbivore-induced plant volatiles in an evolutionary context. Entomol Exp Appl 97:237–249

Dicke MA, Van Poecke RMP (2002) Signaling in plant-insect interactions: signal transduction in direct and indirect plant defence. In: Scheel D, Wasternack C (eds) Plant signal transduction: frontiers in molecular biology. Oxford University Press, Oxford, pp 289–316

Dixon RA, Achnine L, Kota P, Liu CJ, Reddy MSS, Wang LJ (2002) The phenylpropanoid pathway and plant defence – a genomics perspective. Mol Plant Pathol 3:371–390

Drukker B, Bruin J, Jacobs G, Kroon A, Sabelis MW (2000) How predatory mites learn to cope with variability in volatile plant signals in the environment of their herbivorous prey. Exp Appl Acarol 24:881–895

Dudareva N, D'Auria JC, Nam KH, Raguso RA, Pichersky E (1998) Acetyl-CoA: benzylalcohol acetyltransferase – an enzyme involved in floral scent production in *Clarkia breweri*. Plant J 14:297–304

Dudareva N, Negre F, Nagegowda DA, Orlova I (2006) Plant volatiles: recent advances and future perspectives. Crit Rev Plant Sci 25:417–440

Dudareva N, Pichersky E, Gershenzon J (2004) Biochemistry of plant volatiles. Plant Physiol 135:1893–1902

Effmert U, Saschenbrecker S, Ross J, Negre F, Fraser CM, Noel JP, Dudareva N, Piechulla B (2005) Floral benzenoid carboxyl methyltransferases: from in vitro to *in planta* function. Phytochemistry 66:1211–1230

Engelberth J, Alborn HT, Schmelz EA, Tumlinson JH (2004) Airborne signals prime plants against insect herbivore attack. Proc Natl Acad Sci USA 101:1781–1785

Farag MA, Fokar M, Abd, H, Zhang H, Allen RD, Pare PW (2005) (Z)-3-Hexenol induces defense genes and downstream metabolites in maize. Planta 220:900–909

French CJ, Vance CP, Towers GHN (1976) Conversion of *p*-coumaric acid to *p*-hydroxybenzoic acid by cell free extracts of potato tubers and *Polyporus hispidus*. Phytochem 15:564–566

Frey M, Spiteller D, Boland W, Gierl A (2004) Transcriptional activation of Igl, the gene for indole formation in *Zea mays*: a structure-activity study with elicitor-active N-acyl glutamines from insects. Phytochem 65:1047–1055

Frey M, Stettner C, Paré PW, Schmelz EA, Tumlinson JH, Gierl A (2000) An herbivore elicitor activates the gene for indole emission in maize. Proc Natl Acad Sci USA 97:14801–14806

Gang DR, Lavid N, Zubieta C, Chen F, Beuerle T, Lewinsohn E, Noel JP, Pichersky E (2002) Characterization of phenylpropene *O*-methyltransferases from sweet basil: facile change of substrate specificity and convergent evolution within a plant OMT family. Plant Cell 14: 505–519

Gols R, Posthumus MA, Dicke MA (1999) Jasmonic acid induces the production of gerbera volatiles that attract the biological control agent *Phytoseiulus persimilis*. Entomol Exp Appl 93:77–86

Gouinguene S, Pickett JA, Wadhams LJ, Birkett MA, Turlings TCJ (2005) Antennal electrophysiological responses of three parasitic wasps to caterpillar-induced volatiles from maize (*Zea mays* Mays), cotton (*Gossypium herbaceum*), and cowpea (*Vigna unguiculata*). J Chem Ecol 31:1023–1038

Hardie J, Isaacs R, Pickett JA, Wadhams LJ, Woodcock CM (1994) Methyl salicylate and (−)-(*1R, 5S*)-myrtenal are plant-derived repellents for black bean aphid, *Aphis fabae* Scop. (Homoptera: Aphididae) J Chem Ecol 20:2847–2855

Hoballah-Fritsche ME, Tamó C, Turlings TCJ (2002) Differential attractiveness of induced odors emitted by eight maize varieties for the parasitoid *Cotesia marginiventris*: is quality important? J Chem Ecol 28:951–968

Hopke J, Donath J, Blechert S, Boland W (1994) Herbivore-induced volatiles: the emission of acyclic homoterpenes from *Phaseolus lunatus* and *Zea mays* can be triggered by a beta-glucosidase and jasmonic acid. FEBS Lett 352:146–150

James DG, Price TS (2004) Field-testing of methyl salicylate for recruitment and retention of beneficial insects in grapes and hops. J Chem Ecol 30:1613–1627

Jarvis AP, Schaaf O, Oldham NJ (2000) 3-Hydroxy-3-phenylpropanoic acid is an intermediate in the biosynthesis of benzoic acid and salicylic acid but benzaldehyde is not. Planta 212: 119–126

Kaminaga Y, Schnepp J, Peel G, Kish CM, Ben-Nissan G, Weiss D, Orlova I, Lavie O, Rhodes D, Wood K, Porterfield M, Cooper AJL, Schloss JV, Pichersky E, Vainstein A, Dudareva N (2006) Plant phenylacetaldehyde synthase is a bifunctional homotetrameric enzyme that catalyzes phenylalanine decarboxylation and oxidation. J Biol Chem 281:23357–23366

Kant MR, Ament K, Sabelis MW, Haring MA, Schuurink RC (2004) Differential timing of spider mite-induced direct and indirect defenses in tomato plants. Plant Physiol 135:483–495

Kappers IF, Aharoni A, Van Herpen TWJM, Luckerhoff LLP, Dicke MA, Bouwmeester HJ (2005) Genetic engineering of terpenoid metabolism attracts bodyguards to *Arabidopsis*. Science 309:2070–2072

Kapteyn J, Qualley AV, Xie Z, Fridman E, Dudareva N, Gang DR (2007) Evolution of cinnamate/*p*-coumarate carboxyl methyltransferases and their role in the biosynthesis of methylcinnamate. Plant Cell 19:3212–3229

Kessler A, Baldwin IT (2001) Defensive function of herbivore-induced plant volatile emissions in nature. Science 291:2142–2143

Kessler A, Baldwin IT (2004) Herbivore-induced plant vaccination. Part I. The orchestration of plant defenses in nature and their fitness consequences in the wild tobacco *Nicotiana attenuata*. Plant J 38:639–649

Kessler A, Halitschke R, Diezel C, Baldwin IT (2006) Priming of plant defense responses in nature by airborne signaling between *Artemisia tridentata* and *Nicotiana attenuata*. Oecologia 148:280–292

Knudsen JT, Gershenzon J (2006) The chemical diversity of floral scent. In: Dudareva N, Pichersky E (eds) Biology of floral scent. Taylor & Francis, Boca Raton, pp 27–52

Koeduka T, Fridman E, Gang DR, Vassão DG, Jackson BL, Kish CM, Orlova I, Spaaova SM, Lewis NG, Noel JP, Baiga TJ, Dudareva N, Pichersky E (2006) Eugenol and isoeugenol, characteristic aromatic constituents of spices, are biosynthesized via reduction of a coniferyl alcohol ester. Proc Natl Acad Sci USA 103:10128–10133

Koeduka T, Orlova I, Kish CM, Ibdah M, Wilkerson CG, Baiga TJ, Noel JP, Dudareva N, Pichersky E (2007) The multiple phenylpropene synthases in both *Clarkia breweri* and *Petunia hybrida* represent two distinct protein lineages and continue to evolve their product specificity. Plant J (in press)

Leitner M, Boland W, Mithöfer A (2005) Direct and indirect defences induced by piercing-sucking and chewing herbivores in *Medicago truncatula*. New Phytol 167:597–606

Leon J, Shulaev V, Yalpani N, Lawton MA, Raskin I (1995) Benzoic acid 2-hydroxylase, a soluble oxygenase from tobacco, catalyzes salicylic acid biosynthesis. Proc Natl Acad Sci USA 92:10413–10417

Loughrin JH, Manukian A, Heath RR, Tumlinson JH (1995a) Volatiles emitted by different cotton varieties damaged by feeding beet armyworm larvae. J Chem Ecol 21:1217–1227

Loughrin JH, Manukian A, Heath RR, Turlings TCJ (1994) Diurnal cycle of emission of induced volatile terpenoids by herbivore-injured cotton plants. Proc Natl Acad Sci USA 91: 11836–11840

Loughrin JH, Potter DA, Hamilton-Kemp TR (1995b) Volatile compounds induced by herbivory act as aggregation kairomones for the Japanese beetle (*Popilla japonica* Newman). J Chem Ecol 21:1457–1467

Mercke P, Kappers IF, Verstappen FWA, Vorst O, Dicke MA, Bouwmeester HJ (2004) Combined transcript and metabolite analysis reveals genes involved in spider mite induced volatile formation in cucumber plants. Plant Physiol 135:2012–2024

Mithöfer A, Wanner G, Boland W (2005) Effects of feeding *Spodoptera littoralis* on lima bean leaves. II. Continuous mechanical wounding resembling insect feeding is sufficient to elicit herbivory-related volatile emission. Plant Physiol 137:1160–1168

Ngoh SP, Choo LEW, Pang FY, Huang Y, Kini MR, Ho SH (1998) Insecticidal and repellent properties of nine volatile constituents of essential oils against the American cockroach. *Periplaneta americana* (L.). Pestic Sci 54:261–268

Ninkovic V, Ahmed E, Glinwood R, Pettersson J (2003) Effects of two types of semiochemical on population development of the bird cherry oat aphid *Rhopalosiphum padi* in a barley crop. Agric Forest Entomol 5:27–33

Obeng-Ofori D, Reichmuth CH (1997) Bioactivity of eugenol, a major component of essential oil of *Ocimum suave* (Wild.) against four species of stored-product Coleoptera. Int J Pest Manag 43:89–94

Orlova I, Marshall-Colón A, Schnepp J, Wood B, Varbanova M, Fridman E, Blakeslee JJ, Peer WA, Murphy AS, Rhodes DR, Pichersky E, Dudareva N (2006) Reduction of benzenoid synthesis in petunia flowers reveals multiple pathways to benzoic acid and enhancement in auxin transport. Plant Cell 18:3458–3475

Ozawa R, Shimoda T, Kawaguchi M, Arimura G, Horiuchi J, Nishioka T, Takabayashi J (2000) *Lotus japonicus* infested with herbivorous mites emits volatile compounds that attract predatory mites. J Plant Res 113:427–433

Ozawa R, Shiojiri K, Sabelis MW, Arimura G, Nishioka T, Takabayashi J (2004) Corn plants treated with jasmonic acid attract more specialist parasitoids, thereby increasing parasitation of the common armyworm. J Chem Ecol 30:1797–1808

Paré PW, Tumlinson JH (1997) De novo biosynthesis of volatiles induced by insect herbivory in cotton plants. Plant Physiol 114:1161–1167

Paré PW, Tumlinson JH (1998) Cotton volatiles synthesized and released distal to the site of insect damage. Phytochem 47:521–526

Pettersson J, Pickett JA, Pye BJ, Quiroz A, Smart LE, Wadhams LJ, Woodcock CM (1994) Winter host component reduces colonization by bird-cherry-oat aphid, *Rhopalosiphum padi* (L.) (Homoptera, Aphididae), and other aphids in cereal fields. J Chem Ecol 20:2565–2574

Price PW, Bouton CE, Gross P, McPheron BA, Thompson JN, Weis AE (1980) Interactions among 3 trophic levels – influence of plants on interactions between insect herbivores and natural enemies. Ann Rev Ecol System 11:41–65

Radwanski ER, Last RL (1995) Tryptophan biosynthesis and metabolism – iochemical and molecular genetics. Plant Cell 7:921–934

Rodriguez-Saona C, Crafts-Brandner SJ, Cañas LA (2003) Volatile emissions triggered by multiple herbivore damage: beet armyworm and whitefly feeding on cotton plants. J Chem Ecol 29:2539–2550

Röse USR, Manukian A, Heath RR, Tumlinson JH (1996) Volatile semiochemicals released from undamaged cotton leaves. Plant Physiol 111:487–495

Ruther J, Kleier S (2005) Plant-plant signaling: ethylene synergizes volatile emission in *Zea mays* induced by exposure to (*Z*)-3-hexen-1-ol. J Chem Ecol 31:2217–2222

Schmelz EA, Alborn HT, Banchio E, Tumlinson JH (2003) Quantitative relationships between induced jasmonic acid levels and volatile emissions in *Zea mays* during *Spodoptera exigua* herbivory. Planta 216:665–673

Schnee C, Köllner TG, Held M, Turlings TCJ, Gershenzon J, Degenhardt J (2006) The products of a single maize sesquiterpene synthase form a volatile defense signal that attracts natural enemies of maize herbivores. Proc Natl Acad Sci USA 103:1129–1134

Schnitzler JP, Madlung J, Rose A, Seitz HU (1992) Biosynthesis of *p*-hydroxybenzoic acid in elicitor-treated carrot cell cultures. Planta 188:594–600

Shulaev V, Silverman P, Raskin I (1997) Airborne signalling by methyl salicylate in plant pathogen resistance. Nature 385:718–721

Takabayashi J, Dicke MA (1996) Plant-carnivore mutualism through herbivore-induced carnivore attractants. Trends Plant Sci 1:109–113

Takabayashi J, Dicke M, Posthumus MA (1991) Variation in composition in predator attracting allelochemicals emitted by herbivore-infested plants: relative influence of plant and herbivore. Chemoecology 2:1–6

Takabayashi J, Dicke MA, Takahashi S, Posthumus MA, Van Beek TA (1994) Leaf age affects composition of herbivore-induced synomones and attraction of predatory mites. J Chem Ecol 20:373–386

Takabayashi J, Takahashi S, Dicke MA, Posthumus MA (1995) Developmental stage of herbivore *Pseudaletia separata* affects production of herbivore-induced synomone by corn plants. J Chem Ecol 21:273–287

Tieman DM, Loucas HM, Kim JY, Clark DG, Klee HJ (2007) Tomato phenylacetaldehyde reductases catalyze the last step in the synthesis of the aroma volatile 2-phenylethanol. Phytochemistry 68:2660–2669

Tieman D, Taylor M, Schauer N, Fernie AR, Hanson AD, Klee HJ (2006) Tomato aromatic amino acid decarboxylases participate in synthesis of the flavor volatiles 2-phenylethanol and 2-phenylacetaldehyde. Proc Natl Acad Sci USA 103:8287–8292

Ton J, D'Alessandro M, Jourdie V, Jakab G, Karlen D, Held M, Mauch-Mani B, Turlings TCJ (2007) Priming by airborne signals boosts direct and indirect resistance in maize. Plant J 49:16–26

Turlings TCJ, Loughrin JH, McCall PJ, Röse USR, Lewis WJ (1995) How caterpillar-damaged plants protect themselves by attracting parasitic wasps. Proc Natl Acad Sci USA 92:4169–4174

Turlings TCJ, McCall PJ, Alborn HT, Tumlinson JH (1993) An elicitor in caterpillar oral secretions that induces corn seedlings to emit chemical signals attractive to parasitic wasps. J Chem Ecol 19:411–425

Turlings TCJ, Tumlinson JH (1992) Systemic release of chemical signals by herbivore-injured corn. Proc Natl Acad Sci USA 89:8399–8402

Turlings TCJ, Tumlinson JH, Heath RR, Proveaux AT, Doolittle RE (1991) Isolation and identification of allelochemicals that attract the larval parasitoid, *Cotesia-marginiventris* (Cresson), to the microhabitat of one of its hosts. J Chem Ecol 17:2235–2251

Turlings TCJ, Tumlinson JH, Lewis WJ (1990) Exploitation of herbivore-induced plant odors by host-seeking parasitoid wasps. Science 250:1251–1253

Vancanneyt G, Sanz C, Farmaki T, Paneque M, Ortego F, Castanera P, Sanchez-Serrano JJ (2001) Hydroperoxide lyase depletion in transgenic potato plants leads to an increase in aphid performance. Proc Natl Acad Sci USA 98:8139–8144

Van Den Boom CEM, Van Beek TA, Posthumus MA, De Groot AE, Dicke MA (2004) Qualitative and quantitative variation among volatile profiles induced by *Tetranychus urticae* feeding on plants from various families. J Chem Ecol 30:69–89

Van Poecke RMP, Dicke MA (2002) Induced parasitoid attraction by *Arabidopsis thaliana*: involvement of the octadecanoid and the salicylic acid pathway. J Exp Bot 53:1793–1799

Van Poecke RMP, Posthumus MA, Dicke MA (2001) Herbivore-induced volatile production by *Arabidopsis thaliana* leads to attraction of the parasitoid *Cotesia rubecula*: chemical, behavioral, and gene-expression analysis. J Chem Ecol 27:1911–1928

Vassao DG, Gang DR, Koeduka T, Jackson B, Pichersky E, Davin LB, Lewis NG (2006) Chavicol formation in sweet basil (*Ocimum basilicum*): cleavage of an esterified C9 hydroxyl group with NAD(P)H-dependent reduction. Org Biomol Chem 4:2733–2744

Verdonk JC, De Vos CHR, Verhoeven HA, Haring MA, van Tunen AJ, Schuurink RC (2003) Regulation of floral scent production in petunia revealed by targeted metabolomics. Phytochem 62:997–1008

Vet LEM, Dicke MA (1992) Ecology of infochemical use by natural enemies in a tritrophic context. Ann Rev Entomol 37:141–172

Wang J, De Luca V (2005) The biosynthesis and regulation of biosynthesis of Concord grape fruit esters, including 'foxy' methylanthranilate. Plant J 44:606–619

Wang J, Dudareva N, Bhakta S, Raguso RA, Pichersky E (1997) Floral scent production in *Clarkia breweri* (Onagraceae). II. Localization and developmental modulation of the enzyme SAM:(Iso)Eugenol O-methyltransferase and phenylpropanoid emission. Plant Physiol 114:213–221

Watanabe S, Hayashi K, Yagi K, Asai T, Mactavish H, Picone J, Turnball C, Watanabe N (2002) Biogenesis of 2-phenylethanol in rose flowers: incorporation of [^{2}H$_8$]L-phenylalanine into 2-phenylethanol and its beta-D-glucopyranoside during the flower opening of *Rosa* 'Hoh-Jun' and *Rosa damascena*. Mill Biosci Biotechnol Biochem 66:943–947

Weissbecker B, Van Loon JJA, Dicke MA (1999) Electroantennogram responses of a predator, *Perillus bioculatus*, and its prey, *Leptinotarsa decemlineata*, to plant volatiles. J Chem Ecol 25:2313–2325

Wildermuth MC, Dewdney J, Wu G, Ausubel FM (2001) Isochorismate synthase is required to synthesize salicylic acid for plant defence. Nature 414:562–565

Yazaki K, Heide L, Tabata M (1991) Formation of *p*-hydroxybenzoic acid from *p*-coumaric acid by cell free extract of *Lithospermum erythrorhizon* cell cultures. Phytochem 30:2233–2236

Zhu J, Park K (2005) Methyl salicylate, a soybean aphid-induced plant volatile attractive to the predator *Coccinella setempunctata*. J Chem Ecol 31:1733–1746

Chapter 21
Ecological Roles of Vegetative Terpene Volatiles

Jörg Degenhardt

With their enormous number and large structural diversity, terpenes dominate most plant volatile blends. Terpenes were shown to mediate the interactions between plants and many organisms including arthropods, nematodes, and other plants. Often, these interactions are based on volatile mono- and sesquiterpenes which are released from the plant after damage by herbivores. The terpenes attract herbivore enemies that attack the herbivore and thereby may reduce the damage to the plant. This interaction was termed 'indirect defense' and will often benefit the plant, but other organisms like parasitic plants or insects and can utilize the volatile signals to their advantage. Despite the relatively low number of volatile terpene-mediated interactions identified today, many more are likely to be discovered with the advance of volatile collection methods and molecular techniques.

21.1 Introduction

Terpenes form the largest group of volatile compounds among the natural products of plants. Many of the 30,000 terpenes, especially monoterpenes, sesquiterpenes, and irregular terpenes of low molecular weight have high vapor pressures (Connolly and Hill 1991). High vapor pressure results in the emission of these compounds from the plant into the environment. Although terpene volatilization from flowers has been known for many years (Knudsen et al. 2006), these compounds are also emitted from the vegetative tissues of plants. A detailed analysis of emission patterns in plants revealed that most tissues produce specific blends of terpenes (Connolly and Hill 1991; for the model plants *Arabidopsis* and maize, see Chen et al. 2004; Köllner et al. 2004; Tholl et al. 2005). Some of these are emitted from specialized terpene storage organs, for example the glandular trichomes of mints and the resin ducts in conifers, but many volatiles are not emitted from anatomically-specialized storage sites. Less than two decades ago, it was demonstrated that volatiles emitted from

---

J. Degenhardt
Department of Biochemistry, Max Planck Institute for Chemical Ecology, D-07745 Jena, Germany
e-mail: degenhardt@ice.mpg.de

A. Schaller (ed.), *Induced Plant Resistance to Herbivory*,
© Springer Science+Business Media B.V. 2008

plants after herbivore damage can attract enemies of the herbivore (Dicke et al. 1990; Turlings et al. 1990). Under favorable environmental conditions, this attraction of herbivore enemies can limit herbivore damage to the plant. This review will focus on volatile terpene signals from vegetative plant tissues and their roles in plant defense against herbivores.

21.2 Terpenes as Volatile Plant Signals

Unlike fatty acid-derived signals and aromatic volatiles (see contributions by Tumlinson and Engelberth this volume and Qualley and Dudareva this volume), terpenes are commonly emitted in complex blends with a large structural diversity between the compounds. Responsible for most of this terpene diversity are multi-product terpene synthases, the key enzymes of terpene biosynthesis (see Bohlmann this volume; Gershenzor and Kreis 1999). Terpene synthases can produce blends of up to 50 different compounds from one substrate and form mixtures with defined relative ratios of products (Steele et al. 1998). In communication with other organisms, these mixtures might provide a more specific signal than signals consisting of single terpenes.

While low concentrations of volatile terpenes are produced in almost every plant tissue, the highest terpene concentrations from vegetative tissues are released in response to environmental cues. Terpene emission is increased in response to damage and contact with elicitors of herbivores (Mattiacci et al. 1995; Alborn et al. 1997; Halitschke et al. 2001; Spiteller and Boland 2003; Tumlinson and Lait 2005), and abiotic factors like wounding (Howe 2004; Schmelz et al. 2001; Mithöfer et al. 2005), UV-radiation (Johnson et al. 1999), O_3 and CO_2 concentration (Vuorinen et al. 2004a, b; Jasoni et al. 2004; Beauchamp et al. 2005), nutritional status of the plant (Schmelz et al. 2003), as well as temperature and light (Guenther et al. 1993; Takabayashi et al. 1994; Gouinguene and Turlings 2002). In contrast to many compounds of the lipoxygenase pathway, terpenes are usually released after de novo synthesis (Paré and Tumlinson 1997) involving gene transcription (Schnee et al. 2002, 2006; Gomez et al. 2005), and within a time window starting one to two hours after elicitation (Turlings et al. 1998).

Despite the enormous number of terpenes emitted by plants, relatively few terpenes have been clearly identified as signals between organisms to date. In this chapter, we will present some of these interactions involving volatile terpenes and discuss the problems associated with the study of terpene-mediated interactions.

21.2.1 Tritrophic Interactions of Plants with Herbivores and Natural Herbivore Enemies

Almost two decades ago, researchers in the Netherlands and the United States first observed that herbivore damage to certain plants induces the emission of volatile organic compounds that attract natural enemies of the herbivores (Dicke et al. 1990; Turlings et al. 1990). This phenomenon has been reported in more than 15 different plant species after feeding by an assortment of arthropod herbivores and was termed ‚indirect defense' (Dicke 1999; Dicke and van Loon 2000; Kessler and Baldwin 2002; Meiners and Hilker 2000). The herbivore enemies that respond to volatiles from herbivore-damaged plants include various carnivorous arthropods, both predators and parasitoids. Attraction of herbivore enemies has been shown to benefit the plant by reducing subsequent herbivory and increasing reproductive fitness (Hoballah and Turlings 1999; van Loon et al. 2000; Kessler and Baldwin 2001), although such advantages are not realized in all cases (Coleman et al. 1999).

The attraction of the predatory mite *Phytoseiulus persimilis* to lima bean plants infested with the spider mite *Tetranychus urticae* has been studied in detail (Dicke et al. 1990). Olfactory assays testing the attraction of single compounds from the complex volatile blend demonstrated that the predatory mite was not only attracted to the aromatic compound methyl salicylate (De Boer et al. 2004; De Boer and Dicke 2004; see also chapter of Qualley and Dudareva this volume), but also the sesquiterpene alcohol nerolidol. Transgenic *Arabidopsis* overexpressing a nerolidol synthase from strawberry were used as volatile source in olfactometer experiments with *P. persimilis*. Nerolidol-emitting transgenic plants were more attractive to the predator than undamaged wild type plants (Kappers et al. 2005). In addition to an innate attraction to compounds of the plant volatile bouquet, the predatory mites also have the ability to associate odors with host presence (De Boer and Dicke 2006). The homoterpene $(3E,7E)$-4,8,12-trimethyl-1,3,7,11-tridecatetraene, a irregular C_{16} terpene olefin, is released from lima bean in response to feeding by *T. urticae* but not by the non-host organism *Spodoptera exigua* (Fig. 21.1A). After a series of experiences with the presence and absence of the prey, the predatory mite utilized the homoterpene as an indicator for the presence of prey (De Boer et al. 2004). The ability for associative learning may guide the predatory mites to locate their prey under natural conditions in a complex environment (De Boer and Dicke 2006). Not only does this help the predator to identify plants infested with prey, but also allows it to adapt when the prey switch between different host species during the season (Drukker et al. 2000).

Maize plants damaged by larvae of lepidopteran herbivores like *Spodoptera littoralis* emit a complex blend of volatiles dominated by mono- and sesquiterpenes (Turlings et al. 1991). These volatiles attract females of the parasitic wasp *Cotesia marginiventris* which use the lepidopteran larvae as hosts (Turlings et al. 1990). The parasitation might benefit the maize plants under permitting circumstances since the parasitized larvae feed less and will not procreate (Hoballah et al. 2004; Degenhardt et al. 2003). Since the herbivore-induced volatiles of maize consist of a complex blend of compounds, it is difficult to demonstrate which of the compounds are attractive to the parasitic wasp. The major sesquiterpenes of herbivore-induced maize are produced by the terpene synthase TPS10 which is strongly expressed after herbivory by lepidopterans. TPS10 forms (E)-β-farnesene, (E)-α-bergamotene, and other herbivory-induced sesquiterpene hydrocarbons from the substrate farnesyl diphosphate (Fig. 21.1B; Schnee et al. 2006). Overexpression of TPS10 in

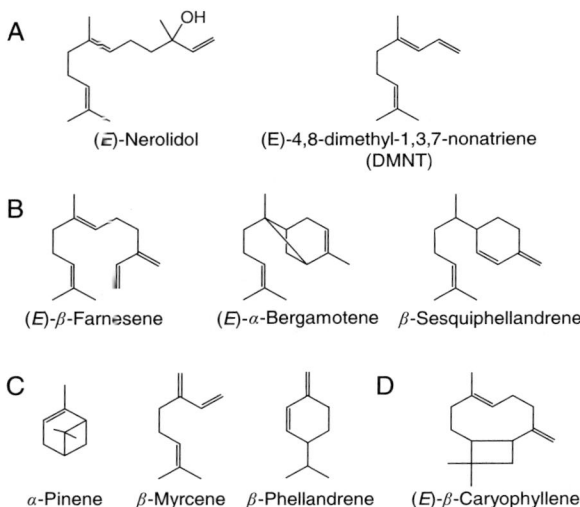

Fig. 21.1 Structures of volatile plant terpenes involved in interactions with other organisms. (**A**) (*E*)-nerolidol and (*E*)-4,8-dimethyl-1,3,7-nonatriene (DMNT) are used by the predatory mite *P. persimilis* to locate spider mites. (**B**) (*E*)-β-farnesene, (*E*)-α-bergamotene and (*E*)-β-sesquiphellandrene attract parasitic wasps to maize damaged by lepidopteran larvae. (**C**) α-pinene, β-myrcene and β-phellandrene attract dodder (*C. pentagona*) to its host plant. (**D**) (*E*)-β-Caryophyllene guides entomophatogenic nematodes to maize roots damaged by *D. v. virgifera*

Arabidopsis thaliana resulted in plants emitting high quantities of TPS10 sesquiterpene products identical to those released by maize. Using these transgenic *Arabidopsis* plants as odor sources in olfactometer assays showed that females of the parasitoid *C. marginiventris* learn to exploit the TPS10 sesquiterpenes to locate their lepidopteran hosts after prior exposure to these volatiles in association with hosts (Schnee et al. 2006). This gene-based dissection of the herbivore-induced volatile blend demonstrates that a single gene such as *tps10* can be sufficient to mediate the indirect defense of maize against herbivore attack. Furthermore, associative learning can also adapt parasitoids to alterations of the herbivore-induced volatile blend by plant species, age and tissue of the plant, and abiotic conditions (Takabayashi et al. 1994; De Moraes et al. 1999; Schmelz et al. 2003; Van den Boom et al. 2004). However, females of *C. marginiventris* are also attracted to the full blend of maize volatiles without prior association, indicating that the blend contains additional attractive compounds that elicit an innate response (Hoballah and Turlings 2005). Bioassay-guided fractionation of the maize volatiles has not yet identified such compounds (D'Alessandro and Turlings 2005, 2006). The combination of both innate and learned responses might allow this generalist parasitic wasp to locate a wide range of hosts on different plant species in a natural, complex environment (Turlings and Wäckers 2004; Degenhardt et al. 2003). Interestingly, the emission of volatiles in response to herbivore damage is not only beneficial for the maize plant since larvae of lepidopteran *Spodoptera frugiperda* use these volatiles as a cue to find their food plants (Carroll et al. 2006). Further studies are required to determine whether the benefit of the volatile signal outweighs its disadvantages under specific environmental conditions.

The function of terpenes as defense signals under field conditions was studied on a wild tobacco species, *Nicotiana attenuata*. The release of terpenes was mimicked by application of a lanolin paste that emitted physiological concentrations of the monoterpene linalool and the sesquiterpene (E)-α-bergamotene (Kessler and Baldwin 2001). The emission of exogenous linalool decreased lepidopteran oviposition rates on *N. attenuata* plants while the release of (E)-α-bergamotene increased egg predation rates by a generalist predator. These observations provided conclusive evidence that indirect, terpene-based plant defenses can reduce the herbivore load of a plant in a natural environment (Kessler and Baldwin 2001).

Terpene-mediated interactions were not only observed in response to damage of the leaves but also in response to root-feeding herbivores. Larvae of the beetle *Diabrotica virgifera virgifera* (Western corn rootworm) are an important pest of maize. In response to feeding by the larvae, maize roots release a signal that strongly attracts the entomopathogenic nematode *Heterorhabditis megidis* (Boff et al. 2001; van Tol et al. 2001). The signal released by the maize roots was identified as (E)-β-caryophyllene, a sesquiterpene olefin (Fig. 21.1D; Rasmann et al. 2005). Most North American maize lines do not release (E)-β-caryophyllene, whereas European lines and the wild maize ancestor, teosinte, do so in response to *D. v. virgifera* attack. Field experiments showed a five-fold higher nematode infection rate of *D. v. virgifera* larvae on a maize variety that produces the signal than on a variety that does not. Spiking the soil near the latter variety with authentic (E)-β-caryophyllene decreased the emergence of adult *D. v. virgifera* to less than half (Rasmann et al. 2005).

Not only feeding, but also oviposition of the herbivore can induce terpene emission in plants. The pine sawfly (*Diprion pini*) lays its eggs on pine twigs and wounds the surface of the needles in the process. The volatiles emitted in response to oviposition attract a wasp that parasitizes saw fly eggs (Hilker et al. 2002). The signal attracting the wasp is the sesquiterpene (E)-β-farnesene which is only recognized in combination with other, constitutively released pine volatiles (Mumm et al. 2003; Mumm and Hilker 2005).

21.2.2 Interference of Plant Volatile Terpenes with Insect Pheromones

Many species of aphids release a sesquiterpene alarm pheromone, (E)-β-farnesene, which lowers the risk of predation of other aphids in the vicinity, either by causing them to move away (Hardie et al. 1999), or by increasing their proportion of winged progeny (Kunert et al. 2005). To determine if plant-produced (E)-β-farnesene can mimic these effects, transgenic *Arabidopsis* plants overexpressing a (E)-β-farnesene synthase from *Mentha x piperita* were generated by Beale et al. (2006).

These transgenic plants emitted high levels of (E)-β-farnesene and elicited potent alarm and repellent responses in the aphid *Myzus persicae*, and an arrestant response in the aphid parasitoid *Diaeretiella rapae* (Beale et al. 2006). In addition, (E)-β-farnesene attracts further predators and parasitoids which are natural enemies of aphids (Molck et al. 1999; Du et al. 1998; Al Abassi et al. 2000; Foster et al. 2005). Thereby, the emission of (E)-β-farnesene after aphid-infestation provides the plant with both direct and indirect defenses. The monoterpene alcohol linalool, albeit no alarm pheromone, also repelled *M. persicae* in experiments with transgenic *Arabidopsis* plants overexpressing a terpene synthase from strawberry (Aharoni et al. 2003).

21.2.3 Terpene-Mediated Interactions Between Plants

Volatile-emitting plants can prime the defense metabolism in adjacent plants. The process involves increased transcription of defense-related genes and allows the plant to respond faster and more vigorously to herbivore attack (Baldwin et al. 2006; Turlings and Ton 2006). Most of these interactions were shown to be based on volatiles derived from the lipoxygenase pathway, the so called 'green leaf volatiles' (Bate and Rothstein 1998; Arimura et al. 2001; Farag and Pare 2002; Engelberth et al. 2004; Farag et al. 2005; Ruther and Kleier 2005). Little is known about the role of terpenes in priming and plant–plant interaction. Only one study in lima bean (*Phaseolus lunatus*) suggests a role of terpenes in this interaction. When these plants were attacked by the spider mite *T. urticae*, the neighboring plants became less susceptible to spider mites and more attractive to predatory mites like *P. persimilis* (Bruin et al. 1992). Lima bean infested by spider mite released a volatile blend dominated by (E)-β-ocimene and (E)-4,8-dimethyl-1,3,7-nonatriene (DMNT), a monoterpene and C_{11} homoterpene, respectively. In neighboring plants, each of these compounds induced the transcript level of pathogen-related proteins and phenylalanine ammonia lyase (Arimura et al. 2000). The volatiles also increased the transcript concentrations of two enzymes involved in terpene biosynthesis: lipoxygenase catalyses an early step in jasmonate biosynthesis, an important regulator of terpene biosynthesis, and farnesyl diphosphate synthase, which is an enzyme of terpene biosynthesis. Interestingly, the transcript levels of these genes were induced more quickly after exposure of lima bean to DMNT and a related homoterpene, $(3E,7E)$-4,8,12-trimethyl-1,3,7,11-tridecatetraene (Arimura et al. 2000). In other plant species, these terpenes were not effective in priming. In maize, for example, no effects were observed after exposure of the plant to exogenous DMNT (Ruther and Fürstenau 2005).

Volatile terpenes also play an important role in host finding by parasitic plants. Seedlings of dodder (*Cuscuta pentagona*) display a host finding behavior that is guided by volatiles of the host plant. (Runyon et al. 2006). The seedling is able to distinguish between the volatiles of tomato (*Solanum lycopersicum*), impatiens (*Impatiens wallerana*) and wheat (*Triticum aestivum*) as well as several synthetic mono- and sesquiterpenes (Fig. 21.1C). This indicates a finely tuned recognition of host plants by the composition of their terpenes emissions (Runyon et al. 2006).

21.3 Perspectives

In the last decades, advances in sensitive, reliable methods for volatile collection and identification have provided us with better insight into terpene-mediated interactions among plants and their enemies. Given the large numbers of terpenes emitted and the large families of terpene synthase genes present in plants, the discovery of many more such interactions is most likely. The advances in molecular biology facilitate the research on genes of terpene biosynthesis and the regulation of terpene emission in response to outside cues. The characterization of terpene biosynthesis genes also provides tools to engineer transgenic plants with specifically altered volatile emission. In the future, such plants will continue to be helpful to unravel the complex ecological functions of terpene emission.

References

Aharoni A, Giri AP, Deuerlein S, Griepink F, de Kogel WJ, Verstappen FWA, Verhoeven HA, Jongsma MA, Schwab W, Bouwmeester HJ (2003) Terpenoid metabolism in wild-type and transgenic *Arabidopsis* plants. Plant Cell 15:866–2884

Al Abassi S, Birkett MA, Pettersson J, Pickett JA, Wadhams LJ, Woodcock CM (2000) Response of the seven-spot ladybird to an aphid alarm pheromone and an alarm pheromone inhibitor is mediated by paired olfactory cells. J Chem Ecol 26:1765–1771

Alborn HT, Turlings TCJ, Jones TH, Stenhagen G, Loughrin JH, Tumlinson JH (1997) An elicitor of plant volatiles from beet armyworm oral secretion. Science 276:945–949

Arimura G, Ozawa R, Shimoda T, Nishioka T, Boland W, Takabayashi J (2000) Herbivory-induced volatiles elicit defence genes in lima bean. Nature 406:512–515

Arimura G, Ozawa R, Nishioka T, Boland W, Koch T, Kuhnemann F, Takabayashi J (2001) Herbivore-induced volatiles induce the emission of ethylene in neighboring lima bean plants. Plant J 29:87–98

Baldwin IT, Halitschke R, Paschold A, von Dahl CC, Preston CA (2006) Volatile signaling in plant–plant interactions: 'talking trees' in the genomics era. Science 311:812–815

Bate NJ, Rothstein SJ (1998) C6-volatiles derived from the lipoxygenase pathway induce a subset of defense-related genes. Plant J 16:561–569

Beale MH, Birkett MA, Bruce TJA, Chamberlain K, Field LM, Huttly AK, Martin JL, Parker R, Phillips AL, Pickett JA, Prosser IM, Shewry PR, Smart LE, Wadhams LJ, Woodcock CM, Zhang YH (2006) Aphid alarm pheromone produced by transgenic plants affects aphid and parasitoid behavior. Proc Natl Acad Sci USA 103:10509–10513

Beauchamp J, Wisthaler A, Hansel A, Kleist E, Miebach M, Niinemets U, Schurr U, Wildt J (2005) Ozone induced emissions of biogenic VOC from tobacco: relationships between ozone uptake and emission of LOX products. Plant Cell Environ 28:1334–1343

Boff MIC, Zoon FC, Smits PH (2001) Orientation of *Heterorhabditis megidis* to insect hosts and plant roots in a Y-tube sand olfactometer. Entomol Exp Appl 98:329–337

Bruin J, Dicke M, Sabelis MW (1992) Plants are better protected against spider-mites after exposure to volatiles from infested conspecifics. Experientia 48:525–529

Carroll MJ, Schmelz EA, Meagher RL, Teal PEA (2006) Attraction of *Spodoptera frugiperda* larvae to volatiles from herbivore-damaged maize seedlings. J Chem Ecol 32:1911–1924

Chen F, Ro DK, Petri J, Gershenzon J, Bohlmann J, Pichersky E, Tholl D (2004) Characterization of a root-specific *Arabidopsis* terpene synthase responsible for the formation of the volatile monoterpene 1,8-cineole. Plant Physiol 135:1956–1966

Coleman RA, Barker AM, Fenner M (1999) Parasitism of the herbivore *Pieris brassicae* L. (*Lep., Pieridae*) by *Cotesia glomerata* L. (*Hym., Braconidae*) does not benefit the host plant by reduction of herbivory. J Appl Entomol Z Ang Entomol 123:171–177

Connolly JD, Hill RA (eds) (1991) Dictionary of terpenoids. Chapman and Hall, London

D'Alessandro M, Turlings TCJ (2005) In situ modification of herbivore-induced plant odors: a novel approach to study the attractiveness of volatile organic compounds to parasitic wasps. Chem Senses 30:739–753

D'Alessandro M, Turlings TCJ (2006) Advances and challenges in the identification of volatiles that mediate interactions among plants and arthropods. Analyst 131:24–32

De Boer JG, Dicke M (2004) The role of methyl salicylate in prey searching behavior of the predatory mite *Phytoseiulus persimilis*. J Chem Ecol 30:255–271

De Boer JG, Dicke M (2006) Olfactory learning by predatory arthropods. Anim Biol 56:143–155

De Boer JG, Posthumus MA, Dicke M (2004) Identification of volatiles that are used in discrimination between plants infested with prey or nonprey herbivores by a predatory mite. J Chem Ecol 30:2215–2230

Degenhardt J, Gershenzon J, Baldwin IT, Kessler A (2003) Attracting friends to feast on foes: engineering terpene emission to make crop plants more attractive to herbivore enemies. Curr Opin Biotechnol 14:169–176

De Moraes CM, Lewis WJ, Pare PW, Alborn HT, Tumlinson JH (1999) Herbivore-infested plants selectively attract parasitcids. Nature 393:570–573

Dicke M (1999) Are herbivore-induced plant volatiles reliable indicators of herbivore identity to foraging carnivorous arthropods? Entomol Exp Appl 91:131–142

Dicke M, Van Beek TA, Posthumus MA, Ben Dom N, Van Bokhoven H, De Groot AE (1990) Isolation and identification of voatile dairomone that affects acarine predator–prey interactions. J Chem Ecol 16:381–396

Dicke M, van Loon JJA (2000) Multitrophic effects of herbivore-induced plant volatiles in an evolutionary context. Ent Exp Appl 97:237–249

Drukker B, Bruin J, Jacobs G, Kroon A, Sabelis MW (2000) How predatory mites learn to cope with variability in volatile plant signals in the environment of their herbivorous prey. Exp Appl Acarol 24:881–895

Du Y, Poppy GM, Powell W, Pickett JA, Wadhams LJ, Woodcock CM (1998) Identification of semiochemicals released during aphid feeding that attract parasitoid *Aphidius ervi*. J Chem Ecol 24:1355–1368

Engelberth J, Alborn HT, Schmelz EA, Tumlinson JH (2004) Airborne signals prime plants against insect herbivore attack. Proc Natl Acad Sci USA 101:1781–1785

Farag MA, Pare PW (2002) C6-Green leaf volatiles trigger local and systemic VOC emissions in tomato. Phytochemistry 61:545–554

Farag MS, Fokar M, Zhang HA, Allen RD, Pare PW (2005) (Z)-3-Hexenol induces defense genes and downstream metabolites in maize. Planta 220:900–909

Foster SP, Denholm I, Thompson R, Poppy GM, Powell W (2005) Reduced response of insecticide-resistant aphids and attraction of parasitoids to aphid alarm pheromone; a potential fitness trade-off. Bull Entomol Res 95:37–46

Gershenzon J, Kreis W (1999) Biosynthesis of monoterpenes, sesquiterpenes, diterpenes, sterols, cardiac glycosides and steroid saponins. In: Wink M (ed) Biochemistry of plant secondary metabolism, Ann Plant Rev, vol 2. Sheffield Academic Press, Sheffield, pp 222–299

Gomez SK, Cox MM, Bede JC, Inoue K, Alborn HT, Tumlinson JH, Korth KL (2005) Lepidopteran herbivory and oral factors induce transcripts encoding novel terpene synthases in *Medicago truncatula*. Arch Insect Biochem Phys 58:114–127

Gouinguene SP, Turlings TCJ (2002) The effects of abiotic factors on induced volatile emissions in corn plants. Plant Physiol 129:1296–1307

Guenther AB, Zimmerman PR, Harley PC, Monson RK, Fall R (1993) Isoprene and monoterpene emission rate variability – model evaluations and sensitivity analyses. J Geophys Res 98:12609–12617

Halitschke R, Schittko U, Pohnert G, Boland W, Baldwin IT (2001) Molecular interactions between the specialist herbivore *Manduca sexta* (Lepidoptera, sphingidae) and its natural host *Nicotiana attenuata* III. Fatty acid–amino acid conjugates in herbivore oral secretions are necessary and sufficient for herbivore-specific plant responses. Plant Physiol 125:711–717

Hardie J, Pickett JA, Pow EM, Smiley DWM (1999) Aphids. In: Hardie J, Minks AK (eds) Pheromones of non-lepidopteran insects associated with agricultural plants. CAB International, Wallingford, pp 227–250

Hilker M, Kobs C, Varama M, Schrank K (2002) Insect egg deposition induces *Pinus sylvestris* to attract egg parasitoids. J Exp Biol 205:455–461

Hoballah MEF, Turlings TCJ (1999) Experimental evidence that plants under caterpillar attack may benefit from attracting parasitoids. Evol Ecol Res 3:553–565

Hoballah ME, Turlings TCJ (2005) The role of fresh versus old leaf damage in the attraction of parasitic wasps to herbivore-induced maize volatiles. J Chem Ecol 31:2003–2018

Hoballah ME, Köllner TG, Degenhardt J, Turlings TCJ (2004) Costs of induced volatile production in maize. OIKOS 105:168–180

Howe GA (2004) Jasmonates as signals in the wound response. J Plant Growth Regul 33:223–237

Jasoni R, Kane C, Green C, Peffley E, Tissue D, Thompson L, Payton P, Paré PW (2004) Altered leaf and root emissions from onion (*Allium cepa* L.) grown under elevated CO_2 conditions. Environ Exp Bot 51:273–280

Johnson CB, Kirby J, Naxakis G, Pearson S (1999) Substantial UV-B-mediated induction of essential oils in sweet basil (*Ocimum basilicum* L.). Phytochemistry 51:507–510

Kappers IF, Aharoni A, van Herpen T, Luckerhoff LLP, Dicke M, Bouwmeester HJ (2005) Genetic engineering of terpenoid metabolism attracts bodyguards to *Arabidopsis*. Science 309:2070–2072

Kessler A, Baldwin IT (2001) Defensive function of herbivore-induced plant volatile emissions in nature. Science 1291:2141–2144

Kessler A, Baldwin IT (2002) Plant responses to insect herbivory: the emerging molecular analysis. Annu Rev Plant Biol 53:299–328

Köllner TG, Schnee C, Gershenzon J, Degenhardt J (2004) The sesquiterpene hydrocarbons of maize (*Zea mays*) form five groups with distinct developmental and organ-specific distribution. Phytochemistry 65:1895–1902

Knudsen JT, Eriksson R, Gershenzon J, Stahl B (2006) Diversity and distribution of floral scent. Bot Rev 72:1–120

Kunert G, Otto S, Röse USR, Gershenzon J, Weisser WW (2005) Alarm pheromone mediates production of winged dispersal morphs in aphids. Ecol Lett 8:596–603

Mattiacci L, Dicke M, Posthumus MA (1995) Beta-glucosidase—an elicitor of herbivore-induced plant odor that attracts host-searching parasitic wasps. Proc Natl Acad Sci USA 92:2036–2040

Meiners T, Hilker M (2000) Induction of plant synomones by oviposition of a phytophagous insect. J Chem Ecol 26:221–232

Molck G, Micha SG, Wyss U (1999) Attraction to odour of infested plants and learning behaviour in the aphid parasitoid Aphelinus abdominalis. J Plant Dis Protect 106:557–567

Mumm R, Schrank K, Wegener R, Schulz S, Hilker M (2003) Chemical analysis of volatiles emitted by *Pinus sylvestris* after induction by insect oviposition. J Chem Ecol 29:1235–1252

Mumm R, Hilker M (2005) The significance of background odour for an egg parasitoid to detect plants with host eggs. Chem Senses 30:337–343

Mithöfer A, Wanner G, Boland W (2005) Effects of feeding *Spodoptera littoralis* on lima bean leaves. II. Continuous mechanical wounding resembling insect feeding is sufficient to elicit herbivory-related volatile emission. Plant Physiol 137:1160–1168

Paré PW, Tumlinson JH (1997) De novo biosynthesis of volatiles induced by insect herbivory in cotton plants. Plant Physiol 114:1161–1167

Rasmann S, Köllner TG, Degenhardt J, Hiltpold I, Toepfer S, Kuhlmann U, Gershenzon J, Turlings TCJ (2005) Recruitment of entomopathogenic nematodes by insect-damaged maize roots. Nature 434:732–737

Runyon JB, Mescher MC, De Moraes CM (2006) Volatile chemical cues guide host location and host selection by parasitic plants. Science 313:1964–1967

Ruther J, Fürstenau B (2005) Emission of herbivore-induced volatiles in absence of a herbivore – response of *Zea mays* to green leaf volatiles and terpenoids. Z Naturforsch 60:743–756

Ruther J, Kleier S (2005) Plant–plant signaling: ethylene synergizes volatile emission in *Zea mays* induced by exposure to (Z)-3-hexen-1-ol. J Chem Ecol 31:2217–2222

Schmelz EA, Alborn HT, Tumlinson JH (2001) The influence of intact-plant and excised-leaf bioassay designs on volicitin- and jasmonic acid-induced sesquiterpene volatile release in *Zea mays*. Planta 214:171–179

Schmelz EA, Alborn HT, Banchio E, Tumlinson JH (2003) Quantitative relationships between induced jasmonic acid levels and volatile emission in *Zea mays* during *Spodoptera exigua* herbivory. Planta 216:665–673

Schnee C, Köllner TG, Gershenzon J, Degenhardt J (2002) The maize gene terpene synthase 1 encodes a sesquiterpene synthase catalyzing the formation of (E)-farnesene, (E)-nerolidol, and (E, E)-farnesol after herbivore damage. Plant Physiol 130:2049–2060

Schnee C, Köllner TG, Held M, Turlings TCJ, Gershenzon J, Degenhardt J (2006) The products of a single maize sesquiterpene synthase form a volatile defense signal that attracts natural enemies of maize herbivores. Proc Natl Acad Sci USA 103:1129–1134

Spiteller D, Boland W (2003) N-(17-acyloxy-acyl)-glutamines: novel surfactants from oral secretions of lepidopteran larvae. J Org Chem 68:8743–8749

Steele CL, Katoh S, Bohlmann J, Croteau R (1998) Regulation of oleoresinosis in grand fir (*Abies grandis*) – differential transcriptional control of monoterpene, sesquiterpene, and diterpene synthase genes in response to wounding. Plant Physiol 116:1497–1504

Takabayashi J, Dicke M, Posthumus MA (1994) Volatile herbivore-induced terpenoids in plant mite interactions – variation caused by biotic and abiotic factors. J Chem Ecol 20:1329–1354

Tholl D, Chen F, Petri J, Gershenzon J, Pichersky E (2005) Two sesquiterpene synthases are responsible for the complex mixture of sesquiterpenes emitted from *Arabidopsis* flowers. Plant J 42:757–771

Tumlinson JH, Lait CG (2005) Biosynthesis of fatty acid amide elicitors of plant volatiles by insect herbivores. Arch Insect Biochem Physiol 58:54–68

Turlings TCJ, Tumlinson JH, Lewis WJ (1990) Exploitation of herbivore-induced plant odors by host-seeking parasitic wasps. Science 250:1251–1253

Turlings TCJ, Tumlinson JH, Heath RR, Proveaux AT, Doolittle RE (1991) Isolation and identification of allelochemicals that attract the larval parasitoid, *Cotesia marginiventris* (Cresson), to the microhabitat of one of its hosts. J Chem Ecol 17:2235–2251

Turlings TCJ, Lengwiler UB, Bernasconi ML, Wechsler D (1998) Timing of induced volatile emissions in maize seedlings. Planta 207:146–152

Turlings TCJ, Wäckers FL (eds) (2004) Recruitment of Predators and parasitoids by Herbivore-damaged plants. Cambridge University Press, Cambridge

Turlings TCJ, Ton J (2006) Exploiting scents of distress: the prospect of manipulating herbivore-induced plant odours to enhance the control of agricultural pests. Curr Opin Plant Biol 9:421–427

Van Den Boom CEM, Van Beek TA, Posthumus MA, De Groot A, Dicke M (2004) Qualitative and quantitative variation among volatile profiles induced by *Tetranychus urticae* feeding on plants from various families. J Chem Ecol 30:69–89

van Loon JJA, De Boer JG, Dicke M (2000) Parasitoid-plant mutualism: parasitoid attack of herbivore increases plant reproduction. Entomol Exp Appl 97:219–227

van Tol RWHM, van der Sommen ATC, Boff MIC, van Bezooijen J, Sabelis MW, Smits PH (2001) Plants protect their roots by alerting the enemies of grubs. Ecol Lett 4:292–294

Vuorinen T, Nerg AM, Holopainen JK (2004a) Ozone exposure triggers the emission of herbivore-induced plant volatiles, but does not disturb tritrophic signalling. Environ Pollut 131:305–311

Vuorinen T, Reddy GVP, Nerg AM, Holopainen JK (2004b) Monoterpene and herbivore-induced emissions from cabbage plants grown at elevated atmospheric CO_2 concentration. Atmos Environ 38:675–682

Abbreviations

AADC - aromatic amino acid decarboxylase
ABA - abscisic acid
ACA - *Amaranthus caudatus* agglutinin
ACC - 1-aminocyclopropane-1-carboxylic acid
ACH - acyl-thioesterase
ACS - 1-aminocyclopropane-1-carboxylate synthase
ACX - acyl-CoA oxidase
ADT - arogenate dehydratase
AHCT - anthocyanin *O*-hydroxycinnamoyltransferase
AMAT - anthraniloyl-coenzyme A:methanol acyltransferase
AOC - allene oxide cyclase
AOS - allene oxide synthase
APA - *Allium porrum* lectin
ARG - ARGINASE
AS - anthranilate synthase
ASAL - *Allium sativum* agglutinin
AtPep - *Arabidopsis* defense-related peptide
BA2H - benzoic acid 2-hydroxylase
BAHD - derived from BEAT, AHCT, HCBT, DAT
BALDH - benzaldehyde dehydrogenase
BAMT - SAM:benzoic acid carboxyl methyltransferase
BBI - Bowman-Birk inhibitor
BEAT - acetyl-CoA:benzyl alcohol acetyltransferase
BPBT - benzoyl-CoA:benzyl alcohol/phenylethanol benzoyltransferase
Bt toxin - *Bacillus thuringiensis* toxin
BSMT - *S*-adenosyl-l-Met:benzoic acid/salicylic acid carboxyl methyltransferase
BZL - benzoyl-coenzyme A ligase
C3H - Coumarate-3-hydroxylase
C4H - cinnamate-4-hydroxylase
CAD - cinnamyl alcohol dehydrogenase
CCMT - *S*-adenosyl-L-Met:coumarate/cinnamate carboxyl methyltransferase
CCoAOMT - caffeoyl-CoA-*O*-methyltransferase

CCR - cinnamoyl-CoA oxidoreductase
CFAT - coniferyl alcohol acetyltransferase
CHI - chalcone isomerase
CHS - chalcone synthase
4-CL - 4-coumaryl-CoA ligase
CM - chorismate mutase
C/N - carbon/nutrient
coi1 - coronatine insensitive 1
COMT - caffeic acid/5-hydroxyferulic acid O-methyltransferase
conA - concanavalin A
CPA - carboxypeptidase A
CPB - carboxypeptidase B
CPC - CAPRICE
CPY - carboxypeptidase Y inhibitor
cpr - constitutive expresser of PR genes
CTS - COMATOSE
DAHP - 3-deoxy-D-*arabino*-heptulosonate-7-phosphate
DAT - deacetylvindoline 4-O-acetyltransferase
DGDG - digalactosyldiacylglycerol
DHS - deoxyhypusine synthase
dnOPDA - dinor-OPDA
DMAPP - dimethylallyl diphosphate
DMNT - (E)-4,8-dimethyl-1,3,7-nonatriene
DOPA - dihydroxy phenylalanine
DXPS - deoxyxylulose phosphate synthase
EA - extracellular alkalinization
eIF5A - eukaryotic initiation factor 5A
EFN - extrafloral nectar
EGL - *ENHANCER OF GLABRA*
EGS - eugenol synthase
12,13-EOT - 12,13(S)-epoxy-octadecatrienoic acid
ESI-MS - electrospray mass spectrometry
EST - expressed sequences tag
ET - ethylene
F5H - ferulate-5-hydroxylase
FAC - fatty acid-amino acid conjugate
FC - fusicoccin
FDP - farnesyl diphosphate
FL-cDNA - full-length cDNA
Fuc - fucose
Gal - galactose
GalNAc - N-acetylgalactosamine
GC-EAG - gas chromatography-electroantennography
GC/MS - gas chromatography / mass spectrometry
GDP - geranyl diphosphate

Abbreviations

GGDP - geranylgeranyl diphosphate
GL - glabrous
GlcNAc - N-acetylglucosamine
GLVs - green leaf volatiles
GNA - *Galanthus nivalis* agglutinin
GOGAT - glutamine 2-oxoglutarate amino transferase
GOX - glucose oxidase
GS - glutamine synthetase
GS-II - *Griffonia simplicifolia* lectin II
HCBT - anthranilate *N*-hydroxycinnamoyl/ benzoyltransferase
HHT - hydroxyfatty acid feruloyl-CoA transferase
HLH - helix-loop-helix
HPL - hydroperoxide lyase
HPLC - high-pressure liquid chromatography
13-HPOT - 13(S)-hydroperoxy-octadecatrienoic acid
HSS - homospermidine synthase
HTH - hydroxycinnamoyl CoA:tyramine hydroxycinnamoyltransferase
HypSys - hydroxyproline-rich glycopeptides
IAA - indole acetic acid
ICS - isochorismate synthase
IDP - isopentenyl diphosphate
IGL - indole-3-glycerol phosphate lyase
IGS - isoeugenol synthase
IL-1β - interleukin-1β
IPL - isochorismate pyruvate lyase
ISR - induced systemic resistance
JA - jasmonic acid
jai1 - jasmonate insensitive 1
jar1 - jasmonic acid resistent 1
JAZ - jasmonate ZIM domain
JIP - jasmonate-inducible protein
JMT - jasmonate methyl transferase
KAT - 3-ketoacyl-CoA thiolase
LOX - lipoxygenase
LRR - leucine-rich repeat
Man - mannose
MAPK - mitogen-activated protein kinase
MeJA - methyl jasmonate
MEP - methylerythritol phosphate
MeSA - methyl salicylate
MEV - mevalonic acid
MFP - multifunctional protein
MGDG - monogalactosyldiacylglycerol
MTI-2 - mustard trypsin inhibitor
NMR - nuclear magnetic resonance

NO - nitric oxide
NOS - nitric oxide synthase
OGAs - oligogalacturonides
OPC-8:0 - 3-oxo-2-(2'(Z)-pentenyl)-cyclopentane-1-octanoic acid
OPCL - OPC-8:0 CoA Ligase
OPDA - oxo-phytodienoic acid
OPR - 12-oxophytodienoate reductase
OS - oral secretion
P450 - P450 dependent monooxygenases
PA - pyrrolizidine alkaloid
PAAS - phenylacetaldehyde synthase
PAL - phenylalanine ammonia-lyase
PC - prohormone convertase
PCI - potato carboxypeptidase inhibitor
PG - polygalacturonase
PI - proteinase inhibitor
PLA_2 - phospholipase A_2
PMA - plasma membrane proton ATPase
PMT - putrescine N-methyl transferase
PNT - prephenate aminotransferase
PP - Polyphenolic parenchyma
PPO - polyphenol oxidase
PR - pathogenesis-related
Prosys - prosystemin
PUFAs - polyunsaturated fatty acids
PT - prenyltransferase
qRT-PCR - quantitative real time PCR
QTL - quantitative trait loci
RIP - ribosome-inactivating protein
ROS - reactive oxygen species
RNAi - RNA interference
RTN - REDUCED TRICHOME NUMBER
SA - salicylic acid
SABATH - derived from SAMT, BAMT, theobromine synthase
SAMT - SAM:salicylic acid carboxyl methyltransferase
SAR - systemic acquired resistance
SIPK - stress-induced MAPK
SnRK1 - SNF1-related kinase 1
SOD - superoxide dismutase
SKTI - soybean Kunitz-type inhibitor
SPR - Suppressed in *35S::prosystemin*-mediated responses
Sys - systemin
TD - threonine deaminase
TPS - terpenoid synthases
TRD - traumatic resin duct

Abbreviations

THT - tyramine hydroxycinnamoyl transferase
TMOF - trypsin modulating oostatic factor
TMV - tobacco mosaic virus
TRY - TRIPTYCHON
TTG1 - TRANSPARENT TESTA GLABRA1
VEG - ventral eversible gland
VIGS - virus-induced gene silencing
VOC - volatile organic compound
VSP - vegetative storage protein
WIPK - wound-induced MAPK
WGA - wheat germ agglutinin
WT - wild type

Subject Index

A

Abiotic stress, 74, 75, 89, 90, 95, 100, 198, 259, 329, 338, 340
Agriculture, 51, 426
Agrochemicals, 107, 109, 119, 299
Alkalinization response, 315, 323
Alkaloids, 14, 15, 22, 36, 38, 41, 62, 69, 71, 90
α-Amylase inhibitor, 237, 294
Allelochemicals, 72, 148
Allene oxide cyclase (AOC), 333, 351, 357, 358, 402
    crystal structure, 351
    stereochemical control, 351
Allocation costs, 62, 63, 71
Anthocyanidin synthase, 204, 206, 207
Antinutritive enzymes, 7
Antixenosis, 2, 12
Apparency, apparence, 63
Arabidopsis defense-related peptides (AtPeps), 321
    expression patterns, 322
    gene family, 322
    receptor protein (AtPepR1), 323
Arginase, 256, 271–276, 279, 281
    biochemical properties, 273
    depletion of Arg, 275
    pH optimum of, 274
    physiological function of, 273
    tissue-specific expression, 273
Aromatic volatiles, 409–426
    biological activity, 415
    biosynthesis of, 410
    derived from chorismate, 414
    derived from Phe, 410
    in plant direct defense, 415
        insecticidal activity, 418
        repellent properties (or activity), 418
    in plant indirect defense, 418
        tritrophic interactions, 418
    in plant–plant interactions, 425
Autotoxicity costs, 62

B

Benzenoid compounds, 410, 412, 414
    via non-β-oxidative pathway, 412
    via β-oxidative pathway, 412
    volatile benzenoids, 412

C

Caeliferin, 395–397, 401, 403
    biological activity of, 396
    chain lengths, 396
    function in the grasshoppers, 396
    origin of the fatty acid moiety, 396
Calcium oxalate, 148, 154, 155
Callose, 12
Cardenolides, 12
(E)-β-Caryophyllene, 347, 436
Cellulose, 12, 96, 111, 113
Chalcone isomerase (CHI), 204–206
Chalcone synthase (CHS), 204–206
Cinnamate-4-hydroxylase (C4H), 194, 200, 201, 204, 411
Cinnamoyl-CoA oxidoreductase (CCR), 136, 201, 204
Community structure, 50
Constitutive resistance, 10, 108, 112
Coronatine, 358, 381
Coronatine-insensitive1 (COI1), 18–21, 23, 350, 358–360
Costs
    of allocation, 62, 63, 70, 71
    of autotoxicity, 62
    of defense, 61, 64, 70, 74, 76, 78
    ecological, 62–64, 70
    evolutionary, 62
    physiological, 62
    of resistance, 66

Coumarate-3-hydroxylase (C3H), 200, 201, 204, 205
4–Coumaroyl-CoA ligase (4CL), 136, 200, 201, 204
Crop protection, 51, 109, 119, 300
Cuticle, 10, 107–120
    chemical composition, 108, 112, 114, 116
    microrelief, 109, 110
    optical properties, 107, 110, 112
    penetration, 119
    solvent extraction, 111
    ultrastructure, 108, 112
Cyanogenic glucosides, 14
Cytochrome P450 dependent monooxygenase (P450), 173–175, 177–183
    biochemical characterization of, 177
    evolution of, 178

D

Defense
    anti-nutritional, 8, 15
    anti-nutritive, 3, 253, 256
    chemical, 7, 13, 15, 61, 62, 96, 148, 157, 175, 176, 196, 214, 217, 223, 225, 285, 389, 390
    direct, 2, 7–23, 31, 32, 36, 39, 63, 68, 70, 107, 112, 114, 116, 120, 183, 192, 259, 350, 371, 374, 375, 381, 399, 418, 410, 415
    ecological, 173
    indirect, 2, 3, 14, 31–52, 63, 70, 115, 117, 118, 120, 181, 183, 330, 339, 374, 375, 379, 389, 399, 400, 415, 418, 425, 433, 435, 436, 438
    morphological, 7, 10
    physical, 7, 10, 14
    structural, 15, 96
    systemic, 16, 18, 22, 349
Deoxyhypusine synthase (DHS), 220–222
Dihydroflavonol, 4–reductase, 204, 206, 207
Dirigent proteins, 353

E

Ecological
    costs, 62–64, 70
    relevance, 48–50, 425
Ecosystem, 8, 47, 50, 374, 403
Elicitation studies, 61, 64–67
Elicitor, 16, 20, 21, 38, 49, 51, 61, 65, 67, 72, 74, 117, 155, 174, 175, 196, 316, 322, 330, 335, 336, 338, 350, 369, 371, 373, 374, 376, 378, 379, 381, 389–395, 399, 401–403, 434
Environmental stress, 74, 76–79, 208

Essential amino acids, 14, 235, 248, 256, 261, 262, 271, 272
Ethylene (ET), 15, 17, 18, 30, 34, 38, 135, 140, 141, 147, 163, 174, 175, 179, 180, 259, 264, 322, 330, 333, 335, 336, 341, 342, 374, 393, 398
Extrafloral nectar (EFN), 2, 31–33, 36, 37, 39, 41–44, 50, 99, 400

F

(E)-β-farnesene, 435–438
Fatty acid–amino acid conjugates, (FACs), 38, 65, 329, 330, 339, 342, 371, 373, 374, 377, 378, 380, 381, 390–396, 403
    common features of, 391
    influenced by the caterpillar diet, 393
    origin of the fatty acid moiety, 393
    role in caterpillar metabolism, 393–395
    synthesis by bacteria, 394
F-box protein, 18
Feeding guild, 10, 34, 37, 38, 62, 95, 97, 330, 339
Fitness
    benefit, 2, 9, 10, 43, 62, 64, 65, 68, 72
    cost, 9, 39, 64–68, 70, 71, 73, 76, 99, 100
    Darwinian, 76, 78
Flavonoids, 110, 111, 119, 156, 189–192, 194, 200, 204–208, 255, 410
Flavonoid 3'-hydroxylase, 204, 206

G

Galactolipids, 359
Gene
    duplication, 178, 220, 279, 314, 315, 320, 335
    neo-functionalization, 94, 178
Genetic
    background, 66, 67, 73
    linkage, 61, 69, 73
    modification, 48, 51, 236
    studies, 64, 78, 94
    variation, 9, 38, 64, 66, 94, 219, 257, 423
Glabrous, 90, 94, 110
Glucosinolate, 9, 10, 14, 15, 36, 42, 66, 111
Grafting, 11, 18, 20, 21, 320
Green leaf volatiles (GLVs), 35–37, 39, 397–403, 438
    activation of the octadecanoid pathway, 400
    biosynthetic pathway for, 397
    induction of JA, 401
    in intra-plant communication, 400
    and plant defense, 398

H

in plant–plant signaling, 398, 400
priming effect of, 399

Headspace, 48, 415, 421, 422
Hemolymph, 64, 224, 247, 297, 300
Herbivore
  performance, 13, 76, 98, 257, 271
  preference, 1, 9, 15
Homospermidine synthase (HSS), 218, 220–222
  evolution of, 220
Host
  acceptance, 118–120
  range, 13, 41, 378–380
  rejection, 120
  searching, 41, 42, 44
Hydrogen peroxide, 16, 17, 44
Hydroxyproline-rich systemin glycopeptides (HypSys), 314, 317, 320, 321, 324, 330, 332, 338
  immunolocalization, 321
  overexpression of, 321
  processing events, 321
Hyperparasitoid, 41, 42, 45, 52, 118

I

Indole, 34, 135, 410, 411, 413, 415, 419, 423–425
Infochemicals, 33, 36, 39, 111, 112
Insect
  associative learning, 435, 436
  biting, chewing, 12
  generalist, 13, 38, 72, 96, 116, 219, 244, 279, 331, 374, 375, 379, 422, 436, 437
  metabolic adaptations, 225
  midgut chemistry, 264
  miners, 95, 97
  monophagous, 13, 113, 224
  phloem-feeding, 37, 291, 292, 295, 300, 321
  piercing–sucking, 12, 36, 189, 192, 193, 291, 292, 330, 331, 336, 337, 370
  polyphagous, 13, 223–226, 395
  specialist, 12, 13, 38, 40, 66, 96, 148, 219, 279, 331, 379, 380, 392
Interaction
  antagonistic, 89
  synergistic, 70, 72, 73
  tritrophic, 2, 36, 52, 89, 97, 100, 397, 409, 419, 421, 423, 434

J

JA-Ile, 19–21, 23, 278, 355, 356, 358–360
Jasmonate ZIM domain proteins (JAZ proteins), 23, 358–360
Jasmonic acid (JA), or jasmonates, 7, 15, 16, 19–23, 30, 33, 35–38, 48, 50, 65, 70, 72, 75, 89, 96, 116, 117, 193, 241, 258, 274–279, 290, 294, 295, 314, 316, 318, 320, 329, 330, 332–339, 342, 349–360, 374, 380, 381, 389, 398–402, 424
  accumulation, 19, 21, 36, 48, 353, 355, 357, 399, 401, 402
  biosynthesis, 11, 17–19, 21, 320, 333, 335, 337, 351, 353–355, 358, 359, 401, 402
  as defense regulators, 350
  metabolites of, 355
  signaling properties, 353, 355

L

Laccase, 207
Latex (latices), 7, 12, 13, 288
  lacticifers, 12
Leaf toughness, 12
Lectins
  amaranthins, 286, 288, 290, 291
  carbohydrate-binding motifs, 299
  as a carrier for toxins, 300
  classical lectins, 285, 296
  Cucurbitaceae phloem lectins, 286–288, 290
  GNA-related lectins, 286–288, 291, 292, 296, 298, 299
  inducible lectins, 285, 287, 290, 294, 298, 299
  insecticidal activity of, 285–301
  Jacalin-related lectins, 286–290, 292, 293, 295
  jasmonate-inducible lectin, 294
  lectins composed of hevein domains, 286–288, 291
  lectins with ricin-B domains, 286, 287, 289, 294
  legume lectins, 286, 287, 289, 293, 296
  mechanism of action, 296
  seed lectins, 286, 288, 292, 293
  snowdrop lectin, 286, 288, 292, 293
  three-dimensional structure, 296
Lignin, lignification, 12, 13, 15, 132, 136, 148, 189, 190, 192, 196, 200, 204, 255, 414
Linkage
  disequilibrium, 69
  functional, 70, 72–74
  genetic, 61, 69, 73
  metabolic, 70–72

Linolenic acid, 17, 18, 34, 35, 48, 314, 319, 336, 351, 377, 379, 389, 391–394
Lipocalin, 353
Lipoxygenase, 50, 351, 397, 434, 438
Locomotion, 107, 108, 112, 115, 117

M

MAP kinase (MAPK), 17, 319, 320, 329–342, 374
MEROPS database, 236, 237, 248
Metabolites (metabolism)
  primary, 13–15, 61, 67, 71, 74, 179, 194, 281
  secondary, 7, 13–15, 16, 22, 35, 61–64, 67, 68, 71, 72, 74, 90, 96, 99, 107, 108, 111, 175, 177, 179, 192, 255, 265, 271, 272, 281, 349, 374, 409, 410, 413
Methylerythritol phosphate (MEP) pathway, 173, 175, 177
Methyl jasmonate (MeJA), 18–21, 34, 35, 48, 49, 51, 65, 67, 70, 71, 78, 155, 174, 175, 179, 180, 256, 259, 274, 276, 295, 316–318, 321, 322, 335, 338, 349, 350, 354, 359, 360, 381, 399, 415
Methyl salicylate (MeSA), 33, 36, 49, 415, 418, 421–423, 425, 435
Mevalonic acid (MEV) pathway, 175–177, 181
Microarray, 14, 15, 19, 22, 181, 190, 194, 208, 238, 322, 331, 338, 378–380, 399
Mimicry, 118
Moving-target model, 76
Mustard trypsin inhibitor (MTI), 237–239, 241, 246
Mutant studies, 61, 64, 67

N

Nectar, 2, 31–33, 36, 42–44, 99, 286, 400
Nicotine, 10, 34, 42, 62, 64, 69, 70
Nictaba, 287, 290
Nitric oxide (NO) synthase (NOS) 273
N-recycling pathway, 198

O

Octadecanoid pathway, 17, 35, 36, 50, 147, 163, 319, 320, 323, 349–351, 356–358, 389, 390, 402
  activation by GLVs, 401
  activation by insect-derived elicitors, 402
  compartmentation in plastids and peroxisomes, 353
  in JA biosynthesis, 319
  post-translational regulation, 358
Octadecanoid signaling, 175, 313, 314, 322, 323, 398, 400, 401

Oleoresin, 173–175, 177–179, 181
Olfactory assays (or olfactometer), 48, 418, 421, 435, 436
Oligogalacturonides (OGAs), 16, 17, 356
Optimal defense theory, 62
Orysata, 290
Oviposition, 7, 10, 39, 44, 48, 51, 96, 108, 112–114, 116–120, 164, 192, 224, 298, 390, 437
Oxophytodienoate reductase (OPR3), 353, 354, 358, 401, 402
  crystal structure, 358
  monomer/dimer equilibrium, 358
  reaction mechanism, 354
  stereospecificity, 354
12–Oxo-phytodienoic acid (OPDA), 19, 34, 35, 49, 314, 319, 335, 338, 351–354, 356, 358, 359, 401, 402
  in complex lipids, 359
  enantiomers, 351, 354
  spontaneous cyclization, 351
β-Oxidation, 18, 352–355, 401, 410, 412
'Oxylipin signature', 36

P

Parasitoid, 3, 32, 39, 41, 42, 44, 45, 47, 48, 63, 64, 89, 98–100, 107, 108, 115, 118–120, 213, 224, 225, 346, 374, 392, 418–420, 422–424, 435, 438
Pathogen invasion, 131, 138, 142
Perfuming, 47
Periderm
  anatomy, 132
  chemical composition, 133
  native, 132–134
Periderm development, 138
  hormonal involvement, 141
  temperature dependence of, 140
Peritrophic membrane, 191, 296, 297
Pharmacophagy, 223
Phellem, 133, 134, 137, 139, 141, 155
  See also Periderm, anatomy
Phellogen, 133, 134, 139–141, 155
  See also Periderm, anatomy
Phenolics
  anti-microbial properties, 192
  anti-oxidative properties, 192
  induced biosynthesis of, 194
  as induced defense, 192
  as pre-formed (constitutive) defense, 192
Phenotypic plasticity, 33, 34, 66, 74, 78, 174
Phenylalanine ammonia-lyase (PAL), 136, 139, 141, 153, 156, 164, 194, 196–198, 200, 410, 411, 413, 425, 438

Subject Index

Phenylpropanoid pathway (or metabolism), 15, 136, 189–208
Phenylpropenes, 409, 414
Pheromone, 117, 157, 182, 183, 224–226, 353, 437, 438
Phospholipase, 17, 314, 336, 358
Pollination, pollinator, 30, 38, 39, 41–43, 46, 47, 51, 52, 66
Polyphenolic parenchyma cells (PP cells), 147
   activation of, 147, 156
   anatomy and development of, 149
   defensive role of, 156
   maturation of, 151
Polyphenol oxidase (PPO), 18, 190, 253–265, 272
   activation, 260
   anti-sense suppression of, 259
   biochemistry of, 254
   defensive functions of, 259
   gene families, 259
   induction of, 259, 261
   mechanisms of action, 261
   overexpression of, 259, 260
   in pathogen defense, 263
   proteolytic processing, 260
   substrates, 255, 261, 264
   tissue-specific expression, 259
Post-ingestive defense, 272, 274, 278, 279
Predator, 2, 31, 33, 36–38, 40, 43, 45–47, 50, 63, 72, 73, 89, 97–100, 107, 108, 115, 118–120, 181, 213, 225, 235, 236, 248, 291, 293, 299, 300, 389, 400, 418–424, 435, 437, 438
Prenyltransferases (PT), 173–176
Priming, 39, 78, 399, 400, 402, 403, 425, 438
Prosystemin (ProSys), 17, 18, 21, 247, 313–320, 324, 334, 335, 337, 357
   amino acid composition of, 316
   compartmentalization, 318
   gene duplication, 314, 315, 320
   gene expression, 316
   proteolytic processing (or cleavage), 315–317
Proteinase inhibitor (PI), 1, 3, 8, 9, 11, 14–18, 20–23, 36, 43, 68, 69, 71, 72, 116, 236, 241–243, 246, 247, 271, 272, 299, 315, 332–335, 338, 339, 350, 376, 399
   antinutritional effects of, 235
   degradation of, 246
   determinants of specificity, 241
   families, 242, 243
   mechanism of inhibition, 241
   MEROPS database, 237, 248
   multimeric associations, 244
   regulation of protease expression, 247
   resistance of insect proteases, 245
   species-specific, 237
   steric occlusion, 244
   structure-function relationship, 241
   tissue specificity, 241
Putrescine N-methyl transferase (PMT), 71
Pyrrolizidine alkaloids (PAs), 39, 213
   detoxification of, 216
   lycopsamine type, 214, 220
   occurrence of, 214, 222
   in plant defense, 219
   profile, 218, 219
   senecionine type, 214–216
   sequestration, 222–224
   site of synthesis, 218
   structural (or chemical) diversity of, 214
   tissue specificity, 219
   toxicity of, 216

R

Reproductive output (or success), 1, 7, 66, 67, 74
Resin
   cavities, 157, 160, 174
   cells, 157, 160, 161, 174
   ducts, 12, 13, 147–153, 157, 159–163, 165, 174, 175, 178, 179, 182, 433
Resistance
   constitutive, 10, 108, 112, 157, 160, 163, 164
   preformed, 7, 10
Resource
   allocation, 70, 71
   mobilization, 22
Ribosome-inactivating proteins (RIPs), 289, 290, 294
Ricin, 286, 287, 289, 290, 294, 299

S

Salicylic acid (or salicylate) (SA), 31, 34, 75, 96, 238, 259, 264, 295, 299, 330, 332–334, 337–339, 342, 374, 381, 393, 411–413, 415, 418, 421–425, 435
Schizogenesis, 161
Scion, 11, 19, 21, 319
Scotophase, 42
Seed production, 42, 65, 67, 71, 74
Selection pressure, 33, 62, 220, 244
Selective advantage, 62, 72, 379
Senecionine, 214–216, 219
Senecionine N-oxygenase, 213, 217–219, 224

454   Subject Index

Shikimate pathway, 194–198, 412, 426
Sib analysis, 66
Sieve element, 19, 318, 357
Silica, 12, 22, 48
Source–sink, 74
Spines, 7, 8, 10
Stone cells, 150, 152, 154, 155
Suberin, 12, 13, 131–133, 135–141, 148, 189, 190, 192, 204, 205
   aliphatic domain, 135–137
   biosynthesis, 136
   as diffusion barrier, 135–138
   phenolic (or aromatic) domain, 138
   purification, 136
   See also Periderm, chemical composition
'Supply-side hypothesis', 63
Systemic acquired resistance (SAR), 67, 264, 314, 338
Systemic wound signaling (systemic wound response), 9, 11, 16, 18–20, 22, 23, 256, 314, 319–321, 324, 356, 357
Systemin (Sys), 7, 11, 16, 17, 19, 21–23, 117, 247, 253, 256, 259, 313–315, 317–321, 323–325, 329, 330, 332–340, 356, 357
   definition of, 323
   evolution, 324
   receptor, 324
   signaling events, 336
   turnover, 316

T

Tannins, 14, 156, 189–194, 198, 200, 204, 205, 207, 208, 271
Terpene (terpenoids), 14, 22, 32, 36, 48, 111, 173–183, 214, 377, 398, 410, 425, 426, 433–435, 437–439
   accumulation of, 175, 178, 179
   biosynthesis, 13, 174–179, 181, 182, 375, 378, 434, 438
      biochemistry of, 175
      subcellular compartmentation, 178
   effects on insects, 180
   induction by MeJA, 180
   manipulation of, 182
   pheromones, 182, 183, 437
Terpene (or terpenoid) synthase (TPS), 47, 174–181, 377, 434, 435, 439
   evolution of, 178, 179
   overexpression of, 438
   reaction mechanism, 176
Terpenoid volatiles, 175, 180–182, 412
   as alarm pheromone, 437
   in host finding, 438
   in priming, 438
   in tritrophic interactions, 434
Thorns, 2, 7, 8, 10, 32, 100, 291, 422
Threonine deaminase (TD), 256, 271, 272, 276, 277
   in anti-insect defense, 278
   biodegradative TDs, 276
   biosynthetic TDs, 276
   evolution of, 279
   expression pattern of, 279
   phylogenetic groups, 277
   proteolytic processing of, 278
   regulatory domain of, 278, 279
   stress-induced TD expression, 277
Tolerance, 2, 10, 15, 16, 23, 62, 63, 89, 90, 160, 265
Traumatic resin ducts (TRDs), 147–150, 153, 157, 160–164, 174, 175, 179, 182
   anatomy and development of, 160
   defensive roles of, 163
Trichome, 7, 9, 10, 12, 50, 89–100, 110, 111, 114–116, 255, 260, 350, 433
   adaptive significance, 99
   density, 9, 12, 49, 89–92, 94, 95, 97–100
   development, 89, 350
   genetic basis of initiation, 94
   genetic variation, 94
   glandular, 12, 90, 94, 95–98, 116, 255, 260, 350, 433
   initiation, 93, 94
   non-glandular, 12, 94–97
   polymorphism, 94
   spacing, 93
Trophic level, 35, 39, 40, 43, 47, 48, 51, 63, 97, 118, 225
Trypsin modulating oostatic factor (TMOF), 247, 317

U

Ubiquitin, 23, 358
α, β-Unsaturated carbonyls, 354

V

Vascular system, 17, 19
Vegetative storage proteins (VSPs), 16, 23, 271, 272, 280, 350
   growth-inhibiting effect of, 280
   phosphatase activity, 271
Volatile aromatic compounds, 410, 411, 425
   See also Aromatic volatiles
Volatile terpenes, 164, 433, 434, 437, 438
   See also Terpenoid volatiles
Volatile blend, 35, 39, 44–51, 392, 398, 400, 421–425, 433, 435, 436, 438

Volicitin, 38, 371, 373, 377, 379, 381, 390–394, 401, 403, 415
   absolute configuration of, 392

W

Wax
   amorphous, 113
   compounds, 109, 110, 114, 136, 140
   crystal, 7, 110, 113, 117
   cuticular, 108, 109, 114, 117, 119, 120
   epicuticular, 10, 108–117, 119, 120
   film, 10, 109

Wound
   callus, 13
   healing, 13, 131, 135, 138–141, 147, 148, 152, 153, 155, 190
   mechanical, 329
   periderm, 13, 131–142, 147, 148, 152, 153, 155, 190
   response, 18, 20, 131, 141, 192, 194, 264, 319, 320, 332–339, 358, 369, 370, 372, 399, 401
      local, 319
      systemic, 18, 20, 22, 314, 356, 357
   signaling, 9, 11, 16, 18–21, 23, 131, 132, 256, 319, 320, 329, 330, 332, 334, 356

X

Xylem mother cells, 163

Taxonomic Index

A

Abies, 148, 158, 162
Abies balsamea, 151, 161
Abies grandis, 161, 176
Acyrthosiphon pisum, 44, 115, 297, 298
Adenostyles alliariae, 223
Agaricus bisporus, 286
Agathis, 158
Alfalfa, 280, 332, 341
Alliaceae, 288, 292
Allium cepa, 291
Allium porrum, 292, 299
Allium sativum, 292, 299
Alloxysta victrix, 42
Alnus incana, 91
Amaranthus caudatus, 291, 299
Amphorophora idaei, 114
Anagasta kuehniella, 293
Angiosperms, 89, 147, 177, 178, 213, 214, 220–222
Antirrhinum (majus), 413
Ants, 2, 31, 32, 39, 40, 98, 115, 118, 224
Aphidius ervi, 45, 115
Aphids, 34, 40, 42, 44, 45, 99, 113–116, 118, 119, 223, 255, 291–293, 295, 297, 330, 337, 339, 340, 370, 398, 418, 422, 437, 438
Aphis fabae, 45, 416, 418
Aphis glycines, 419, 420, 422
Aphis gossypii, 291
Aphis jacobaeae, 223
Apium graveolens, 116
Apocheima pilosaria, 198
Apocynaceae, 214, 216, 221
Apple, 98, 118, 259, 291, 422
Arabidopsis, 9, 15, 23, 48, 89, 93, 94, 96, 165, 197, 198, 201, 202, 206, 236–239, 246, 264, 276, 277, 280, 298, 313–318, 321–323, 329–332, 338, 339, 350, 351, 354–356, 358, 359, 379, 397, 398, 401, 413, 422, 433, 435–438
Arabidopsis, thaliana, 37, 47, 49, 51, 67, 69, 89, 91, 93, 94, 96, 222, 236, 321, 412, 413, 419, 436
Arabidopsis lyrata, 90, 94, 413
Araceae, 288, 292
Arachis hypogeae, 293
Araucaria, 154, 158
Araucariaceae, 148, 149, 157, 158
Archaea, 220
Arctiidae, 222, 223
Arctiid moths, 223, 224
Arisaema jacquemontii, 292
Arthropods, 18, 29, 32, 33, 36, 40, 41, 47, 49, 96, 113, 238, 372, 373, 375, 400, 421, 423, 424, 433, 435
Artocarpus integrifolia, 288
Arum maculatum, 292
Asclepiadacea, 12
Asclepias, 97
Asclepias syriaca, 97
Ascogaster quadridentata, 118
Asteraceae, 111, 214, 216, 220, 223
Atherigona soccata, 113

B

Bacillus thuringiensis, 300
Bactrocera cucurbitae, 292
Bald Cypress, 151
Banana, 254, 413
Bark beetles, 12, 156, 157, 163, 164, 181–183
Basil, 413, 414, 418
Bauhinia monandra, 293
Bauhinia purpurea, 293
Bees, 42

457

Beetle, 1, 12, 43, 46, 79, 95, 97, 100, 113, 114, 117, 156, 157, 163, 164, 181–183, 223, 224, 257, 262, 294, 374, 418, 422, 424, 437
Bemisia tabaci, 424
Beta vulgaris, 116, 118
Betula pendula, 116
Betula pubescens, 91
Birch, 116, 198, 258
Bluestain fungus (or fungi), 156
Bombyx mori, 294, 370, 377
Boraginaceae, 214, 216, 218–221, 223
Botrytis cinerea, 398
Brasilian soldier beetle, 223
Brassicaceae, 111
Brassica nigra, 91
Brassica oleracea, 32, 46, 51
Brassica rapa, 66
Bromeliaceae, 288
Brussels sprouts, 46
Bumble bees, 42
Butterflies, 39, 223, 369, 422

C

Cabbage, 38, 46, 113–115, 422
Cabbage looper, 279, 374
Callosobruchus chinensis, 294
Callosobruchus maculatus, 291, 293, 294, 297, 298
Calocedrus, 159
Camponotus crassus, 118
Camptotheca acuminata, 200
Canavalia ensiformis, 293
Capsicum annuum, 416, 422
Cardiochiles nigriceps, 45, 379, 419
Carrot, 412
Cathaya, 158, 160
Cedrus, 159, 162
Centaurium erythraea, 412
Cepaea hortensis, 424
Cephalotaxaceae, 157, 158
Ceratocysis polonica, 152, 153, 161
Ceroplastes albolineatus, 223
Chamaecyparis, 159
Chauliognathus falax, 223
Chickpea, 245, 276, 277, 298
Choristoneura occidentalis, 175
Chrysomelidae, 113, 114, 117, 422
Chrysopa nigricornis, 422
Cicadellidae, 292
Cicer arietinum, 298
Cinara pinea, 45
Cinnamomum camphora, 294

Clarkia breweri, 413, 414
Clavigralla tomentosicollis, 293
Cnaphalocrocis medinalis, 300
Cnidoscolus texanus, 92
Coccinelids, 99, 224
Coccinella septempunctata, 419, 422
Coffee, 255, 257
Coleoptera, 91, 92, 113, 114, 117, 223, 257, 258, 290, 422
Concord grape, 414
Conifers, 12, 13, 147–151, 153–158, 160, 161, 163–165, 173–183, 433
Convolvulaceae, 214, 216
Cotesia congregata, 64
Cotesia glomerata, 32, 44, 46, 47
Cotesia kariyai, 46, 419, 420, 423
Cotesia marginiventris, 45, 47, 48, 400, 419, 420, 423–425, 435, 436
Cotesia plutellae, 44
Cotesia rubecula, 419, 422
Cotton, 44, 45, 94, 257, 291, 292, 294, 295, 297, 376, 377, 423, 424
Cowpea weevil, 291–294, 298
Crotalariae, 214
Cruciferae, 237
Cryptomeria, 159
Ctenarytaina spatulata, 113
Cucumber, 412, 421, 422
Cucumis sativus, 412, 419, 422
Cucurbitaceae, 237, 286–288, 290
Culex pipines pallens, 294
Cupressaceae, 148, 149, 157, 158
Cupressus, 159
Cuscuta pentagona, 118
Cydia pomonella, 118
Cynoglossum officinale, 218, 219, 222

D

Datura stramonium, 416, 422
Datura wrightii, 94
Delphacidae, 292
Dendrocerus carpenteri, 42
Dendroctonus ponderosa, 182
Deraeocoris brevis, 422
Diabrotica undecimpunctata, 291, 293
Diabrotica virgifera virgifera, 436, 437
Diaeratiella rapae, 115, 119, 438
Diatraea saccharalis, 292
Dicots, 20, 263, 332
Dicyphus minimus, 419, 422
Diprion pini, 183, 437
Diptera, 113, 114, 257, 290, 403
Dodder, 436, 438

Taxonomic Index 459

Douglas fir, 174, 175, 177, 178
Drosophila, 216, 389, 391, 392
Duckweed, 7

E

Elm, 46
Empoasca fabae, 293
Eoreuma loftini, 292
Epitrix hirtipennis, 422
Erwinia carotovora, 138
Estigmene acrea, 224
Eucalyptus, 8, 120
Eukarya, 220
Euneura augarus, 41, 42, 45
Eupatorieae, 214, 216, 220
European cinnabar moth, 222

F

Fabaceae, 214–216, 257, 287, 289
Fall armyworm, 295, 371, 374, 377, 392
Fire ant, 98
Flea beetle, 113, 223, 422
Flies, 41, 42, 117, 415
Fusarium sambucinum, 138

G

Galanthus nivalis, 288, 300
Geocoris pallens, 73, 419, 422
Gerbera, 422, 423
Gerbera jamesonii, 419, 422
Glechoma hederacea, 289
Glycine max, 416, 419, 420, 422
Gossypium birsutum, 423
Gossypium herbaceum, 416, 417
Grammia geneura, 224
Grand fir, 161, 176–178
Griffonia simplicifolia, 293, 295, 297
Gryllidae, 392
Gypsy moth, 193, 198

H

Harmonia axyridis, 422
Helianthus tuberosus, 293
Heliothis (Helicoverpa) armigera, 245, 246, 260, 291, 292, 294, 295, 374, 375
Heliothis (Helicoverpa) zea, 44, 198, 245, 256–259, 370, 371, 375–377, 380, 382, 393
Heliothis subflexa, 377, 379, 393
Heliothis virescens, 45, 244, 247, 294, 375, 377–379, 393, 419
Heliotropium indicum, 218, 222
Hemerobius sp, 422
Hemiptera, 223
Hessian fly, 114, 295

Heteroptera, 97, 114, 257, 422
Heterorhabditis megidis, 43, 437
Hevea brasiliensis, 288
Homoptera, 113–116, 119, 257, 290–293
Hoya carnosa, 413
Humulus lupulus, 416, 422
Hymenoptera, 115, 118, 119
Hypericum androsaemum, 412

I

Ilex aquifolium, 112
Impatiens wallerana, 438
Insecta, 1
Ips pini, 182
Iridaceae, 288

J

Jack bean, 289
Jack fruit, 288
Jasmine, 349
Jerusalem artichoke, 200

K

Keteleeria, 158
Koelreuteria paniculata, 293

L

Lacanobia oleracea, 292, 293, 298, 300
Lamiaceae, 287, 289
Largus rufipennis, 223
Larix, 148, 158, 160
Leguminosae, 237
Lepidium virginicum, 92
Lepidoptera (or lepidopteran), 38, 65, 72, 91, 92, 96, 113, 114, 118, 213, 238, 247, 253, 256–258, 261, 262, 264, 271, 272, 274, 279, 290, 292, 293, 332, 369, 370, 380, 381, 389, 391–394, 396, 403, 422, 435–437
Leptinotarsa decemlineata, 114, 257, 258, 262, 420
Leucoptera coffeella, 257
Liliaceae, 288
Lilioceris lilii, 117
Lilioceris merdigera, 117
Liliopsida, 288
Lima bean, 33, 35, 37, 39, 46, 50, 392, 400, 421, 423–425, 435, 438
Lipaphis erysimi, 291, 292, 297
Lithospermum erythrorhizon, 412
Loblolly pine, 177, 178, 180
Locusta migratoria, 114
Longitarsus, 223
Lotus japonicus, 422
Lucilia cuprina, 291
Lycopersicon hirsutum, 114

Lycopersicon peruvianum 30, 315, 317, 336, 337
Lymantria dispar, 259, 260, 262
Lysibia nana, 41

M

Macaranga, 115, 120
Maclura pomifera, 292, 293
Macrosiphum euphorbiae, 42, 258
Maize, 38, 43, 45, 114, 137, 165, 277, 354, 374, 377, 381, 415, 423–425, 433, 435–438
Malacosoma disstria, 257, 260
Malus sp., 417, 418
Manduca quinquemaculata, 259, 419, 422
Manduca sexta, 23, 64, 68, 72, 73, 118, 258, 274–279, 332, 337, 339, 342, 356, 372, 374, 377–380, 392, 394, 399, 419
Marchantia polymorpha, 288
Mayetiola destructor, 295
Mechanitis polymnia, 118
Medicago sativa, 114
Medicago truncatula, 37, 198, 375, 377, 416–418, 421
Melampsora medusae, 263
Melanopus sanguinipes, 118
Meligethes aeneus, 294
Mentha x piperita, 437
Mesobuthus tamulus, 300
Metasequoia, 159
Microplitis rufiventris, 48, 424
Milkweed, 12, 50, 51, 97
Miridae, 97, 422
Mnesampela privata, 113
Monocots, 220, 236–238, 263, 288, 291, 332
Moraceae, 287, 292, 293
Moths, 42, 46, 113, 118, 193, 198, 222–224, 292, 293, 369
Musa sp., 413
Mustard, 43, 113, 237, 238, 241, 291, 292, 297
Mycorrhiza, 52
Myzus nicotianae, 294
Myzus persicae, 119, 291–294, 300, 438

N

Nematodes, 43, 44, 238, 331, 337, 433, 436, 437
Nepenthes, 120
Nicotiana, 64, 70, 94, 208
Nicotiana attenuata, 10, 15, 20, 23, 38, 48–50, 62, 63, 65, 68–74, 276–278, 338, 342, 350, 356, 374, 378–380, 412, 419, 422, 437
Nicotiana plumbaginifolia, 276
Nicotiana suaveolens, 413

Nicotiana sylvestris, 70
Nicotiana tabacum, 46, 75, 222, 290, 375, 378
Nilaparvata lugens, 291, 292, 297, 300
Norway spruce, 149–155, 160–165, 174, 176–178, 180, 181, 205
Nothotsuga, 158

O

Ocimum basilicum, 413
Ophiostoma clavigerum, 182
Ophiostoma ips, 182
Orchidaceae, 214, 216, 218, 219, 288
Oreina, 223
Orgyia leucostigma, 260
Orius tristicolor, 423
Orthoptera, 114, 118, 223, 257, 258, 262, 392, 395, 403
Oryza sativa, 236
Ostrinia nubilalis, 291, 293, 297

P

Pacific yew, 154
Panonychus ulmi, 417
Papaver, 111
Paropsis charybdis, 113
Parsley, 200, 332
Parsnip, 208
Pastinaca sativa, 66
Peanut, 247, 248, 289, 293, 423
Perilampus fulvicornis, 118
Perillus bioculatus, 420, 424
Petunia hybrida, 412
Phaedon cochleariae, 113, 117
Phalaenopsis, 218, 219, 222
Phaseolus lunatus, 416–421, 438
Phaseolus vulgaris, 294
Phratora vulgatissima, 97, 100
Physalis angulata, 377, 379, 393
Physcomitrella patens, 178
Phytolacca americana, 291
Phytoseiulus persimilis, 46, 47, 419–422, 435, 436, 438
Picea, 148, 158, 160
Picea abies, 152, 161, 174
Picea glauca, 163, 174
Picea sitchensis, 174
Pieris rapae, 32, 44, 47, 372, 380, 419, 422
Pinaceae, 148–151, 154, 155, 157, 158, 160, 173, 193, 198, 437
Pine, *Pinus*, 45, 148, 158, 160, 163, 164, 173, 175, 177, 178, 180, 182, 183
Pineapple, 237
Pinellia ternata, 292
Pinus longaeva, 148

Pinus sylvestris, 183, 193
Pinus taeda, 198
Pissodes strobi, 163, 175
Pisum, 120
Pittocaulon (ex *Senecio*) *praecox*, 223
Plantago major, 119
Platyphora, 223
Plutella xylostella, 44, 47, 113, 114, 119, 246
Podocarpaceae, 148, 149, 157, 158
Podocarpus, 159
Pokeweed, 255
Popilla japonica, 417–419
Poplar, 9, 15, 192, 193, 196, 198, 200, 254–257, 259–261, 263, 277, 322, 331
Populus tremuloides, 255
Potato, 1, 8, 9, 16, 42, 79, 98, 114, 131–141, 192, 194, 196, 198, 204, 205, 237, 238, 245, 257–259, 262, 264, 272, 276–278, 280, 286, 288, 291–294, 297, 315, 317, 321, 322, 324, 325, 355, 374, 398, 412, 424
Predatory mites, 33, 46, 49, 98, 415, 421–423, 425, 435, 436
Prostephanus truncates, 418
Pseudaletia separata, 45, 374, 419, 420, 423
Pseudolarix, 158, 162
Pseudomonas syringae, 67, 241, 260, 263, 264, 381
Pseudotsuga, 158, 160
Pseudotsuga menziesii, 174
Psophocarpus tetragonolobus, 293
Pythium irregulare, 323, 356

R

Radish, 9, 42, 50, 65, 97, 99, 379
Raphanus raphanistrum, 92, 99
Raphanus sativus, 99
Red cotton bug, 297
Red oak, 208, 263
Reduviidae, 73
Rhododendron, 114
Rhopalosiphum maidis, 416, 417
Rhopalosiphum padi, 295, 418
Rice, 165, 236, 237, 277, 286, 290–292, 300, 331, 354
Ricinus communis, 289
Robinia pseudo-acacia, 422
Rubus idaeus, 114
Ruscaceae, 288

S

Sagebrush, 43, 50, 258, 399, 400
Salix borealis, 92

Salix cinerea, 29, 95, 97
Salix viminalis, 92, 97
Sambucus nigra, 289
Sawfly, 119, 183, 183, 437
Schistocerca americana, 395, 396
Scirpophaga incertulas, 300
Scolytidae, 157
Scots pine, 164, 183
Segestria florentina, 300
Senecio, 111, 213, 217–220, 222–224
Senecio erucifolius, 218
Senecio jacobaea, 39, 218–220, 222
Senecioneae, 214, 220
Senecio vernalis, 218, 222
Senecio vulgaris, 218
Sequoiadendron, 159
Sequoia sempervirens, 147
Sesamina noonagrioides, 198
Sipha flava, 116
Sitka spruce, 15, 174–178, 180–182, 194, 198, 200, 204, 205, 207
Sitobion avenae, 42
Sitodipolis mosellana, 198
Sitophilus granaries, 418
Sitophilus zeamais, 418
Solanaceae (or solanaceous), 22, 114, 257, 258, 277, 313–315, 321, 323, 324
Solanum berthaultii, 255
Solanum lycopersicum, 273, 413, 419, 438
Solanum nigrum, 22, 380
Solanum tuberosum, 67, 134, 420
Sorghum, 113, 114, 116
Sorghum halepense, 116
Soybean, 98, 198, 235, 247, 248, 257, 274, 280, 286, 289, 322, 350, 377, 422
Sphenostylis stenocarpa, 293
Spider mite, 33, 35–37, 46, 258, 421, 423–425, 435, 436, 438
Spodoptera exigua, 45, 72, 258, 294, 373, 375, 377, 379, 390, 394, 416, 417, 419, 421, 423, 424, 435
Spodoptera frugiperda, 114, 243, 295, 373, 374, 377, 437
Spodoptera littoralis, 198, 292, 295, 373, 374, 377, 437
Spodoptera litura, 114, 260, 375
Spruce, 15, 149–155, 160–165, 173–182, 194, 198, 200, 204, 205, 207
Spruce budworms, 175
Squash bug, 99
Stephanitis pyrioides, 114
Stephanotis floribunda, 413
Stethorus punctum picipes, 422

Strawberry, 48, 435, 438
Sweet potato, 196, 198, 230
Symphytum officinale, 213, 222
Syrphid flies, 42

T
Taxaceae, 111, 149, 157, 158
Taxodium distichum, 151
Taxodium mucronatum, 147
Taxus brevifolia, 154
Teosinte, 437
Tephritid flies, 117
Tetranychus urticae, 46, 416–422, 424, 425, 435, 438
Therioaphis maculata, 114
Thuja, 43, 159
Thuja occidentalis, 43
Tiger moth, 223
Tobacco, 34, 42, 43, 45, 47, 51, 64, 200, 236, 256, 258, 259, 274, 276, 290–292, 294, 295, 299, 314, 315, 317–321, 323, 324, 332–336, 339
Tobacco hornworm, 64, 274
Tomato, 1, 7–9, 11, 16–20, 23, 34, 36, 50, 194, 196, 253, 257–261, 263–265, 273–279, 286, 292, 313–321, 332, 335–340
Tomato moth, 292
Torreya, 159
Tribolium castaneum, 418
Trichogramma ostriniae, 224
Trichoplusia ni, 258, 279
Tsuga, 158, 162

Tyria jacobaeae, 40, 222–224

U
Ulmus minor, 46
Urtica dioica, 92, 96
  dioica, 92
  sondenii, 92
Utetheisa ornatrix, 224

V
Vicia faba, 45, 416, 418
Vitis labrusca, 414
Vitis vinifera, 416, 422

W
Wasps, 32, 33, 41, 44, 45, 64, 115, 183, 224, 300, 400, 415, 423, 425, 435, 437
Weevil, 43, 163, 164, 175, 180–182, 194, 200, 204, 205, 207, 246, 291, 293, 294, 298
Western corn rootworm, 437
White pine weevil, 163, 164, 175
White spruce, 163, 164, 174, 176, 177, 182
Willow, 95, 97, 100, 119
Willow warblers, 119
Wireworms, 44
Wollemia, 158

X
Xanthogaleruca lutteola, 46

Z
Zea mays, 44–47, 419, 420
Zonocerus variegatus, 223